Advanced Analysis of Marine Structures

Advanced Analysis of Marine Structures

Editors

Bin Liu
Kun Liu
Chenfeng Li

MDPI

Basel • Beijing • Wuhan • Barcelona • Belgrade • Novi Sad • Cluj • Manchester

Editors

Bin Liu
Wuhan University of Technology
Wuhan
China

Kun Liu
Jiangsu University of Science and Technology
Zhenjiang
China

Chenfeng Li
Harbin Engineering University
Harbin
China

Editorial Office
MDPI AG
Grosspeteranlage 5
4052 Basel, Switzerland

This is a reprint of articles from the Special Issue published online in the open access journal *Journal of Marine Science and Engineering* (ISSN 2077-1312) (available at: https://susy.mdpi.com/special_issue/process/1209307/search).

For citation purposes, cite each article independently as indicated on the article page online and as indicated below:

Lastname, A.A.; Lastname, B.B. Article Title. *Journal Name* **Year**, *Volume Number*, Page Range.

ISBN 978-3-7258-1689-7 (Hbk)
ISBN 978-3-7258-1690-3 (PDF)
doi.org/10.3390/books978-3-7258-1690-3

Contents

Journal of
Marine Science and Engineering

Editorial

Advanced Analysis of Marine Structures

Bin Liu [1], Kun Liu [2] and Chenfeng Li [3,*]

1 Green & Smart River-Sea-Going Ship, Cruise and Yacht Research Centre, Wuhan University of Technology, Wuhan 430063, China
2 School of Naval Architecture and Ocean Engineering, Jiangsu University of Science and Technology, Zhenjiang 212100, China
3 College of Shipbuilding Engineering, Harbin Engineering University, Harbin 150001, China
* Correspondence: lichenfeng@hrbeu.edu.cn

In the analysis and design of marine structures, one of the key issues is the accurate prediction of their strength under various load conditions, particularly impact and ultimate and fatigue strength. The advanced analysis of marine structures involves the mechanical analyses of advanced materials such as alloys and composite materials; it also involves the strength analyses of novel structures such as sandwich structures, in order to render marine structures lightweight, safe, and economical throughout their lifetimes.

The present Special Issue contains 19 articles; in total, 11 papers belong to the area of impacts and dynamic responses of ships and offshore structures, 3 papers are related to the fatigue strength of marine structures, 3 papers cover the various aspects of the strength assessment of ships and offshore structures, 1 paper revolves around the analysis of gravity anchors, and 1 paper deals with the sound absorption capability of marine equipment.

The fracture prediction of steel-plated structures is a key issue in impact strength analysis. Cerik and Choung [Contributions 1] carried out a validation study on a recently proposed rate-dependent shell element fracture model using both quasi-static and dynamic impact tests on square hollow sections made from offshore high-tensile-strength steel. A rate-dependent forming limit curve was used to predict the membrane loading-dominated failure, while a rate-dependent ductile fracture locus was applied for predicting the failure that is governed by bend loading.

The cruciform structure is a typical form in the double-hull ships. Liu et al. [Contributions 2] presented an experimental, numerical, and analytical study of the cruciform structure subjected to a local in-plane load, in order to investigate its crushing deformation and resistance.

For reducing accidental consequences, specific consideration is required to optimize the crashworthiness design. Qiu et al. [Contributions 3] proposed a novel method that addresses multi-working conditions and combines orthogonal testing with a backpropagation neural network to establish an efficient surrogate model for collision optimization.

Ice loading is a hot topic in the field of impact engineering, focusing on the safety of polar navigation. Liu et al. [Contributions 4] selected two numerical benchmark tests to validate the fluid–structure coupling approach and its program. During the ice-breaking simulation, the generation and propagation of radial and circular cracks in level ice are modeled, and the influence of ship speed and ice thickness on the ice load are investigated and analyzed. Chen et al. [Contributions 5] analyzed the dynamic response of a submarine during surface navigation in floating ice channels under special conditions. The fluid–structure coupling method was employed to simulate the structure–ice interaction of a submarine.

In polar engineering, the low temperature affects a ship's safety, focusing on the changing of the mechanical properties of steel materials. Zhang et al. [Contributions 6] performed a series of quasi-static and dynamic tests to investigate the behavior of EH36

Citation: Liu, B.; Liu, K.; Li, C. Advanced Analysis of Marine Structures. *J. Mar. Sci. Eng.* **2024**, *12*, 1111. https://doi.org/10.3390/jmse12071111

Received: 19 June 2024
Accepted: 28 June 2024
Published: 2 July 2024

steel at temperatures ranging from 20 °C to −60 °C; the variations of yield and ultimate tensile stress are given.

Impact analysis has been widely used to evaluate the strength of various types of structures, such as floating pontoons. Khalifa et al. [Contributions 7] analyzed the dynamic response of a steel ferry exposed to side explosions with different explosive charges at certain stand-off distances. An innovative mitigation system has been proposed to dissipate the blast energy of the explosion based on the scientific theory of impedance using sacrificial cladding.

Box girders can be seen as simplified ship structures that are able to withstand a vertical bending moment. Shi et al. [Contributions 8] analyzed the dynamic elastic–plastic response and ultimate strength of a box girder under a bending moment. Based on published experimental results, the nonlinear finite element analysis method is validated to determine the ultimate strength of a box girder under a bending moment.

Hydroelastic analysis is often used to analyze the hydrodynamic and structural response characteristics of marine structures, an example of which is a large multi-body floating offshore structure. Gu et al. [Contributions 9] used a method based on a CFD-FEA (computation fluid dynamics–finite element analysis) coupling simulation to evaluate the hydroelastic response of the flexible floating structures.

The dynamic response of steel catenary risers used in offshore platforms is difficult to analyze due to their complicated loadings. Yu et al. [Contributions 10] proposed a novel experimental platform to conduct dynamic loading tests on a truncated-model steel catenary riser within the touchdown zone. The facilities of the platform include a soil tank, a loading system, and a soil stirring system.

Codends are the posterior components of trawl nets; they collect the catch, and play a crucial role in the selectivity process. Zhang et al. [Contributions 11] investigated the effects of various catch configurations on the hydrodynamic characteristics, geometrical profile, and fluttering motions of a codend in a flume tank. A codend structure was designed and tested using various catch configurations, including grooved-type configurations (canvas, green canvas, and basketballs) and spherical configurations (table tennis balls filled with water, balloons filled with water, and balls made of twine) in the flume tank.

Fatigue failure is one of the main fracture modes in ships and offshore structures. To comprehensively understand the distribution pattern of the hot spot stress method at spatial tubular joints, Wang et al. [Contributions 12] selected to research the typical joints by performing hot spot stress testing.

Low cycle fatigue damage has received attention in relation to ship structures experiencing high stresses and large deformations. Qin et al. [Contributions 13] carried out experimental and numerical studies on the crack propagation behavior of cracked plates under low cycle fatigue loads, in order to explain the fatigue crack propagation mechanism. The effect of the stress ratio and maximum applied load on the crack propagation behavior was investigated.

The lumping block equivalent method has been widely used to reduce the computational effort in the fatigue damage assessment of offshore structures. Guo et al. [Contributions 14] proposed a novel wave energy equivalence-based lumping block equivalent method to accurately determine the wave parameters of the representative sea states. The novelty of the proposed method is that a compact relationship between the input wave energy component and mooring lines' fatigue damage is derived.

The superstructure of cruise ships is designed to be plumper with numerous decks and complex structural forms. To control the weight and the center of gravity, the bending stiffness of the superstructure is always designed to be weaker than that of the main hull, resulting in a stiffness step. Pei et al. [Contributions 15] conducted an experimental analysis to investigate the stiffness step between the main hull and the superstructure of a typical cruise ship. By comparing the longitudinal stress distribution characteristics both with and without the stiffness step with the theoretical results, the influence of the stiffness step on the longitudinal strength is investigated.

Fires and explosions in ships and offshore platforms cause serious human, economic, and environmental damage. Zong et al. [Contributions 16] experimentally and numerical evaluated the fire resistance of A-60-class ship bulkheads and decks through two sets of standard fire resistance tests.

The inflated membrane structures satisfy the demands of maritime salvage and military transportation for long-distance delivery and rapid response. Ye et al. [Contributions 17] comprehensively investigated the bending and failure behavior of inflated membrane structures.

Moreover, gravity anchors are widely used in ships and offshore platforms. Less attention has been paid to the bearing properties of gravity anchors in clay. Yu et al. [Contributions 18] investigated gravity anchors by conducting scaled model tests; the bearing process of gravity anchors in clay was simulated through the finite element method.

For marine equipment, their low-frequency broadband sound absorption capability is an important characteristic. Hu et al. [Contributions 19] designed a novel sound absorption structure with cavities by adding resonators and honeycombs to traditional sound absorption structures with cavities. The approximate multilayer sound absorption theoretical model based on the modified transfer matrix method is used to verify the accuracy of finite element calculations.

In summary, the articles presented in this Special Issue cover broad research topics related to the advanced analysis of marine structures, guiding readers through the best analysis approach.

Author Contributions: Writing—original draft preparation, B.L.; writing—review and editing, K.L. and C.L. All authors have read and agreed to the published version of the manuscript.

Funding: This research received no external funding.

Conflicts of Interest: The authors declare no conflicts of interest.

List of Contributions

1. Cerik, B.C.; Choung, J. Fracture Prediction of Steel-Plated Structures under Low-Velocity Impact. *J. Mar. Sci. Eng.* **2023**, *11*, 699. https://doi.org/10.3390/jmse11040699.
2. Liu, H.; Liu, K.; Wang, X.; Gao, Z.; Wang, J. On the Resistance of Cruciform Structures during Ship Collision and Grounding. *J. Mar. Sci. Eng.* **2023**, *11*, 459. https://doi.org/10.3390/jmse11020459.
3. Qiu, W.; Liu, K.; Liu, H.; Zong, S.; Wang, J.; Gao, Z. Crashworthiness Optimization Method of Ship Structure under Multi-Working Conditions. *J. Mar. Sci. Eng.* **2023**, *11*, 1335. https://doi.org/10.3390/jmse11071335.
4. Liu, R.; Xue, Y.; Lu, X. Coupling of Finite Element Method and Peridynamics to Simulate Ship-Ice Interaction. *J. Mar. Sci. Eng.* **2023**, *11*, 481. https://doi.org/10.3390/jmse11030481.
5. Chen, Z.; Zhao, W.; Chen, Z.; Feng, G.; Ren, H.; Gui, H. Dynamic Response Analysis of Submarines Based on FEM-ALE Coupling Method in Floating Ice Conditions. *J. Mar. Sci. Eng.* **2023**, *11*, 1560. https://doi.org/10.3390/jmse11081560.
6. Zhang, J.; Kang, X.; Shi, X.; Guedes Soares, C.; Song, M. Low Temperature Effect on the Mechanical Properties of EH36 with Strain Rates. *J. Mar. Sci. Eng.* **2023**, *11*, 678. https://doi.org/10.3390/jmse11030678.
7. Khalifa, Y.A.; Lotfy, M.N.; Fathallah, E. Effectiveness of Sacrificial Shielding for Blast Mitigation of Steel Floating Pontoons. *J. Mar. Sci. Eng.* **2023**, *11*, 96. https://doi.org/10.3390/jmse11010096.
8. Shi, G.-J.; Wang, D.-Y.; Wang, F.-H.; Cai, S.-J. Analysis of Dynamic Response and Ultimate Strength for Box Girder under Bending Moment. *J. Mar. Sci. Eng.* **2023**, *11*, 373. https://doi.org/10.3390/jmse11020373.
9. Gu, N.; Liang, D.; Zhou, X.; Ren, H. A CFD-FEA Method for Hydroelastic Analysis of Floating Structures. *J. Mar. Sci. Eng.* **2023**, *11*, 737. https://doi.org/10.3390/jmse11040737.
10. Yu, Y.; Xu, S.; Yu, J.; Xu, L.; Liu, X.; Liu, P. Experimental Study of the Dynamic Performance of Steel Catenary Riser within the Touchdown Zone. *J. Mar. Sci. Eng.* **2023**, *11*, 151. https://doi.org/10.3390/jmse11010151.
11. Zhang, F.; Tang, H.; Thierry, N.N.B.; Liu, W.; Sun, Q.; Zhu, M.; Zhang, C.; Guo, X.; Shan, C.; Xu, L.; et al. The Oscillating Behavior of Trawl Codends Including Various Geometric Configurations of Simulated Catch. *J. Mar. Sci. Eng.* **2023**, *11*, 1026. https://doi.org/10.3390/jmse11051026.

12. Wang, K.; Wang, Y.; Gao, S.; Wang, B.; Chen, J.; Luo, P. Experimental Study of Hot Spot Stress for Spatial CHS KK-Joints. *J. Mar. Sci. Eng.* **2023**, *11*, 1432. https://doi.org/10.3390/jmse11071432.
13. Qin, D.; Geng, X.; Jie, Z.; Yaoyu, H. Experimental and Numerical Study on Crack Propagation of Cracked Plates under Low Cycle Fatigue Loads. *J. Mar. Sci. Eng.* **2023**, *11*, 1436. https://doi.org/10.3390/jmse11071436.
14. Guo, Y.; Wang, S.; Guo, H.; Song, X. A Novel Wave Energy Equivalence Based Lumping Block Method for Efficiently Predicting the Fatigue Damage of Mooring Lines. *J. Mar. Sci. Eng.* **2023**, *11*, 1679. https://doi.org/10.3390/jmse11091679.
15. Pei, Z.; Yang, B.; Liu, G.; Wu, W. Experimental Research on the Stiffness Step between the Main Hull and Superstructure of Cruise Ships. *J. Mar. Sci. Eng.* **2023**, *11*, 1264. https://doi.org/10.3390/jmse11071264.
16. Zong, S.; Liu, K.; Qiu, W.; Gao, Z.; Wang, J. Numerical and Experimental Analysis of Fire Resistance for Bulkhead and Deck Structures of Ships and Offshore Installations. *J. Mar. Sci. Eng.* **2023**, *11*, 1200. https://doi.org/10.3390/jmse11061200.
17. Ye, Y.; Gan, J.; Liu, H.; Guan, Q.; Zheng, Z.; Ran, X.; Gao, Z. Experimental and Numerical Studies on Bending and Failure Behaviour of Inflated Composite Fabric Membranes for Marine Applications. *J. Mar. Sci. Eng.* **2023**, *11*, 800. https://doi.org/10.3390/jmse11040800.
18. Yu, J.; Liu, P.; Yu, Y.; Liu, X.; Li, H.; Sun, R.; Zong, X. Research on Bearing Characteristics of Gravity Anchor in Clay. *J. Mar. Sci. Eng.* **2023**, *11*, 505. https://doi.org/10.3390/jmse11030505.
19. Hu, J.; Lin, Y.; Zhou, Z.; Cao, X.; Chi, Q.; Wu, W. Study on the Low-Frequency and Broadband Sound Absorption Perfor-mance of an Underwater Anechoic Layer with Novel Design. *J. Mar. Sci. Eng.* **2023**, *11*, 409. https://doi.org/10.3390/jmse11020409.

Journal of
Marine Science
and Engineering

MDPI

Article

A Novel Wave Energy Equivalence Based Lumping Block Method for Efficiently Predicting the Fatigue Damage of Mooring Lines

Yuanzhi Guo [1], Shuqing Wang [1], Haiyan Guo [1] and Xiancang Song [1,2,*]

[1] College of Engineering, Ocean University of China, Qingdao 266100, China; bengyz@163.com (Y.G.); shuqing@ouc.edu.cn (S.W.); hyguo@ouc.edu.cn (H.G.)
[2] Shandong Provincial Key Laboratory of Marine Environment and Geological Engineering, Ocean University of China, Qingdao 266100, China
* Correspondence: songxiancang@ouc.edu.cn

Abstract: The lumping block equivalent method (LBEM) is widely used to reduce the computational effort in the fatigue damage assessment of offshore structures, and the wave parameters of the representative sea states (RSSs) resulting from LBEM are of vital importance for the accurate prediction of offshore structures' fatigue damage. In this study, a novel wave energy equivalence (WEE)-based LBEM is proposed to determine the wave parameters of the RSS accurately. The novelty of the proposed method is that a compact relationship between the input wave energy component and mooring lines' fatigue damage is derived, and the modified statistical relationships between the wave parameters and spectral moments are provided by incorporating the effects of the peak enhancement factor of the input wave spectrum, the number of original sea states (OSSs) and the equivalence bandwidth of the OSSs. Based on the compact relationship, the wave energy component of the RSS can be determined from the wave energy component of the OSSs for each wave frequency from the viewpoint of the fatigue damage equivalence criterion. The wave energy distribution of the RSS can be accurately characterized with the wave energy distribution of the OSSs, and the spectral moments of the RSS can be calculated by its energy distribution directly, without any approximation. Moreover, the wave parameters of the RSS can be determined from the modified statistical relationships easily. The effectiveness of the proposed WEE LBEM is numerically investigated with a moored semi-submersible platform. Results show that the proposed WEE LBEM is robust, efficient and accurate within engineering expectations, and it outperforms the conventional LBEMs both in accuracy and robustness.

Keywords: floating offshore structure; fatigue damage assessment; lumping block equivalent method; wave energy equivalence; mooring system

Citation: Guo, Y.; Wang, S.; Guo, H.; Song, X. A Novel Wave Energy Equivalence Based Lumping Block Method for Efficiently Predicting the Fatigue Damage of Mooring Lines. *J. Mar. Sci. Eng.* **2023**, *11*, 1679. https://doi.org/10.3390/jmse11091679

Academic Editor: José António Correia

Received: 3 August 2023
Revised: 22 August 2023
Accepted: 24 August 2023
Published: 25 August 2023

1. Introduction

A floating offshore structure generally composed of large-scaled floaters and slender mooring lines is the infrastructure used for the exploitation of natural resources in deep and ultra-deep water [1,2]. When the floating offshore structure is in operational conditions, the stochastic wave acting on the structure can be considered as the cyclic hydrodynamic load, which can lead the mooring lines to experience fatigue failure, even if the mooring line's tension is much smaller than its minimum breaking load [3,4]. Therefore, the mooring line's fatigue damage is one of governing factors for the design of floating offshore structures [5–7].

Generally, through an integration account of efficiency and safety requirements, the classical frequency- and time-domain fatigue assessment methods are the most commonly used methods in the design stage of floating offshore structures [8,9]. However, due to the nonlinearities inherent in the system, the dynamic responses of the floating offshore structure are non-Gaussian processes, and the classical frequency-domain fatigue assessment method, which is based on the Rayleigh distribution, therefore becomes inapplicable [3,10]. The time-domain fatigue assessment method can consider the nonlinearities of the system and the hydrodynamic loads accurately via a coupled dynamic analysis model, and the dynamic response can be converted into a series of response ranges and cycle numbers with the aid of the rain-flow counting algorithm [11]. As a result, it can be used as the benchmark for other methods [12,13]. However, the time-domain method has a remarkable computational cost. On the one hand, the short-term variability contributed by the stochastic wave necessitates multiple realizations to acquire statistical convergence. On the other hand, the simulation duration should be long enough to obtain sufficient low-frequency cycles to accurately estimate the mooring line's fatigue damage [14–16].

To reduce the computational cost, the lumping block equivalent method (LBEM), utilizing one representative sea state (RSS) to replace a group of original sea states (OSSs) to estimate the offshore structure's fatigue damage, is widely adopted in engineering practice [5,17]. The main challenge of LBEM is to select an RSS for each lumping block that can predict the fatigue damage resulting from the OSSs accurately [18]. The methodologies available to determine the wave parameters of the RSS broadly fall into three categories. The DNV and Sheehan LBEMs are the representative methods of the first category, where the wave parameters of the RSS can be determined from the wave parameters of the OSSs directly [19,20]. Referring to DNV LBEM, the summation of the interval and the largest wave height of the OSSs in the block are set to the wave height of the RSS, and the summation of the half interval and averaging wave period of the OSSs in the block are set to the wave period of the RSS [19]. In Sheehan LBEM, the largest wave height of the OSSs in the block is selected as the wave height of the RSS, and the probabilistic average of the wave period of the OSSs in the block is set to the wave period of the RSS [20]. The fatigue damage of offshore structures therefore will be overestimated significantly as the wave height of the RSS is much larger than the needed one.

In the second category, the OSSs with the same wave height/period are lumped into a block, and the wave period/height of the RSS can be calculated from the correlation of the wave period/height, the sea state's occurrence probability and a fatigue parameter, such as in the Mittendorf and Burton LBEMs [21,22]. In the Mittendorf LBEM, the OSSs with the same wave height are lumped into a block, and the probabilistic average of the wave period of the OSSs in the block is set to the wave period of the RSS [21]. In the Burton LBEM, the OSSs with the same wave period are lumped into a block, and the wave height of the RSS can be determined from the correlation of the wave height, the sea state's occurrence probability and the exponent of the S-N curve [22]. Obviously, the wave parameters of the RSS can be easily calculated for a specific lumping block case, but the accuracy of these LBEMs becomes very poor if the OSSs comprise several wave height/period intervals. Different from the Mittendorf and Burton LBEMs, the OSSs with different wave heights and periods are lumped into a block in the Jia LBEM [23]. The correlation utilized in the Burton LBEM is employed to calculate the wave height of the RSS, and the probabilistic average period principle used in the Mittendorf LBEM is adopted to determine the wave period of the RSS [23]. It has been proven that the offshore structure's fatigue damage resulting from the RSS related to the Jia LBEM method is smaller than that resulting from the OSSs.

In the third type of method, the wave parameters of the RSS are determined from the wave power spectral density (PSD) of the OSSs from the viewpoint of a fatigue damage equivalence criterion, such as in the Seidel and Song LBEMs [17,24]. In the Seidel LBEM, a compact equation between the input wave energy and fatigue damage equivalent loads on the large-diameter monopile wind turbine is derived and the relationship between the wave energy of the RSS and the OSSs can be established based on the compact equation from the viewpoint of the fatigue damage equivalence criterion. The wave height and period of the RSS can be obtained based on the equivalent wave energy and quasi-static considerations. It has been proven that the Seidel LBEM can give excellent fatigue prediction results for the structure components at the bottom of large-diameter monopile wind turbines. However, its accuracy becomes poor if the resonance part of the response is governing or the quasi-static response is of higher importance [24]. In the Song LBEM, the relationship between the sea state's energy and the structure's fatigue damage is provided by considering the influence of the structural fatigue parameter. From the viewpoint of the fatigue damage equivalence criterion, the spectral moments of the RSS can be determined from the spectral moments of the OSSs. The wave height and period of the RSS can be obtained from the statistical relationships between the spectral moments and wave parameters in a straightforward manner [17]. The effectiveness of the Song LBEM has been proven by many researchers. However, the second spectral moment of the RSS is obtained by an analogy method rather than the fatigue damage equivalence criterion, which has an influence on the wave period of the RSS and causes this method to slightly underestimate the structure's fatigue damage in some cases [25].

The purpose of this study is to propose a wave energy equivalence-based lumping block method to efficiently and accurately predict mooring lines' fatigue damage induced by stochastic wave loads. In the proposed method, a relationship between the input wave energy and the mooring line's fatigue damage is first established by considering the effect of the mooring line's nonlinearities and the fatigue parameter, and the relationship between the wave energy distribution of the RSS and the OSSs is further derived from the fatigue damage equivalence viewpoint. Two modified statistical relationships between the spectral moments and wave parameters are provided by incorporating the effect of the spectrum's peak enhancement factor, the sea state's number in the block and the sea state's equivalence bandwidth. The spectral moments of the RSS can be determined from its wave energy distribution in a straightforward manner, and the proposed method has direct physical significance compared to the conventional LBEMs. To present the theories of the LBEMs and the results of the investigation, the manuscript is structured as follows. The details of the conventional and proposed LBEMs are introduced in Sections 2 and 3. In Section 4, the environmental conditions and the numerical model are provided. The effectiveness of the proposed method is validated with a series of case studies as presented in Section 5. Finally, the conclusions are summarized in Section 6.

2. Conventional LBEMs

To clearly illustrate the theories of the LBEMs, it is assumed that the wave scatter diagram consists of N OSSs, they are lumped into N_b blocks and there are n_j OSSs in the jth block. The significant wave height and mean up-crossing period, as well as the occurrence probability of the RSS related to the jth block, are characterized by H_{srj}, T_{zrj} and p_{rj}, respectively.

For different LBEMs, the equivalent occurrence probability of the RSS can be calculated with the same principle, which is identical to the sum of the occurrence probability of the OSSs in the block, as shown in Equation (1).

$$p_{rj} = \sum_{k=1}^{n_j} p_{kj} \tag{1}$$

where p_{kj} is the occurrence probability of the kth OSS in the jth block.

2.1. DNV LBEM

In DNV LBEM, the H_{srj} and T_{zrj} of the RSS can be calculated based on the following expressions [20]:

$$H_{srj} = \max_{1 \le k \le n_j} \left\{ H_{skj} \right\} + \Delta H_s \tag{2}$$

$$T_{zrj} = \frac{\sum_{k=1}^{n_{T_{zj}}} T_{zkj}}{n_{T_{zj}}} + \frac{\Delta T_z}{2} \tag{3}$$

where $max\{\cdot\}$ represents the maximum value in the set; ΔH_s and ΔT_z are the interval of the wave height and period in the block; $n_{T_{zj}}$ is the parameter related to the number of T_z encompassed by the jth block, and it can be given as

$$n_{T_{zj}} = \max_{1 \le i \le n_{H_{sij}}} \left[n_{T_{zij}} \right] \tag{4}$$

where $n_{T_{zij}}$ is the number of T_z with the same H_{sij} in the jth block; $n_{H_{sij}}$ is the number of H_{sij} encompassed by the jth block.

2.2. Sheehan LBEM

Referring to the Sheehan LBEM, the H_{srj} of the RSS is identical to the largest H_{skj} of the OSSs in the block. The T_{zrj} of the RSS is identical to the probabilistic average of the OSS's T_{zkj} in the block, and it yields [21]

$$H_{srj} = \max_{1 \le k \le n_j} \left\{ H_{skj} \right\} \tag{5}$$

$$T_{zrj} = \frac{\sum_{k=1}^{n_j} T_{zkj} p_{kj}}{p_{rj}} \tag{6}$$

2.3. Jia LBEM

According to the theory of the Jia LBEM, the H_{srj} of the RSS can be calculated based on the relation of the fatigue parameter m, H_s and p_k of the OSSs, while the T_{zrj} of the RSS can be calculated by the probabilistic average period of the OSSs and yields [16]

$$H_{srj} = \left(\frac{\sum_{k=1}^{n_j} H_{skj}^{m} p_{kj}}{p_{rj}} \right)^{\frac{1}{m}} \tag{7}$$

$$T_{zrj} = \frac{\sum_{k=1}^{n_j} T_{zkj} p_{kj}}{p_{rj}} \tag{8}$$

where m is the exponent of the fatigue curve.

2.4. Spectral Moment Equivalence (SME)-Based LBEM

In the SME LBEM, the relation of the input spectral moment and structural fatigue damage is provided, and it can be given as [18]

$$Da_k = \kappa m_{0k}^{\lambda m} \tag{9}$$

where Da_k is the cumulative fatigue damage contributed by the kth OSS; m_{0k} is the zeroth spectral moment of the kth OSS; m is the exponent of the fatigue curve; λ is the exponent of the zeroth spectral moment obtained via the regression algorithm, and $\lambda = 2/m$; κ is the coefficient related to the zeroth spectral moment of the input sea state and the fatigue damage of the offshore structure.

Based on the relation, the cumulative fatigue damage caused by the RSS and OSSs related to the jth block can be given as

$$Da_{rj} = \kappa m_{0rj}^{\lambda m} p_{rj} \tag{10}$$

$$Da_j = \kappa \sum_{k=1}^{n_j} Da_{kj} p_{kj} = \kappa \sum_{k=1}^{n_j} m_{0kj}^{\lambda m} p_{kj} \tag{11}$$

where Da_{rj} and Da_j are the cumulative fatigue damage resulting from the RSS and OSSs related to the jth block; m_{0rj} is the zeroth spectral moment of the RSS related to the jth block; m_{0kj} is the zeroth spectral moment of the kth OSS in the jth block.

The relation of the zeroth spectral moments of the RSS and OSSs can be set up according to the fatigue equivalence criterion and yields

$$m_{0rj} = \left(\frac{\sum_{k=1}^{n_j} m_{0kj}^{\lambda m} p_{kj}}{p_{rj}} \right)^{\frac{1}{\lambda m}} \quad (12)$$

The second spectral moment of the RSS can be determined from the second spectral moments of the OSSs by analogy with Equation (12), and it can be given as

$$m_{2rj} = \left(\frac{\sum_{k=1}^{n_j} m_{2kj}^{\lambda m} p_{kj}}{p_{rj}} \right)^{\frac{1}{\lambda m}} \quad (13)$$

where m_{2rj} represents the second spectral moment of the RSS related to the jth block, and m_{2kj} represents the second spectral moment of the kth OSS in the jth block.

The H_{srj} and T_{zrj} of the RSS can be determined from its spectral moments directly based on the following formula:

$$H_{srj} = 4.0\delta_j \sqrt{m_{0rj}} \quad (14)$$

$$T_{zrj} = 2\pi \sqrt{\frac{m_{0rj}}{m_{2rj}}} \quad (15)$$

where δ_j is a correction factor adopted to consider the effect of the OSS's number in the jth block, and it has the following expression:

$$\delta_j = (n_j)^{\frac{1}{4mn_j}} \quad (16)$$

where n_j is the number of OSSs in the jth block, and m is the exponent of the fatigue curve.

3. The Novel Wave Energy Equivalence (WEE)-Based LBEM

The SME LBEM simply depends on the zeroth and second spectral moments and the occurrence probability of the OSSs, and it is therefore very easy to achieve for practicing engineers. However, the formula adopted to calculate the second spectral moment of the RSS is derived by the analogy method rather than the fatigue equivalence criterion, which makes the SME LBEM slightly non-conservative for some lumping block cases.

In this study, a novel wave energy equivalence (WEE)-based LBEM is proposed to improve the applicability of the SME LBEM. In the proposed WEE LBEM, a compact relationship between the input wave energy and the mooring lines' fatigue damage is established with the aid of a regression algorithm, and the relationship between the wave energy distribution of the RSS and OSSs is further derived from the viewpoint of the fatigue equivalence criterion. The spectral moments, including the zeroth, first and second spectral moments, of the RSS are then calculated from its wave energy distribution directly, and the wave parameters of the RSS are finally determined from the modified statistical relationships between the spectral moments and wave parameters. The flow-chart of the proposed WEE LBEM is illustrated in Figure 1.

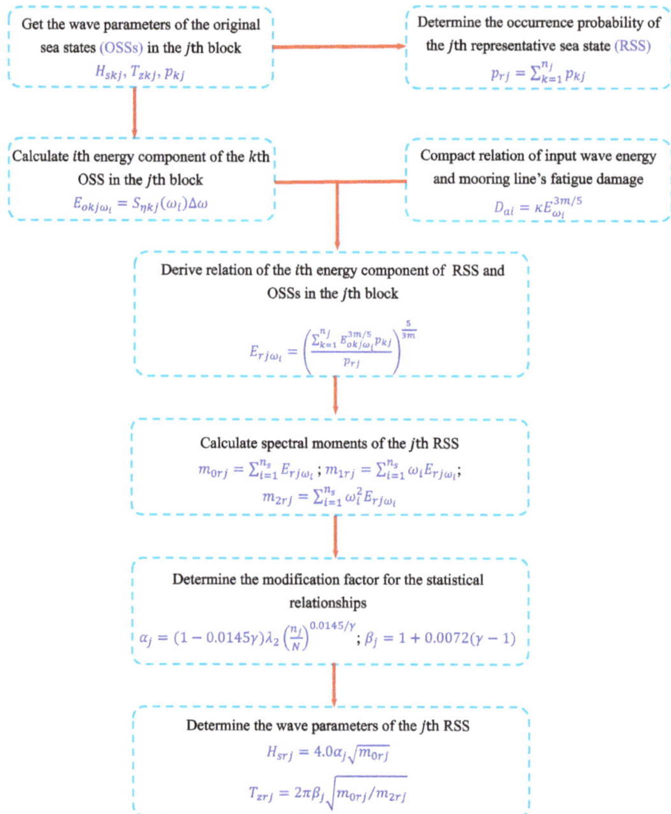

Figure 1. Flow-chart of the proposed WEE LBEM.

3.1. The Relationship between Input Wave Energy and Mooring Lines' Fatigue Damage

Based on the Longuet-Higgins wave model, the stochastic wave can be converted into a series of regular waves with different amplitudes, frequencies and phases. This means that the wave power spectral density (PSD) of the input sea state can be discretized into a series of energy components, and each energy component can be used to construct the regular wave with different frequencies. Therefore, the relationship between the input wave energy component and the mooring lines' fatigue damage can be established for a specific wave frequency. The flow-chart of the wave PSD discretization and regular wave construction is illustrated in Figure 2. As the dynamic responses of the mooring system are very sensitive to both the H_s and T_z of the input sea state, the sea states with non-zero occurrence probability in the wave scatter diagram are selected. Furthermore, to ensure that the discrete wave frequency intervals of the wave PSD are the same for different sea states, the lower and upper limits of the wave frequency and the number of discrete wave energy components should be identical to each other. The lower and upper limits of the wave frequency can be determined based on the wave parameters of the selected sea states, and they yield

$$\omega_L = \left(-\frac{3.11}{[min(H_{sk})]^2 ln(0.002)}\right)^{0.25} \tag{17}$$

$$\omega_H = \left(-\frac{3.11}{[max(H_{sk})]^2 ln(0.998)}\right)^{0.25} \tag{18}$$

where $min(*)$ and $max(*)$ represent the minimum and maximum values in the set, respectively.

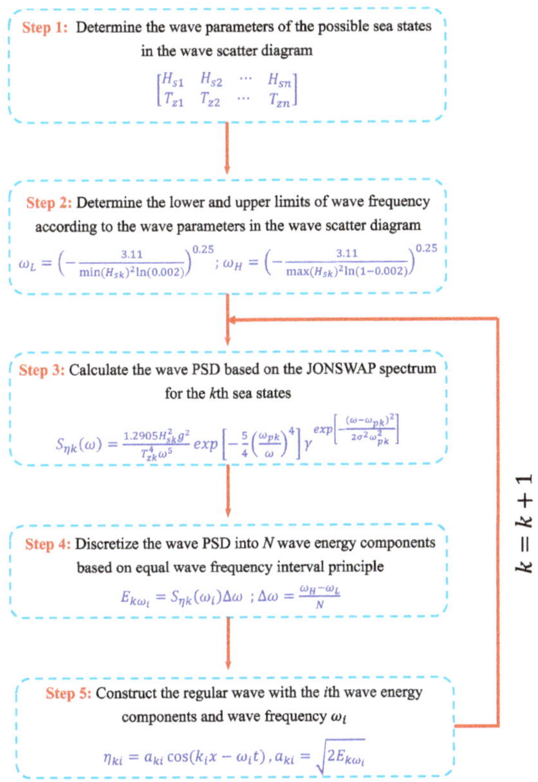

Figure 2. Flow-chart of the wave PSD discretization and regular wave construction.

In this paper, the JONSWAP spectrum is adopted to depict the wave energy distribution of the selected sea state, and the wave PSD of the kth sea state can be given as

$$S_{\eta k}(\omega) = \frac{1.2905H_{sk}^2 g^2}{T_{zk}^4\omega^5} exp\left[-\frac{5}{4}\left(\frac{\omega_{pk}}{\omega}\right)^4\right]\gamma^{exp[-\frac{(\omega-\omega_{pk})^2}{2\sigma^2\omega_{pk}^2}]} \tag{19}$$

$$\begin{cases} \omega \leq \omega_{pk} & \sigma = 0.07 \\ \omega > \omega_{pk} & \sigma = 0.09 \end{cases} \tag{20}$$

$$\omega_{pk} = \frac{2\pi}{1.407(1-0.287ln\gamma)^{1/4}T_{zk}} \tag{21}$$

The wave PSD of the kth sea state is further discretized into N wave energy components according to the equal wave frequency interval principle, and the wave energy for the ith wave frequency ω_i can be given as

$$E_{k\omega_i} = S_{\eta k}(\omega_i)\Delta\omega \tag{22}$$

$$\Delta\omega = \frac{\omega_H - \omega_L}{N} \tag{23}$$

where N is the number of discretized wave energy components, and it is 50 in this paper.

The regular wave elevation associated with the ith wave energy component of the kth sea state can be given as

$$\eta_{ki} = a_{ki} \cos(k_i x - \omega_i t) \tag{24}$$

where a_{ki} is the amplitude of the regular wave and $a_{ki} = \sqrt{2E_{k\omega_i}}$; k_i is the wave number, which can be calculated based on the dispersion relation.

Repeating steps 3–5, all of the regular waves related to the selected sea states can be constructed. It should be mentioned that the mooring lines' fatigue damage under all of the constructed regular waves are estimated, but, due to space limitations, this paper presents results for only ten wave frequencies, covering almost all of the wave frequencies that may be encountered by mooring lines, as illustrated in Figures 3 and 4.

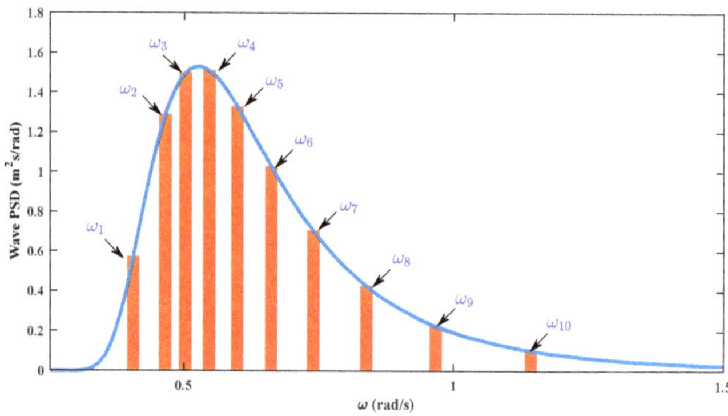

Figure 3. Illustration of wave PSD discretization.

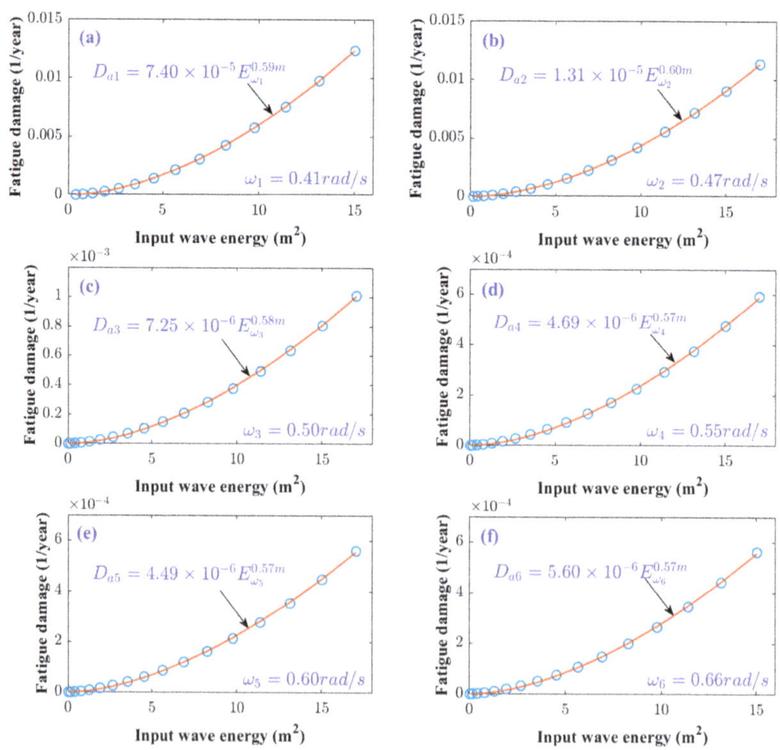

Figure 4. *Cont.*

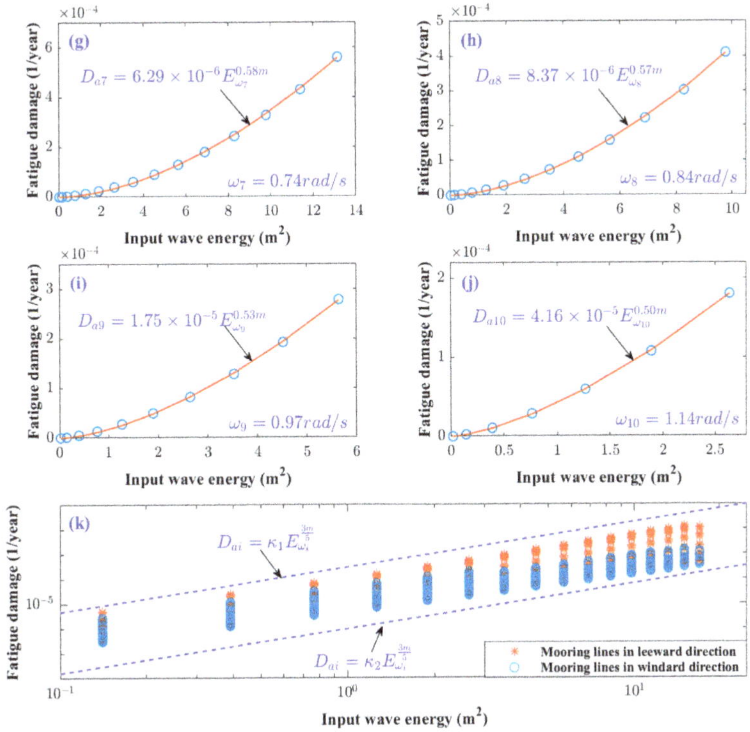

Figure 4. The relationships between the input wave energy and mooring lines' fatigue damage.

Mooring lines' tension resulting from a series of regular waves with different amplitudes and frequencies is first estimated with the fully coupled analysis model, and the mooring lines' fatigue damage is then estimated with the time-domain fatigue assessment method to fully consider the effect of the mooring line's nonlinearities. The relationships between the input wave energy and the mooring lines' fatigue damage are set up for different input sea states with the aid of a regression algorithm, and the corresponding results are presented in Figure 4a–j. In the figure, the circles represent the mooring lines' fatigue damage contributed by the regular waves related to different input sea states, while the line denotes the fitted results with the regression algorithm.

From the figure, one can find that the mooring lines' fatigue damage is proportional to the λmth of the input wave energy for different wave frequencies, and the parameter m is the exponent of the fatigue curve. It is notable that the exponent λ decreases slightly as the wave frequency increases, and the maximum value of the exponent λ is not larger than 3/5, as illustrated in Figure 4k. Therefore, the relationship between the input wave energy and mooring lines' fatigue damage can be characterized by a formula with a constant exponent, and the exponent λ can be set to its maximum value (i.e., 3/5) for conservative considerations.

The compact formula adopted to depict the relationship between the input wave energy and the mooring lines' fatigue damage for a specific wave frequency ω_i can be given as

$$D_{ai} = \kappa_i E_{\omega_i}^{3m/5} \tag{25}$$

where D_{ai} is the mooring lines' fatigue damage resulting from the ith wave energy component E_{ω_i} related to wave frequency ω_i; κ_i is the coefficient related to wave frequency ω_i; m is the exponent of the T-N curve, which is 3 for mooring lines.

3.2. The Relationship between the Wave Energy of RSS and OSSs

The fatigue damage caused by the ith wave energy component related to frequency ω_i in the RSS should be identical to the sum of the fatigue damage caused by the ith energy component with the same frequency in all of the OSSs in the block. Therefore, the ith wave energy component of the RSS can be obtained from the ith wave energy component of the OSSs in the block.

As illustrated in Figure 5, for a three-OSS block case, the relationship between the ith wave energy component of the RSS and the OSSs for wave frequency ω_i can be given as

$$E_{rj\omega_i} = \left(\frac{\sum_{k=1}^{n_j} E_{okj\omega_i}^{3m/5} p_{kj}}{p_{rj}} \right)^{\frac{5}{3m}} \tag{26}$$

where $E_{rj\omega_i}$ is the ith wave energy component related to wave frequency ω_i of the jth RSS; $E_{okj\omega_i}$ is the ith wave energy component related to wave frequency ω_i of the kth OSS in the jth block, and $E_{okj\omega_i} = S_{\eta kj}(\omega_i)\Delta\omega$, where $S_{\eta kj}$ is the wave PSD of the kth OSS in the jth block and $\Delta\omega$ is the wave frequency discretization interval of the wave PSD; p_{kj} is the occurrence probability of the kth OSS in the jth block; n_j is the number of OSSs in the jth block; p_{rj} is the occurrence probability of the jth RSS (see Equation (1)).

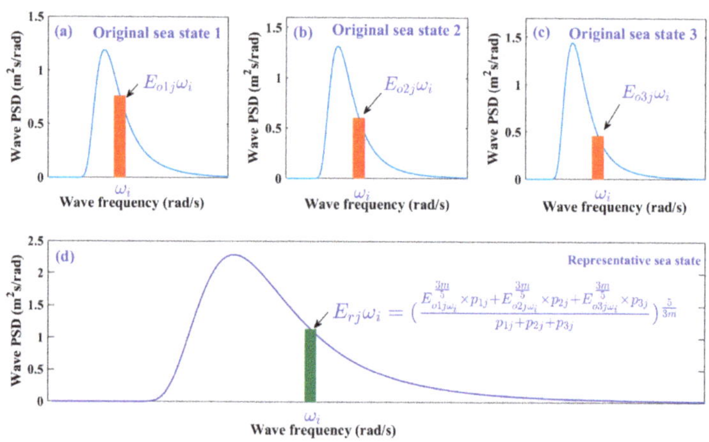

Figure 5. Illustration of the relationship between wave energy of the RSS and OSSs related to wave frequency ω_i. (**a**) Wave PSD for original sea state 1, (**b**) Wave PSD for original sea state 2, (**c**) Wave PSD for original sea state 3, (**d**) Wave PSD for representative sea state.

The wave energy distribution of the RSS can be obtained by applying Equation (18) to all of the wave frequencies of the OSS in the block. Then, the spectral moments of the jth RSS, including the zeroth, first and second spectral moments, can be calculated based on its wave energy distribution directly, yielding

$$m_{0rj} = \sum_{i=1}^{n_s} E_{rj\omega_i} \tag{27}$$

$$m_{1rj} = \sum_{i=1}^{n_s} \omega_i E_{rj\omega_i} \tag{28}$$

$$m_{2rj} = \sum_{i=1}^{n_s} \omega_i^2 E_{rj\omega_i} \tag{29}$$

where m_{0rj}, m_{1rj} and m_{2rj} are the zeroth, first and second spectral moments of the RSS, and n_s is the number of discretization intervals of the RSS's wave PSD.

3.3. The Modified Statistical Relationships between Spectral Moments and Wave Parameters

The wave parameters of the RSS, including the H_{srj} and T_{zrj}, can be calculated based on the statistical relationships between the spectral moments and wave parameters. However, many researchers have shown that the fatigue damage contributed by the RSS is closely related to the number of OSSs in the block and the bandwidth of the RSS [17,25]. According to the commonly utilized statistical relationship, the significant wave height is only related to the zeroth spectral moment, which can be determined from the energy distribution. Therefore, the parameters that have an influence on the energy distribution of the RSS should be introduced to construct the correction factor for the significant wave height. Firstly, the wave energy distribution of the OSS is closely related to the wave spectrum and the spectrum's enhancement factor, and the spectrum's enhancement factor has a significant influence on the energy distribution of the RSS in turn. Therefore, the effect of the spectrum's peak enhancement factor should be taken into consideration.

Secondly, the lumping block usually contains several OSSs, and the energy distribution is different for OSSs with different significant wave heights and up-crossing periods. The energy distribution of the RSS determined from the energy distribution of the OSS is usually a broadband spectrum, which cannot be depicted by the standard wave spectrum. The difference between the calculated wave energy distribution and the theoretical wave energy distribution depicted by a standard wave spectrum can be characterized by the equivalent bandwidth. Therefore, the effect of the equivalent bandwidth of the RSS should be considered.

Thirdly, the significant wave heights and up-crossing periods encompassed by the block increase as the number of OSSs in the block increases, and the fatigue damage contributed by the low-frequency cycles associated with the OSS with very long periods, and the high-frequency cycles associated with the OSSs with very small periods, will be suppressed by the representative sea state. Therefore, a factor related to the number of original sea states in the block should be introduced to amplify the wave energy distribution to obtain conservative results.

Since it is a modification of the commonly utilized statistical relationship, the correction factor should vary slightly with the input parameters. In this paper, the correction factor α_j related to the spectrum's peak enhancement factor γ, the number of OSSs n_j and the bandwidth of the RSS λ_{2j} are introduced for the H_{srj} of the RSS in exponent form, and it can be given as

$$\alpha_j = (1 - 0.0145\gamma)\lambda_{2j}\left(\frac{n_j}{N}\right)^{0.0145/\gamma} \tag{30}$$

where N is the number of OSSs in the wave scatter diagram; λ_{2j} is the equivalent bandwidth of the jth RSS and $\lambda_{2j} = \frac{\sqrt{m_{0rj}m_{2rj}}}{m_{1rj}}$; the coefficient "0.0145" is obtained from the simulation data via the regression algorithm.

As illustrated in Figure 6, the correction factor α_j increases slightly as the number of OSSs in the jth block increases. In addition, the H_{srj} of the RSS calculated based on the modified statistical relationship is slightly larger than those calculated based on the commonly utilized statistical relationship (i.e., $H_{srj} = 4.0\sqrt{m_{0rj}}$), and the discrepancies between them increase as the number of OSSs increases. It should be mentioned that the correction factor α_j will approach 1.0 as the number of OSSs decreases to 1, and the H_{srj} of the RSS will approach the H_{skj} of the OSS.

The T_{zrj} of the RSS calculated based on the commonly utilized statistical relationship (i.e., $T_{zrj} = 2\pi\sqrt{m_{0rj}/m_{2rj}}$) is close to the target ones when the spectrum's enhancement factor γ is identical to 1.0, as presented in Figure 7a. However, the T_{zrj} of the RSS determined from the commonly utilized statistical relationship is smaller than the target ones when the spectrum's enhancement factor γ is larger than 1.0, and the discrepancies between them increase as the parameter γ increases, as illustrated in Figure 7b. As the up-crossing period is very close to the cycle number of the response, which is very important for fatigue damage, the spectrum's peak enhancement factor should be incorporated into the correction factor to allow the calculated up-crossing period to match the target ones.

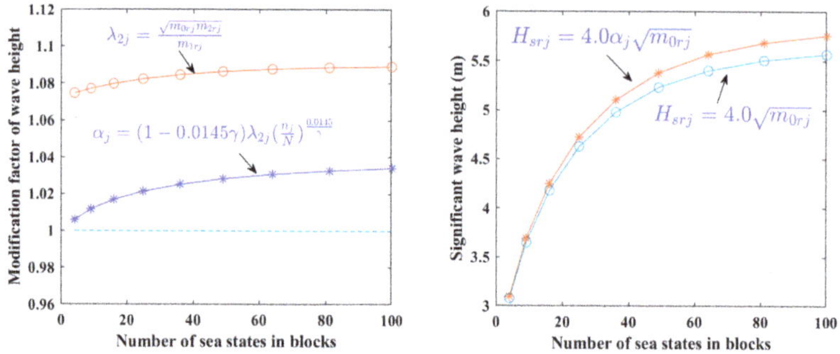

Figure 6. Relation of the correction factor α_j and the number of OSSs in the block.

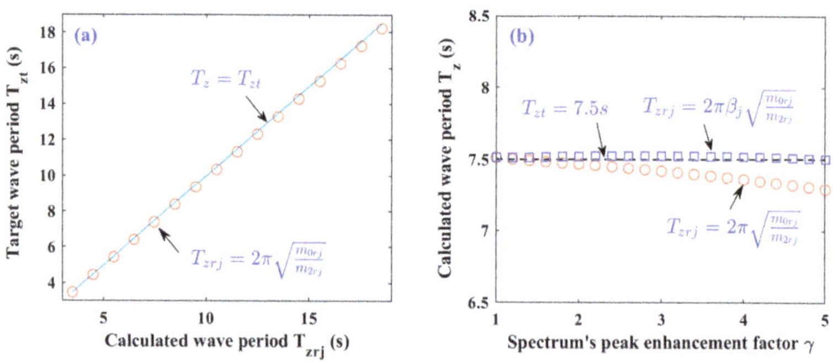

Figure 7. The relation of correction factor β_j and spectrum's peak enhancement factor. (**a**) Wave period calculated by the commonly utilized statistical relationship with γ of 1.0, (**b**) Wave period calculated by the commonly utilized and modified statistical relationships with different γ value.

In this study, a correction factor β_j related to the spectrum's peak enhancement factor γ is introduced for the T_{zrj} of the RSS, and it can be given as

$$\beta_j = 1 + 0.0072(\gamma - 1) \tag{31}$$

where coefficient "0.0072" is obtained from the simulation data via the regression algorithm.

As illustrated in Figure 7b, after considering the effect of the spectrum's peak enhancement factor γ, the T_{zrj} of the RSS determined from the modified statistical relationship are in perfect agreement with the target ones, and the correction factor will approach 1.0 as the spectrum's enhancement factor decreases to 1.0.

3.4. Wave Parameters of the RSS

The H_{srj} and T_{zrj} of the RSS can be determined from the following formulae:

$$H_{srj} = 4.0\alpha_j\sqrt{m_{0rj}} \tag{32}$$

$$T_{zrj} = 2\pi\beta_j\sqrt{m_{0rj}/m_{2rj}} \tag{33}$$

The WEE LBEM simply depends on the wave energy and occurrence probability of the OSSs in the block, which is very easy achieve for practicing engineers. In addition, the spectral moments adopted to calculate the wave parameters of the RSS can be obtained from its wave energy distribution directly, which gives the proposed LBEM more physical significance compared to the conventional LBEMs.

4. Numerical Model and Environmental Conditions

4.1. Numerical Model

The semi-submersible (SEMI) platform adopted in Song and Wang's research is adopted in this study to investigate the effectiveness of the proposed WEE LBEM [11,18]. As shown in Figure 8a, the SEMI is a deep-draft platform where four large columns are connected together with a ring pontoon. The SEMI operates at a water depth of 1828.8 m and is positioned with 12 mooring lines. These mooring lines are composed of three different components and grouped into four bundles that connect to the SEMI with the fairleads at each corner, as illustrated in Figure 8b. The main specifications of the SEMI and the mooring lines are summarized in Tables 1 and 2, respectively.

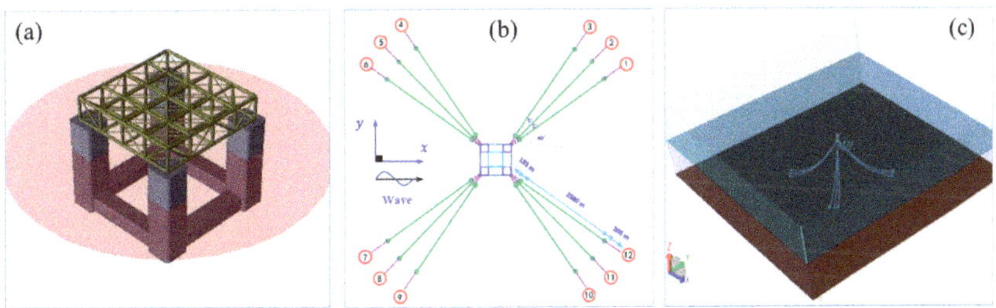

Figure 8. Illustration of the SEMI: (**a**) semi-submersible platform; (**b**) layout of mooring system; (**c**) dynamic analysis model.

Table 1. Main specifications of the SEMI.

Item	Value	Item	Value
Column side (m)	17.069	Pontoon length (m)	48.768
Column height (m)	59.131	Pontoon width (m)	12.192
Column corner radius (m)	2.438	Pontoon height (m)	9.754
Operation draft (m)	37.795	Pontoon corner radius (m)	1.219
Freeboard (m)	21.336	Radius of gyration in roll (m)	36.354
Vertical of COG (m)	24.465	Radius of gyration in pitch (m)	36.354
Total displacement (t)	63,811	Radius of gyration in yaw (m)	45.785

The numerical simulations are conducted in the DNV-SESAM software with the coupled dynamic analysis model, as illustrated in Figure 8c. In the numerical simulation, the head wave is defined as propagating along the *x*-axis, and the simulation duration is set to 10,800 s (i.e., 3 h) with a time step of 0.1 s for each study case.

Table 2. Main specifications of mooring lines.

Item	Platform Chain	Wire Rope	Anchor Chain
Length of lines (m)	183	2560	305
Diameter of lines (m)	0.147	0.273	0.147
Wet density of lines (kg/m)	360.47	57.58	360.47
Axial stiffness of lines (kN/m)	1,845,400	81,237	1,845,400
Minimum breaking load (kN)	19,089	12,705	19,089

4.2. Environmental Conditions

In engineering practice, the wind and current loads on SEMI are usually simplified as a constant load, which generally causes the mean offset of the SEMI, and they have little contribution to the structure's fatigue damage [4]. Therefore, only the wave hydrodynamic loads on the SEMI are of concern in this study. To estimate the cumulative fatigue damage of the mooring lines occurring at the fairleads of SEMI, the long-term wave condition should be considered, which is usually discretized into a series of short-term sea states represented by the wave scatter diagram in engineering practice. A typical wave scatter diagram in the North Sea, which includes 197 short-term sea states, is illustrated in Table 3.

Table 3. A typical wave scatter diagram in the North Sea.

$H_s(m)$	$T_z(s)$																	
	1.5	2.5	3.5	4.5	5.5	6.5	7.5	8.5	9.5	10.5	11.5	12.5	13.5	14.5	15.5	16.5	17.5	18.5
0.5	0	0	13	1337	8656	11,860	6342	1863	369	56	7	1	0	0	0	0	0	0
1.5	0	0	0	293	9860	49,760	77,380	55,697	23,757	7035	1607	305	51	8	1	0	0	0
2.5	0	0	0	22	1975	21,588	62,300	74,495	48,604	20,660	6445	1602	337	63	11	2	0	0
3.5	0	0	0	2	349	6955	32,265	56,750	50,991	28,380	11,141	3377	843	182	35	6	1	0
4.5	0	0	0	0	60	1961	13,543	32,885	38,575	26,855	12,752	4551	1309	319	69	13	2	0
5.5	0	0	0	0	10	510	4984	16,029	23,727	20,083	11,260	4636	1509	410	97	21	4	1
6.5	0	0	0	0	2	126	1670	6903	12,579	12,686	8259	3868	1408	422	109	25	5	1
7.5	0	0	0	0	0	30	521	2701	5944	7032	5249	2767	1117	367	102	25	6	1
8.5	0	0	0	0	0	7	154	979	2559	3506	2969	1746	776	277	84	22	5	1
9.5	0	0	0	0	0	2	43	332	1019	1599	1522	992	483	187	61	17	4	1
10.5	0	0	0	0	0	0	12	107	379	675	717	515	273	114	40	12	3	1
11.5	0	0	0	0	0	0	3	33	133	266	314	247	142	64	24	7	2	1
12.5	0	0	0	0	0	0	1	10	44	99	128	110	68	33	13	4	1	0
13.5	0	0	0	0	0	0	0	3	14	35	50	46	31	16	7	2	1	0
14.5	0	0	0	0	0	0	0	1	4	12	18	18	13	7	3	1	0	0
15.5	0	0	0	0	0	0	0	0	1	4	6	7	5	3	1	1	0	0
16.5	0	0	0	0	0	0	0	0	0	1	2	2	2	1	1	0	0	0

4.3. Wave Scatter Diagram Discretization

To examine the applicability of the proposed WEE LBEM to different lumping block partition cases, the OSSs in the wave scatter diagram illustrated in Table 3 are lumped into 57, 29, 15 and 6 blocks according to a specific rule, respectively. Taking the 57-block case, for example, OSSs in adjacent rows and columns are lumped into a block, as illustrated in Figure 9a. Similarly, OSSs in adjacent 3 rows and 3 columns, 4 rows and 4 columns, 6 rows and 8 columns are lumped into a block for 29, 15 and 6 block cases, as illustrated in Figure 9b–d, respectively. Compared with the full wave scatter diagram, the computational effort of the fatigue damage assessment can be reduced by 3.46, 6.79, 13.13 and 32.83 times when the OSSs in the wave scatter diagram are lumped into 57, 29, 15 and 6 blocks, respectively.

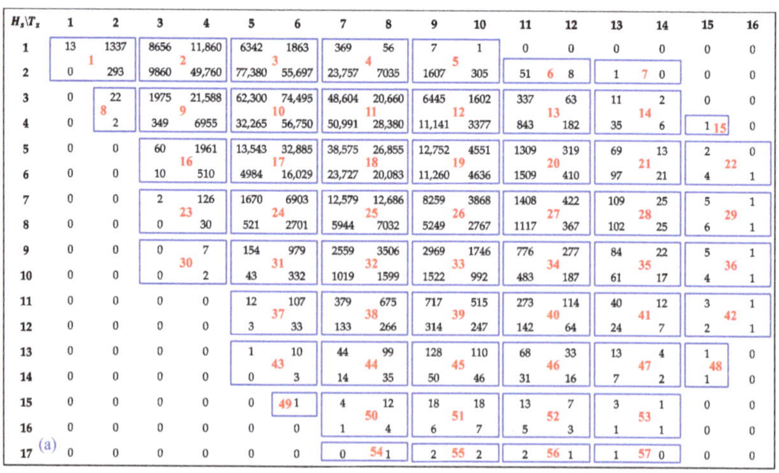

Figure 9. *Cont.*

(b)

$H_s \backslash T_z$	1	2	3	4	5	6	7	8	9	10	11	12	13	14	15	16
1	13	1337	8656	11,860	6342	1863	369	56	7	1	0	0	0	0	0	0
2	0	293	9860	49,760	77,380	55,697	23,757	7035	1607	305	51	8	1	0	0	0
3	0	22	1975	21,588	62,300	74,495	48,604	20,660	6445	1602	337	63	11	2	0	0
4	0	2	349	6955	32,265	56,750	50,991	28,380	11,141	3377	843	182	35	6	1	0
5	0	0	60	1961	13,543	32,885	38,575	26,855	12,752	4551	1309	319	69	13	2	0
6	0	0	10	510	4984	16,029	23,727	20,083	11,260	4636	1509	410	97	21	4	1
7	0	0	2	126	1670	6903	12,579	12,686	8259	3868	1408	422	109	25	5	1
8	0	0	0	30	521	2701	5944	7032	5249	2767	1117	367	102	25	6	1
9	0	0	0	7	154	979	2559	3506	2969	1746	776	277	84	22	5	1
10	0	0	0	2	43	332	1019	1599	1522	992	483	187	61	17	4	1
11	0	0	0	0	12	107	379	675	717	515	273	114	40	12	3	1
12	0	0	0	0	3	33	133	266	314	247	142	64	24	7	2	1
13	0	0	0	0	1	10	44	99	128	110	68	33	13	4	1	0
14	0	0	0	0	0	3	14	35	50	46	31	16	7	2	1	0
15	0	0	0	0	0	1	4	12	18	18	13	7	3	1	0	0
16	0	0	0	0	0	0	1	4	6	7	5	3	1	1	0	0
17	0	0	0	0	0	0	0	1	2	2	2	1	1	0	0	0

(c)

$H_s \backslash T_z$	1	2	3	4	5	6	7	8	9	10	11	12	13	14	15	16
1	13	1337	8656	11,860	6342	1863	369	56	7	1	0	0	0	0	0	0
2	0	293	9860	49,760	77,380	55,697	23,757	7035	1607	305	51	8	1	0	0	0
3	0	22	1975	21,588	62,300	74,495	48,604	20,660	6445	1602	337	63	11	2	0	0
4	0	2	349	6955	32,265	56,750	50,991	28,380	11,141	3377	843	182	35	6	1	0
5	0	0	60	1961	13,543	32,885	38,575	26,855	12,752	4551	1309	319	69	13	2	0
6	0	0	10	510	4984	16,029	23,727	20,083	11,260	4636	1509	410	97	21	4	1
7	0	0	2	126	1670	6903	12,579	12,686	8259	3868	1408	422	109	25	5	1
8	0	0	0	30	521	2701	5944	7032	5249	2767	1117	367	102	25	6	1
9	0	0	0	7	154	979	2559	3506	2969	1746	776	277	84	22	5	1
10	0	0	0	2	43	332	1019	1599	1522	992	483	187	61	17	4	1
11	0	0	0	0	12	107	379	675	717	515	273	114	40	12	3	1
12	0	0	0	0	3	33	133	266	314	247	142	64	24	7	2	1
13	0	0	0	0	1	10	44	99	128	110	68	33	13	4	1	0
14	0	0	0	0	0	3	14	35	50	46	31	16	7	2	1	0
15	0	0	0	0	0	1	4	12	18	18	13	7	3	1	0	0
16	0	0	0	0	0	0	1	4	6	7	5	3	1	1	0	0
17	0	0	0	0	0	0	0	1	2	2	2	1	1	0	0	0

(d)

$H_s \backslash T_z$	1	2	3	4	5	6	7	8	9	10	11	12	13	14	15	16
1	13	1337	8656	11,860	6342	1863	369	56	7	1	0	0	0	0	0	0
2	0	293	9860	49,760	77,380	55,697	23,757	7035	1607	305	51	8	1	0	0	0
3	0	22	1975	21,588	62,300	74,495	48,604	20,660	6445	1602	337	63	11	2	0	0
4	0	2	349	6955	32,265	56,750	50,991	28,380	11,141	3377	843	182	35	6	1	0
5	0	0	60	1961	13,543	32,885	38,575	26,855	12,752	4551	1309	319	69	13	2	0
6	0	0	10	510	4984	16,029	23,727	20,083	11,260	4636	1509	410	97	21	4	1
7	0	0	2	126	1670	6903	12,579	12,686	8259	3868	1408	422	109	25	5	1
8	0	0	0	30	521	2701	5944	7032	5249	2767	1117	367	102	25	6	1
9	0	0	0	7	154	979	2559	3506	2969	1746	776	277	84	22	5	1
10	0	0	0	2	43	332	1019	1599	1522	992	483	187	61	17	4	1
11	0	0	0	0	12	107	379	675	717	515	273	114	40	12	3	1
12	0	0	0	0	3	33	133	266	314	247	142	64	24	7	2	1
13	0	0	0	0	1	10	44	99	128	110	68	33	13	4	1	0
14	0	0	0	0	0	3	14	35	50	46	31	16	7	2	1	0
15	0	0	0	0	0	1	4	12	18	18	13	7	3	1	0	0
16	0	0	0	0	0	0	1	4	6	7	5	3	1	1	0	0
17	0	0	0	0	0	0	0	1	2	2	2	1	1	0	0	0

Figure 9. Wave scatter diagram discretization: (**a**) 57-block case; (**b**) 29-block case; (**c**) 15-block case; (**d**) 6-block case.

5. Results and Discussion

To examine the effectiveness of the proposed WEE LBEM, two different scenarios are analyzed. The first scenario is that the JONSWAP spectrum is adopted to depict the wave energy distribution of the input sea state, where the spectrum's peak enhancement factor is set to 3.3, and the performance of the conventional and proposed LBEMs are evaluated. The second scenario is that the Pierson–Moskowitz (P-M) spectrum is adopted to depict the wave energy distribution of the input sea state, where the spectrum's peak enhancement factor is set to 1.0, and the applicability of the proposed WEE LBEM to various spectra is investigated.

5.1. Wave Parameters of the RSS

The wave parameters of the RSS are first calculated based on these five LBEMs, and the six-block case is selected to demonstrate the discrepancies in the wave parameters determined from different LBEMs. The comparison results related to the JONSWAP spectrum with γ of 3.3 are listed in Tables 4 and 5, respectively.

Table 4. H_{srj} of the RSS resulting from different LBEMs.

Blocks	DNV LBEM/m	Sheehan LBEM/m	Jia LBEM/m	SME LBEM/m	WEE LBEM/m
Block 1	8.500	7.500	3.237	3.561	3.584
Block 2	8.500	7.500	4.717	4.993	5.238
Block 3	8.500	7.500	6.014	6.206	6.275
Block 4	15.500	14.500	9.044	9.240	9.291
Block 5	17.500	16.500	9.448	9.610	10.276
Block 6	17.500	16.500	10.061	10.289	10.506

Table 5. T_{zrj} of the RSS resulting from different LBEMs.

Blocks	DNV LBEM/s	Sheehan LBEM/s	Jia LBEM/s	SME LBEM/s	WEE LBEM/s
Block 1	6.500	7.692	7.692	7.899	8.083
Block 2	12.000	10.291	10.291	10.217	10.681
Block 3	17.000	14.792	14.792	14.544	14.905
Block 4	8.000	8.363	8.363	8.200	8.368
Block 5	12.000	11.164	11.164	10.786	11.271
Block 6	17.000	14.921	14.921	14.648	15.044

As can be seen from Table 4, the discrepancies in the significant wave heights resulting from the different LBEMs are remarkable for a specific block. Taking block 1, for example, the significant wave heights determined from the DNV, Sheehan, Jia, SME and WEE methods are 8.500 m, 7.500 m, 3.237 m, 3.561 m and 3.584 m, respectively. The significant wave height determined from the DNV LBEM is slightly larger than that resulting from the Sheehan LBEM, but both of them are much larger than those calculated by the other three LBEMs. While the significant wave height determined from the Jia LBEM is slightly smaller than that calculated by the SME LBEM, both of them are slightly smaller than that resulting from the WEE LBEM. Similar variation trends can be found for the other five blocks.

From Table 5, one can find that the up-crossing periods determined from the DNV LBEM are smaller than those determined from the Sheehan LBEM for the block 1 and 4 cases, but larger than those resulting from the Sheehan LBEM for the block 2, 3, 5 and 6 cases. The up-crossing periods calculated by the Sheehan LBEM are identical to those calculated by the Jia LBEM for the same theories used to calculate the up-crossing periods. The up-crossing period calculated by the Sheehan and Jia LBEMs is slightly smaller than that calculated by the SME LBEM for the block 1 case, but larger than those calculated by the SME LBEM for the block 2–6 cases. It is notable that the up-crossing periods resulting from the WEE LBEM are slightly larger than those resulting from the Sheehan, Jia and SME LBEMs.

To examine the applicability of the proposed WEE LBEM to various wave spectra, the wave parameters of the RSS related to the JONSWAP and P-M spectra are calculated, and the results are presented in Figure 10. From the figure, one can find that the spectrum's peak enhancement factor γ has a slight influence on the wave parameters of the RSS. The H_{srj} of the RSS calculated by the proposed WEE LBEM with $\gamma = 1.0$ is slightly larger than that calculated by the proposed WEE LBEM with $\gamma = 3.3$. Nevertheless, the T_{zrj} of the RSS calculated by the proposed WEE LBEM with $\gamma = 1.0$ is slightly smaller than that calculated by the proposed WEE LBEM with $\gamma = 3.3$. Similar variation trends are found for the RSSs for the 57-, 29- and 15-block cases, but the results are not shown here due to space limitations.

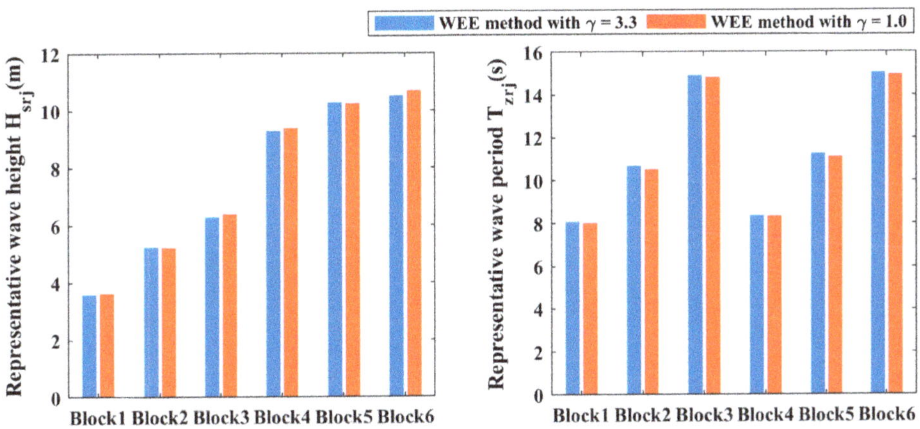

Figure 10. Wave parameters of the RSSs calculated by the proposed WEE LBEM related to different wave spectra.

5.2. Fatigue Damage of Mooring Lines Related to the JONSWAP Spectrum

The fatigue damage of mooring lines accumulated at the fairleads of SEMI under the RSSs and OSSs is estimated via the time-domain fatigue assessment method. To clearly illustrate the performance of the LBEMs, the mooring lines' fatigue damage contributed by the RSSs is normalized to the fatigue damage contributed by all the OSSs in the wave scatter diagram. The normalized fatigue damage of mooring lines 4 and 6 related to the JONSWAP spectrum is summarized in Tables 6 and 7, respectively.

Table 6. Normalized fatigue damage of mooring line 4 resulting from different LBEMs related to JONSWAP spectrum.

Block Partition Cases	Method				
	DNV LBEM	Sheehan LBEM	Jia LBEM	SME LBEM	WEE LBEM
57 Blocks	2.4776	1.2933	0.9639	1.0407	1.0480
29 Blocks	3.5750	1.6673	0.9235	0.9802	1.0807
15 Blocks	4.6778	2.2025	0.9147	0.9909	1.1366
6 Blocks	14.1194	6.1259	0.8611	0.9812	1.1228

Due to the normalization, the analysis indicates that the LBEM underestimates the mooring line's fatigue damage if the normalized fatigue damage is smaller than one. Otherwise, this means that the LBEM overestimates the mooring line's fatigue damage. It is worth noting that the fatigue damage estimated by the T-N curve may be smaller than the actual fatigue damage. If the fatigue damage is further underestimated by the lumping block equivalent method, the offshore structure will be in a dangerous condition. The underestimation of fatigue damage is undesirable in engineering practice.

Table 7. Normalized fatigue damage of mooring line 6 resulting from different LBEMs related to JONSWAP spectrum.

Block Partition Cases	Method				
	DNV LBEM	Sheehan LBEM	Jia LBEM	SME LBEM	WEE LBEM
57 Blocks	2.4801	1.3005	0.9652	1.0460	1.0487
29 Blocks	3.5769	1.6851	0.9277	0.9888	1.0856
15 Blocks	4.7424	2.2180	0.9104	0.9920	1.1318
6 Blocks	14.8135	6.3071	0.8624	0.9861	1.1241

It can be seen from the tables that the fatigue damage of the mooring lines resulting from these five LBEMs shows similar characteristics for various mooring lines. The mooring lines' fatigue damage resulting from the DNV LBEM is much larger than the benchmark value. Taking mooring line 4, for example, the overestimation level of the DNV LBEM is approximately 147.76% for the 57-block case, and it reaches approximately 1311.94% for the six-block case. As a result, the DNV LBEM overestimates the mooring lines' fatigue damage significantly. In addition, the mooring lines' fatigue damage resulting from the Sheehan LBEM is much larger than the benchmark value, and the overestimation level increases from 29.33% to 512.59% as the number of partitioned blocks in the wave scatter diagram decreases from 57 to 6. Although the Sheehan LBEM overestimates the mooring lines' fatigue damage remarkably as well, its overestimation level is much smaller than that of the DNV LBEM. The reason for this phenomenon is that the effect of the sea state's occurrence probability is taken into consideration in the process of determining the wave parameters of the RSS and a smaller H_{srj} is selected.

Different from the DNV and Sheehan LBEMs, the mooring lines' fatigue damage resulting from the Jia LBEM is consistently smaller than the benchmark value. This means that the Jia LBEM always underestimates the mooring lines' fatigue damage. When the number of OSSs in the block increases, the underestimation level of the Jia LBEM increases significantly, and the largest underestimation level can reach 13.89%. In contrast to the Jia LBEM, the mooring line fatigue damage related to the SME LBEM is slightly larger than the benchmark value for the 57-block case, but it is slightly smaller than the benchmark value for the 29-, 15- and 6-block cases. Taking mooring line 4, for example, the overestimation and underestimation levels of the SME LBEM are 4.07%, 1.98%, 0.91% and 1.88% for the 57-, 29-, 15- and 6-block cases, respectively. The underestimation level of the SME LBEM is much smaller than that of the Jia LBEM. This means that the SME LBEM gives more accurate fatigue damage predictions than the Jia LBEM.

A noteworthy observation is that the mooring lines' fatigue damage contributed by the proposed WEE LBEM is close to and always larger than the benchmark values. Different from the DNV and Sheehan LBEMs, the overestimation level of the proposed WEE LBEM increases slightly as the number of OSSs in the block increases, and the largest overestimation level is approximately 10%. In contrast to the Jia and SME LBEMs, the mooring lines' fatigue damage related to the proposed WEE LBEM is consistently larger than the benchmark value, and it can maintain its accuracy for different lumping block cases. These characteristics show that the proposed WEE LBEM yields the most accurate fatigue damage prediction, and it has robustness to different lumping block partitions.

5.3. Fatigue Damage of Mooring Lines Related to the P-M Spectrum

To fully investigate the applicability of the proposed WEE LBEM, the mooring lines' fatigue damage related to the P-M spectrum is further estimated with the time-domain fatigue assessment method. Similar to the results related to the JONSWAP spectrum, the fatigue damage of mooring lines resulting from the RSSs is normalized to the fatigue damage contributed by all of the OSSs in the wave scatter diagram, and the normalized fatigue damage for mooring lines 4 and 6 is listed in Tables 8 and 9, respectively.

Table 8. Normalized fatigue damage of mooring line 4 resulting from different LBEMs related to P-M spectrum.

Block Partition Cases	Method				
	DNV LBEM	Sheehan LBEM	Jia LBEM	SME LBEM	WEE LBEM
57 Blocks	2.4485	1.2663	0.9639	1.0501	1.0193
29 Blocks	3.4050	1.5877	0.9169	0.9761	1.0361
15 Blocks	4.7910	2.0190	0.8777	0.9280	1.0539
6 Blocks	13.6143	5.3221	0.7469	0.8262	1.0043

Table 9. Normalized fatigue damage of mooring line 6 resulting from different LBEMs related to P-M spectrum.

Block Partition Cases	Method				
	DNV LBEM	Sheehan LBEM	Jia LBEM	SME LBEM	WEE LBEM
57 Blocks	2.4491	1.2754	0.9645	1.0546	1.0222
29 Blocks	3.4184	1.6093	0.9187	0.9807	1.0394
15 Blocks	4.7586	2.0653	0.8833	0.9387	1.0640
6 Blocks	14.2487	5.5122	0.7554	0.8371	1.0137

There are five features worthy of attention in Tables 8 and 9. First, the mooring lines' fatigue damage contributed by the five LBEMs shows similar variation trends for different mooring lines. Second, the DNV and Sheehan LBEMs still overestimate the mooring lines' fatigue damage remarkably. When the number of OSSs in the block increases, the overestimation level increases dramatically, showing similar characteristics to the results under the JONSWAP spectrum.

Third, the Jia LBEM still underestimates the mooring lines' fatigue damage. When the number of OSSs in the block increases, the underestimation level increases significantly, and the largest underestimation level is approximately 25.31%, which is much larger than that related to the JONSWAP spectrum. Fourth, the SME LBEM overestimates the mooring lines' fatigue damage for the 57-block case, but it underestimates the mooring line fatigue damage for the 29-, 15- and 6-block cases. When the number of OSSs in the block increases, the underestimation level increases, and the largest underestimation level is approximately 17.38%, which is much smaller than the results related to the JONSWAP spectrum.

The most important feature is that the proposed WEE LBEM yields the most accurate and smallest conservative fatigue damage prediction for the mooring lines among these five LBEMs. When the number of OSSs in the block increases, the overestimation level of the proposed WEE LBEM increases slightly. Taking mooring line 6 as an example, the overestimation levels of the proposed WEE LBEM are 2.22%, 3.94%, 6.40% and 1.37% for the 57-, 29-, 15- and 6-block cases, respectively.

To further examine the applicability of the proposed WEE LBEM to various wave spectra, the normalized fatigue damage of mooring lines related to the JONSWAP and P-M spectra are compared, and the results are presented in Figure 11.

It can be seen from the figure that the fatigue damage of the leeward mooring lines (e.g., mooring lines 1 and 3) related to the P-M spectrum is slightly larger than that related to the JONSWAP spectrum for the 57- and 6-block cases, while the fatigue damage of the leeward mooring lines related to the P-M spectrum is slightly smaller than that related to the JONSWAP spectrum for the 29- and 15-block cases. However, the discrepancies in the mooring lines' fatigue damage related to the JONSWAP and P-M spectra are negligible.

One can also find that the fatigue damage of the windward mooring lines (e.g., mooring lines 4 and 6) related to the P-M spectrum is slightly smaller than the results related to the JONSWAP spectrum for all the lumping block cases. When the number of OSSs in the block increases, the discrepancies in the fatigue damage related to the P-M and JONSWAP spectra increase slightly. This indicates that the correction factor related to the spectrum's peak enhancement factor adopted in the proposed WEE LBEM makes it applicable to different spectra, and these qualities make the proposed WEE LBEM a useful tool for the fatigue damage assessment of the mooring system in its preliminary design stage.

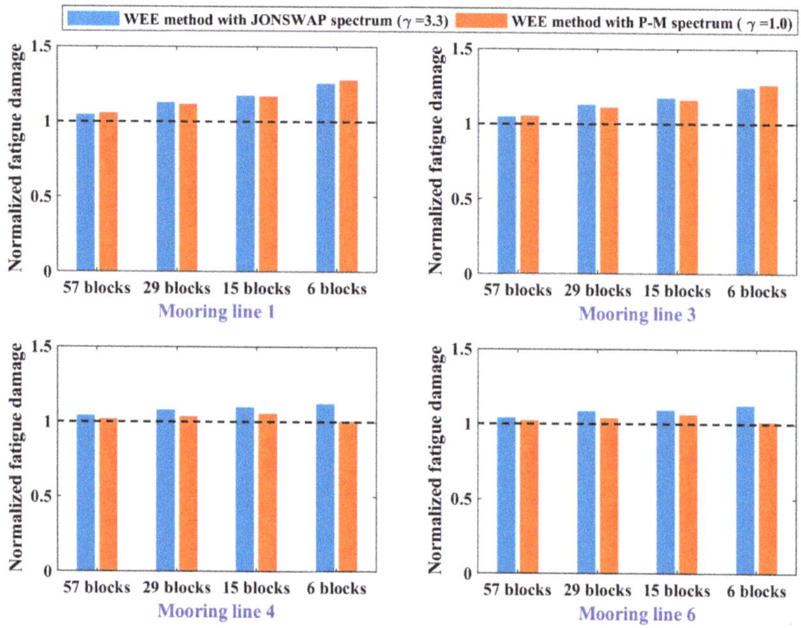

Figure 11. Normalized fatigue damage of mooring lines contributed by the proposed WEE LBEM related to the JONSWAP and P-M spectra.

6. Conclusions

This paper presents a novel LBEM from the viewpoint of wave energy equivalence to efficiently and accurately estimate a mooring line's fatigue damage at the preliminary design stage. In the proposed method, a compact relationship between the input wave energy and the mooring line's fatigue damage and a modified statistical relationship between the wave parameters and spectral moments are provided via the regression algorithm. The wave energy distribution of the RSS can be obtained from the wave energy distribution of the OSS based on the compact relationship, the spectral moment of the RSS can be calculated from its wave energy distribution directly, and the wave parameters of the RSS can be determined from the modified statistical relationships easily. The effectiveness of the proposed WEE LBEM has been numerically investigated with the moored SEMI. According to this study, several conclusions can be drawn as follows.

(1) The DNV and Sheehan LBEMs consistently overestimate mooring lines' fatigue damage significantly, and the overestimation level increases dramatically as the number of OSSs in the block increases. The overestimation level of the DNV LBEM is much larger than that of the Sheehan LBEM, and the largest overestimation level of the DNV LBEM can reach 1311.94%.

(2) The Jia LBEM consistently underestimates mooring lines' fatigue damage, while the SME LBEM underestimates mooring lines' fatigue damage for some cases. The underestimation level of these two methods increases as the number of OSSs in the block increases, but the underestimation level of the SME LBEM is much smaller than that of the Jia LEMB for the same lumping block case. The largest underestimation level of the Jia LBEM can reach 13.89%.

(3) The proposed WEE LBEM can yield the most accurate but slightly conservative fatigue damage predictions, and the largest overestimation level is approximately 10% for all the cited scenarios. It has excellent performance for different input wave spectra and is applicable to different lumping block cases, and it outperforms the conventional LBEM both in accuracy and robustness.

The proposed WEE LBEM will be a powerful tool for mooring line fatigue damage assessment in the preliminary stage of design, where a parameter study may be required and the costs of a time-domain fatigue assessment for a full wave scatter diagram are prohibitive.

Author Contributions: Conceptualization, Y.G. and X.S.; methodology, Y.G.; software, Y.G. and X.S.; validation, S.W. and H.G.; formal analysis, Y.G. and X.S.; writing—original draft preparation, Y.G. and X.S.; writing—review and editing, S.W. and H.G.; supervision, S.W.; project administration, S.W. and X.S.; funding acquisition, S.W. and X.S. All authors have read and agreed to the published version of the manuscript.

Funding: This research was funded by the National Natural Science Foundation of China (52088102), the Natural Science Foundation of Shandong Province (ZR2021QE101) and the Major Research Development Program of Shandong Province (2021CXGC010701).

Institutional Review Board Statement: Not applicable.

Informed Consent Statement: Not applicable.

Data Availability Statement: The data presented in this study are all included in this article.

Conflicts of Interest: The authors declare no conflict of interest.

References

1. Gemilang, G.M.; Reed, P.A.S.; Sobey, A.J. Low-cycle fatigue assessment of offshore mooring chains under service loading. *Mar. Struct.* **2021**, *76*, 102892. [CrossRef]
2. Zhao, Y.; Dong, S. Comparison of environmental contour and response-based approaches for system reliability analysis of floating structures. *Struct. Saf.* **2022**, *94*, 102150. [CrossRef]
3. Yue, J.; Yang, K.; Peng, L.; Guo, Y. A frequency-time domain method for ship fatigue damage assessment. *Ocean Eng.* **2021**, *220*, 108154. [CrossRef]
4. Guellec, C.; Doudard, C.; Levieil, B.; Jian, L.; Ezanno, A.; Calloch, S. Parametric method for the assessment of fatigue damage for marine shaft lines. *Mar. Struct.* **2023**, *87*, 103325. [CrossRef]
5. Katsikogiannis, G.; Sørum, S.H.; Bachynski, E.E.; Amdahl, J. Environmental lumping for efficient fatigue assessment of large-diameter monopile wind turbines. *Mar. Struct.* **2021**, *77*, 102939. [CrossRef]
6. Zhang, X.; Ni, W.; Sun, L. Fatigue analysis of the oil offloading lines in FPSO System under wave and current loads. *J. Mar. Sci. Eng.* **2022**, *10*, 225. [CrossRef]
7. Takeuchi, T.; Osawa, N.; Tatsumi, A.; Inoue, T.; Hirakawa, S.; Seki, N.; Yoshida, T.; Miratsu, R.; Ikeda, S. Fatigue assessment of ship structures based on equivalent wave probability (EWP) concept (1st report): Proposal of EWP concept and its verification by 8600TEU container ship's onboard hull monitoring. *Mar. Struct.* **2023**, *91*, 103476. [CrossRef]
8. Chang, A.; Li, H.; Wang, S.; Du, J. Probabilistic analysis and fatigue damage assessment of offshore mooring system due to non-Gaussian bimodal tension processes. *J. Ocean Univ. China* **2017**, *16*, 585–601. [CrossRef]
9. Zheng, T.S.; Chen, N.Z. Time-domain fatigue assessment for blade root bolts of floating offshore wind turbine (FOWT). *Ocean Eng.* **2022**, *262*, 112201. [CrossRef]
10. Song, X.; Wang, S.; Hu, Z.; Li, H. A hybrid Rayleigh and Weibull distribution model for the short-term motion response prediction of moored floating structures. *Ocean Eng.* **2019**, *182*, 126–136. [CrossRef]
11. Pham, Q.H.; Gagnon, M.; Antoni, J.; Tahan, A.; Monette, C. Rainflow-counting matrix interpolation over different operating conditions for hydroelectric turbine fatigue assessment. *Renew. Energ* **2021**, *172*, 465–476. [CrossRef]
12. Katsikogiannis, G.; Hegseth, J.M.; Bachynski-Polić, E.E. Application of a lumping method for fatigue design of monopile-based wind turbines using fully coupled and simplified models. *Appl. Ocean Res.* **2022**, *120*, 102998. [CrossRef]
13. Yuan, K.; Jin, H.; Chai, W. Development of a new spectral method for fatigue damage assessment in bimodal and trimodal Gaussian random processes. *Ocean Eng.* **2023**, *267*, 113273. [CrossRef]
14. He, W.T.; Xie, L.J.; Wang, S.Q.; Hu, Z.Q.; Xie, D.; Wang, C.Z. Preliminary assessment of the mooring fatigue performance of a semi-submersible platform in time-domain utilizing fracture mechanics-based approach. *Appl. Ocean Res.* **2022**, *129*, 103407. [CrossRef]
15. Han, C.S.; Mo, C.G.; Tao, L.B.; Ma, Y.L.; Bai, X. An efficient fatigue assessment model of offshore wind turbine using a half coupling analysis. *Ocean Eng.* **2022**, *263*, 112318. [CrossRef]

16. Low, Y.M.; Huang, X. Long-term extreme response analysis of offshore structures by combining importance sampling with subset simulation. *Struct. Saf.* **2017**, *69*, 79–95. [CrossRef]
17. Song, X.; Wang, S. A novel spectral moments equivalence based lumping block method for efficient estimation of offshore structural fatigue damage. *Int. J. Fatigue* **2019**, *118*, 162–175. [CrossRef]
18. Haselibozchaloee, D.; Correia, J.; Mendes, P.; Jesus, A.D.; Berto, F. A review of fatigue damage assessment in offshore wind turbine support structure. *Int. J. Fatigue* **2022**, *164*, 107145. [CrossRef]
19. DNV. *SESAM User's Manual—Deep Water Coupled Floater Motion Analysis*; DNV: Oslo, Norway, 2010.
20. Sheehan, J.M.; Grealish, F.W.; Harte, A.M.; Smith, R.J. Characterizing the wave environment in the fatigue analysis of flexible risers. *J. Offshore Mech. Arct. Eng.* **2006**, *128*, 108–118. [CrossRef]
21. Mittendorf, K.E. Joint description methods of wind and waves for the design of offshore wind turbines. *Mar. Technol. Soc. J.* **2009**, *43*, 23–33. [CrossRef]
22. Burton, T.; Jenkins, N.; Sharpe, D.; Bossanyi, E. *Wind Energy Handbook*; John Wiley & Sons: Hoboken, NJ, USA, 2011.
23. Jia, J. An efficient nonlinear dynamic approach for calculating wave induced fatigue damage of offshore structures and its industrial applications for lifetime extension. *Appl. Ocean Res.* **2008**, *30*, 189–198. [CrossRef]
24. Seidel, M. Wave induced fatigue loads: Insights from frequency domain calculations. *Stahlbau* **2014**, *83*, 535–541. [CrossRef]
25. Hou, H.M.; Dong, G.H.; Xu, T.J. An improved lumping block equivalent method for predicting fatigue damage of mooring system for fish cage. *Ocean Eng.* **2019**, *193*, 106567. [CrossRef]

Journal of
*Marine Science
and Engineering*

MDPI

Article

Dynamic Response Analysis of Submarines Based on FEM-ALE Coupling Method in Floating Ice Conditions

Zhongyu Chen [1], Weidong Zhao [1,2,*], Zhanyang Chen [1], Guoqing Feng [2], Huilong Ren [2] and Hongbin Gui [1]

1 School of Ocean Engineering, Harbin Institute of Technology, Weihai 264209, China
2 College of Shipbuilding Engineering, Harbin Engineering University, Harbin 150009, China
* Correspondence: weidong.zhao@hit.edu.cn

Abstract: To address global challenges, research on the safety of polar navigation is indispensable. However, most of studies focus on traditional surface vessels, with few research studies on submarine. The dynamic response of submarine during surface navigation in floating ice channels under special conditions is studied in this work. Firstly, a model of the submarine incorporating an intact internal frame was established. Subsequently, the FEM-ALE coupled method was employed to simulate the structure-ice interaction, and the obtained results was verified by the Colbourne method. Then, the parametric study (navigation speed, ice thickness, and floating ice size) were analyzed from the perspectives of ice resistance, stress and plastic strain. Finally, an empirical equation suitable for the interaction between submarine and floating ice during surface navigation is improved based on the Colbourne method.

Keywords: structure-ice collision; FEM-ALE; submarines; ice resistance; floating ice condition

Citation: Chen, Z.; Zhao, W.; Chen, Z.; Feng, G.; Ren, H.; Gui, H. Dynamic Response Analysis of Submarines Based on FEM-ALE Coupling Method in Floating Ice Conditions. *J. Mar. Sci. Eng.* **2023**, *11*, 1560. https://doi.org/10.3390/jmse11081560

Academic Editor: Md Jahir Rizvi

Received: 26 June 2023
Revised: 29 July 2023
Accepted: 1 August 2023
Published: 7 August 2023

1. Introduction

With the increasing degree of globalization and the ever-growing energy shortage problem, the Arctic region harbors paramount developmental significance [1]. The Arctic region has a transcontinental route linking Asia and Europe, and it contains a large amount of untapped resources [2]. However, numerous natural obstacles, such as floating ice, layer ice, and ridges, as well as extreme environmental conditions, pose a serious threat to the safety of polar navigation. Therefore, safety assessments of structures such as icebreakers, commercial vessels, and submarines that navigate in the Arctic region are particularly important [3]. Current researches mainly focuses on traditional surface vessels, such as icebreakers and commercial vessels, with limited research on submarines.

The dynamic response during ship-ice interaction process is affected by the shape of the ship, structural strength, and ice properties [4]. Analytical methods that are conventional have difficulty comprehensively considering multiple factors. Therefore, researchers commonly use empirical equation methods, simulation methods, and experimental methods. The experimental method can be divided into full-scale ship tests and model tests, which yield more accurate results but are costly and time consuming. Empirical equations are derived from the summary and integration of data from considerable of tests, and often apply only to a certain type of operating condition. A relatively lower cost is associated with the simulation method, rendering it applicable to a majority of operational conditions [5].

In the early nineteenth century, experimental methods were primarily employed for the accumulation of a significant volume of crucial data, from which empirical equations were derived. The earliest ice resistance equation, presented by Runeberg [6], took into consideration the influences of frictional forces and bow trim angles. Ship design parameters were initially considered by Shimanskii [7], while Kashteljan [8] pioneered the partitioning of ice resistance into breaking resistance, overturning and submergence resistance, and damaged floating ice resistance. Jones [9] introduced semi-empirical methods into the study of continuous ice breaking. Based on the aforementioned research and the accumulation of a substantial amount of data, numerous empirical equations with notable effectiveness were developed. For level ice conditions, methods such as the Lindqvist [10] approach, which incorporated considerations of friction and ship hull geometry, as well as the Riska [11] method derived from the modified Lindqvist approach, had been established. Conversely, for floating ice conditions, the relevant empirical equations were predominantly compiled and summarized from experimental data. Examples include the Bronnikov method [12] based on ship tests and the Mellor method based on the Mohr-Coulomb failure criterion [13].

In the experimental aspect, Daley [14] conducted observations on the failure process of the ice layer through full-scale and model-scale experiments and proposed a contact model for the ice layer's edge. Jeong [15] conducted ice tank experiments with square-shaped floating ice in three different channel widths and analyzed the impact of ice size on resistance, while proposing a rapid method for calculating ice concentration. Kim [16] conducted model-scale ice resistance experiments in a simulated ice environment using triangular ice elements made of paraffin instead of real ice. Several operating conditions were set based on different velocities, ice concentration, and ship hull waterline entry angles. The experimental results were compared with numerical simulation results. Jeong [17] and others conducted model-scale ship tests in a frozen ice region at the MOERI ice tank, resulting in the prediction of planar ice resistance for various thicknesses and bending strengths.

With the advancement of computer technology and numerical simulation methods, researchers have begun to employ simulation techniques to study the physical processes and mechanical characteristics of ice collisions. Among these, FEM, SPH, DEM, and their combined usage are the most common research approaches employed today.

The application of the FEM to simulate ice loads was pioneered by Määttänen and Hoikkanen in 1990, while Evgin et al. subsequently successfully utilized the DEM for icebreaking simulations [18]. Munjiza [19] proposed the FEM-DEM method, which integrated the advantages of both approaches, using FEM to simulate ice fracture and DEM to simulate ice accumulation. Huang et al. [20] based on the combination of CFD and DEM methods, investigated fragmented ice channels and discovered a linear relationship between the thickness and diameter of floating ice and ice resistance. Based on the ice shell collision mechanism and fundamental icebreaking characteristics, Karl [21] introduced a simplified numerical model to predict the ice impact forces acting on the vessel under horizontal ice conditions. The model addresses two critical failure modes, namely local crushing and flexural fracture. Andrei [22] constructed a CFD numerical model based on the DEM-BEM theory, suitable for simulating the interaction between floating structures and fragmented ice. The potential flow theory was employed to predict the flow field around the ship hull and the surrounding fragmented ice. Finally, the computational results were compared with the ice tank test results.

Zhang [23] performed numerical simulations of the collision between a vertical cylindrical structure and layered ice by utilizing LS-DYNA finite element software. Two different thicknesses of cohesive element ice models were constructed, and the S-ALE fluid–structure coupling method was employed. Song, Kim, and others [24] utilized the ALE method within the LS-DYNA finite element software to simulate the mutual interaction between floating ice and marine structures. They accounted for the fluid–structure interaction problem and validated the correctness of the ALE method and the coupling algorithm in the interaction between fluid, ice, and structures.

Apart from those, Su [25] incorporated numerical models to investigate the overall and local ice loads on ship structures. Through two case studies, simulated ice loads on the vessel were examined, analyzing ship performance, statistical framework loads induced by ice, and the spatial distribution of ice loads around the ship. A comparison with field measurements was conducted as well. Chai [26] applied probability methods and models to seek the correlation between ice-induced load statistics and major ice conditions in the field of ship and ocean engineering. Zhao [27] performed the probability-based fatigue damage assessment of vessels traversing horizontal ice fields. A novel procedure utilizing numerical simulation had been developed for the evaluation of fatigue damage, and the process was demonstrated through a hypothetical scenario involving the icebreaking vessel Snow Dragon 2. The sensitivity of the procedure to key analysis parameters, such as sample size and initial crack size, was also taken into consideration by Zhao [28]. The impact of low temperature on the computational results was analyzed as well.

The icebreaking research mentioned above primarily focuses on surface vessels, with limited experimental studies and insufficient available data conducted on submarines, thus few related empirical equations have been proposed for the ice resistance of submarines. However, they may encounter collisions with small-scale floating ice when they need to surface during certain special emergency tasks. This paper aims to investigate the dynamic response of submarines navigating through the surface of floating ice-field using the FEM-ALE coupling approach. The effects of navigation speed, ice thickness, and floating ice size on the polar navigation safety of submarines are studied. Special emphasis is placed on the ice resistance force-time history, as well as the stress–strain distribution on both the surface and interior of the structures, to propose a reference condition for safe navigation. The novelty of this study lies in the fact that the submarines are not modeled as empty shells, but as fully integrated bow structures with internal plates, frames, and trusses. Based on the numerical results, critical conditions and empirical equations are proposed.

2. Theoretical Background

2.1. The Theory of Ship-Ice Interaction

The collision between ships and ice involves a significant amount of non-linear behaviors, such as material non-linearity, geometric non-linearity, and contact non-linearity. Material non-linearity refers to the complex stress–strain relationship of sea ice material, which is influenced by multiple factors. Geometric non-linearity refers to the large deformation of sea ice, which is not suitable for the small deformation principle. Contact non-linearity refers to the unstable contact between two substances, resulting in a large number of non-linear changes. In this work, comprehensive consideration is given to the three non-linearities, where both the sea ice and structure are simplified as elastic-plastic materials, and the contact is computed using an automatic face-to-face contact algorithm. In terms of the algorithm, the simulation calculation of the ship collision process essentially involves the numerical solution of collision equations, which is accomplished using the explicit central difference method [29].

When the results of $0, \ldots, t_n$, time steps are known:

$$M\ddot{U}(t_n) = P(t_n) - F^{\text{int}}(t_n) + H(t_n) - C\dot{U}(t_n) \tag{1}$$

where: $P(t)$—external force vector;

$F^{\text{int}}(t_n)$—internal force vector;

$F^{\text{int}}(t_n) = \sum \int_{\Omega} B^T \sigma_n d\Omega + F^{contact}$, Nodal element internal force;

$H(t_n)$—hourglass resistance.

Inverting the equation, it can get:

$$\ddot{U}(t_n) = M^{-1}\left[P(t_n) - F^{\text{int}}(t_n) + H(t_n) - C\dot{U}(t_n)\right] \tag{2}$$

According to the definition of finite difference method, the displacement, velocity, and acceleration at time t_{n+1} can be solved by the following equation:

$$\dot{U}(t_{n+\frac{1}{2}}) = \dot{U}(t_{n-\frac{1}{2}}) + \ddot{U}(t_n)\Delta t_n \tag{3}$$

$$U(t_{n+1}) = U(t_n) + \dot{U}(t_{n+\frac{1}{2}})\Delta t_{n+\frac{1}{2}} \tag{4}$$

where: $\Delta t_{n+\frac{1}{2}} = \frac{\Delta t_n + \Delta t_{n+1}}{2}$, $\Delta t_{n-\frac{1}{2}} = \frac{\Delta t_n + \Delta t_{n-1}}{2}$.

After a series of calculations, the mass influence coefficients are transformed into a diagonal matrix, achieve decoupling, and enable independent computations. In this calculation process, the influence of the stiffness matrix is assumed to be neglected, and only the central single-point integration is considered, significantly reducing the computational steps and decreasing computation time.

2.2. *The Material Model*

Sea ice is a typical nonlinear material whose physical properties are influenced by various factors, including load strain, temperature, ice age, and salinity. Describing the constitutive model of sea ice accurately in simulation calculations is challenging, and existing research often replaces sea ice materials with ice load models.

At present, there is no perfect ice material model due to the complex physical properties and microstructure of ice. Based on experimental and simulation results, researcher [30] have proposed several reliable ice load models, including elasto-plastic models based on plasticity theory, viscoplastic models, improved models, anisotropic failure models, and visco-elastic-plastic models based on the particle flow theory of sea ice, as well as crushable foam models.

For the elastic-plastic model [31], it is assumed that the material undergoes only elastic deformation when the stress is relatively low. However, when the stress exceeds the yield stress, the stress–strain curve deviates from the elastic stage and becomes a sloping line. Once the yield stage is reached, the material starts exhibiting irreversible plastic deformation, and as the stress continues to increase, the material may fail and undergo damage.

Based on the existing experiments, the stress-strain curve of ice floe is shown in red line in Figure 1. However, elastoplastic models are often used instead of actual ice floes in the simulation of structure-ice interaction. The curve is approximated as two straight lines, the former part is the elastic stage, and the latter part is the plastic stage. The failure modes of sea ice can be simplified into four types: local fragmentation caused by compression failure, instability caused by buckling failure, cracks generated by shear failure, and fractures caused by bending failure. This study focuses on maximum floating ice sizes of only 6 m × 6 m, which are relatively small and rarely experience the latter three failure modes. Typically, only compression failure occurs, specifically in the parts directly in contact with the structure. In this work, the elastic-plastic model is utilized, which effectively captures most properties of the floating ice and enables failure simulation through mesh removal.

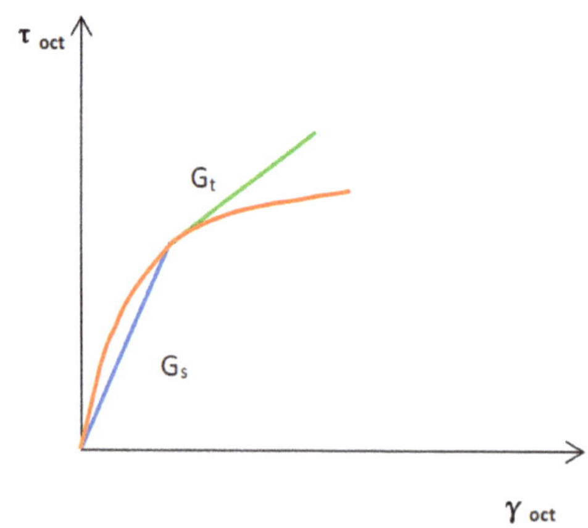

Figure 1. Simplified schematic diagram of elastic-plastic model [31].

2.3. FEM-ALE Coupling Method

Finite element method includes Lagrangian, Eulerian, and ALE methods [32]. In the Lagrangian method, element points are identical to material points, and the element is fixed to the object, moving along with the model nodes. In the Eulerian method, element points are spatial points, and the element is fixed in space, unaffected by the movement of material points. In the ALE description, a reference configuration independent of the actual and initial configurations is introduced, with element points as reference points. The motion of the element in space is arbitrary, independent of the Lagrangian and Eulerian coordinate systems, and can be chosen as needed. When the element motion velocity is reasonable, this method can accurately simulate object deformation and track object motion, making it suitable for modeling nonlinear and large deformation changes, particularly in fluid domains. As shown in Figure 2, A, B and C are grid movements of Lagrangian, Euler, and ALE methods respectively.

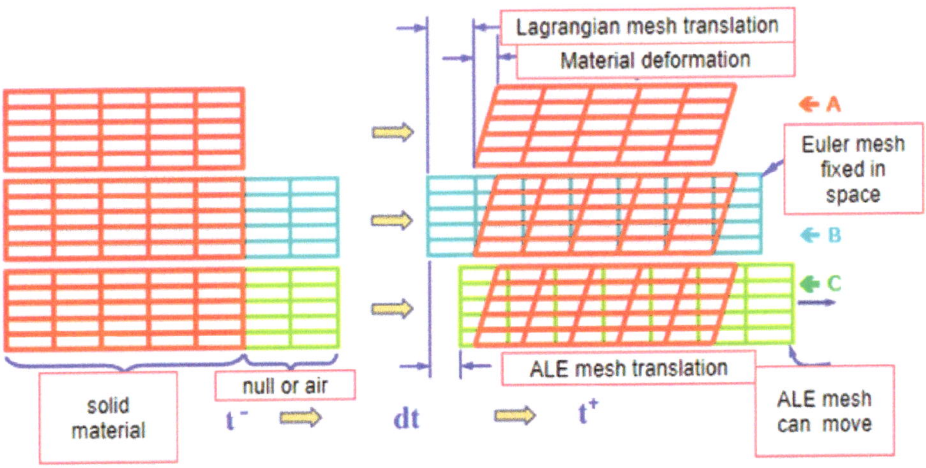

Figure 2. Schematic diagram of mesh movement.

In the traditional Eulerian conservation equations, the conventional terms have been replaced by relative velocity yielding the conservation equations in an arbitrary Lagrangian form, as expressed below [33]:

Mass conservation equation:

$$\frac{\partial \rho}{\partial t}|_x + c \cdot \nabla \rho = -\rho \nabla \cdot v \tag{5}$$

Momentum conservation equation:

$$\rho\left(\frac{\partial \rho}{\partial t}|_x + (c \cdot \nabla)\nu\right) = \nabla \cdot \sigma + \rho b \tag{6}$$

Energy conservation equation:

$$\rho\left(\frac{\partial \rho}{\partial t}|_x + c \cdot \nabla E\right) = \nabla \cdot (\sigma \cdot v) + v \cdot \rho b \tag{7}$$

Internal energy conservation equation:

$$\rho\left(\frac{\partial \rho}{\partial t}|_x + c \cdot \nabla e\right) = \sigma \cdot \nabla^S \nu \tag{8}$$

At each computational instant, two distinct stages are encompassed. In the first stage, material does not flow across the boundaries of the grid, and there is no material overflow throughout the entire calculation process, ensuring mass conservation. In the second stage, material flows across the edges of the elements, referred to as convection. The ALE method computes the transport quantities, internal energy, and momentum of the various physical properties as the material passes through the element boundaries. Unlike the first stage of the Lagrangian method, the second stage involves the generation of an independent motion in the grid, separate from the material. The grid's independent relative motion allows it to return to its original position or any other position that facilitates more accurate calculations. ALE does not support implicit time integration, dynamic relaxation, and contact. However, large deformation motion can be well described by ALE, which is applicable for fluid–structure interaction in fluid dynamics.

The main models in this work includes the submarine, floating ice, and fluid domains, with defined contacts between floating ice and the submarine, and coupling between each model and the fluid domain. Both floating ice and structures use an elastoplastic material model, representing contact between structures and floating ice as contact between elastic bodies. A penalty function algorithm is employed for contact, which checks if nodes penetrate the master surface at each time step and applies a large interface contact force at the penetrated locations. The magnitude of this force is proportional to the penetration depth and the stiffness of the master surface, ensuring high efficiency and applicability.

The collision between the ship and ice in this work is highly complex and involves erosion double-sided contact, where the structure and floating ice act as master and slave surfaces, respectively [34]. The fluid–structure interaction utilizes a penalty function algorithm, and the ALE moving mesh method is employed to calculate fluid motion. Due to the LS-DYNA explicit integration scheme utilizing the central difference method, the accuracy of calculations is affected by the hourglass deformation of quadrilateral meshes. For fluids, hourglass control based on a viscous equation is applied to suppress hourglass deformation, while for solids, hourglass control based on a stiffness equation is used to counteract hourglass deformation by deforming in the opposite direction.

2.4. Colbourne Method

The maximum ice resistance is often employed for evaluating structural strength, whereas the average ice resistance is generally utilized for calculating economic benefits. The validation of the maximum ice resistance results is carried out in this work. The Colbourne method [35], which treats the ice resistance as a unified entity by disregarding infrequent ice fracture occurrences, is adopted to validate the predicted maximum ice resistance based on the numerical simulation. This approximation aligns with the simulation settings in the computational software. According to Colbourne method the total resistance is divided into open water resistance and ice resistance, as expressed by the following equation:

$$\begin{aligned} R_{OW} &= C_{OW}V^2 \\ R_P &= C_P\rho_i g B h_i V^2 C^n \end{aligned} \tag{9}$$

where:

C_{OW}—the open water resistance coefficient,
C_P—the drag coefficient of ice floe,
ρ_i—the density of ice,
h_i—the ice thickness,
g—the acceleration of gravity,
V—the ship speed,
C—the concentration of ice floe.

The drag coefficient of floating ice is a dimensionless value, defined as follows:

$$C_P = \frac{R_P}{\rho_i g B h_i V^2 C^n} \tag{10}$$

where:

B—maximum beam for the structure

Similarly, the non-dimensionalization of navigation speed is also performed in the Colbourne method. It is considered that the Froude number is related to the ice concentration. The ice Froude number is defined as follows:

$$Frp = \frac{V}{\sqrt{gh_iC}} \tag{11}$$

Based on the Colbourne method, it is considered that there exists a linear relationship between the natural logarithm of Cp and the natural logarithm of Fr_p. However, in the case of this study, the structure is unique and lacks relevant experimental data for validation. Therefore, utilizing multiple simulated data obtained from our own simulations, the natural logarithm of Cp and the natural logarithm of Fr_p are calculated, and a corresponding curve can be plotted. The study aims to evaluate the accuracy by inversely verifying the error in the resistance results based on this curve.

3. The Interaction between Submarine and Floating Ice

3.1. The Numerical Setup

A typical submarine is chosen for the structure-ice interaction in this work. To simplify the computational process and ensure accuracy, only the bow part which includes an internally intact plate-frame structure, is utilized for collision analysis, The length of the bow of the structure is 9.2 m, the width is 7.8 m, the height above water is 2.1 m, and the height below water is 6.7 m.

The collision between the structure and the floating ice primarily occurs with the first ice floe. To reduce computational time while maintaining accuracy, the mesh size of the first ice floe is refined to 0.1 m × 0.1 m × 0.1 m, while the remaining ice floes have a mesh size of 0.2 m × 0.2 m × 0.2 m. The ice concentration is set at 50%, and several rectangular ice floes with cross-sectional dimensions of 6 m × 6 m, 4 m × 4 m, and 2 m × 2 m are considered. The structure-ice interaction model and the investigated operating conditions are presented in Figure 3 and Table 1, respectively.

Figure 3. The numerical model for structure-ice interaction.

Table 1. Working condition parameter.

No.	Speed (kn)	Ice Thickness (m)	The Size of the First Floating Ice (m × m)
1	6	1.2	6 × 6
2	4	1.2	6 × 6
3	2	1.2	6 × 6

Table 1. *Cont.*

No.	Speed (kn)	Ice Thickness (m)	The Size of the First Floating Ice (m × m)
4	1	1.2	6 × 6
5	6	1.0	6 × 6
6	6	0.8	6 × 6
7	6	0.5	6 × 6
8	6	1.2	4 × 4
9	6	1.2	2 × 2

3.2. *The Material Parameters*

3.2.1. The Fluid Domain

The fluid domain (sea water and air) adopts the null hydrodynamic material type in LS DYNA. The related parameters are given in Table 2.

Table 2. The material parameters for the fluid domain.

	Density (kg/m^3)	Cut off Pressure (Pa)	Viscosity Coefficient (Pa·s)
Sea water	1.03	−1000	1.79×10^{-3}
Air	1018	−100	1.75×10^{-5}

The effects of seawater on the behavior of the ship and ice can be considered by application of the ALE method. In LS-DYNA, the computation of fluids necessitates the definition of materials and EOS. An EOS such as the Gruneisen model is suggested to simulate water in the current FEM solver. The Gruneisen EOS with cubic shock velocity-particle velocity defines the pressure for a compressed material.

$$P = \frac{\rho_0 C^2 \mu\left[1 + \left(1 - \frac{\gamma_0}{2}\right) - \frac{\alpha}{2}\mu^2\right]}{\left[1 - (S_1 - 1)\mu - S_2\frac{\mu^2}{\mu+1} - S_3\frac{\mu^3}{(\mu+1)^2}\right]^2} + (\gamma_0 + \alpha\mu)E_0 \tag{12}$$

where:

E—the internal energy per initial volume,

C—the intercept of the $\mu_s - \mu_p$ curve (speed of sound in water),

S_1, S_2 and S_3—the coefficients of the slope of the $\mu_s - \mu_p$ curve,

γ_0—the Gruneisen gamma parameter,

α—the first-order volume correction to γ_0.

The seawater characteristics based on the relevant parameters are given in Table 3.

Table 3. The EOS parameters for sea water.

	C	S1	S2	GAMAO	E$_0$ (J)	V$_0$ (m^3)
Sea water	1480.000	1.920	−0.096	0.350	2.895×10^5	1.000

The air domain adopts the linear equation EOS, and the specific equation is shown in Equation (13). The related parameters are given in Table 4.

$$P = C_0 + C_1\mu + C_2\mu^2 + C_3\mu^3 + \left(C_4 + C_5\mu + C_6\mu^2\right)E \tag{13}$$

Table 4. Table The EOS parameters for air.

	C0	C1	C4	C5	E_0 (J)	V_0 (m^3)
Air	0.000	0.000	0.400	0.400	2.530×10^5	1.000

3.2.2. The Material Parameters for Submarines

In this work, an elastic-plastic model more suitable for high-strength steel is adopted which defines the structural deformation during the collision process of structures. Related material parameters are given in Table 5.

Table 5. The material parameters for submarines.

Density (kg/m^3)	Elastic Modulus (Pa)	Possion's Ratio	Yield Strength (Pa)	Tangent Modulus (Pa)	SRC	SRP	Plastic Failure Strain
7850.0000	2.060×10^{11}	0.300	3.900×10^8	1.180×10^9	3200	5.00	0.28

Where: SRC—Strain rate parameter C; SRP—Strain rate parameter *p*.

3.2.3. The Material Parameters for Floating Ice

The ice material employed in this work utilizes the previously mentioned isotropic elastic failure model. This model exhibits elastic deformation when subjected to low stress levels. However, plastic deformation occurs when the stress exceeds the yield strength. In the event that the plastic strain surpasses 0.002, the corresponding mesh elements are automatically deleted due to failure. The related parameters of the model are given in Table 6.

Table 6. The material parameters for floating ice.

Density (kg/m^3)	Tangent Modulus (Pa)	Yield Strength (Pa)	Hardening Modulus (Pa)	Plastic Failure Strain	Failure Pressure (Pa)	Bulk Modulus (Pa)
920.00	2.20×10^9	2.12×10^6	4.26×10^9	0.002	-4.00×10^6	5.26×10^9

3.3. The Verification for Numerical Model

3.3.1. The Verification for Pressure Gradient with Single Floating Ice

Before the numerical simulation of structure-ice interaction, it is necessary to validate the stability of the fluid pressure and the ability of a single ice floe to float stably. The air pressure validation is conducted by examining the variations in air pressure gradient and the temporal changes in air pressure at monitored nodes, as shown in Figure 4.

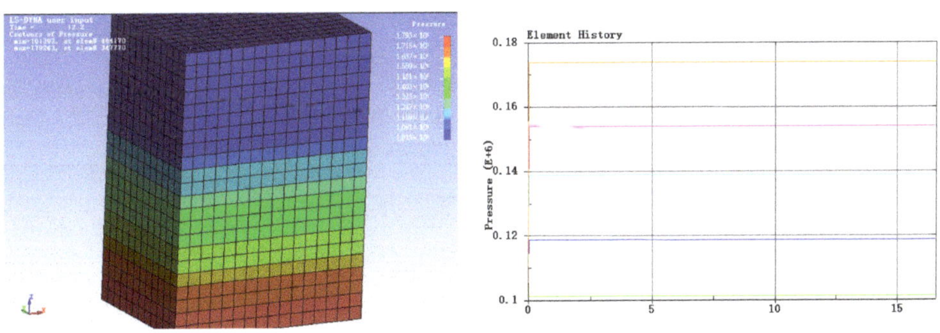

(a) (b)

Figure 4. (**a**) The distribution of pressure gradient; (**b**) the pressure at monitored nodes.

It can be seen from Figure 4 that the pressure gradually increases with depth, and the pressure at the nodes, showing in color lines, remains stable. A floating ice floe is placed within the region, and the results animation is observed and the time history of the ice floe's acceleration is shown in Figure 5. As shown in Figure 5, the acceleration of the individual ice gradually stabilizes around zero, indicating floating stability.

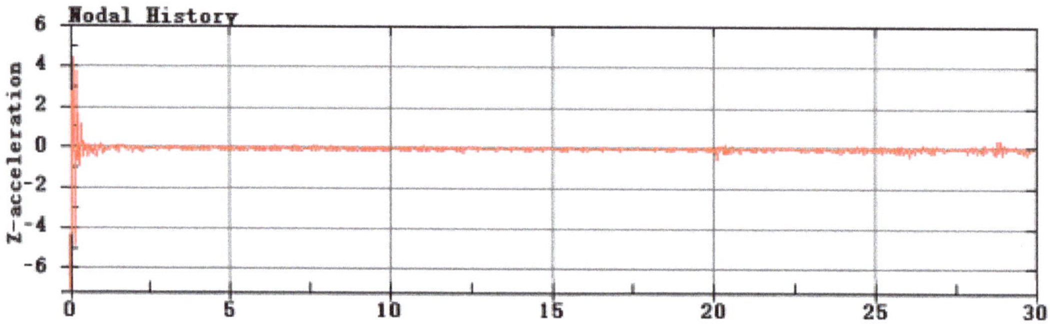

Figure 5. The time history of the ice floe's acceleration.

3.3.2. The Verification for Ice Resistance

There are four different navigation speeds considered in this research, i.e., 1 knot, 2 knots, 4 knots, and 6 knots. In the numerical calculation, a 120-core parallel computation is employed, and the 1 kn working condition requires approximately 36 h. The simulation results provide ice resistance data, excluding the open water resistance. The specific data is shown in Table 7.

Table 7. The numerical results based on structure-ice interaction.

Navigation Speed (kn)	1	2	4	6
Maximum ice resistance (N)	2.44×10^5	4.00×10^5	4.01×10^5	6.54×10^5

By substituting the average ice resistance into Equations (10) and (11), $\ln Fr_p$ and $\ln Cp$ can be obtained, as shown in Table 8.

Table 8. $\ln Fr_p$ and $\ln Cp$ data.

Navigation Speed (kn)	1	2	4	6
$\ln Cp$	3.679	2.787	1.403	1.081
$\ln Fr_p$	−1.551	−0.858	−0.165	0.241

The relation between Fr_p and $\ln Cp$ is presented in Figure 6. Based on the least squares method, the trendline equation is derived as $\ln Cp = -1.526 \ln Fr_p + 1.3474$. Applying the operating conditions of a velocity of 6 knots, ice thickness of 1.2 m, and the dimensions of the initial ice floe as 6 m × 6 m, the ice Froude number is obtained. Then, the natural logarithm of Cp is obtained. Finally, the theoretical ice resistance is calculated, resulting in a value of 5.91×10^5 N. In comparison, the simulation yields a predicted resistance of 6.54×10^5 N, indicating a discrepancy of 10.65%. Therefore, the obtained results are deemed reliable.

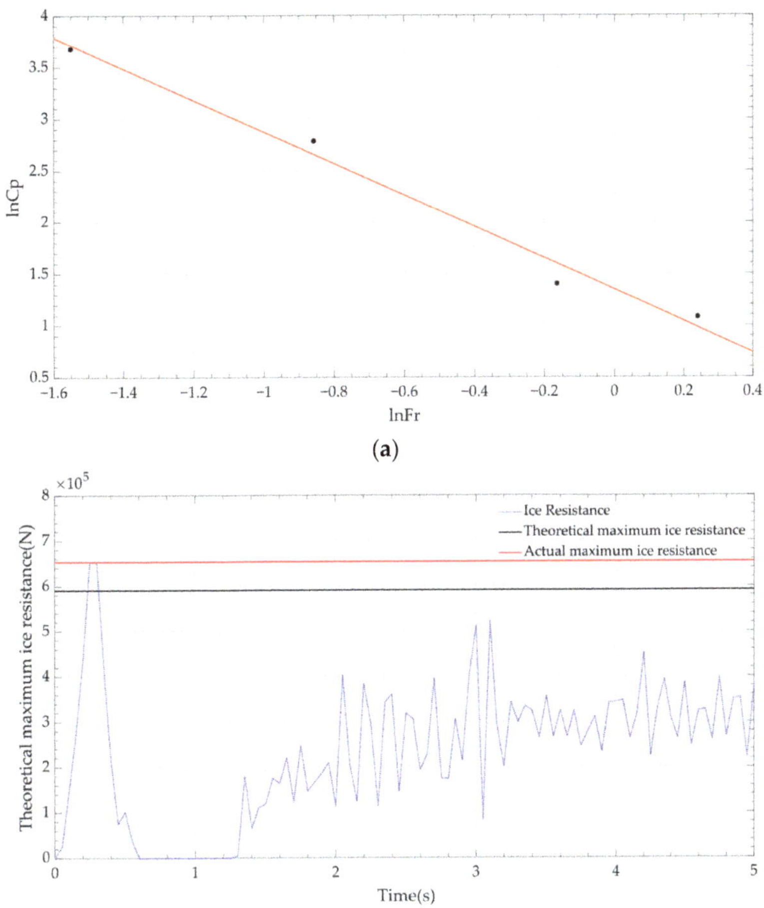

Figure 6. (**a**) The relation between $\ln Fr_p$ and $\ln Cp$; (**b**) the comparison of ice resistance between theory and numerical simulation.

4. Results and Discussion

4.1. The Structural Response of Structure-Ice Interaction

The work conditions in the previous section are subjected to in-depth analysis. Combined with the animation and time history curve, some mechanism of submarine collisions with ice floes was discovered. Different from traditional icebreakers due to the similar size between the submarine and the ice floes, the submarine almost only collides with the first ice floe, and the remaining floes are pushed aside without accumulating. Following impact, the ice floe momentarily attains higher velocity than the structure due to the latter's significantly greater mass, subsequently decelerating until the next impact occurs, as shown in Figure 7. This repetitive loading–unloading process leads to multiple peaks in the response. The location where the structure encounters collisions and the internal circular plate experience significant stress, making them susceptible to plastic deformation.

Figure 7. (a) loading stage; (b) unloading stage.

4.2. The Parametric Study of Structure-Ice Interaction

4.2.1. The Effect of Ship Speed on Structural Response

To study the effect of navigation speed on ice resistance, the ice thickness is maintained at 1.2 m, and the dimensions of the initial ice floe remain constant at 6 m × 6 m. The results are computed for velocities of 6 knots, 4 knots, 2 knots, and 1 knot, respectively. The corresponding results are shown in Figures 8 and 9.

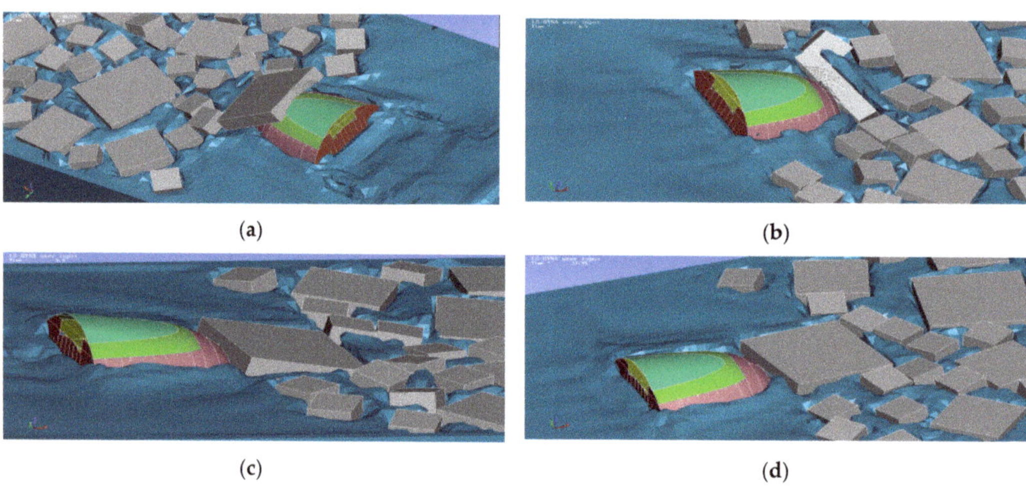

Figure 8. (**a**) The structure-ice interaction at 6 kn; (**b**) the structure-ice interaction at 4 kn; (**c**) the structure-ice interaction at 2 kn; (**d**) the structure-ice interaction at 1 kn.

It can be seen from Figure 8 that when the structure is moving at a speed of 1 knot, a small amount of fragmentation occurs in the first ice floe, and it is predominantly translating along the x-direction. As the submarines move forward, the ice resistance acting on the structure gradually decreases. When the structure sails at speeds of 2 knots, 4 knots, and 6 knots the arc surface at the head of the structure causes the ice floe to turn over, thereby affecting the load characteristics. The maximum and average values of the ice resistance over time are extracted and listed in Table 9.

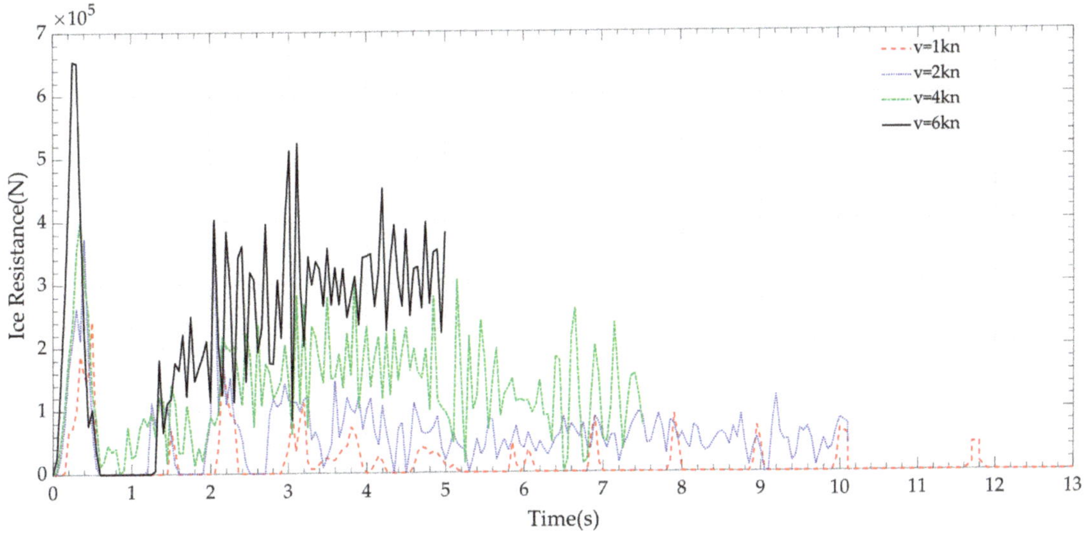

Figure 9. The ice resistance at different navigation speeds.

With the increase in navigation speed, both the average and maximum ice resistance increase. Specifically, there is a slight difference in the maximum ice resistance between 4 knots and 2 knots. It can be seen from Figure 8 that there is a direct connection between the relative position of interaction between ice resistance and structural response in the floating ice field.

Table 9. The ice resistance at different navigation speeds.

Navigation Speed (kn)	Average Ice Resistance (N)	Maximum Ice Resistance (N)
1	1.376×10^4	2.44×10^5
2	6.154×10^4	4.00×10^5
4	1.329×10^5	4.01×10^5
6	2.306×10^5	6.54×10^5

The navigation speed can affect the motion of ice floe, especially the first floe. There is some overturning of ice floe that can be observed at 1 knot and 2 knots. When the speed exceeds 4 knots, the ice floe flips upward, reaching the top of the structure, resulting in a significantly different form of interaction compared to the previous two scenarios.

Furthermore, the effective stresses and plastic strains of the structure at different navigation speeds are shown in Figure 10.

The plastic strains occur can be found in Figure 10. Both effective stress and structural deformation increase as the vessel speed increases. Although the maximum ice resistance is similar at 2 knots and 4 knots, plastic deformation occurs when the speed exceeds 4 knots, while no plastic deformation occurs when the speed is below 2 knots. Thus, when navigating in icy waters with ice floe sizes not smaller than 6 m × 6 m and ice thicknesses not less than 1.2 m, it is recommended to maintain a navigation speed below 4 knots.

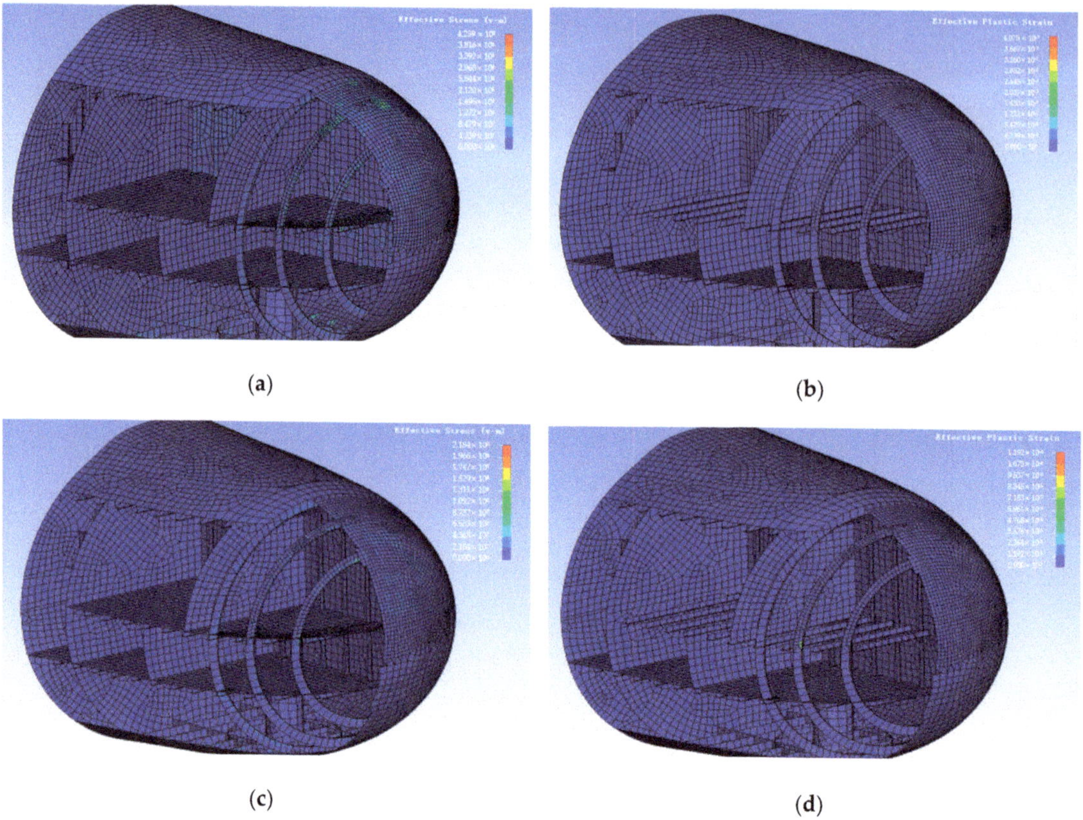

Figure 10. (**a**) The effective stress at 6 kn; (**b**) the structural deformation at 6 kn; (**c**) the effective stress at 4 kn; and (**d**) the structural deformation at 4 kn.

4.2.2. The Effect of Ice Floe Size on Structural Response

To study the influence of ice floe size with an ice thickness of 1.2 m and a vessel speed of 6 knots, the first ice floe is modeled in three different sizes: 6 m × 6 m, 4 m × 4 m, and 2 m × 2 m. Related results are shown in Figure 11.

It can be seen from Figures 11 and 12 that at speed of 6 knots, all three sizes of ice floes experience upturning, indicating a similar mode of motion. The unloading time is influenced by the size of the floating ice. After the collision, smaller ice floes attain higher velocities, leading to easy detachment from the structure, and the subsequent re-contact occurs later. The ice resistances with different ice floe sizes are shown in Table 10.

It can be seen from Table 10 that as the ice floe size increase, both the maximum ice resistance and average ice resistance increase. For the structural stress and plastic deformation, it is observed from Figure 13 that plastic deformation only occurs in the 6 m × 6 m size scenario, while the other two scenarios do not exhibit plastic deformation. Therefore, when the structure is sailing at a speed of 6 knots with an ice thickness of 1.2 m, collisions with ice floes of size 6 m × 6 m should be avoided as much as possible.

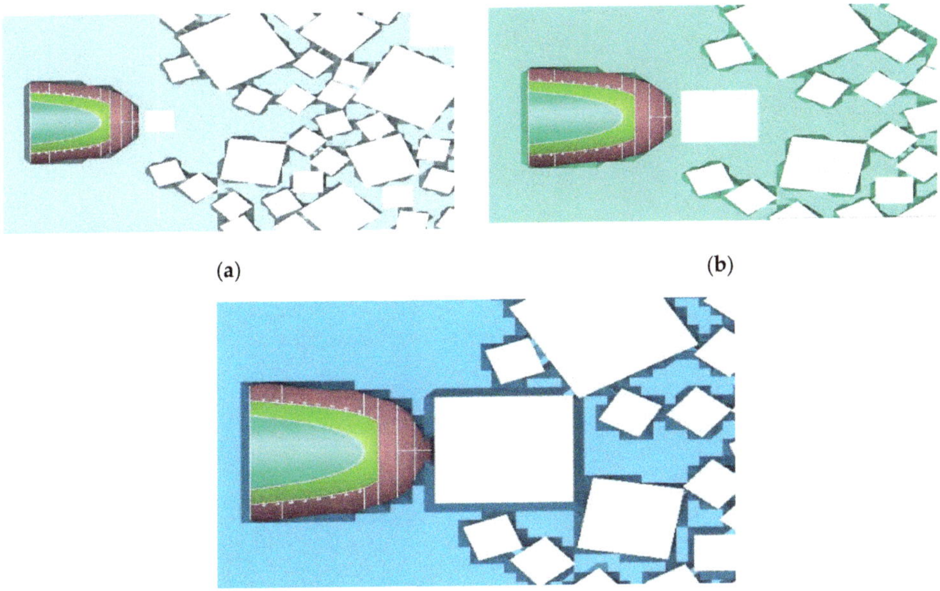

(a) (b) (c)

Figure 11. (**a**) The structure-ice interaction with size 2 m × 2 m of the ice floe; (**b**) the structure-ice interaction with size 4 m × 4 m of the ice floe; (**c**) the structure-ice interaction with size 6 m × 6 m of the ice floe.

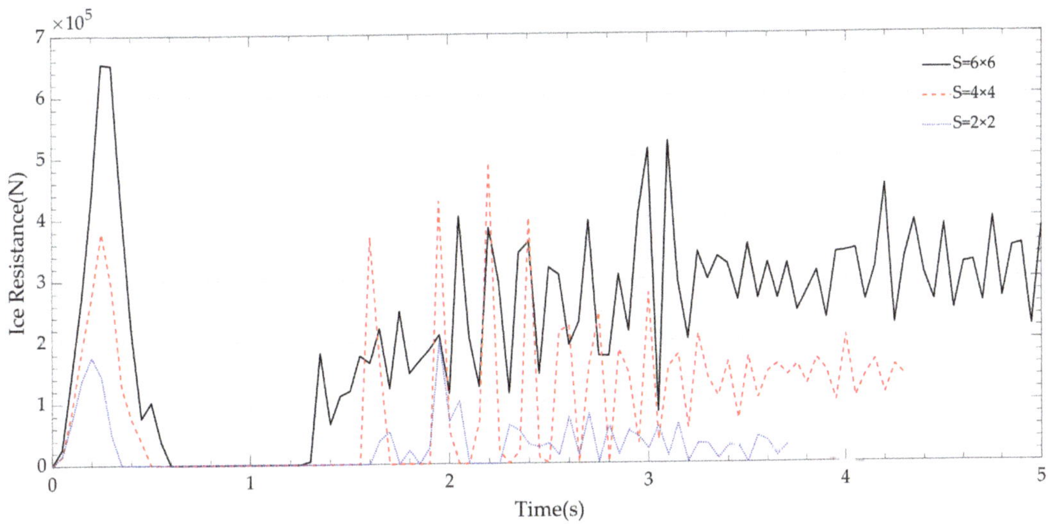

Figure 12. The ice resistance with different ice floe size.

Table 10. The ice resistances with different ice floe size.

Ice Floe Size (m × m)	2 × 2	4 × 4	6 × 6
Maximum ice resistance (N)	2.01×10^5	4.9×10^5	6.54×10^5
Average ice resistance (N)	2.755×10^4	9.845×10^4	2.306×10^5

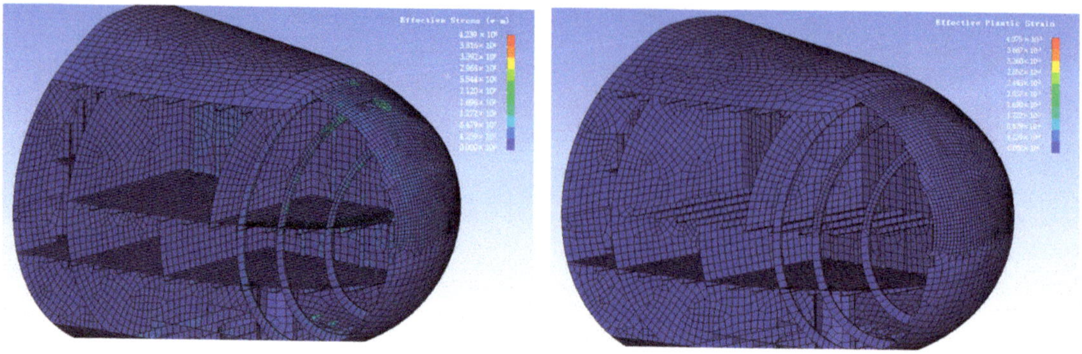

(a) (b)

Figure 13. (**a**) The effective stress at 6 m × 6 m; (**b**) the structural deformation at 6 m × 6 m.

4.2.3. The Effect of Ice Thickness on Structural Response

When studying the influence of ice thickness, while maintaining a vessel speed of 6 knots and a first ice floe size of 6 m × 6 m, the results are calculated for ice thicknesses of 0.5 m, 0.8 m, 1.0 m, and 1.2 m, as shown in Figures 14 and 15.

(a) (b)

Figure 14. (**a**) The result animation of 0.5 m working condition t = 1.15 s; (**b**) the resulting animation of 0.5 m working condition t = 4.40 s.

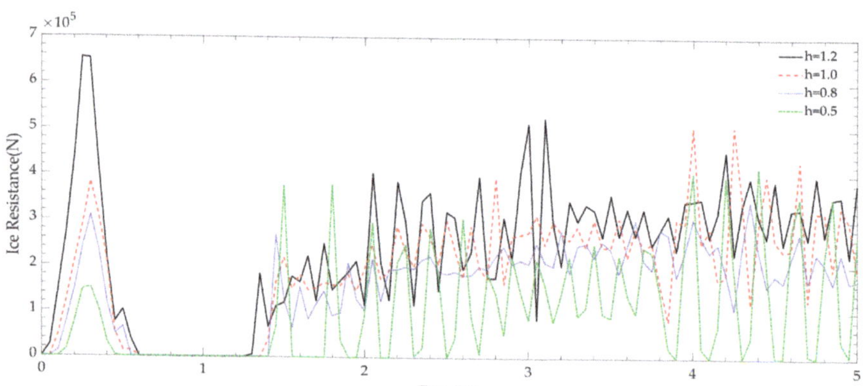

Figure 15. The ice resistance at different ice thickness.

The process of structure-ice interaction with thicknesses of 0.8 m, 1.0 m, and 1.2 m is similar. However, when the ice thickness is 0.5 m, the ice floes are light and easily flip to the top of the structure, undergoing multiple impacts and rebounds. Combined with the analysis of Figure 15, the influence of ice thickness is similar to that of ice floe size, whereby a smaller ice thickness results in a longer duration of separation from contact after impact. The extracted the ice resistance data are shown in Table 11.

Table 11. The ice resistance at different ice thickness.

Ice Thickness (m)	Average Ice Resistance (N)	Maximum Ice Resistance (N)
0.5	9.922×10^4	4.18×10^5
0.8	1.555×10^5	3.43×10^5
1.0	1.913×10^5	5.06×10^5
1.2	2.306×10^5	6.54×10^5

The structure-ice interaction at the four ice thicknesses shows similar behavior. When the ice thickness increases, both the maximum ice resistance and average ice resistance gradually increase. By extracting the structural stress and plastic deformation data, it is observed from Figure 16 that plastic deformation only occurs when the ice thickness is 1.2 m, while the other three scenarios do not exhibit plastic deformation. Therefore, when the structure is sailing at a speed of 6 knots with a floe size of 6 m × 6 m, collisions with ice floes of 1.2 m thickness should be avoided as much as possible.

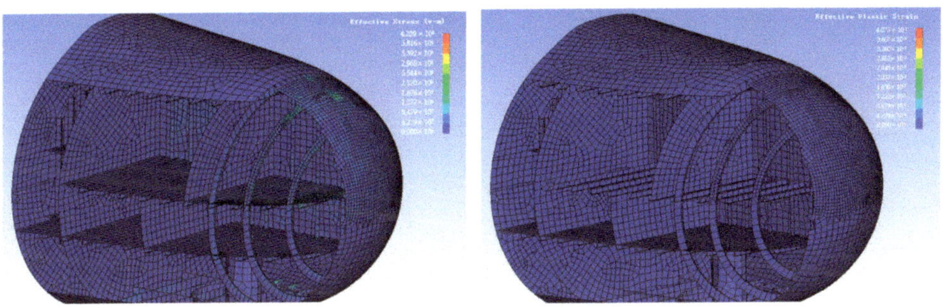

(**a**) The effective stress at 1.2 m (**b**) The structural deformation at 1.2 m

Figure 16. The structural response at different ice thicknesses.

4.3. The Updated Ice Resistance Method

The effect of three factors (navigation speed, ice floe size, ice thickness) on the interaction between submarine and floating ice was studied in this work. The employed Colbourne method, which is only suitable for floating ice conditions, is solely dependent on navigation speed, ice thickness, and ice concentrations, disregarding the size of the floating ice, which is inconsistent with reality and contradicts the simulation results. This work attempts to incorporate the parameter S to represent the size of the floating ice cross-sectional area and introduces a new ice Fourier number, denoted as Equation (14).

$$Fr_p = \frac{V}{\sqrt{gh_iCS}} \tag{14}$$

where, S—the size of the floating ice cross-sectional area

The definition of Cp remains unchanged. Condition 3 represents the critical scenario of both flipped and unflipped floating ice, exhibiting a complex behavior. Condition 7 involves thinner floating ice that flips upwards and undergoes multiple loading–unloading events, where the ice in contact with the bow is not solely limited to the initial ice floe. By excluding the data from these two particular scenarios, the results for the remaining seven conditions are calculated and plotted in Figure 17 and Table 12, yielding the following outcomes:

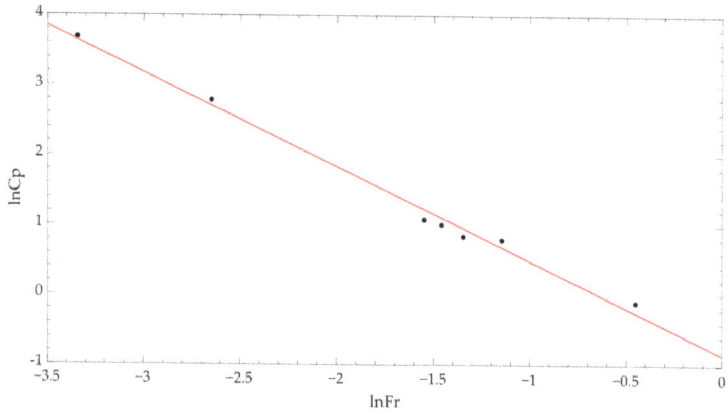

Figure 17. The relation between $\ln Fr_p$ and $\ln Cp$.

Table 12. Ice resistance results after equation improvement.

Number	Case Condition	Ice Resistance Based on Numerical Simulation	$\ln Fr_p$	$\ln Cp$	Ice Resistance Based on the Updated Method	Error
1	6 kn, 6 × 6, 1.2 m	6.54×10^5 N	−1.551	1.081	7.60×10^5 N	13.99%
2	1 kn, 6 × 6, 1.2 m	2.44×10^5 N	−3.343	3.679	2.33×10^5 N	−4.42%
3	2 kn, 6 × 6, 1.2 m	4.00×10^5 N	−2.650	2.787	3.69×10^5 N	−8.45%
6	6 kn, 6 × 6, 0.8 m	3.43×10^5 N	−1.348	0.841	3.87×10^5 N	11.19%
7	6 kn, 6 × 6, 1.0 m	5.06×10^5 N	−1.460	1.007	5.61×10^5 N	9.76%
8	6 kn, 2 × 2, 1.2 m	2.01×10^5 N	−0.452	−0.099	1.74×10^5 N	−15.41%
9	6 kn, 4 × 4, 1.2 m	4.90×10^5 N	−1.146	0.792	4.41×10^5 N	−11.02%

Where, Error = (theoretical − simulation)/theoretical × 100%.

To verify the rationality of the updated ice resistance method, this study computes the case where the initial ice floe size is 5 m × 5 m, navigation speed is 6 knots, and ice thickness is 1.2 m. The maximum ice resistance obtained is 5.43×10^5 N, while the theoretical value calculated using the equation is 5.95×10^5 N, resulting in an error of 8.80%, which falls within an acceptable range. Based on these results, an updated ice resistance method is proposed in this work, which takes into account parameters such as navigation speed, ice thickness, and floating ice size comprehensively, making it a suitable empirical equation for this submarine.

5. Conclusions

In this work, the dynamic response of submarines navigating in floating ice field was conducted based on the FEM-ALE coupled method. Different from traditional icebreakers, the submarine almost only collides with the first ice floe, and the remaining floes are pushed aside without accumulating. The collision is a continuous loading–unloading process, and its frequency is negatively correlated with the three working condition factors. The intensity of dynamic response of the submarine is generally positively correlated with the three working condition factors, but it is locally influenced by the relative position of the collision. All three working condition factors affect the degree of ice floe flipping, thereby influencing the relative collision position.

In this work, plastic deformation of the structure only occurred under the conditions of an ice thickness of 1.2 m, floating ice size of 6 m by 6 m, and navigation speeds of 6 knots and 4 knots. Therefore, it is recommended that the structure should try to avoid more severe conditions than an ice thickness of 1.2 m, a floating ice size of 6 m by 6 m, and a navigation speed of 4 knots when navigating on floating ice field. Based on the simulation results in this paper, an updated ice resistance method was developed by introducing the floating ice size to the Colbourne method, which was validated using a 5 m $\times$ 5 m condition.

Due to the extensive computational time required for each scenario in this study, the mechanism of the impact of floating ice flipping on collisions was not quantitatively investigated. The critical working conditions currently proposed are estimated values that require further refinement into more accurate values through additional research. The empirical equation improvement scheme presented in this paper, based on simulation results, needs to be validated through model tests.

Author Contributions: Conceptualization, Z.C. (Zhanyang Chen) and H.G.; software, Z.C. (Zhongyu Chen); writing—original draft preparation, Z.C. (Zhongyu Chen); writing—review and editing, W.Z. and H.R.; supervision, W.Z.; funding acquisition, W.Z. and G.F. All authors have read and agreed to the published version of the manuscript.

Funding: This paper was supported by the Shandong Provincial Natural Science Foundation, Fund No. ZR2022QE092 and the National Natural Science Foundation of China, Fund No. 52171301, China Postdoctoral Science Foundation, No. 2023M730829.

Institutional Review Board Statement: Not applicable.

Informed Consent Statement: Not applicable.

Data Availability Statement: Data available on request due to restrictions eg privacy or ethical. The data presented in this study are available on request from the corresponding author. The data are not publicly available due to the structure of the submarine is relatively sensitive.

Conflicts of Interest: The authors declare no conflict of interest.

Nomenclature

ALE	Arbitrary Lagrangian-Eulerian
B	maximum beam for the structure
C	the intercept of the $\mu_s - \mu_p$ curve (speed of sound in water)
CFD	Computational Fluid Dynamics
C_{OW}	The open water resistance coefficient
C_P	The drag coefficient of ice floe
DEM	Discrete Element Method
E	The internal energy per initial volume
EOS	Equations Of State
FEM	Finite Element Method
$F^{\mathrm{int}}(t_n)$	Internal force vector
Fr_p	The ice Froude number
g	The acceleration of gravity
h_i	The ice thickness
$H(t_n)$	Hourglass resistance
M	Mass matrix
$P(t)$	External force vector
R_{ow}	Open water resistance
R_p	Ice resistance
S	The size of the floating ice cross-sectional area
SPH	Meshless Method
SRC	Strain rate parameter C
SRP	Strain rate parameter p
S_1	The coefficients of the slope of the $\mu_s - \mu_p$ curve
V	The navigation speed

α	The first-order volume correction to γ_0
γ_0	The Gruneisen gamma parameter
ρ_i	The density of ice

References

1. Huang, L.; Li, Z.; Ryan, C.; Ringsberg, J.W.; Pena, B.; Li, M.; Ding, L.; Thomas, G. Ship resistance when operating in floating ice floes: Derivation, validation, and application of an empirical equation. *Mar. Struct* **2021**, *79*, 103057. [CrossRef]
2. Ghosh, S.; Rubly, C. The emergence of Arctic shipping: Issues, threats, costs, and risk-mitigating strategies of the Polar Code. *Aust. J. Marit. Ocean Aff.* **2015**, *7*, 171–182. [CrossRef]
3. Xue, Y.; Liu, R.; Li, Z.; Han, D. A review for numerical simulation methods of ship-ice interaction. *Ocean Eng.* **2020**, *215*, 107853. [CrossRef]
4. Li, F.; Huang, L. A review of computational simulation methods for a ship advancing in broken ice. *J. Mar. Sci. Eng.* **2022**, *10*, 165. [CrossRef]
5. Huang, L.; Li, M.; Igrec, B.; Cardiff, P.; Stagonas, D.; Thomas, G. Simulation of a ship advancing in floating ice floes. In Proceedings of the Port and Ocean Engineering under Arctic Conditions (POAC), Delft, The Netherlands, 9–13 June 2019.
6. Runeberg, R. On steamers for winter navigation and ice-breaking (including plates at back of volume) Minutes of the Proceedings. *Thomas Telford* **1889**, *97*, 277–301.
7. Shimansky, J.A. Conditional standards of ice qualities of a ship. *ISA Trans.* **1938**, *130*, 1–381.
8. Kashteljan, V.I.; Poznjak, I.I.; Ryvlin, A.J. *Ice Resistance to Motion of a Ship*; Sudostroenie: Leningrad, Russia, 1969.
9. Jones, S.J. Ships in ice-a review. In Proceedings of the 25th Symposium on Naval Hydrodynamics, Washington DC, USA, 8–13 August 2004; National Academies Press: Washington, DC, USA, 2004; Volume 3, pp. 78–85.
10. Lindqvist, G. A straightforward method for calculation of ice resistance of ships. In Proceedings of the POAC, Luleå, Sweden, 12–16 June 1989; pp. 722–735.
11. Riska, K. *Performance of Merchant Vessels in Ice in the Baltic*; Sjafartsverket: Norrköping, Sweden, 1997.
12. Bronnikov, A.V. Analysis of resistance of cargo ships going through pack ice. *Trans. Leningr. Shipbuild. Inst.* **1959**, *24*, 13.
13. Mellor, M. Ship resistance in thick brash ice. *Cold Reg. Sci. Technol.* **1980**, *3*, 305–321. [CrossRef]
14. Daley, C.G. Ice Edge contact-A brittle failure process model. *Loads* **1991**, *144*, 1–92.
15. Cho, S.R.; Jeong, S.Y.; Lee, S. Development of effective model test in pack ice conditions of square-type ice model basin. *Ocean Eng.* **2013**, *67*, 35–44. [CrossRef]
16. Kim, M.C.; Lee, W.J.; Shin, Y.J. Comparative study on the resistance performance of an icebreaking cargo vessel according to the variation of waterline angles in pack ice conditions. *Int. J. Nav. Archit. Ocean Eng.* **2014**, *6*, 876–893. [CrossRef]
17. Jeong, S.Y.; Choi, K. A review on ice resistance prediction equation for icebreaking vessels. In Proceedings of the 19th International Offshore and Polar Engineering, Conference, Chennai, India, 20–24 September 2009; pp. 587–592.
18. Wang, C.; Sun, W. A review of the resistance forecast method of icebreaker. *Ship Sci. Technol.* **2015**, *37*, 1–8.
19. Xiang, J.; Munjiza, A.; Latham, J.P.; Guises, R. On the validation of DEM and FEM-DEM models in 2D and 3D. *Eng. Comput.* **2009**, *26*, 673–687. [CrossRef]
20. Huang, L.; Tuhkuri, J.; Igrec, B.; Li, M.; Stagonas, D.; Toffoli, A.; Cardiff, P.; Thomas, G. Ship resistance when operating in floating ice floes: A combined CFD&DEM approach. *Mar. Struct.* **2020**, *74*, 102817.
21. Zhang, M.; Garme, K.; Burman, M.; Zhou, L. A Numerical Ice Load Prediction Model Based on Ice-Hull Collision Mechanism. *Appl. Sci.* **2020**, *10*, 692. [CrossRef]
22. Tsarau, A.; Lubbad, R.; Løset, S. A numerical model for simulation of the hydrodynamic interactions between a marine floater and broken sea ice. *Cold Reg. Sci. Technol.* **2014**, *103*, 1–14. [CrossRef]
23. Ji, H.; Zhang, J. Fluid-structure interaction method for ice-column collision based on cohesive element method. *Shock Vib.* **2023**, *42*, 41–47+74.
24. Song, M.; Kim, E.; Amdahl, J.; Greco, M.; Souli, M. Numerical Investigation of Fluid-Ice-Structure Interaction During Collision by an Arbitrary Lagrangian Eulerian Method. In Proceedings of the International Conference on Ocean, Offshore and Arctic Engineering Omae, Busan, Republic of Korea, 18–24 June 2016.
25. Su, B.; Riska, K.; Moan, T. Numerical study of ice-induced loads on ship hulls. *Mar. Struct.* **2011**, *24*, 132–152. [CrossRef]
26. Chai, W.; Leira, B.J.; Naess, A. Short-term extreme ice loads prediction and fatigue damage evaluation for an icebreaker. *Ships Offshore Struct.* **2018**, *13* (Suppl. S1), 127–137. [CrossRef]
27. Zhao, W.; Leira, B.J.; Kim, E.; Feng, G.; Sinsabvarodom, C. A probabilistic framework for the fatigue damage assessment of ships navigating through level Ice fields. *Appl. Ocean Res.* **2021**, *111*, 102624. [CrossRef]
28. Zhao, W.; Leira, B.J.; Feng, G.; Gao, C.; Cui, T. A reliability approach to fatigue crack propagation analysis of ship structures in polar regions. *Mar. Struct.* **2021**, *80*, 103075. [CrossRef]
29. Zheng, Y. Research on Calculation Method of Ice Load for Polar Heavy Icebreaker. Ph.D. Thesis, Harbin Engineering University, Harbin, China, 2019.
30. Gao, Y. An Elastic-Plastic Ice Material Model and Iceberg Shape Sensitivity Analysis for Ship-Iceberg Collision Scenario. Ph.D. Thesis, Shanghai Jiao Tong University, Shanghai, China, 2015.
31. Lu, S.; Shen, H.T. Constitutive laws for river ice dynamics. *Ice Surf. Wate* **1998**, *32*, 109–116.

32. Mao, F. Study on the Bow Configuration of Polar Sailing Ship Based on Icebreaking Capacity. Master's Thesis, Harbin Engineering University, Harbin, China, 2017.
33. Zhang, X. Numerical Simulation of Fluid-Structure Interaction Arising from Bottom Slamming. Master's Thesis, Dalian University of Technology, Dalian, China, 2007.
34. Ni, B.Y.; Huang, Q.; Chen, W.S. Numerical simulation of ice resistance of ship turning in level ice zone considering fluid effects. *J. Ship Res.* **2020**, *15*, 1–7.
35. Colbourne, D.B. Scaling pack ice and iceberg loads on moored ship shapes. *Ocean. Eng. Int.* **2000**, *4*, 39–45.

Journal of
Marine Science and Engineering

MDPI

Article

Experimental and Numerical Study on Crack Propagation of Cracked Plates under Low Cycle Fatigue Loads

Dong Qin [1], Xu Geng [2,*], Zhao Jie [3] and Hu Yaoyu [4]

1 Key Laboratory of High Performance Ship Technology, Wuhan University of Technology, Ministry of Education, Wuhan 430063, China
2 School of Civil Engineering and Architecture, Wuhan Polytechnic University, Wuhan 430048, China
3 School of Naval Architecture and Navigation, Wuhan Technical College of Communications, Wuhan 430065, China; strlab@whut.edu.cn
4 Marine Equipment and Technology Institute, Jiangsu University of Science and Technology, Zhenjiang 212000, China; huyaoyu@whut.edu.cn
* Correspondence: xugeng_1988@163.com

Abstract: The traditional study on fatigue strength for ship structures usually focuses on high cycle fatigue and ignores low cycle fatigue. However, given the recent trend towards large-scale ship development, the stress and deformation experienced by ship structures are becoming increasingly significant, leading to greater attention being paid to low cycle fatigue damage. Therefore, experimental and numerical studies on crack propagation behavior of cracked plates under low cycle fatigue loads were carried out in this paper, in order to explain the fatigue crack propagation mechanism. The effect of the stress ratio and maximum applied load on the crack propagation behavior was investigated by conducting experimental research on the cracked plate of AH32 steel. The experimental results show that an increasing maximum applied load and decreasing stress ratio will shorten the fatigue life of the cracked plate. Meanwhile, based on the finite element method, the distribution of the stress–strain field at the crack tip and the effect of crack closure were evaluated. The influencing factors such as the stress ratio and crack length were considered in numerical studies, which provided a new way to study the low cycle fatigue crack propagation behavior.

Keywords: low cycle fatigue; crack growth; cracked plate; experimental study

Citation: Qin, D.; Geng, X.; Jie, Z.; Yaoyu, H. Experimental and Numerical Study on Crack Propagation of Cracked Plates under Low Cycle Fatigue Loads. *J. Mar. Sci. Eng.* **2023**, *11*, 1436. https://doi.org/10.3390/jmse11071436

Academic Editor: José António Correia

Received: 29 June 2023
Revised: 13 July 2023
Accepted: 14 July 2023
Published: 18 July 2023

1. Introduction

Our foreign trade has flourished in recent years, and its dependence on the marine environment and resources has increased substantially. Developing the marine economy and building "maritime power" has become an inevitable choice for our economy to reach a new level. The trend of large-scale development of ships is inevitable. However, given the recent trend towards large-scale ship development, the stress and deformation experienced by ship structures are becoming increasingly significant, leading to greater attention being paid to low cycle fatigue damage.

The origins of the low cycle fatigue problem date back to the 1960s, when Manson and Coffin [1] established the low cycle fatigue theory, respectively. Their research indicated that plastic deformation of materials under high-stress amplitude loading was the primary cause of fatigue damage in structures. As further research was conducted in this area, both domestic and international scholars began to recognize the significance of low cycle fatigue on ship structures [2–8]. In 1965, Dunham [2] conducted fatigue tests on large submarine structures with full-scale models, observing crack propagation behavior under low cycle fatigue loading. Later, Jordan and Cochran [3,4] investigated the fatigue damage phenomenon of 86 ships in operation, finding that structural fatigue damage was a crucial factor leading to the failure of ships. Munse [5] summarized the findings of Jordan's research, developing S-N fatigue curves for locations prone to fatigue damage and proposing a fatigue strength assessment applicable to different hull components. Radek [6] conducted low cycle fatigue tests using miniature test samples, which demonstrated that cyclic parameters produced consistent results regardless of the varying sample volumes.

In the 1980s, researchers in domestic academia began to focus on studying the problem of low cycle fatigue in ship structures. In 1991, Chen [9] conducted tests on large submarine structures using scale models and employed similar theoretical approaches to analyze the results. Building on Urm's [10] work on low cycle fatigue in oil tankers, Cui [11] investigated the causes of low cycle fatigue damage in ship structures and developed a nonlinear model for calculating accumulative fatigue damage to predict their low cycle fatigue life. In 2008, Liu [12] carried out experiments on 921A steel plates to study the behavior of crack propagation in damaged ship structures under cyclic tensile loading, which served as a foundation for subsequent predictions of low cycle fatigue in ship structures.

The cracked plate will gradually propagate under cyclic loading, and the stress and strain field near the crack tip will change accordingly. The fracture parameters, such as the stress intensity factor K [13–19], crack tip opening displacement (CTOD) [20,21], the J-integral [22], the size of the reversed plastic zone [23–25], etc., are important factors to assess the fatigue crack propagation life. Fatigue crack propagation research has been primarily conducted by scholars from the perspective of the crack tip stress distribution [26,27] and crack closure effect [28–30], etc., leading to several valuable conclusions [31–40]. When a severe stress concentration occurs, plastic deformation arises in the surrounding area, which gradually accumulates, eventually resulting in crack propagation. As the crack tip passes through the plastic zone, a plastic trailing zone forms at the original crack tip, creating a significant residual compressive stress at the crack tip. This residual compressive stress causes crack closure, which affects the stress–strain field distribution near the crack tip, ultimately reducing the effective stress intensity factor range and the crack growth rate. Therefore, the paper will carry out experimental and numerical studies on the crack propagation behavior of cracked plates under low cycle fatigue loads, in order to explain the fatigue crack propagation mechanism. The effect of the stress ratio and maximum applied load on the crack propagation behavior was investigated by conducting experimental research on a cracked plate of AH32 steel. Meanwhile, based on the finite element method, the distribution of the stress–strain field at the crack tip and the effect of crack closure were evaluated. The influencing factors such as the stress ratio and crack length were considered in numerical studies, which provided a new way to study the low cycle fatigue crack propagation behavior.

2. Experimental Investigations on the Crack Growth Behavior of Cracked Plates under Low Cycle Fatigue Loads

2.1. Experimental Setup

The experiments on low cycle fatigue crack growth were conducted on a 3 mm thick steel plate with a central crack, and the standard used for the tests is ASTNE 647-05 [41]. The dimensions of the plate and crack are shown in Figure 1, where the initial crack length is a_0 = 16 mm. The material used for the tests was AH32 steel, which is a high-strength steel commonly used in the manufacturing of hull and deck components for ocean-going, coastal, and inland waterway shipping vessels. The basic mechanical properties of the material are listed in Table 1. The chemical composition of the material is as follows (in weight percentage): Carbon (C) 0.18%, Aluminum (Al) 0.02%, Silicon (Si) 0.32%, Manganese (Mn) 1.4%, Phosphorus (P) 0.04%, Sulfur (S) 0.04%, and Niobium (Nb) 0.02%.

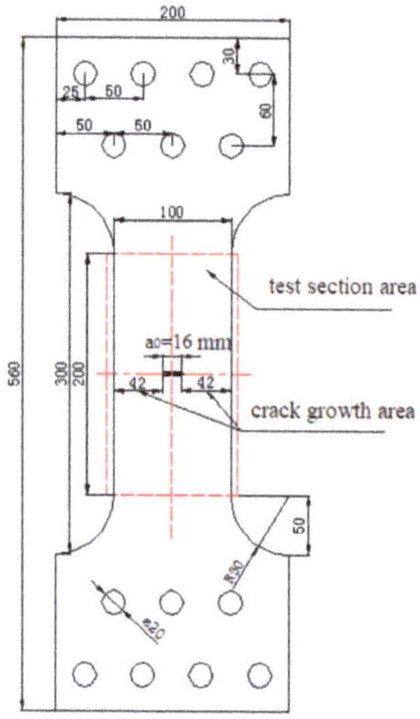

Figure 1. The geometric dimensions of central cracked plate.

Table 1. Mechanical behavior for AH32 steel.

Yield Stress (MPa)	Ultimate Tensile Strength (MPa)	Elastic Modulus (GPa) E	Poisson's Ratio ν
345	510	206	0.3

The experiments were conducted at room temperature of approximately 20–25 °C and in the air. A computer-controlled servo-hydraulic test machine MTS322 (±250 kN) was used in the testing process, as shown in Figure 2. Each specimen was subjected to load control at a frequency of 1 Hz using a sine wave, with a maximum applied load of 95 kN and stress ratios of 0, 0.1, and 0.2, respectively. The applied cyclic loads used in the present experiment are categorized as low cycle fatigue loads due to the fact that the fatigue lives of smooth specimens subjected to the same cyclic loads are shorter than 10,000 cycles. The crack length was determined by a metallographic microscope connected to a video travelling microscope, while an electronic extensometer was installed on both sides of the crack surface to measure the crack opening displacement (COD). The test procedure for evaluating the behavior of a cracked plate can be divided into three distinct steps. Firstly, the area around the crack is carefully sanded to ensure a smooth surface, and strain gauges are affixed to collect data on the crack tip strain value. After the preparation work, the cracked plate is fixed to the MTS by bolts. Then, a single loading is required before cyclic loading to remove the initial deformation of the cracked plate. Finally, the loading equipment program is adjusted, and cyclic loading is performed while collecting data on COD using an extensometer and observing the crack growth process using a metallographic microscope. The MTS is set up with displacement protection to prevent excessive displacement caused by fracture failure of the cracked plate to prevent damage to the machine. The loading conditions and experimental results of AH32 steel are summarized in Table 2.

Table 2. Loading conditions and experimental results of AH32 steel.

Specimen No.	P_{max}/kN	Stress Ratio R	Fatigue Life
01	95	0	2142
02	95	0.1	2455
03	95	0.2	4670
04	100	0.2	3425

Figure 2. Scheme of fatigue crack growth test.

2.2. Experimental Result and Discussion

In order to study the impact of the maximum applied load and stress ratio on fatigue crack growth behavior in cracked plates, a series of low cycle fatigue tests were performed. As depicted in Figure 3, the results indicate that an increase in the maximum applied load at a constant stress ratio leads to an increase in crack growth rate, resulting in shorter fatigue life. Conversely, when the maximum applied load remains constant, an increase in the stress ratio leads to a decrease in the crack growth rate, thereby prolonging the material's fatigue life. Figure 4 illustrates how crack opening displacement (COD) varies under different stress ratios, revealing that the effect of various loads on COD differs. Specifically, as the stress ratio increases, the value of COD does not change significantly during the early stages, displaying an approximately linear relationship. However, once the crack has propagated to a certain extent, the COD value begins to increase significantly until the plate fractures. Notably, the change in COD value accurately represents the crack growth rate.

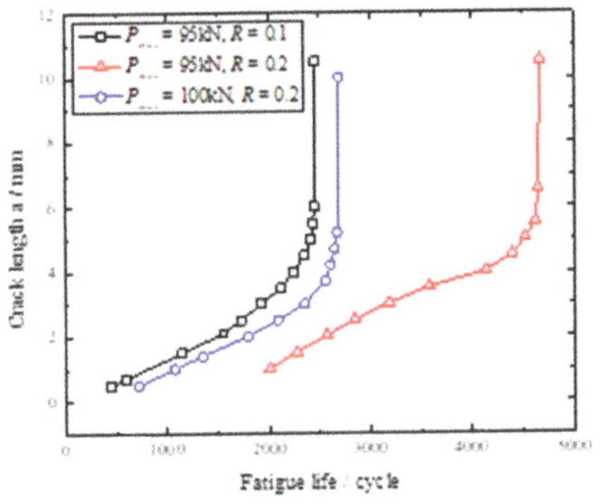

Figure 3. The fatigue life vs. crack length.

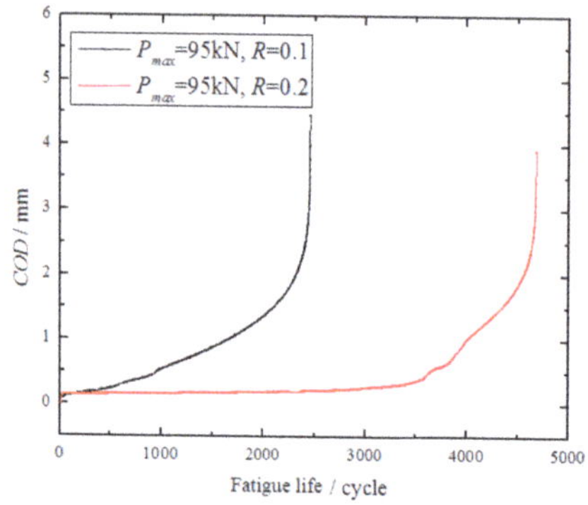

Figure 4. The fatigue life vs. COD.

2.3. *Observation of Fracture Morphology*

In Figure 5, the fracture morphology of an AH32 steel specimen examined during a fatigue test with a stress ratio R = 0.1 is presented. The prefabricated crack area has a relatively smooth surface because it was produced using wire-cut EDM. It is noticeable that there is a significant accumulation of plastic deformation near the crack tip region, and the crack growth zone exhibits plastic slip, which is relatively smoother than the accelerated growth zone. This phenomenon occurs due to the continuous stretching and extrusion of the upper and lower surfaces of the crack during its growth process. As the crack enters the accelerated growth zone, it primarily stretches instead of being extruded, resulting in a comparatively rough section in this part. This section is similar to the fracture surface observed in cases of static load damage.

Figure 5. The fracture morphology of cracked plate.

3. Numerical Analysis and Discussion

This paragraph describes a finite element (FE) analysis performed on a central cracked plate under constant amplitude loading. To simulate high stress and strain gradients near the crack tip more accurately, a singular element is adopted at the crack tip and a refined mesh is used in the region around the crack, as shown in Figure 6. A 1/4 model is considered due to symmetry with dimensions of length L = 100 mm, width w = 50 mm, and initial crack length a_0 = 8 mm. Four-node plane-stress elements with full integration (CPS4) are used in the FE mesh model. To simulate crack growth behavior better, a dynamic crack growth approach is employed. This involves periodic release nodes, which allow the crack to propagate gradually. Contacts and constraints are used to simulate the boundary conditions. Specifically, contacts are arranged on both the upper and lower crack surfaces and a rigid line is set at the partition interface to prevent the upper and lower surfaces from crossing each other. The master surface refers to the rigid line, while the slave surface pertains to the crack surface. Since there is no inclination for sliding to occur between the contact surfaces, a friction coefficient of zero has been assigned. The process begins with the bonding of nodes located on the contact surface, followed by the gradual release of these nodes through low cycle fatigue loads to grow the crack. In this finite element analysis, appropriate boundary conditions have been implemented by setting X-symmetric constraints on the Y-axis and Y-symmetric constraints on the X-axis to account for symmetry. The Chaboche kinematic hardening model has been utilized to accurately capture the elastic-plastic stress–strain behavior of the component with a crack. This model is particularly suitable for materials exhibiting Masing-type behavior. The cyclic plasticity model parameters have been optimized based on fatigue experimental data from AH-32 steel and are presented in Table 3.

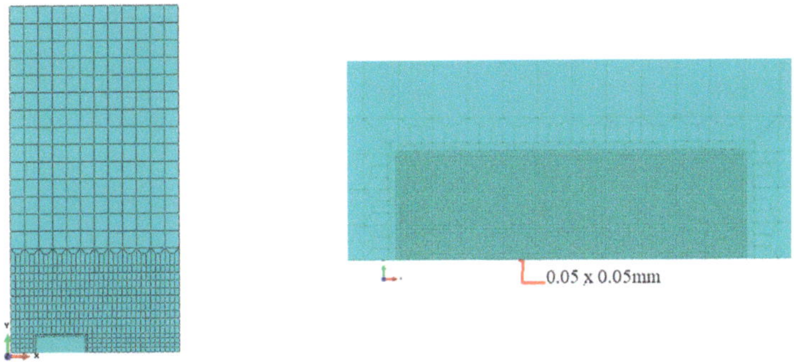

(a) The FE model (b) The refined region

Figure 6. FE model and refined mesh.

Table 3. Parameters of the Chaboche model.

E = 206 GPa	V = 0.3	Q = 72 MPa	k = 8 MPa
C_1 = 314,310	C_2 = 28,071	C_3 = 1950	
α_1 = 800	α_2 = 321	α_3 = 0	

To balance between accuracy and computational time, it is necessary to select an appropriate refined mesh size. In this study, three mesh sizes of 0.1 × 0.1 mm, 0.05 × 0.05 mm, and 0.02 × 0.02 mm were used to simulate fatigue crack propagation under a constant amplitude cyclic load with a maximum load of P_{max} = 95 kN and a stress ratio of R = 0.1. As can be seen from Table 4, there types of mesh were considered to calculate the crack closure parameter U, and the difference between the three mesh sizes was relatively small. While the 0.1 × 0.1 mm mesh may provide an overall assessment of a good economy, it falls short in accurately simulating the stress distribution at the crack tip. To address this limitation, a refined mesh of 0.05 × 0.05 mm is employed in the FE model. This refined mesh offers the dual advantage of efficient computation time and accurate simulation of stress distribution.

Table 4. Results of convergence study.

Mesh Size	Crack Closure Parameter U	Difference/%
0.1 × 0.1 mm	0.725	1.11
0.05 × 0.05 mm	0.717	-
0.02 × 0.02 mm	0.713	0.56

To validate the rationality of the numerical model, a finite element analysis was performed with a stress ratio of 0.1. The obtained stress–strain curve from the analysis was then compared with the experimental stress–strain curve, as shown in Figure 7. From Figure 7, it can be observed that the experimental and numerical stress–strain curves are in good agreement, indicating that the chosen material parameters in the FE model accurately reflect the stress–strain behavior of the material and the rationality of the numerical model.

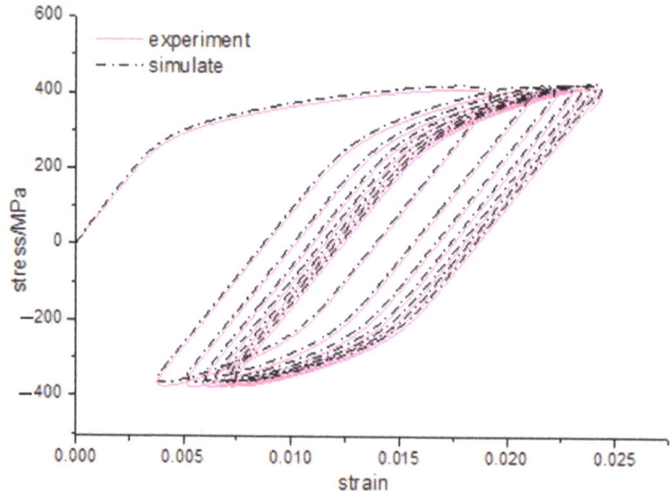

Figure 7. Comparison of stress–strain curves.

3.1. Analysis of Crack Opening Displacement

The opening and closing of crack opening displacement (COD) can better reflect the change in the crack surface during the crack propagation process. Figure 8 shows the variation of COD under different stress ratios and a maximum applied load of 95 kN. It is evident that the crack is closed at the minimum applied load, with increasing stress ratios leading to larger closure effects. This implies that higher stress ratios tend to bring the crack surfaces closer together, thereby hindering crack propagation. Furthermore, the COD is reduced at the maximum applied load with an increasing stress ratio, primarily due to a decrease in effective driving stress. As the stress ratio increases, the proportion of the applied load that contributes to driving the crack further decreases. Consequently, the crack growth rates are reduced due to the reduced driving force. This finding indicates that higher stress ratios exacerbate the crack closure phenomenon, resulting in reduced crack growth rates, and ultimately causing a retardation effect.

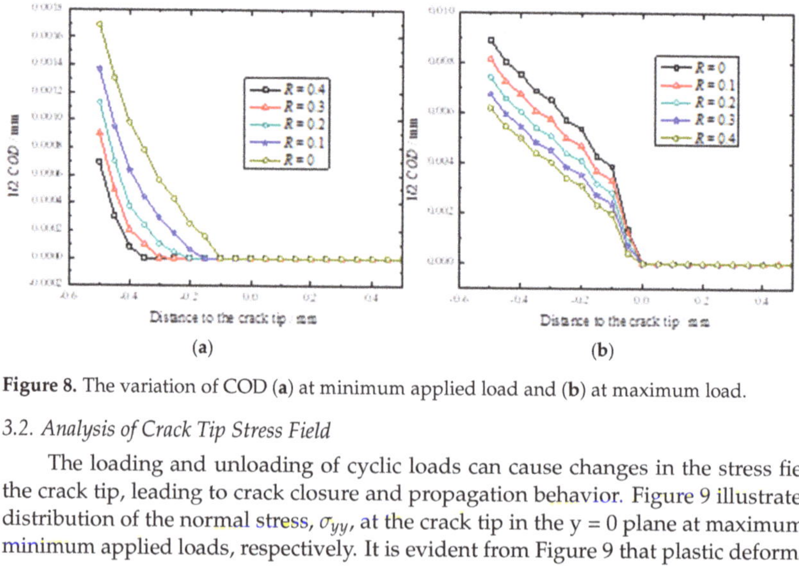

Figure 8. The variation of COD (**a**) at minimum applied load and (**b**) at maximum load.

3.2. Analysis of Crack Tip Stress Field

The loading and unloading of cyclic loads can cause changes in the stress field at the crack tip, leading to crack closure and propagation behavior. Figure 9 illustrates the distribution of the normal stress, σ_{yy}, at the crack tip in the y = 0 plane at maximum and minimum applied loads, respectively. It is evident from Figure 9 that plastic deformation occurs behind the crack tip as the load gradually decreases to the minimum applied load, resulting in a compressive stress field that partially closes the crack. As the crack length increases, the residual compressive stress at the crack tip remains almost constant. However, the stress concentration at the crack tip can cause the material to enter a reverse yielding state. Upon loading to the maximum applied load, the crack opens and enters the forward-yielding state, resulting in tensile plastic strain. This analysis indicates that during cyclic loading and unloading, the crack tip experiences reciprocal action between forward and reverse yielding states, leading to the gradual accumulation of plastic strain and prompting crack propagation.

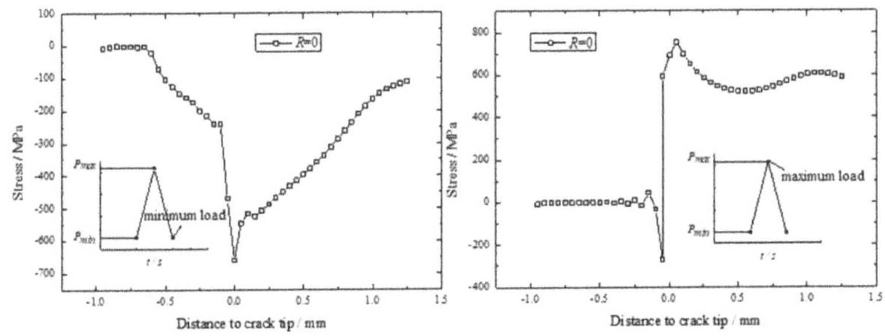

(a) (b)

Figure 9. The distribution of crack tip stress field (**a**) at minimum applied load and (**b**) at maximum load.

3.3. *Analysis of Crack Closure Parameter*

The load ratio parameter U, which is commonly employed to characterize the development of crack closure during fatigue crack propagation, can be expressed as:

$$U = \frac{K_{\max} - K_{op}}{K_{\max} - K_{\min}} \tag{1}$$

where $K_{\max}$ is the maximum intensity factor, $K_{\min}$ is the minimum intensity factor, and K_{op} is the intensity factor with respect to the crack opening load. The load parameter takes values between 0, which indicates complete crack closure, and 1, which corresponds to the fully open crack condition.

Figure 10 illustrates the load ratio parameter U for different stress ratios, with a comparison between the numerical simulation and test results, where the solid lines represent the FE results and the solid circles represent the test results. The results show that as the stress ratio increases, the load ratio parameter also increases while the crack closure effect weakens. This is because larger stress ratios increase the minimum stress level, which offsets some of the residual compressive stress at the crack tip that promotes crack closure. Consequently, the crack closure effect diminishes. While Figure 10 shows noticeable differences between the experimental and numerical results, the latter demonstrates a more prominent crack closure effect. The average errors in the load ratio parameter for increasing stress ratios are 3.6%, 3.9%, 4.1%, and 4.4%, respectively. Possible reasons for these errors include the idealized numerical simulation of the crack without accounting for blunting at the crack tip, which affects the level of crack closure. Additionally, the extensometer's relatively large distance from the crack tip in the experiment may have resulted in small measured results due to its insensitivity. Lastly, processing procedures and the initial defects of the specimen could also contribute to measurement errors.

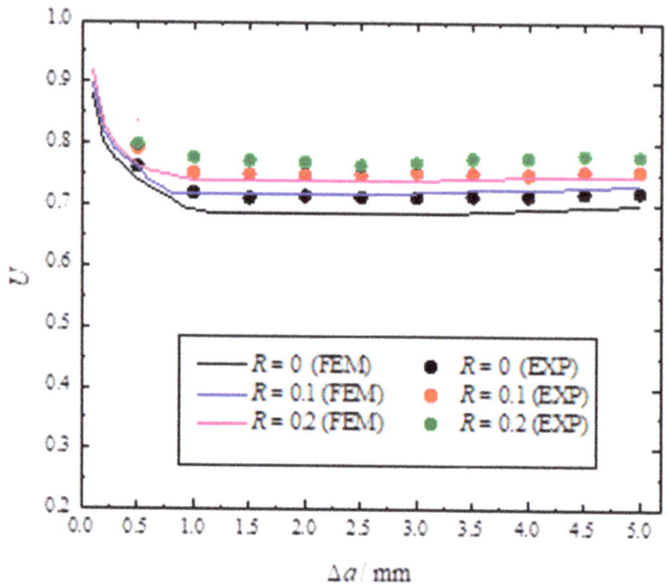

Figure 10. The crack closure level under different stress ratios.

3.4. Relationship between Accumulative Plastic Strain and Crack Closure Parameter

3.4.1. Study of Crack Closure Mechanism and Evolution of Accumulative Plastic Strain

The phenomenon of crack closure is closely linked to the accumulative plastic strain at the crack tip. When a crack propagates, the plastic zone ahead of the crack tip affects the formation of the trailing zone, while the crack closure effect influences the stress–strain distribution in the vicinity of the crack tip, thus slowing down the crack growth rate. The accumulative plastic strain at the crack tip has two impacts on the crack closure phenomenon. Firstly, it affects the current crack length's crack closure effect, and secondly, it influences the residual compressive stress field in the trailing zone of the new crack tip after crack propagation due to the irrecoverable deformation formed at the previous crack tip, which in turn affects the crack closure effect.

Based on Figure 11, it can be observed that the accumulative plastic strain at the crack tip experiences a peak value, followed by a decreasing trend, which is primarily influenced by the preceding hardening process [42]. The initial peak value of the accumulative plastic strain plays a significant role in facilitating crack closure, while the residual trailing zone gradually forms as the crack propagates. As a result, the residual compressive stress in the trailing zone becomes the primary source of crack closure. During the initial stage of crack propagation, the accumulative plastic strain at the crack tip first decreases and then stabilizes, with a corresponding decrease and stabilization of the crack closure parameter. This indicates that the stability of the accumulative plastic strain is closely related to the crack closure effect. As the crack continues to propagate, the accumulative plastic strain at the crack tip slowly increases, while the crack closure parameter remains unchanged. Once the crack length reaches a certain point, the accumulative plastic strain at the crack tip becomes larger, causing the crack closure parameter to exhibit a slow increase. In a related study [43], the reduction of the crack closure effect typically occurs only when the crack length is longer, but their study involves low load levels. In contrast, the present study involves higher low cycle fatigue loads, suggesting that the reduction in the closure effect may occur earlier under higher load levels.

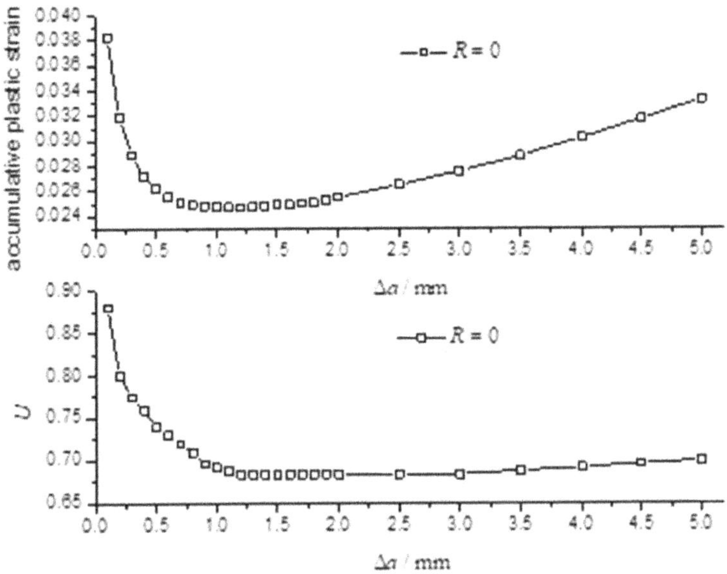

Figure 11. Relationship between accumulative plastic strain and crack closure parameter.

3.4.2. The Evolution of Residual Compressive Stress Field

The portion of the crack surface that is closed can bear some of the load, which reduces the stress–strain field at the crack tip and consequently lowers the level of accumulative plastic strain. Conversely, an increase in accumulative plastic strain can result in a lower crack contact stress, which leads to a reduced crack closure effect. These two factors have a reciprocal influence on each other. The residual compressive stress field in the trailing zone is the primary determinant of the level of crack closure. Therefore, investigating the relationship between the accumulative plastic strain at the crack tip and the residual compressive stress field in the trailing zone can indirectly reveal the connection between the accumulative plastic strain at the crack tip and the crack closure effect.

The stress distribution near a crack tip was studied at minimum load for a stress ratio of R = 0, with different crack lengths. The relationship between the accumulative plastic strain and the residual compressive stress field at the crack tip was analyzed in Figure 12. Figure 12a indicates that as the crack length increased from 0.6 mm to 1.2 mm, the accumulative plastic strain decreased initially, but then stabilized. Meanwhile, the residual compressive stress increased consistently, which aligns with the observation that the crack closure parameter decreased and the crack closure effect increased. In Figure 12b, when the crack length increased from 1.2 mm to 1.8 mm, there was no significant change in the accumulative plastic strain or the corresponding residual compressive stress field. As a result, there was no notable effect on the crack closure effect, and the crack closure parameter remained stable. Figure 12c showed that as the crack length grew from 1.8 mm to 4.5 mm, the accumulative plastic strain increased continuously, while the residual compressive stress decreased slowly. This decrease is a crucial factor behind the reduction in the crack closure effect.

According to the above analysis, it is concluded that the residual compressive stress field in the trailing zone is more indicative of the level of crack closure. As the accumulative plastic strain at the crack tip increases, the residual compressive stress decreases, which in turn reduces the effectiveness of crack closure.

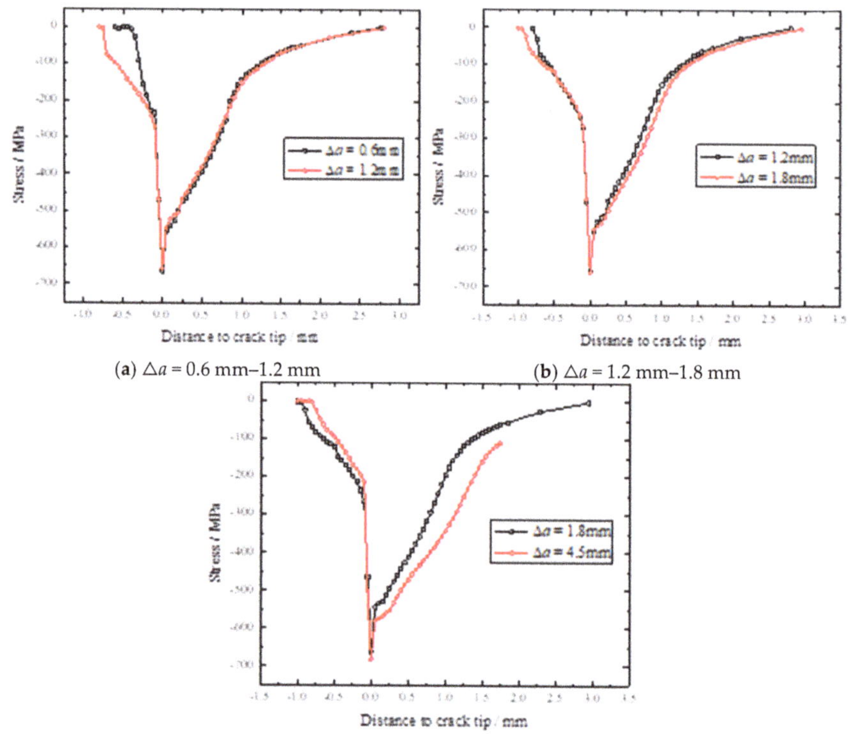

(**a**) Δa = 0.6 mm–1.2 mm (**b**) Δa = 1.2 mm–1.8 mm (**c**) Δa = 1.8 mm–4.5 mm

Figure 12. The stress distribution at minimum applied load.

3.4.3. The Effect of the Stress Ratio

Figure 13 shows the effect of the stress ratio on accumulative plastic strain and crack closure parameters. It can be observed that when the stress ratios are 0.1, 0.3, and 0.4, the corresponding crack closure parameters at steady state are 0.717, 0.771, and 0.82, respectively. The corresponding range of decrease in accumulative plastic strain is 1.29%, 1.04%, and 0.95%, respectively. It is evident that a smaller stress ratio results in a smaller crack closure parameter, as well as a greater reduction in the range of accumulative plastic strain. The data suggest that an increase in the crack closure effect can lead to a decrease in accumulative plastic strain, with both factors tending to stabilize at a later stage. Additionally, the length of the crack has an influence on its propagation behavior. During the final stages of crack propagation, an increase in accumulative plastic strain may hinder the crack closure effect, causing the crack closure parameter to gradually increase. This trend is primarily due to a decline in the residual compressive stress field behind the crack tip.

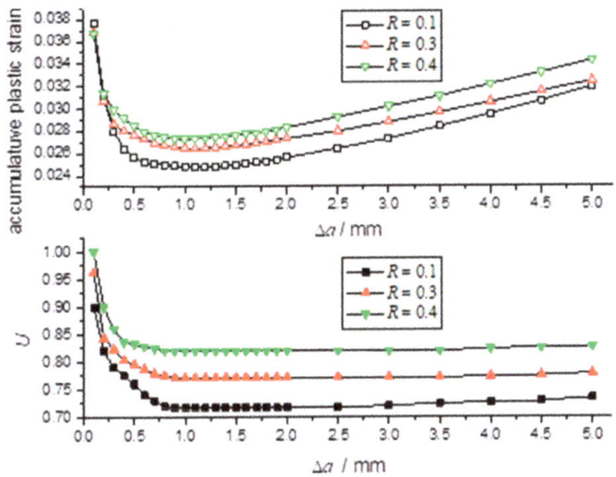

Figure 13. The effect of stress ratio.

4. Conclusions

A study was conducted to examine the behavior of crack propagation of cracked plates subjected to low cycle fatigue loads. Both experimental and numerical methods were employed to investigate the effect of the stress ratio and the initial crack length on the crack propagation process. The findings shed new light on the low cycle fatigue crack propagation behavior and offer a fresh perspective for future research. Some conclusions can be drawn:

(1) The test was conducted on AH32 steel plates to investigate the effect of maximum applied load and stress ratio on crack growth behavior. The results show that the increasing maximum applied load and decreasing stress ratio will shorten the fatigue life of cracked plates.

(2) A numerical analysis of a cracked plate was carried out to study the distribution of the stress–strain field and crack closure parameter. The study found that an increase in the stress ratio led to a rise in the residual compressive stress in the trailing zone, resulting in lengthened crack closure and an amplified effect. Additionally, the crack closure effect and accumulative plastic strain at the crack tip influenced the crack growth rate.

(3) The crack closure effect is influenced by the residual compressive stress in the trailing zone. Higher residual compressive stress enhances the effect, while crack closure weakens the stress–strain field at the crack tip, suppressing accumulative plastic deformation. As accumulative plastic deformation increases, the compressive stress in the trailing zone decreases, weakening the closure effect. This interaction is important for studying low cycle fatigue.

Author Contributions: Study design and manuscript writing D.Q., Literature search and article translation, X.G.; Making charts and data analysis, Z.J.; Revision, H.Y. All authors have read and agreed to the published version of the manuscript.

Funding: This research was funded by the National Natural Science Foundation of China (grant no. 51909198), The Natural Science Foundation of Hubei Province (Grant No.2022CFB331), the Fundamental Research Funds for the Central Universities (WUT: 3120622721), The Natural Science Foundation of the Jiangsu Higher Education Institutions of China (23KJB58005).

Institutional Review Board Statement: Not applicable.

Informed Consent Statement: Not applicable.

Data Availability Statement: Due to the nature of this research, participants of this study did not agree for their data to be shared publicly, so supporting data is not available.

Acknowledgments: The authors acknowledge the financial support provided by The National Natural Science Foundation of China (Grant No.51909198), The Natural Science Foundation of Hubei Province (Grant No.2022CFB331), the Fundamental Research Funds for the Central Universities (WUT: 3120622721), and The Natural Science Foundation of the Jiangsu Higher Education Institutions of China (23KJB58005).

Conflicts of Interest: The authors declare no conflict of interest.

References

1. Coffin, L. A Study of the Effects of Cyclic Thermal Stresses on a Ductile Metal. *J. Fluids Eng.* **1954**, *76*, 931–949. [CrossRef]
2. Dunham, F.W. Fatigue testing of large-scale models of submarine structural details. *Mar. Technol.* **1965**, *1*, 299–307. [CrossRef]
3. Jordan, C.R.; Cochran, C.S. In-service performance of structural details. *Ship Struct. Comm.* **1978**, *272*, 1–9.
4. Jordan, C.R.; Cochran, C.S. Further survey of in-service performance of structural details. *Ship Struct. Comm.* **1980**, *294*, 2–10.
5. Munse, W.; Wilbur, T.; Tellalian, M.; Nicoll, K.; Wilson, K. Fatigue characterization of fabricated ship detail for design. *Ship Struct. Comm.* **1982**, *318*, 1–10.
6. Radek, P.; Jan, D. Low cycle fatigue properties assessment for rotor steels with the use of miniaturized specimens. *Int. J. Fatigue* **2022**, *154*, 106555.
7. Morishita, T.; Itoh, T.; Sakane, M.; Nakamura, H.; Takanashi, M. Multiaxial fatigue property of Ti–6Al–4V using hollow cylinder specimen under push-pull and cyclic inner pressure loading. *Int. J. Fatigue* **2016**, *87*, 370–380. [CrossRef]
8. Itoh, T.; Nakamura, H.; Takanashi, M.; Wu, M. Multiaxial low cycle fatigue life of Ti-6Al-4V under non-proportional loading with mean strain. *Theor. Appl. Fract. Mech.* **2017**, *90*, 165–173. [CrossRef]
9. Chen, L.; Chen, X. The low cycle fatigue tests on submarine structures. *Ship Sci. Technol.* **1991**, *2*, 19–20.
10. Urm, H.; Yoo, I.; Heo, J. Low cycle fatigue strength assessment for ship structures. In Proceedings of the 9th Symposium on Practical Design of Ships and Other Floating Structures, Travemuende, Germany, 12–17 September 2004; pp. 774–781.
11. Han, Y.; Cui, W.; Huang, X.; Wu, Y. Fatigue strength assessment of large-scale ship structures. *Shipbuild. China* **2007**, *48*, 60–67.
12. Liu, Y.; Zhu, X.; Huang, X. Experimental research on low frequency fatigue crack propagation rate of 921A hull steel structure. *J. Nav. Univ. Eng.* **2008**, *20*, 69–74.
13. Wang, Q.; Huber, N.; Liu, X.; Kashaev, N. On the analysis of plasticity induced crack closure in welded specimens: A mechanism controlled by the stress intensity factor resulting from residual stresses. *Int. J. Fatigue* **2022**, *162*, 106940. [CrossRef]
14. Masoud, S.; Otmar, K. A novel approach for determining the stress intensity factor for cracks in multilayered cantilevers. *Eng. Fract. Mech.* **2022**, *266*, 108386.
15. Hossein, N.; Pooya, R. Stress concentration factors in tubular T-joints reinforced with external ring under in-plane bending moment. *Ocean Eng.* **2022**, *266*, 112551.
16. Hossein, N.; Pooya, R. Probabilistic analysis of the SCFs in tubular T/Y-joints reinforced with FRP under axial, in-plane bending, and out-of-plane bending loads. *Structures* **2022**, *35*, 1078–1097.
17. Hossein, N.; Pooya, R. Stress concentration factors in tubular T/Y-connections reinforced with FRP under in-plane bending load. *Mar. Struct.* **2021**, *76*, 102871.
18. Hossein, N.; Pooya, R. Static capacity of tubular X-joints reinforced with fiber reinforced polymer subjected to compressive load. *Eng. Struct.* **2021**, *236*, 112041.
19. Hossein, N.; Pooya, R. Stress concentration factors in tubular T/Y-joints strengthened with FRP subjected to compressive load in offshore structures. *Int. J. Fatigue* **2020**, *140*, 105719.
20. Dong, Q.; Yang, P.; Xu, G. Low cycle fatigue crack growth analysis by CTOD under variable amplitude loading for AH32 steel. *Mar. Struct.* **2019**, *63*, 257–268. [CrossRef]
21. Dong, Q.; Yang, P.; Deng, J.; Wang, D. The theoretical and numerical research on CTOD for ship plate under cyclic loading considering accumulative plastic strain. *J. Ship Mech.* **2015**, *19*, 1507–1516.
22. Dowling, N.E. Geometry effects and the J-integral approach to elastic-plastic fatigue crack growth. In *Cracks and Fracture*; Swedlow, J., Williams, M., Eds.; ASTM STP 601; American Society for Testing and Materials: West Conshohocken, PA, USA, 1976; pp. 19–32.
23. Gonzales, G.L.G.; González, J.A.O.; Antunes, F.V.; Neto, D.M.; Díaz, F.A. Experimental determination of the reversed plastic zone around fatigue crack using digital image correlation. *Theor. Appl. Fract. Mech.* **2023**, *125*, 103901. [CrossRef]
24. Chen, J.; Huang, Y.; Dong, L.; Li, Y. A study on evaluation method of crack tip reverse plastic zone size for the center cracked steel plate model under tension-compression cyclic loading. *Eng. Fract. Mech.* **2015**, *133*, 138–151. [CrossRef]
25. Zhang, W.; Liu, Y. Plastic zone size estimation under cyclic loadings using in situ optical microscopy fatigue testing. *Fatigue Fract. Eng. Mater. Struct.* **2011**, *34*, 717–727. [CrossRef]
26. Huang, M.; Cai, L. Unified theoretical solutions for describing the crack-tip stress fields of finite specimens with mode-I crack under fully plastic conditions. *Int. J. Solids Struct.* **2022**, *254–255*, 111846. [CrossRef]
27. Xu, M.; Liu, Y.; Yuan, H. Characterization of crack-tip fields for elasto-plastic fatigue crack growth Part I: Analysis of the J-integral. *Eng. Fract. Mech.* **2022**, *275*, 108847. [CrossRef]
28. Ding, X.; Yan, X.; Guo, Z.; Guo, K. A combined low- and high-cycle life prediction model considering the closure effect of micro-defects. *Fatigue Fract. Eng. Mater. Struct.* **2022**, *45*, 2058–2071. [CrossRef]

29. Mansor, N.; Abdullah, S.; Ariffin, A. Effect of loading sequences on fatigue crack growth and crack closure in API X65 steel. *Mar. Struct.* **2019**, *65*, 181–196. [CrossRef]
30. Ferreira, S.; Castro, J.; Meggiolaro, M.A.; de Oliveira Miranda, A.C. Crack closure effects on fatigue damage ahead of crack tips. *Int. J. Fatigue* **2019**, *125*, 187–198. [CrossRef]
31. Ma, T.; Gao, N.; Chang, L.; He, X.; Zhou, C. Low-cycle fatigue behavior and life prediction of CP-Ti under non-proportional and multiaxial loading. *Eng. Fract. Mech.* **2021**, *254*, 107930. [CrossRef]
32. Deng, J.; Tu, W.; Xiong, K.; Yang, P.; Dong, Q. Analysis of biaxial proportional low-cycle fatigue and biaxial accumulative plasticity of hull inclined-crack plate. *Int. J. Nav. Archit. Ocean Eng.* **2022**, *14*, 100423. [CrossRef]
33. Song, Y.; Yang, P.; Peng, Z.; Jiang, W. Low-Cycle Fatigue Crack Propagation Behavior of Cracked Steel Plates Considering Accumulative Plastic Strain. *Int. J. Steel Struct.* **2020**, *20*, 538–547. [CrossRef]
34. Jiang, W.; Yang, P.; Luo, B.; Xu, Z. Responses of cracked stiffened plates to low-cycle fatigue loads. *Ocean Eng.* **2021**, *241*, 109986. [CrossRef]
35. Dong, Q.; Yang, P.; Xu, G.; Deng, J. Mechanisms and modeling of low cycle fatigue crack propagation in a pressure vessel steel Q345. *Int. J. Fatigue* **2016**, *89*, 2–10. [CrossRef]
36. Tola, A.; Eatherton, M.; Koutromanos, L. Experimental program for characterization of ultra-low cycle fatigue fracture in structural steel. *Eng. Fract. Mech.* **2022**, *276*, 108873. [CrossRef]
37. Chen, J.; Xiao, X.; Xiong, S.; Wang, J.; Huang, H.; Yang, B. Low cycle fatigue behavior of Cu-Cr-Zr alloy with different cold deformation. *Fatigue Fract. Eng. Mater. Struct.* **2023**, *46*, 3–16. [CrossRef]
38. Xiong, K.; Deng, J.; Pei, Z.; Tang, H.; Tu, W. Analysis of accumulative plasticity and fracture behavior of hull cracked plates subjected to biaxial low cycle fatigue loading. *J. Ship Mech.* **2022**, *26*, 113–124.
39. Dong, Q.; Rong, M.; Xu, G. Study of crack closure effect of hull plate under low cycle fatigue. *J. Mar. Sci. Eng.* **2022**, *10*, 1557. [CrossRef]
40. Gan, J.; Sun, D.; Deng, H.; Wang, Z.; Wang, X.; Yao, L.; Wu, W. Fatigue characteristics of designed T-type specimen under two-step repeating variable amplitude load with low-amplitude load below the fatigue limit. *J. Mar. Sci. Eng.* **2021**, *9*, 107. [CrossRef]
41. *E647-05*; Standard Test Method for Measurement of Fatigue Crack Growth Rates. ASTM: West Conshohocken, PA, USA, 2005.
42. Rodrigues, D.; Antunes, F. Finite element simulation of plasticity induced crack closure with different material constitutive models. *Eng. Fract. Mech.* **2009**, *76*, 1215–1230. [CrossRef]
43. Zhou, L.; Wang, J.; Wang, Y.; Li, X.; Chai, Y. The enriched finite element method-virtual crack closure technique for cracked structures. *Thin-Walled Struct.* **2023**, *187*, 110756. [CrossRef]

Journal of
Marine Science and Engineering

MDPI

Article

Experimental Study of Hot Spot Stress for Spatial CHS KK-Joints

Kang Wang [1,2], Yuhang Wang [1], Shan Gao [3,4], Bin Wang [3,4], Jinzhong Chen [3,4] and Pengjun Luo [5,*]

1 School of Civil Engineering, Chongqing University, Chongqing 400045, China
2 Department of Civil Engineering, Osaka University, Suita 567-0047, Osaka, Japan
3 Key Laboratory of Far-Shore Wind Power Technology of Zhejiang Province, Hangzhou 311122, China
4 Powerchina Huadong Engineering Corporation Limited, Hangzhou 311122, China
5 School of Civil Engineering, Chongqing Jiaotong University, Chongqing 400074, China
* Correspondence: pjluo123@cqjtu.edu.cn

Abstract: In offshore structures such as offshore wind turbine jacket foundations, prolonged exposure to wind and wave loads can lead to fatigue failure, especially at the joint connections. Currently, international codes primarily evaluate the fatigue performance of tubular joints using the hot spot stress (HSS) method, with stress concentration factor (SCF) being the most crucial parameter. Numerous studies have investigated the SCF of tubular joints and proposed corresponding calculation formulas. However, most of these studies have focused on simple basic loading cases. To comprehensively understand the distribution pattern of the HSS method at spatial tubular joints, this study selects the spatial CHS KK-joints as the research subject and performs the HSS testing on spatial KK-joints. The research indicates that, in CHS joints, the distribution of the HSS on the chord side remains consistent under tensile or compressive loading conditions along the brace axis. Additionally, within spatial joints, the loading conditions of neighboring braces near the reference brace significantly affect the distribution of HSS on the chord side and exhibit varying spatial effects on the SCF. Furthermore, this study applies recommended calculation formulas from codes to analyze the experimental results. The findings reveal significant discrepancies between the SCF calculation formulas recommended by CIDECT and the test results, rendering it inadequate for accurately calculating the SCF of spatial KK-joints. Therefore, further research is required to develop suitable calculation formulas for SCFs in spatial KK-joints.

Keywords: spatial CHS KK-joints; hot spot stress (HSS); multi-planar interaction; stress concentration factor (SCF)

Citation: Wang, K.; Wang, Y.; Gao, S.; Wang, B.; Chen, J.; Luo, P. Experimental Study of Hot Spot Stress for Spatial CHS KK-Joints. *J. Mar. Sci. Eng.* **2023**, *11*, 1432. https://doi.org/10.3390/jmse11071432

Academic Editor: Vincenzo Crupi

Received: 20 June 2023
Revised: 14 July 2023
Accepted: 15 July 2023
Published: 18 July 2023

1. Introduction

To address increasingly dire environmental and energy challenges, the development of clean energy has emerged as a pivotal direction for sustainable progress worldwide. Wind power, as a form of clean energy, has garnered escalating attention. Due to the abundant wind resources available at sea and the minimal utilization of land, the installed capacity of offshore wind power has been steadily rising [1].

The jacket structure (Figure 1) serves as a commonly employed fixed foundation configuration for offshore wind turbine systems, comprising a steel tower section, a jacket section, and pile foundations. The jacket section typically consists of multiple vertically oriented chord members interconnected by bracing members. The bracing members intersect at a single point on the chord member, forming a joint. Due to the prolonged exposure of offshore wind structures to the cyclic loads of wind and waves, fatigue effects become apparent [2], particularly at the joint where the unique structural configuration tends to induce stress concentration, thereby further amplifying the influence of fatigue loads.

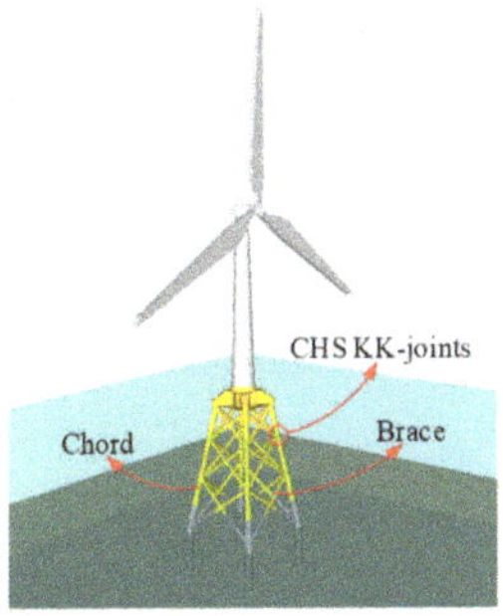

Figure 1. Jacket foundation for offshore wind turbines.

Currently, the commonly used fatigue design method for joints in international codes is the S-N curve method based on hot spot stress (HSS) [3]. Research has revealed that HSS mainly occurs at the weld between the braces and the chords. Therefore, a thorough understanding of stress concentration phenomena at intersecting joints is crucial for the fatigue design of structures.

In numerous previous studies and codes, the stress concentration factor (SCF) is defined as the ratio of HSS at the weld toe to the nominal stress of the brace. Conversely, if the SCFs of the joints are known, multiplying them by the corresponding fatigue load allows for the convenient determination of the HSS at the weld toe of the joints. This method is known as the HSS method. Evidently, it is crucial to identify the corresponding SCFs for the HSS method. Researchers have conducted extensive investigations in this regard. Mohamed et al. [4] conducted SCF tests and fatigue experiments on three T-joints, subjecting three identical T-joints to heat treatment. The results indicate that fatigue cracks initiate at the saddle point of the joint, and the higher the specimen temperature, the longer the fatigue life. Furthermore, a comparison of the experimental results with CIDECT code [5] and API code [6] reveals that the calculated values of CIDECT tend to be unsafe, while those of API tend to be conservative. Lie et al. [7] and N′ Diaye et al. [8] performed HSS analysis on the tubular T-joints using the finite element (FE) method. Gho et al. [9] established an FE model consisting of 192 fully overlapped tubular K-joints and conducted parameter analysis. The results demonstrate that the SCF extreme point occurs at the saddle point of the overlapping joints, leading to the derivation of the SCF calculation formula for overlapped tubular K-joints based on the FE results. In addition, researchers have separately investigated the fatigue performance of intersecting joints in rectangular steel tubes [10–12]. They have conducted fatigue and HSS tests on T-, X-, and K-joints, combined with FE analysis to perform parameter analysis, and proposed corresponding SCF calculation formulas for tubular joints. In addition, Nassiraei [13] conducted a parameter analysis on the local joint flexibility of CHS T/Y-connections strengthened with a collar plate and proposed a recommended design method.

Chen et al. [14], Musa et al. [15], and Kim et al. [16] have conducted SCF experiments on circular concrete-filled steel tubular (CFST) joints and comparative specimens of circular hollow steel (CHS) joints. The forms of the joints include Y-, K-, and KT-joints. These research results all indicate that filling the chord of the steel joint with concrete can achieve a more uniform distribution of stiffness at the intersection, alleviate stress concentration phenomena, and lead to varying degrees of decrease in the SCFs. Nassiraei et al. [17] investigated HSS distribution in tubular T/Y-joints reinforced with FRP and proposed the corresponding formula for calculating SCF.

From the previous studies, it can be observed that there has been extensive research on the HSS testing of joints under basic loads. However, there is still a lack of experimental studies on the HSS testing of tubular joints with spatial joints. Marine structures, during their operational phase, confront a highly complex load environment [18]. The force conditions of tubular joints are inevitably subjected to multiple operating conditions, necessitating research on the spatial interactions of these joints.

In this study, a spatial loading test setup for static and fatigue loading of spatial tubular joints was designed. By conducting HSS tests on spatial CHS KK-joints under various operating conditions, the influence of each condition on the distribution of the HSS at the joints was investigated. Furthermore, by comparing test results with existing codes, recommendations are proposed for subsequent design work.

2. Hot Spot Stress Test

2.1. Specimen Design

To investigate the distribution of the HSS in KK-joints in different combinations of loads, the creators of this study designed and fabricated a spatial CHS KK-joints. The dimensions of the specimen were obtained by scaling down the jacket foundation structure for offshore wind turbines, with a scaling ratio of 1:10. The notations used are given in Table 1.

Table 1. Notation list.

Parameters and Sets	
D	Chord diameter
t_c	Chord wall thickness
L	Chord length
d	Brace diameter
t_b	Brace wall thickness
l	Brace length
θ	Angle between brace and chord
Φ	Angle between two planes formed by the braces
β	d/D
γ	$D/(2T)$
τ	t/T
α	$2L/D$
g	Gap between two braces
ζ	g/D
σ_y	Steel yield stress
E	Steel Young modulus
r_c	Chord radius
r_b	Brace radius

The diameter of the chord member D is 219.1 mm, with a wall thickness t_c of 6.4 mm and a length L of 2000 mm, approximately nine times the diameter of the chord member, while excluding the influence of end constraints on the joint region [19]. The diameter of the brace member d is 114.0 mm, with a wall thickness t_b of 6.0 mm, and the bracing member length l is set at 700 mm (Figure 2). Reinforcing ribs were installed at the ends of both the chord and brace members to enhance the end sections and prevent premature failure due to stress concentration.

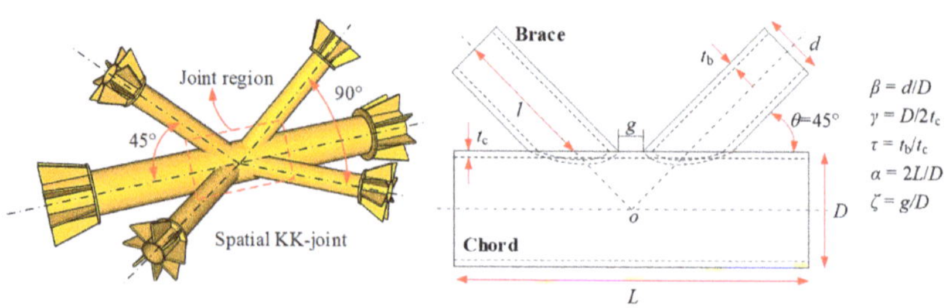

Figure 2. Diagram of specimen.

The specimen is composed of four braces and one chord, with the brace ends cut into intersecting lines and directly welded to the surface of the chord. The axis of each brace forms a 45-degree angle with the axis of the chord, and the two planes formed by the braces have a 90-degree angle between them. The four braces axes intersect at a single point on the chord axis. In offshore wind turbine jacket structures, the quality requirements for welded joints are high due to long-term exposure to combined wind, wave, and current loads. The specimens in this study meet the welding requirements specified by the AWS code [20].

As marine structures continue to evolve, the demands placed on structural steel are increasing. However, for the purposes of this study, commonly used low-carbon steel still suffices to meet the testing requirements for the HSS tests. The selected steel material in this study is Q355B steel. Additionally, to ensure the overall integrity of the structure and prevent the influence of unnecessary welds on the test results, seamless steel tubes are used for all the members. Material property tests were conducted on the steel samples, and the obtained characteristics (yield stress σ_y and Young modulus E of the steel) of the specimens are presented in Table 2.

Table 2. Material properties of specimen.

Members	Thickness (mm)	σ_y (MPa)	E (MPa)
Chord	6.4	365	219.59
Brace	6.0	397	222.82

2.2. Loading and Measurement System

To investigate the spatial interaction mechanism of spatial KK-joints, it is necessary to employ a loading method that combines multiple working conditions, specifically requiring the individual loading of the four braces. Therefore, a spatial loading test setup suitable for statically and dynamically loading spatial tube joints is developed in this study, as illustrated in Figure 3. This test setup primarily consists of four independently controlled MTS loading actuators, supporting columns, oblique support devices, actuator supports, and specimen supports, among other components. The stability of the connections between the various parts is ensured using high-strength friction-type bolts. Additionally, all components placed on the reaction floor are connected to the floor with prestressed steel bars, preventing any relative slippage between the parts. During the installation of the specimen, it is necessary to first utilize a laser calibration instrument to position the specimen at the center of the loading device and align the axes of the braces and actuators, ensuring that the axes of the actuators and the braces coincide, thus preventing any additional loads during the loading process. Furthermore, prior to applying the load to the specimen, it is advisable to conduct a preliminary loading test, which serves to assess the alignment between the specimen and the test setup, as well as the stability of the measurement system.

Due to the effect of notching in the weld joint, direct measurement of stress at the weld toe is not feasible during testing. The prevailing approach commonly employed is the indirect method, which is currently known as the extrapolation method. In accordance with the recommendations provided by CIDECT [5] and IIW codes [21], the extrapolation method for tubular joints primarily encompasses linear extrapolation and quadratic extrapolation. Linear extrapolation involves linear calculations based on two measurement points in the extrapolation region to determine the stress at the weld toe, whereas quadratic extrapolation entails secondary calculations utilizing multiple measurement points. The codes suggest that the linear extrapolation method should be used for circular steel tubular joints. Consequently, the linear extrapolation method for measuring HSS is utilized in the tests for this study.

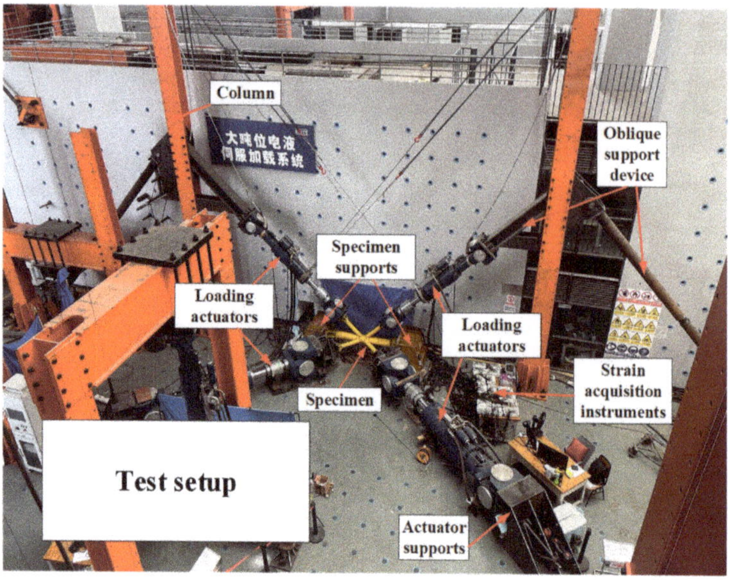

Figure 3. Diagram of test setup.

The arrangement of measurement points in the extrapolation region for the linear extrapolation method is illustrated in Figure 4. As depicted in the diagram, the linear extrapolation method requires the placement of only two strain gauges in the extrapolation region of the weld toe. CIDECT [5] has provided recommendations regarding the positioning of strain gauges, as shown in Table 3. In the table, r_c represents the radius of the chord, while r_b denotes the radius of the brace. Finally, in welded tubular joints, the crown point refers to the highest point in the joints, while the saddle point refers to the lowest point in the joints.

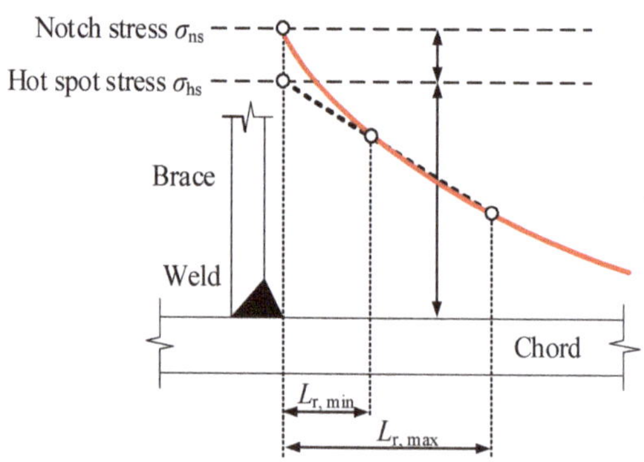

Figure 4. Illustration of linear extrapolation method.

Table 3. Arrangement of strain gauges in linear extrapolation method.

Distance from Weld Toe	Chord		Brace	
	Saddle	Crown	Saddle	Crown
$L_{r,min}$ *	$0.4 \cdot t_c$		$0.4 \cdot t_b$	
$L_{r,max}$ **	$0.09 r_c$	$0.4 \times \sqrt[4]{r_c t_c r_b t_b}$	$0.65 \times \sqrt{r_b t_b}$	

* Minimum value for $L_{r,min}$ is 4 mm. ** Minimum value for $L_{r,max}$ is $L_{r,min} + 0.6\ t_b$.

In various international codes, there are some discrepancies regarding the stress (strain) component of SCFs. The question arises whether the principal stress (strain) or the stress (strain) perpendicular to the weld toe should be utilized. IIW [21] argues in favor of adopting the maximum principal stress (strain), while AWS [20] and API [6] advocate for using the stress (strain) perpendicular to the weld toe. Although in theory, the fatigue crack propagation direction in tubular joints tends to expand towards the direction perpendicular to the maximum principal stress, the limitations of current testing techniques prevent the precise measurement of the maximum principal stress. Additionally, the direct superposition of maximum principal stress is not feasible under different combinations of loads. Moreover, it is more convenient to arrange the strain gauges perpendicular to the weld toe in experimental measurements. Furthermore, CIDECT [5] suggests that the differences between these two types of stresses near the weld toe are not significant. Therefore, CIDECT [5] recommends measuring the strain solely using strain gauges positioned perpendicular to the weld toe, without the need for employing strain gauges that measure the principal strain. In this study, the measurement of strain perpendicular to the weld is chosen.

Hence, the formula for calculating the hot spot strain at the weld toe based on linear extrapolation is as follows:

$$\varepsilon_{\perp W} = \frac{L_{r,max}\varepsilon_{\perp E1} - L_{r,min}\varepsilon_{\perp E2}}{L_{r,max} - L_{r,min}} \tag{1}$$

where $\varepsilon_{\perp W}$ is the strain at the weld toe obtained by linear calculation; $\varepsilon_{\perp E1}$ and $\varepsilon_{\perp E2}$ represent the perpendicular strain measured at the first and second measurement points, respectively, in the extrapolation region, perpendicular to the weld toe; $L_{r,min}$ and $L_{r,max}$ denote the distances from the weld toe to the first and second extrapolation points in the extrapolation region, as specified in Table 3. In the study, $L_{r,min}$ is set to 4 mm on both the chord side and the brace side. $L_{r,max}$ is determined to be 9.9 mm at the saddle point and 8.9 mm at the crown point on the chord side, while it is uniformly 12 mm on the brace side. Interpolation is employed for the points between the crown point and the brace point.

Furthermore, the strain concentration factor (SNCF) of the joints could be obtained by dividing the hot spot strain by the nominal strain.

$$\mathrm{SNCF} = \frac{\varepsilon_{\perp W}}{\varepsilon_N} \tag{2}$$

where ε_N is the nominal strain of the braces, which is mesured in this study by strain gauges.

Furthermore, according to Hooke's Theorem, there exists a relationship between the SCF and the SNCF as expressed by the following equation:

$$\mathrm{SCF} = \mathrm{SNCF} \times (1 + v\varepsilon_{\parallel W}/\varepsilon_{\perp W})/(1 - v^2) \tag{3}$$

where $\varepsilon_{\parallel W}$ represents the strain parallel to the weld toe, calculated using the same method as $\varepsilon_{\perp W}$, and v denotes the Poisson's ratio of the steel material.

$$\varepsilon_{\parallel W} = \frac{L_{r,max}\varepsilon_{\parallel E1} - L_{r,min}\varepsilon_{\parallel E2}}{L_{r,max} - L_{r,min}} \tag{4}$$

where $\varepsilon_{\parallel E1}$ and $\varepsilon_{\parallel E2}$ represent the perpendicular strain measured at the first and second measurement points, respectively, in the extrapolation region, parallel to the weld toe.

Therefore, in order to obtain the SCF more accurately, this study employed a strain gauge parallel to the weld toe beside each strain gauge perpendicular to the weld toe. Additionally, due to the impracticality of measuring all positions along the weld toe during the experimental process, strategically positioned strain measurement points were selected in this study. In particular, the strain distribution around the weld toe of the joints under spatial effects needs to be considered. Hence, eight measurement points are placed for every brace region. The first measurement point is located at the 0° position, which is at the crown point near the side of the chord. Eight measurement points are distributed at 45° intervals along the weld toe, with the coordinate origin being the intersection of the brace axis and the surface of the chord. The positions of the four braces corresponded consistently to the strain gauges. The arrangement diagram and numbering of the strain gauges are illustrated in Figure 5. It should be noted that the points at 0°/360° and 180° are referred to as the crown points of the joints, whereas the points at 90° and 270° are referred to as the saddle points of the joints. The actual arrangement of the gauges on the specimen is shown in Figure 6.

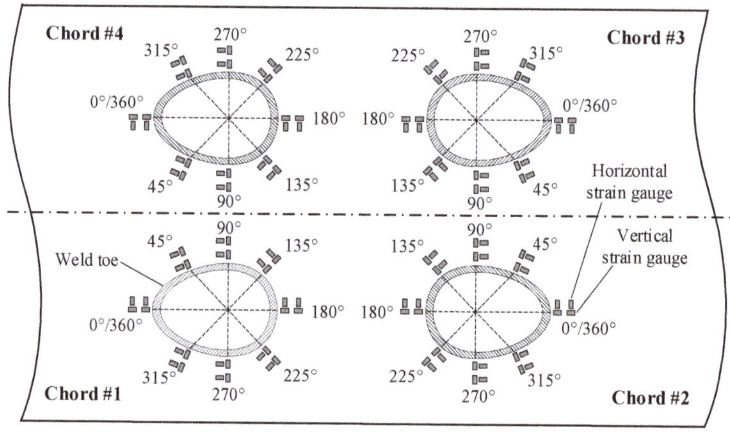

Figure 5. Arrangement and schematic diagram of strain gauges placement and strain gauges numbering on the chord side.

Figure 6. Experimental strain gauge layout diagram.

In order to obtain the coefficient of hot spot stress, it is necessary to further measure the axial strain of the brace as a nominal strain. In this study, four uniformly distributed circumferential axial strain gauges are placed at the midpoint of the brace to avoid the influence of stress concentration at the node and end regions on the test results.

2.3. Load Design Rules

Four actuators on the test setup enable independent loading, providing various load combinations for the joints. Within the IIW code [18], three reference combinations are provided for spatial KK-type joints: the first is force-symmetric loading, the second is unilateral loading, and the third is force anti-symmetric loading, as illustrated in Figure 7.

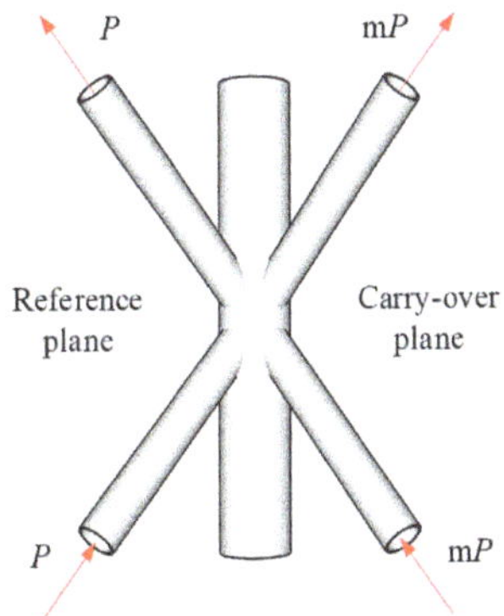

m	Referred to as
1	Symmetrical loading
0	Reference-plane loading
−1	Anti-symmetrical loading

Figure 7. Definition of load combinations for spatial KK-joints in IIW codes.

To fully understand the stress conditions of spatial tubular joints, five loading conditions were proposed for HSS testing, as shown in Table 4. Furthermore, to present a clearer depiction of the loading rules for each loading condition, the loading rules from the table are depicted in Figure 8. It should be noted that the listed loading conditions in Table 4 are implemented by first selecting a reference brace and applying the corresponding load condition to that reference brace. Given the symmetrical construction of the specimen, any brace can serve as the reference brace. Additionally, since the steel material remains in an elastic working state throughout the HSS tests, each brace is repeatedly subjected to the load conditions listed in Table 4. It is important to note that in general, marine structures subjected to long-term loading rarely experience fatigue loads that cause the material to enter the plastic working stage. Therefore, it is necessary to ensure that the stress on the steel material during the HSS tests does not exceed the yield strength of the steel.

Table 4. Loading conditions for specimen.

Load Case	B1		B2		B3		B4	
	T *	C	T	C	T	C	T	C
L1	1 **	−1	0	0	0	0	0	0
L2	1	−1	0	0	0	0	−1	1
L3	1	−1	−1	1	0	0	0	0
L4	1	−1	−1	1	1	−1	−1	1
L5	1	−1	−1	1	−1	1	1	−1

* T and C respectively represent the tensile or compressive conditions of the brace in the axial direction. ** One unit of load.

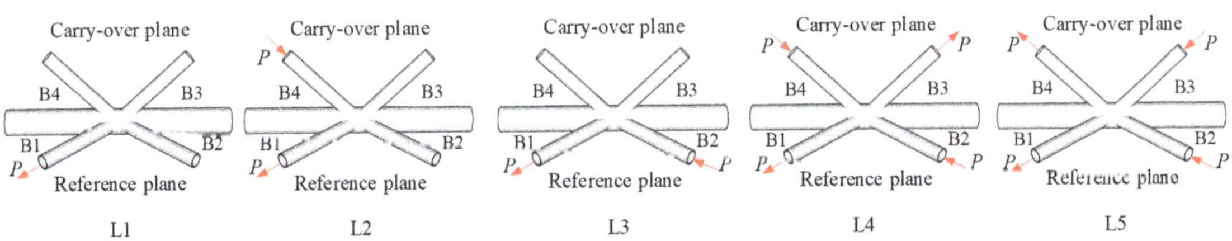

Figure 8. Diagram of loading conditions for specimen.

3. Test Results

3.1. Data Processing

Before processing the experimental results, it is necessary to make certain assumptions:

1.Nominal strain gauges were installed at the midpoint positions of each brace in the test. During the preloading process prior to the start of the test, it is necessary to ensure that the values of the four nominal strain gauges under the same brace do not differ significantly. Ultimately, the nominal strain of that brace is calculated according to Equation (5) and compared with the theoretically calculated results (calculated by Equation (6)) to ensure the accuracy of the measured outcome.

$$\varepsilon_N = \frac{\varepsilon_1 + \varepsilon_2 + \varepsilon_3 + \varepsilon_4}{4} \tag{5}$$

$$\varepsilon_N = \frac{\sigma_N}{E} = \frac{4F_A}{\pi E\left[d^2 - (d - 2t_b)^2\right]} \tag{6}$$

where ε_1, ε_2, ε_3 and ε_4 are nominal strains of the four measuring points along the brace.

2. During the testing process, each brace was used as a reference brace for identical load conditions. Therefore, it can be postulated that the specimen underwent four repeated tests for each loading condition. In actuality, some errors are inevitably encountered during the manufacturing and welding processes of the specimen. Furthermore, in the context of large-scale structural testing, it is unfeasible to guarantee perfect alignment of the loading apparatus. Consequently, the mean value of the results from the four repeated tests was adopted as the ultimate outcome.

3. The existing research generally holds the view that the HSS of the chord in the CHS joint is significantly higher than that of the brace. In the HSS test, the focal point is primarily on the distribution of the HSS on the side of the chord. Consequently, this study mainly focuses on the HSS distribution on the chord side of the spatial KK-joints.

3.2. Discussion

The SCF distribution curves obtained from the test are summarized as shown in the Figure 9. The following distribution patterns can be derived from the figures.

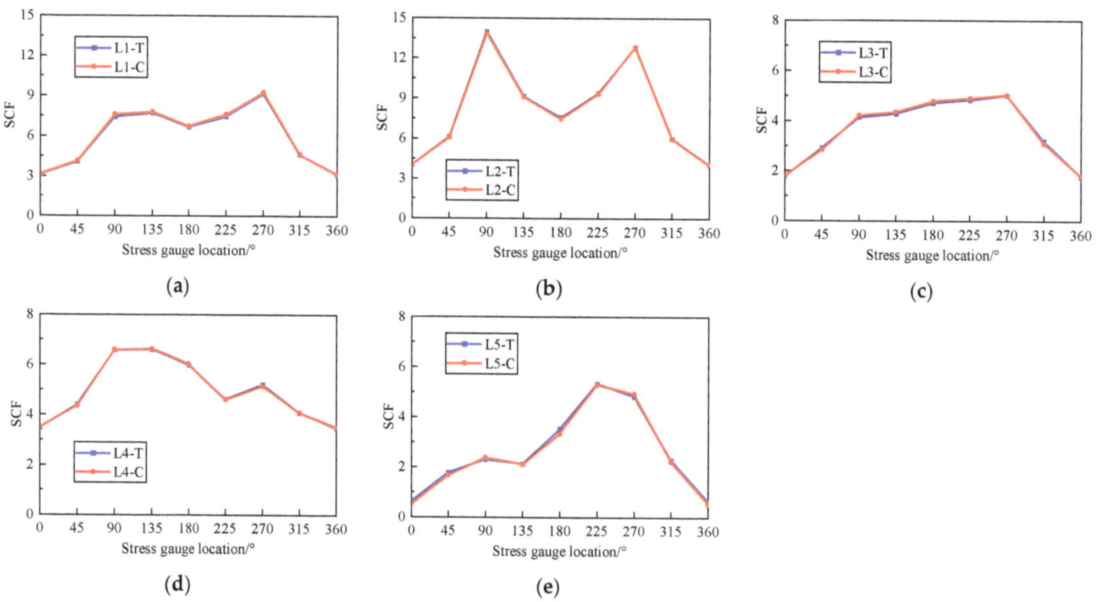

Figure 9. SCF distribution curves on the chord side under different loading conditions. (**a**) Loading case L1. (**b**) Loading case L2. (**c**) Loading case L3. (**d**) Loading case L4. (**e**) Loading case L5.

1. From the overall situation, it can be observed that for the spatial CHS KK-joints, the HSS on the chord side are almost the same under both tension and compression conditions of the brace subjected to various loading conditions. This indicates that for

the CHS KK-joint, a single set of calculation formulas is sufficient to determine the distribution of HSS in the chord under axial loading of the brace.

2. From Figure 9a, it can be observed that when a single brace is subjected to axial loading, the SCF is approximately symmetric about the crown point. However, due to the presence of other braces at the 90° position, the stiffness at that location is greater than at the 270° position. Therefore, the SCF value at the 270° position of the chord will be larger than the SCF value at the 90° position. When a single brace is subjected to axial loading, the SCF values at all measurement points on the chord are greater than 1, with the maximum value occurring at the outer side of the chord at the 270° position, which is 9.26.
3. Upon comparing Figure 9a,b, it can be observed that the stress patterns in these two scenarios resemble those of Y-joints and spatial YY-joints. It is evident that the axial load on the reference brace, as indicated by the SCF values, is significantly influenced by the axial load on the other brace. Particularly noteworthy is the substantial increase in the SCF values at the 90° and 270° positions, which correspond to the saddle points on both sides of the chord. At the 90° position, the SCF is 13.93, representing an improvement of 84.99%, while at the 270° position, the SCF is 12.80, reflecting a 38.98% enhancement. Through force analysis, it can be deduced that in spatial tubular joints, when one side of the reference brace is subjected to load, a lateral force is exerted on the chord, thereby altering the strain distribution on one side of the chord. Additionally, the CHS tube undergoes localized deformation under the load applied by the brace. Hence, under the influence of the other brace, the strain and SCF at the 90° and 270° positions of the chord side increase.
4. When comparing Figure 9a,c, it can be observed that the stress patterns in these two scenarios resemble those of Y-joints and K-joints. Similar to the previous finding, the inclusion of load on other brace alters the distribution of SCF on the reference brace. However, unlike before, in the case of K-joints under loading conditions, the SCF of the brace experiences a significant reduction. Through force analysis, it can be inferred that when the other side brace is subjected to axial force, it acts in the opposite direction to the axial load on the reference brace. In the axial load mode of Y-joints, the chord experiences a lateral force, resulting in lateral deformation. However, in the axial action mode of the K-joints brace, the perpendicular components of the axial forces of the two braces in relation to the chord's direction will cancel each other out, thereby reducing the lateral deformation of the chord at the reference brace and consequently decreasing the SCFs.
5. The SCF distribution of spatial KK-joints under the axial forces of braces is depicted in Figure 9d,e. The loading conditions in Figure 9d,e correspond to the spatial loading scheme of Figure 9c, representing the three force modes specified in the IIW codes [21] for spatial KK-joints. The two braces loaded in Figure 9c are designated as the reference plane. It can be observed that the loading conditions illustrated in Figure 9d,e involve the application of spatial loads on one side of the reference plane. Although the SCF in Figure 9c is not perfectly symmetrical about the crown point, the SCF values on either side of the crown point are similar. Similar to Y-joints, the SCF at 90° is smaller than that at 270° due to structural issues. Through force analysis, it can be deduced that when axial forces are applied to the brace adjacent to the reference brace of Y-joints, the resulting axial force components cause greater deformation in the chord compared to the Y-joints, thereby increasing the SCF of the chord. However, in spatial KK-joints, the axial force components of the two braces on the opposite plane balance each other, reducing the lateral deformation of the chord and consequently decreasing the chord's SCF. Nevertheless, the local deformation of the chord under the force state of the braces on the opposite plane persists. When the axial load direction on the reference brace aligns with that of the brace on its side, it will reduce the SCF of the chord at the 90° location. Conversely, when the axial load direction on the

reference brace opposes that of the brace on its side, it will increase the SCF of the chord member at the 90° location.

To facilitate viewing, the SCF values obtained from the calculations of various loading conditions are consolidated in Table 5. It can be observed that among all the conditions, the highest SCF value occurs at the 90° position of the chord under the L2 loading condition, with tensile and compressive SCF values of 14.00 and 13.85, respectively.

Table 5. Value of SCFs on the chord side under different loading conditions.

Location	L1-T	L1-C	L2-T	L2-C	L3-T	L3-C	L4-T	L4-C	L5-T	L5-C
0	3.15	3.16	4.04	4.04	1.78	1.81	3.48	3.51	0.66	0.53
45	4.10	4.15	6.11	6.06	2.93	2.86	4.42	4.37	1.79	1.67
90	7.45	7.62	14.00	13.85	4.15	4.22	6.59	6.58	2.30	2.38
135	7.72	7.79	9.16	9.12	4.29	4.36	6.60	6.62	2.13	2.11
180	6.69	6.73	7.58	7.48	4.72	4.80	6.00	6.03	3.55	3.35
225	7.48	7.62	9.40	9.33	4.85	4.91	4.63	4.62	5.35	5.30
270	9.16	9.26	12.79	12.81	5.03	5.03	5.21	5.15	4.83	4.93
315	4.64	4.65	6.02	5.98	3.23	3.12	4.09	4.08	2.29	2.23
360	3.15	3.16	4.04	4.04	1.78	1.81	3.48	3.51	0.66	0.53

4. Comparison with Empirical Formulas

Currently, various international codes such as CIDECT [5], API [6], IIW [21], and DNV [22] typically employ the Efthymiou formula. The application range of this formula is as follows: $0.2 \leq \beta \leq 1.0$, $0.2 \leq \tau \leq 1.0$, $8.0 \leq \gamma \leq 32$, and $20^\circ \leq \theta \leq 90^\circ$.

Under the influence of axial force in the brace, the calculation formula of SCF for the T/Y-joints is as follows:

The saddle point of the chord:

$$\mathrm{SCF}_{T/Y} = F_1 \gamma \tau^{1.1} [1.11 - 3(\beta - 0.52)^2] \sin^{1.6}\theta \tag{7}$$

The crown point of the chord:

$$\mathrm{SCF}_{T/Y} = \gamma^{0.2} \tau [2.65 + 5(\beta - 0.65)^2] + \tau\beta(0.25\alpha - 3)\sin\theta \tag{8}$$

When α is greater than or equal to 12, $F_1 = 1$; when α is less than 12, F_1 is calculated using Equation (9).

$$F_1 = 1 - (0.83\beta - 0.56\beta^2 - 0.02)\gamma^{0.23} \exp(-0.21\gamma^{-1.16}\alpha^{2.5}) \tag{9}$$

Under the influence of axial force in the brace, the calculation formula of SCF for K-joints is as follows:

$$\begin{aligned} \mathrm{SCF}_K = {} & \left[\tau^{0.9}\gamma^{0.5}(0.67 - \beta^2 + 1.16\beta)\sin\theta\right] \left(\frac{\sin\theta_{max}}{\sin\theta_{min}}\right)^{0.30} \times \left(\frac{\beta_{max}}{\beta_{min}}\right)^{0.30} \\ & \times \left[1.64 + 0.29\beta^{-0.38}\arctan(8\zeta)\right] \end{aligned} \tag{10}$$

Furthermore, within the CIDECT code [5], a modified formula based on Equation (10) for calculating the SCF of spatial CHS KK-joints is proposed. The calculation formula is presented as follows.

$$\mathrm{SCF}_{KK} = f_{\mathrm{geom}}\, f_{\mathrm{load}}\, \mathrm{SCF}_K \tag{11}$$

where the values of f_{geom} and f_{load} are determined by the geometric parameters and the load conditions. In the code, a singular parameter *MCF* has been employed to enhance Equation (11).

$$\mathrm{SCF}_{KK} = MCF \cdot \mathrm{SCF}_K \tag{12}$$

where the value of *MCF* is typically associated with the angle Φ between the plane where the braces are located. Based on reference to the code, *MCF* is assigned a value of 1.25 for the L4 loading case and a value of 1 for the L5 loading case in this study.

The calculated values from the formulas in Equations (7)–(12) are summarized and compared to the test results, and the results are consolidated in Table 6.

Table 6. Comparison between experimental maximum values of SCF and recommendation formula values.

Load Case		Test	Position	Recommendation Formula	Position	Error (%)
L1	T	9.19	270°	10.16	90°/270°	10.6
	C	9.26	270°			9.7
L2	T	14.00	90°			27.4
	C	13.85	90°			26.6
L3	T	5.03	270°	5.70	—	13.3
	C	5.03	270°			13.3
L4	T	6.60	135°	7.13	—	8.0
	C	6.62	135°			7.6
L5	T	5.35	225°	5.70	—	6.5
	C	5.30	225°			7.5

From the comparison results, it can be observed that the existing codes have relatively low precision in calculating the recommended SCFs for spatial KK-joints. Particularly in the L2 loading case, the calculation formula for this loading configuration is not proposed by the code. If the calculation formula for uniplanar Y-joints is used to calculate the SCF under the L2 condition, the maximum error will reach 27.4%. In the code, the calculated results of Equation (12) (obtained by correcting Equation (10)) have an error of between 6.5% and 8% from the test results.

Through the HSS testing for spatial KK-joints, it is evident that the distribution of SCFs for the joints varies significantly under different loading conditions. Additionally, the existing codes could not adequately account for the spatial effects of spatial joints. If the SCF calculation formula for joints under the single force mode is employed to calculate the SCF of spatial joints in actual structures, precise results are often unattainable. Therefore, it is necessary in subsequent work to propose corresponding SCF calculation formulas for spatial joints under different conditions.

5. Conclusions

An experimental investigation of HSS for spatial CHS KK-joints was proposed in this study. Initially, the test setup was proposed for loading spatial joints, followed by a detailed description of the HSS loading conditions and measurement device. Lastly, the HSS of the joints was tested under both individual and combined loading conditions, and the results are compared with the outcomes derived from existing recommended calculation formulas. The main conclusions are as follows:

1. Under axial loading conditions in the brace direction, the distribution of the SCFs on the chord side of the CHS KK-joints is essentially similar to that under axial tensile forces in the brace direction.
2. When the joints are subjected to spatial loads, the distribution of SCFs on the chord side, with reference to the brace, is greatly influenced by other force-bearing braces.
3. The existing codes for calculating the SCFs of spatial joints primarily rely on the application of corrective parameters to the basic formulas. However, this approach leads to significant discrepancies in the calculated results and fails to accurately predict the SCFs of spatial joints. Therefore, it is necessary to propose suitable SCF calculation formulas specifically designed for spatial joints in order to enhance calculation accuracy.

Author Contributions: Conceptualization, Y.W. and B.W.; methodology, K.W.; software, S.G.; validation, P.L. and J.C.; formal analysis, K.W.; investigation, Y.W.; resources, S.G.; data curation, K.W.; writing—original draft preparation, K.W.; writing—review and editing, P.L.; visualization, J.C.; supervision, B.W., P.L. and Y.W.; project administration, S.G. and B.W.; funding acquisition, J.C. All authors have read and agreed to the published version of the manuscript.

Funding: This research was funded by National Natural Science Foundation of China (Nos. 52008063, 52071301, 51939002, 52101333), Chinese Scholarship Council (Grant No. 202206050093), and Zhejiang Provincial Natural Science Foundation of China (LQ21E090009).

Institutional Review Board Statement: Not applicable.

Informed Consent Statement: Not applicable.

Data Availability Statement: Not applicable.

Conflicts of Interest: The authors declare no conflict of interest.

References

1. Li, Y.; Zhang, Y.; Wang, W.; Li, X.; Wang, B. Influence of Corrosion Damage on Fatigue Limit Capacities of Offshore Wind Turbine Substructure. *J. Mar. Sci. Eng.* **2022**, *10*, 1011. [CrossRef]
2. Gan, J.; Sun, D.; Deng, H.; Wang, Z.; Wang, X.; Yao, L.; Wu, W. Fatigue Characteristic of Designed T-Type Specimen under Two-Step Repeating Variable Amplitude Load with Low-Amplitude Load below the Fatigue Limit. *J. Mar. Sci. Eng.* **2021**, *9*, 107. [CrossRef]
3. Zhao, W.; Hsu, W.T. Revaluation of Fatigue Thickness Effect Based on Fatigue Test Database. *J. Mar. Sci. Eng.* **2020**, *8*, 895. [CrossRef]
4. Mohamed, H.S.; Gao, F.; Guan, X.-Q.; Zhu, H.P. Experimental investigation on the fatigue behaviour of heat-treated tubular T-joints. *KSCE J. Civ. Eng.* **2018**, *22*, 2451–2463. [CrossRef]
5. *CIDECT Design Guide No.8, Design Guide for Circular and Rectangular Hollow Section Welded Joints under Fatigue Loading*; TÜV-Verlag: Cologne, Germany, 2000.
6. American Petroleum Institute. *Planning, Designing and Constructing Fixed Offshore Platforms—Working Stress Design*, 22nd ed.; API RP 2A-WSD (R2020); American Petroleum Institute: Washington, DC, USA, 2014.
7. Lie, S.-T.; Chiew, S.-P.; Lee, C.-K.; Huang, Z.-W. Fatigue Performance of Cracked Tubular T Joints under Combined Loads. II: Numerical. *J. Struct. Eng.* **2004**, *130*, 572–581. [CrossRef]
8. N'diaye, A.; Hariri, S.; Pluvinage, G.; Azari, Z. Stress concentration factor analysis for welded, notched tubular T-joints under combined axial, bending and dynamic loading. *Int. J. Fatigue.* **2009**, *31*, 367–374. [CrossRef]
9. Gho, W.M.; Gao, F. Parametric equations for stress concentration factors in completely overlapped tubular K(N)-joints. *J. Constr. Steel Res.* **2004**, *60*, 1761–1782. [CrossRef]
10. Feng, R.; Young, B. Stress concentration factors of cold-formed stainless steel tubular X-joints. *J. Constr. Steel Res.* **2013**, *91*, 26–41. [CrossRef]
11. Tong, L.W.; Zheng, H.Z.; Mashiri, F.R.; Zhao, X.L. Stress-Concentration Factors in Circular Hollow Section and Square Hollow Section T-Connections: Experiments, Finite-Element Analysis, and Formulas. *J. Struct. Eng.* **2013**, *139*, 1866–1881. [CrossRef]
12. Huang, F.; Cheng, B.; Li, C.; Zhao, X.L. Fatigue behavior of bird-beak SHS X-joints under cyclic brace axial forces. *J. Constr. Steel Res.* **2020**, *169*, 106021. [CrossRef]
13. Nassiraei, H. Local joint flexibility of CHS T/Y-connections strengthened with collar plate under in-plane bending load: Parametric study of geometrical effects and design formulation. *Ocean Eng.* **2020**, *202*, 107054. [CrossRef]
14. Chen, J.; Chen, J.; Jin, W.L. Experiment investigation of stress concentration factor of concrete-filled tubular T joints. *J. Constr. Steel Res.* **2010**, *66*, 1510–1515. [CrossRef]
15. Musa, I.A.; Mashiri, F.R.; Zhu, X.; Tong, L. Experimental stress concentration factor in concrete-filled steel tubular T-joints. *J. Constr. Steel Res.* **2018**, *150*, 442–451. [CrossRef]
16. Kim, I.-G.; Chung, C.-H.; Shim, C.-S.; Kim, Y.-J. Stress concentration factors of N-joints of concrete-filled tubes subjected to axial loads. *Int. J. Steel Struct.* **2014**, *14*, 1–11. [CrossRef]
17. Nassiraei, H.; Rezadoost, P. Stress concentration factors in tubular T/Y-joints strengthened with FRP subjected to compressive load in offshore structures. *Int. J. Fatigue.* **2020**, *140*, 105719. [CrossRef]
18. Ahola, A.; Lipiäinen, K.; Lindroos, J.; Koskimäki, M.; Laukia, K.; Björk, T. On the Fatigue Strength of Welded High-Strength Steel Joints in the As-Welded, Post-Weld-Treated and Repaired Conditions in a Typical Ship Structural Detail. *J. Mar. Sci. Eng.* **2023**, *11*, 644. [CrossRef]
19. Cheng, B.; Huang, F.; Duan, Y.; Chen, M.-T. Fatigue Performance of Bird-Beak SHS Gap K-Joints under Brace In-Plane Force. *J. Struct. Eng.* **2021**, *147*, 04021167. [CrossRef]
20. American Welding Society (AWS). *Structural Welding Code—Sheet Steel*; ASW: Miami, FL, USA, 2015.

21. *IIW Doc. XV-1035-99, Recommendations for IIW Sub Commission XV-E*; Abington Publishing: Northampton, UK, 2000.
22. DNVGL. *Fatigue Design of Offshore Steel Structures*; DNVGL-RP-C203; DNVGL: Oslo, Norway, 2016.

Journal of
*Marine Science
and Engineering*

MDPI

Article

Crashworthiness Optimization Method of Ship Structure under Multi-Working Conditions

Weijian Qiu, Kun Liu *, Hewei Liu, Shuai Zong, Jiaxia Wang and Zhenguo Gao

School of Naval Architecture and Ocean Engineering, Jiangsu University of Science and Technology, Zhenjiang 212100, China; 15751774907@163.com (W.Q.); wsslslsw1@163.com (H.L.); 220110101103@stu.just.edu.cn (S.Z.); jxwang66@yeah.net (J.W.); zhenguo.gao@just.edu.cn (Z.G.)

* Correspondence: kunliu@just.edu.cn; Tel.: +86-135-1169-2085; Fax: +86-0511-8444-6543

Abstract: Numerous collision conditions can occur during ship operations, resulting in various consequences that require specific consideration for optimizing crashworthiness design. Existing studies have investigated crashworthiness design in ship structures; however, they often focus on single working conditions and do not comprehensively consider the diverse scenarios encountered during ship operations. To overcome this drawback, this paper proposes a novel method that addresses multi-working conditions and combines orthogonal testing with a backpropagation neural network (BPNN) to establish an efficient surrogate model for collision optimization. The accuracy of the BPNN was improved by introducing the genetic algorithm and Adam algorithm. The technique for order preference by similarity to ideal solution (TOPSIS) is introduced to formulate a multi-working condition optimization function. The crashworthiness of the ship structure is optimized using the sparrow search algorithm (SSA) while considering the constraint of lightweight design. The results demonstrate a substantial reduction in the objective functions for the optimized collision conditions. Moreover, the BPNN predicted values are in good agreement with the finite element simulation results, affirming the effectiveness of the proposed method in improving the crashworthiness of the ship structure and providing valuable guidance for engineering design.

Keywords: ship collision; crashworthiness optimization; surrogate model; multi-working conditions

Citation: Qiu, W.; Liu, K.; Liu, H.; Zong, S.; Wang, J.; Gao, Z. Crashworthiness Optimization Method of Ship Structure under Multi-Working Conditions. *J. Mar. Sci. Eng.* **2023**, *11*, 1335. https://doi.org/10.3390/jmse11071335

Academic Editor: Joško Parunov

Received: 31 May 2023
Revised: 27 June 2023
Accepted: 29 June 2023
Published: 30 June 2023

1. Introduction

With the rapid development of the shipping industry, the number of ships sailing at sea is increasing, leading to a rise in ship accidents. In-depth research on ship collision and crashworthiness optimization is necessary to reduce the losses caused by collision accidents and ensure safe navigation at sea. However, the optimization design of ship structures involves finding the optimal solution in a complex and vast search space, which cannot be completed in a short time using traditional optimization methods and may result in search results exploding. Thus, finding efficient intelligent optimization algorithms for engineering optimization problems with high nonlinearity, constraint, and modeling complexity has become a significant research area in related disciplines [1]. Since engineering optimization is typically accompanied by a large number of finite element calculations, the direct application of intelligent optimization algorithms to solve such problems inevitably requires a substantial amount of calculation time. At this point, a surrogate model can be employed to approximate the output of the original model, thereby reducing computational costs. The surrogate model can rapidly predict system responses or evaluate design performance based on input design variables [2,3]. According to the literature [4], compared to the real value, the error rate using the surrogate model is only 1%. These models include the response surface model (RSM), Kriging model, radial basis function (RBF), and BP neural network, all of which have been applied in structure design [5–8].

Currently, the commonly adopted approach is to integrate surrogate models with optimization algorithms. Lin et al. [9] used the twin-stern fin-fishing vessel as the research object, constructed the sample set of the Kriging model by optimal Latin hypercube sampling (OLHS), and applied the multi-objective evolutionary algorithm (NSGA-II) to achieve the minimum resistance design of the fishing vessel. Prebeg et al. [10] established the surrogate model of tanker crashworthiness with the ultimate strength of the hull girder and ship crashworthiness quality measures as optimization objectives. Through multi-objective optimization, the safety of the structure was improved. Chen et al. [11] combined the adaptive RSM with a multi-objective genetic algorithm to investigate the dynamic response and design optimization of sandwich plates with layered gradients. It can be observed that introducing surrogate models into intelligent optimization algorithms to replace finite element simulation calculations in the optimization process can significantly enhance the optimization efficiency of intelligent optimization algorithms.

Furthermore, for such multi-objective optimization problems, it is worth discussing how to handle conflicts among optimization objectives. Some scholars have adopted the Pareto concept in addressing this issue, which can effectively identify better trade-off solutions in the solution space [12–14]. There are also scholars who have directly optimized all the optimization objectives using a weighted sum method [15,16]. This method, compared to the former one, is simpler and more intuitive in terms of methodology. It also allows for the convenient adjustment of the weights assigned to each optimization objective. However, it is only suitable for problems with lower complexity. When there are multiple conflicts among the optimization objectives, the adoption of the Pareto concept should be considered.

It should be noted that due to the inherent complexity and challenges associated with ship collision issues, current research on optimizing ship structural crashworthiness primarily focuses on a single specific working condition. Compared to other domains, there is relatively limited research on optimizing the structural crashworthiness of ships under multiple working conditions [17–19]. In fact, ship structure design often needs to balance the crashworthiness requirements imposed by various working conditions. When only conducting crashworthiness optimization designs for individual working conditions, conflicts can arise between these conditions. This is because, with identical design variables for the ship structure, the structural damage, deformation, collision forces, and energy absorption differ among these working conditions. While it is possible to demonstrate through verification that the optimization results for one working condition also meet the design requirements for other conditions, this is not the most effective method, especially when two working conditions significantly impact the optimization results.

Therefore, to address the crashworthiness requirements of ship structures under various collision conditions, this paper proposes a multi-working condition crashworthiness optimization method for ship structures based on BPNN, TOPSIS, and SSA (BP-TSSA). To address the issue of insufficient accuracy in traditional BPNN, an improved approach is proposed by incorporating the genetic algorithm (GA) and Adam algorithm to establish a high-precision crashworthiness optimization surrogate model. Different collision positions are selected to carry out multi-working condition crashworthiness optimization analysis on the structural areas mainly concerned by the struck ship, so as to realize the comprehensive optimization effect of multi-working conditions, multi-parameters, and multi-objectives, which can effectively support the engineering design.

It is worth noting that this method still has certain limitations. While it can address crashworthiness optimization problems under different collision locations, angles, and velocities, it cannot simultaneously consider factors such as collision location, collision angle, and collision velocity. Further research is therefore needed to address this limitation.

This paper is organized as follows. Following this introduction, the optimization method, including the improvement of the neural network surrogate model, optimization algorithm, and multi-working optimization functions, is introduced in Section 2. Section 3 describes the application examples of the crashworthiness optimization of the selected structure under two different collision conditions based on BP-TSSA. The discussion of the results is presented in Section 4, and major conclusions are drawn in Section 5.

2. Multi-Working Condition Optimization Method

This paper considers the randomness of hull collision positions in ship collision accidents by selecting two different positions located at the stern of a ship for collision analysis. The first position is at the center of the plate and the second position is at the intersection of the plate and frame, named Condition 1 and Condition 2, respectively (as illustrated in Figure 1). The crashworthiness optimization design process under multi-working conditions is presented in Figure 2. The employed optimization method is BP-TSSA, which utilizes BPNN to establish a collision-resistant optimization surrogate model. Subsequently, the TOPSIS is employed to construct a multi-working condition optimization function for SSA. Finally, the SSA is utilized to optimize the function. The primary optimization process is as follows:

1. Definition of the design variables, constraints, and objective function of the optimization problem and establishment of the mathematical model;
2. Utilization of orthogonal design to construct test sample points in the design domain, with each sample point corresponding to one numerical simulation case. The structural responses corresponding to each sample point are obtained through finite element analysis;
3. Normalization of the orthogonal table and finite element simulation results and input into BP neural network (for different collision conditions, corresponding BPNN is established, respectively);
4. SSA optimization to obtain ideal points of each optimization objective under different collision conditions and establish a multi-working condition evaluation function;
5. SSA optimization to obtain the optimal solution that can make each working condition relatively optimal based on the above evaluation function.

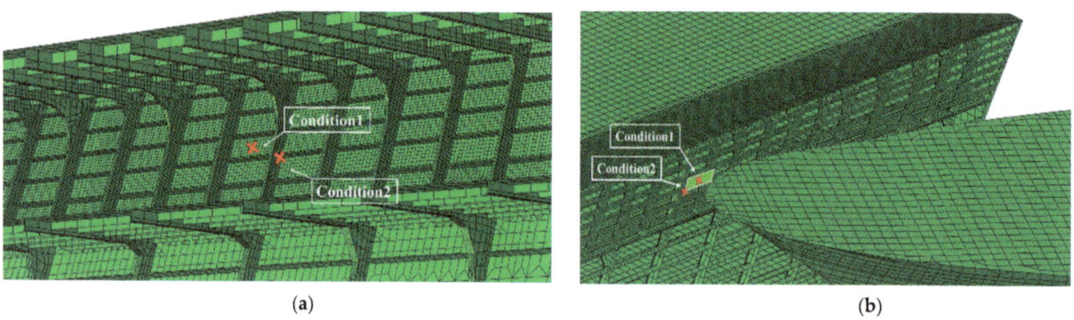

Figure 1. (**a**) Collision positions from interior perspective; (**b**) collision positions from exterior perspective.

2.1. Mathematical Model for Crashworthiness Optimization

2.1.1. Single Working Condition

For conventional single-working condition optimization problems, what needs to be established is a multi-objective optimization function. Multi-objective optimization design can achieve a balance in overall design performance. For instance, optimization objectives such as weight, energy absorption, and deformation are simultaneously considered, aiming to attain a design solution that can optimize these objectives in a relative sense. Mathematically, multi-objective optimization problems can be formulated as follows [20]:

$$\begin{cases} \min & f(\mathbf{X}) = f(x_1, x_2, \cdots, x_n) \\ s.t. & g_j(\mathbf{X}) \leq 0 \qquad\qquad j = 1, 2, \cdots, J \\ & x_{i,min} \leq x_i \leq x_{i,\max} \qquad i = 1, 2, 3, \cdots, n \end{cases} \tag{1}$$

where $\mathbf{X} = (x_1, x_2, \cdots, x_n)$ is a design variable, $f(\mathbf{X})$ is a multi-objective function, J is the total number of constraints, $g_j(\mathbf{X})$ is the jth constraints, x_i is the ith design variable, and $x_{i,\max}$ and $x_{i,\min}$ are the upper and lower limits, respectively.

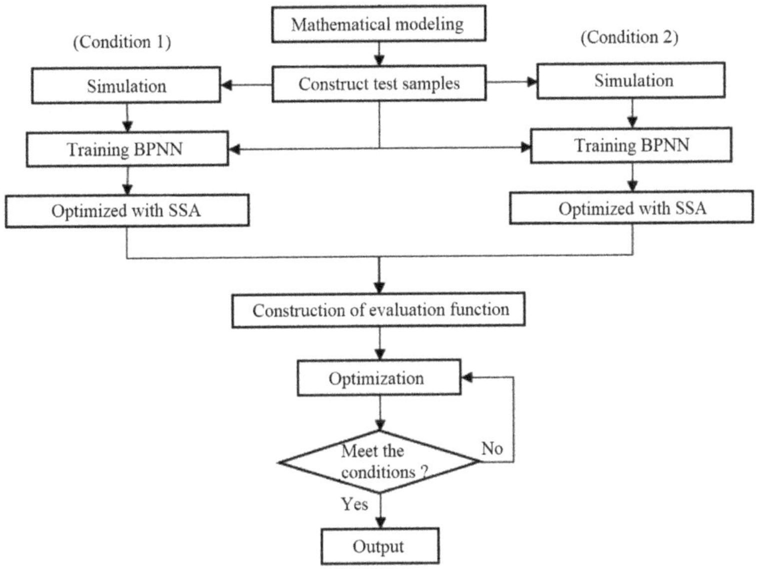

Figure 2. Crashworthiness optimization process.

In this paper, the linear weighting method of constructing the evaluation function is used to address multi-objective optimization problems. The linear weighting method aims to transform a multi-objective optimization problem into a single-objective optimization problem. This is accomplished by assigning weight coefficients to each objective based on their relative importance in the optimization problem. The evaluation function utilizing the linear weighting method can be expressed as follows [21]:

$$F(\mathbf{X}) = \sum_{i=1}^{n} \alpha_i f_i(\mathbf{X}) \tag{2}$$

where α_i is the weight coefficient of ith optimization objective. Each weight coefficient is between 0 and 1, and the sum of all the ownership weight coefficients is equal to 1. $f_{(i)}(\mathbf{X})$ is the ith optimization objective after normalization using Equation (3) [22].

$$Objective_{\text{membership}} = \frac{Objective_{\text{value}} - Objective_{\text{min}}}{Objective_{\text{max}} - Objective_{\text{min}}} \tag{3}$$

where $Objective_{\text{value}}$ is the normalized value of each index and $Objective_{\text{max}}$ and $Objective_{\text{min}}$ are the maximum and minimum values corresponding to each indicator in the orthogonal design sample. It should be noted that $Objective_{\text{max}}$ and $Objective_{\text{min}}$ do not represent optimized maximum and minimum values in the traditional sense, as they are only used for data normalization. They can be found directly in the orthogonal design sample.

In multi-objective optimization problems, certain objectives require maximization, while others require minimization. It is known that the maximum value of a function is equivalent to the minimum value of its negative counterpart. This paper incorporates this concept to enhance the multi-objective optimization function:

$$F(\mathbf{X}) = \sum_{i=1}^{n} \alpha_i f_i(\mathbf{X}) - \sum_{j=n+1}^{m} \alpha_j f_j(\mathbf{X}) \tag{4}$$

where α_i is the optimization objective weight coefficient that requires the minimum value and α_j is the optimization objective weight coefficient that requires the maximum value.

This paper considers several indicators that can comprehensively evaluate the crashworthiness of hull structures as optimization objectives, including weight, peak collision force, collision depth, and plastic energy absorption. Weight is an indispensable factor for realizing ship lightweight. Collision depth, collision force, and plastic energy absorption can, to a certain extent, reflect the crashworthiness of the structure. Regarding the collision depth, this paper primarily focuses on the plastic deformation of the struck ship structure. Generally, the greater the collision depth, the more significant the kinetic energy loss of the struck ship. The lost kinetic energy will be transformed into the plastic energy absorption of the struck ship structure during the collision process, hence increasing the plastic energy absorption of the struck ship structure. In order to minimize damage to people and cargo, it is essential to reduce the plastic deformation of the collision zone of the crashed ship as much as possible or to maintain the overall damage deformation within a narrow range, thus enhancing the safety of the structure. Collision force signifies the hull structure's ability to resist collision loads. In this paper's crashworthiness optimization research, the peak value of the collision force in the collision process is of primary concern. The higher the peak value, the more robust the hull structure's ability to resist collision loads.

It is evident that the following relationships exist among the four comprehensive crashworthiness indicators mentioned above:

1. Increasing the structural size can significantly enhance the crashworthiness of the ship but it will also increase the total weight of the ship;
2. Blindly pursuing structural lightweight can lead to a decrease in the crashworthiness of the ship. Specifically, the peak value of the collision force decreases and the collision depth increases, which is detrimental to the safety of the structure;
3. The coupling relationship between collision depth and plastic energy absorption is strong, considering both the collision depth and plastic energy absorption can effectively reduce the collision depth or collision depth range.

Equation (5) represents the multi-objective evaluation function of ship structure crashworthiness optimization under a single working condition. By assigning weight coefficients to each crashworthiness index and transforming it into the index's minimum value, the conflict between the indexes can be effectively resolved:

$$F = \alpha_M \cdot f(M) + \alpha_D \cdot f(D) - \alpha_P \cdot f(P) + \alpha_E \cdot f(E) \tag{5}$$

where the dimensionless quantity, $f()$, is obtained by normalizing each crashworthiness index using the formula presented in Equation (3). α_M, α_D, α_P, and α_E are the weight coefficients of weight, collision depth, collision force, and energy absorption, respectively. The optimal crashworthiness of a structure is achieved when its weight is minimized, collision force is maximized, and collision depth and plastic energy absorption are minimized. However, the coupling relationship between collision depth and energy absorption should also be considered to avoid overemphasis on optimization. Setting the weight coefficient too high may result in over-optimization, while setting it too low may not effectively improve the crashworthiness of the structure. This paper takes into account the relationships and relative importance of the four crashworthiness indicators and proposes the weight coefficients denoted as: $\alpha_M = 0.5$, $\alpha_D = 0.15$, $\alpha_P = 0.2$, and $\alpha_E = 0.15$, which achieves a balanced optimization.

When the optimization problem has constraints, the constrained problem is transformed into an unconstrained problem through the penalty function [23]:

$$F(\boldsymbol{X}) = \begin{cases} F(\boldsymbol{X}) & if \quad \boldsymbol{X} \in \boldsymbol{K} \\ 1 & if \quad \boldsymbol{X} \notin \boldsymbol{K} \end{cases} \tag{6}$$

where $\boldsymbol{X}$ is the design variable and $\boldsymbol{K}$ is the feasible region of the optimization problem.

2.1.2. Multi-Working Conditions

For the multi-working condition crashworthiness optimization problem addressed in this paper, it is imperative to not only establish a reasonable multi-objective evaluation function for each working condition but also develop a multi-working condition evaluation function to achieve relatively optimal performance across all conditions.

Technique for Order Preference by Similarity to Ideal Solution (TOPSIS), also known as the distance method of superior and inferior solutions, is a multi-objective decision-making method initially proposed by C.L. Hwang and K. Yoon [24]. The fundamental principle behind TOPSIS is to rank alternatives by calculating their distances to both the ideal and worst solutions. By determining the closest proximity to the ideal solution and the farthest distance from the worst solution, the method identifies the best alternative. For multi-objective optimization problems, if it is possible to assign a value f_i^* to each optimization objective such that $f_i^* \leq \min f_i(\boldsymbol{X})$ holds, then these values are referred to as "ideal points" for these optimization objectives.

In this paper, when dealing with multi-working condition optimization problems, the ideal point for the ith optimization objective is taken as the minimum value achieved when optimizing that objective function independently. Therefore, the multi-working condition optimization problem can be transformed into finding a solution that minimizes the weighted sum of distances between each optimization objective and its corresponding minimum value across all working conditions. This can be expressed as follows:

$$F(\boldsymbol{X}) = \sum_{i=1}^{n} \omega_i \cdot [W_1 \cdot |f_{i1}(\boldsymbol{X}) - f_{i1}^*(\boldsymbol{X})| + W_2 \cdot |f_{i2}(\boldsymbol{X}) - f_{i2}^*(\boldsymbol{X})|] \quad (7)$$

where n is the number of optimization objectives, ω_i is the weighting coefficient for each optimization objective, and $|f_{i1}(\boldsymbol{X}) - f_{i1}^*(\boldsymbol{X})|$ and $|f_{i2}(\boldsymbol{X}) - f_{i2}^*(\boldsymbol{X})|$ are the distances between each optimization objective and its corresponding ideal point under two different collision conditions, respectively. Here, $f_{i1}(\boldsymbol{X})$, $f_{i1}^*(\boldsymbol{X})$, $f_{i2}(\boldsymbol{X})$, and $f_{i2}^*(\boldsymbol{X})$ are calculated using the normalization function described earlier in the text. W_1 and W_2 are the weighting coefficients for the two collision conditions. This paper assumes that both collision conditions are equally important; thus, W_1 and W_2 are both set to 0.5.

2.2. BPNN Surrogate Model

The surrogate model serves as a means of emulating the correlation between input variables and output responses within the optimization problem. Typically, the surrogate model is established by either defining the input and output of the model with or without prior knowledge of the internal computation process. The selection of initial sample points determines the amount of original information captured within the model, whereby a greater number of initial sample points results in a higher degree of original information; however, this comes at the expense of increased computational time. To mitigate this, this present paper employs a scientifically rigorous experimental design method (orthogonal design) to systematically arrange tests, thereby reducing the number of tests required and shortening the test cycle [25]. To eliminate the influence of the different dimensions and units of sampling data on the analysis results, all the sample data are normalized to the range of 0~1.

In this paper, BPNN [26] is used as the surrogate model of the optimization problem, and orthogonal design is used to provide initial sample points for the surrogate model. Figure 3 shows the three-layer structure of BPNN. Specifically, the input layer comprises n nodes, the hidden layer contains g nodes, and the output layer encompasses m nodes. The connection weights between nodes are denoted by ω, while θ represents the threshold that affects the nodes.

The training algorithm used for the BPNN is the error backpropagation algorithm, commonly known as the BP algorithm. The objective of backpropagation is to minimize the error during the training process by estimating the error before the output layer using the output error and subsequently estimating the error of the previous layer. This process involves gradually transmitting the error from the output layer to the input layer in the opposite direction of input transmission while gradually updating the weights to approximate the expected output.

However, the BPNN still has the following two shortcomings:

1. It is difficult to determine the initial weights and thresholds of the network, and it is prone to get stuck in a local optimum;
2. The fixed learning rate can result in slow network convergence or system instability.

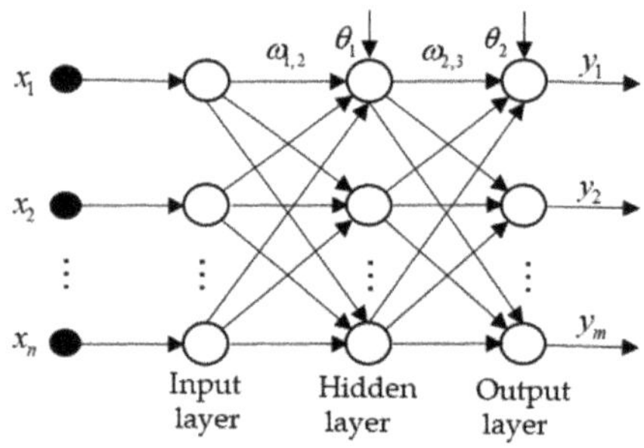

Figure 3. Three-layer BP neural network.

To overcome these limitations, this paper employs both the genetic algorithm and the Adam algorithm to enhance the performance of the BPNN.

2.2.1. Modification of BPNN Surrogate Model

The genetic algorithm was initially proposed by Holland in the 1970s [27]. In this paper, the genetic algorithm is utilized to optimize the initial parameters (weights and thresholds) of the BP neural network (BPNN). The optimization process involves the following steps:

1. Converting the parameters to be optimized into chromosomes using real number coding;
2. Using the network error as the objective function and the reciprocal of the error as the fitness function to evaluate the adaptability of the chromosomes;
3. Performing genetic operations such as selection, crossover, and mutation on the current population to update the population and select the optimal chromosome;
4. Outputting the optimal combination of initial parameters that minimize the error at the end of the iteration.

The fitness function of the genetic algorithm is given as:

$$F = \frac{1}{E + \lambda} \tag{8}$$

where E is the loss function (error) and λ is a constant close to 0. When F exceeds a predetermined upper limit, it is truncated to that upper limit to prevent the influence of F approaching infinity on the optimization process.

The Adam algorithm, introduced by Kingma [28], derives its name from "adaptive moment estimation". It is a network learning algorithm that accelerates network convergence by adaptively adjusting the learning rate and is capable of finding smaller errors than the traditional gradient descent method. As such, it can be used to train BPNN. The Adam algorithm introduces the concept of momentum, where first-order momentum suppresses oscillations in the network training process of the gradient descent method, while second-order momentum provides different learning rates for the network at different stages of training, balancing the adjustment of each parameter update. The formulas for first-order momentum and second-order momentum are given by Equations (9) and (10), respectively:

$$m_t = \beta_1 \cdot m_{t-1} + (1 - \beta_1) \cdot dk \tag{9}$$

$$v_t = \beta_2 \cdot v_{t-1} + (1 - \beta_2) \cdot dk^2 \tag{10}$$

where t is the current iteration number and dk is the gradient of the error function with respect to the weight or threshold. m and v are gradient first-order momentum and second-order momentum, respectively. β_1 and β_2 are the exponential decay rates of m and v, respectively, which are used to emphasize the gradients closer to the current moment, meaning that the impact of gradients decreases as they move farther away from the current moment.

By combining the first-order and second-order momentum of the gradient, the parameter change can be calculated under the current learning rate, and the parameters can be updated accordingly:

$$\eta_t = \alpha \cdot \frac{1}{\sqrt{v_t} + \delta} \tag{11}$$

$$\omega_t = \omega_{t-1} - \eta_t \cdot m_t \tag{12}$$

$$\theta_t = \theta_{t-1} - \eta_t \cdot m_t \tag{13}$$

where α is the initial learning rate, η_t is the learning rate of the current parameter calculated, and δ is a constant used to prevent the denominator from being 0. ω_t and θ_t are the updated weight and threshold values, respectively.

The principles of the Adam algorithm indicate that it utilizes the first-order momentum of the gradient to regulate the direction of model updates and the second-order momentum of the gradient to dynamically adjust the learning rate. This approach not only avoids network oscillation during training but also allocates distinct learning rates to different parameters at various stages, significantly simplifying the process of model adjustment.

The procedures of training BPNN with the Adam algorithm are as follows:

1. Randomly initialize network parameters (weights and thresholds);
2. Combine the sample data with the initialized parameters to calculate the output of each layer in the network;
3. Calculate the gradients by taking derivatives of the loss function with respect to the parameters;
4. Update the first and second moment estimates of the gradient and update the learning rate and parameters accordingly;
5. If the desired number of training iterations is reached, terminate the training process; otherwise, repeat steps (2) to (4).

2.2.2. Parameter Settings for BPNN Surrogate Model

The parameter settings for BPNN include determining the network structure and the algorithm parameters.

While having more hidden layers is commonly believed to improve the model's accuracy, it also increases its complexity. This paper adopts a three-layer backpropagation neural network with only one hidden layer as the primary approach, and the number of nodes in the hidden layer is determined by Kolmogorov's theorem. According to Kolmogorov's theorem, the network model achieves optimal accuracy when the relationship between the m nodes of the input layer and the n nodes of the hidden layer is approximately $n = 2m + 1$.

To introduce nonlinear characteristics to the backpropagation neural network, the Sigmoid function is selected as the activation function. The Sigmoid function can smoothly map the real field to a range of 0 to 1, and its functional form is given by:

$$f(x) = \frac{1}{1 + e^{-x}} \tag{14}$$

The mean square error function (MSE) is utilized as the loss function to assess the discrepancy between the predicted value and the true value of the network model. The loss value is also referred to as the mean square error, and the smaller the value, the higher the model's accuracy. The formula for the mean square error function is:

$$\mathrm{MSE} = \frac{1}{n}\sum_{i=1}^{n}|y - y'|^2 \quad (15)$$

where n is the total number of data output from the output layer, y is the true value, and y' is the predicted value.

BPNN is trained based on normalized sample data, and the preset parameters for the algorithm are set for BPNN in this section. Specifically, the genetic algorithm employs a population size of 50 and performs 5000 iterations, while the Adam algorithm is trained for 8000 iterations with an initial learning rate α of 0.001. Additionally, the first-order momentum exponential decay rate of the gradient β_1 is 0.9, the second-order momentum exponential decay rate of the gradient β_2 is 0.99, and the δ value is set to 10^{-8}.

2.3. Improved Sparrow Search Algorithm

As one of the swarm intelligence optimization algorithms, the sparrow search algorithm shares the same optimization mechanism as other algorithms in this category [29]. It relies on a population-based search, where each individual in the population represents a set of design schemes for the optimization problem at hand. The sparrow search algorithm achieves optimization by imitating the behavior of sparrows, including foraging, anti-predation, and investigation.

In the sparrow search algorithm, the population of sparrows can be divided into two main categories: explorers and followers. Explorers are responsible for locating food sources and providing suitable foraging areas for the followers. The fitness value of a sparrow represents the quality of its current location. When a follower discovers a better position, it competes with the explorers and brings along its own explorers, resulting in dynamic changes within the population.

To mitigate the risk of getting trapped in local optima, the population also includes scout sparrows. These scouts compare their fitness with the sparrows occupying the optimal position, thereby reducing the likelihood of falling into local optima.

Although the sparrow search algorithm generally yields favorable results in addressing most function optimization problems, a study by Yan et al. [30] revealed that the algorithm can still encounter challenges when dealing with complex multi-peak problems. Therefore, there is a need for further enhancements to the sparrow search algorithm.

Opposition-based learning (OBL) is a population generation strategy introduced by Tizhoosh et al. [31] in 2005, specifically designed for swarm intelligence optimization algorithms. OBL aims to enhance the uniformity of the population distribution through an opposite search approach. This strategy effectively addresses the limitations of swarm intelligence optimization algorithms that may suffer from insufficient population diversity due to the random generation of initial populations.

In OBL, if the j-dimensional position information of the ith sparrow in the new population is represented by $x_{i,j}$, the corresponding update formula for the opposite position information $x^*_{i,j}$ can be expressed as follows:

$$x^*_{i,j} = k(lb_j + ub_j) - x_{i,j} \quad (16)$$

where ub_j and lb_j are the upper and lower bounds of the j-dimensional design domain, respectively, and k is a random number between 0 and 1.

To enhance the population diversity in the algorithm, the search strategy of the Bird Swarm algorithm [32] is employed to enhance the location update formula for both explorers and followers. The position update for explorers is based on a Gaussian random distribution. The update formula for follower locations is adjusted to increase the likelihood of approaching the current optimal explorers. This modification is performed with a certain probability:

$$x_{i,j}^{t+1} = \begin{cases} x_{i,j}^{t} + x_{i,j}^{t} \cdot p & if \quad R_2 < ST \\ x_{i,j}^{t} + Q \cdot \boldsymbol{L} & if \quad R_2 \geq ST \end{cases} \tag{17}$$

where $x_{i,j}$ is the location information of the *i*th sparrow in the *j* dimension and *p* is Gaussian distribution with a mean value of 0 and standard deviation of 1. R_2 and ST are alarm value and security value, respectively; Q is a random number subject to normal distribution; and L is a matrix of $1 \times d$, in which each element is 1.

$$x_{i,j}^{t+1} = \begin{cases} Q \cdot \exp\left(\frac{x_{worst}^{t} - x_{i,j}^{t}}{i^2}\right) & if \quad i > n/2 \\ x_{p}^{t+1} + \left(x_{i,j}^{t} - x_{p}^{t+1}\right) \cdot FL \cdot k & if \quad i \leq n/2 \end{cases} \tag{18}$$

where n is the total number of sparrows, i is the follower number except the explorer in the sparrow population, x_p is the current position of the optimal explorer, x_{worst} is the worst position among all the current sparrows, Q is a random number subject to normal distribution, k is a random number between 0 and 1, and $FL \in [0,2]$ is the probability followers approaching the explorer.

Levy flight is a strategy that enhances the local search capability of the algorithm [33]. It is characterized by both small and large step lengths. When the algorithm is searching near the optimal position, the Levy flight strategy effectively prevents it from getting trapped in local optima. Thus, Levy flight can be incorporated into the position update formula for investigators, improving the control parameters of step size, denoted as β and K. A larger step size benefits the global search of the algorithm, while a smaller step size benefits the local search:

$$x_{i,j}^{t+1} = \begin{cases} Levy \cdot x_{best}^{t} + \beta \cdot \left|x_{i,j}^{t} - Levy \cdot x_{best}^{t}\right| & if \quad f_i > f_g \\ x_{i,j}^{t} + K \cdot \left(\frac{\left|x_{i,j}^{t} - x_{worst}^{t}\right|}{(f_i - f_w) + \varepsilon}\right) & if \quad f_i = f_g \end{cases} \tag{19}$$

$$\beta = f_g - (f_g - f_w) \cdot \left(\frac{iter_{\max} - t}{iter_{\max}}\right)^{1.5} \tag{20}$$

$$K = (f_g - f_w) \cdot \exp(-20 \cdot \tan\left(\frac{t}{iter_{\max}}\right)^{2}) \cdot (2c - 1) \tag{21}$$

$$Levy = 0.01 \cdot \frac{r_1 \cdot \sigma}{|r_2|^{1/\lambda}} \tag{22}$$

$$\sigma = \left\{\frac{\Gamma(1+\lambda) \cdot \sin(\pi\lambda/2)}{\Gamma[(1+\lambda)/2]\lambda \cdot 2^{(\lambda-1)/2}}\right\}^{1/\lambda} \tag{23}$$

where t is the current iteration number; $iter_{\max}$ is the maximum number of iterations; x_{best} is the current optimal location; f_i is the fitness value of the current sparrow; f_g is the fitness value of the current optimal position; f_w is the fitness value of the current worst position; ε is a constant used to make the denominator not zero; c is a random number between 0 and 1; Γ is the gamma function; λ is a constant between 0 and 2, generally, 1.5; and r_1 and r_2 are random numbers between 0 and 1.

The main SSA process is as follows:

1. Randomly initialize the sparrow population;
2. Calculate the fitness of each sparrow and sort them in descending order based on their fitness values;
3. Update the location of the explorer and follower sparrows alternately. This step involves adjusting the positions of these sparrows based on specific rules or algorithms;

4. Randomly select investigator sparrows and update their locations using a predefined strategy or method;
5. Check if there is a better position found during the previous steps. If a better position is found, update the optimal position accordingly;
6. Check if the termination condition of the algorithm is satisfied. If it is, output the optimal solution obtained so far. Otherwise, repeat steps (2) to (5) to continue the execution of the algorithm.

3. Example Analysis of Multi-Working Conditions Crashworthiness Optimization

3.1. Ship Collision Scenario and Its Primary Optimization Components

The collision scenario selected for this paper involves the bow of the striking ship colliding with the tail of the struck ship at a collision angle of 150°. Figure 4 illustrates a diagram of the ship collision. The coordinate system's origin for the struck ship is situated at the stern baseline, with the x-axis running from stern to bow, the y-axis running from starboard to port, and the z-axis running from baseline to deck. The coordinate origin for the striking ship is located at the baseline position of the striking ship's stern. When the collision occurs, the striking ship and the struck ship's course align with their respective x-axis coordinates, and there is no rotation.

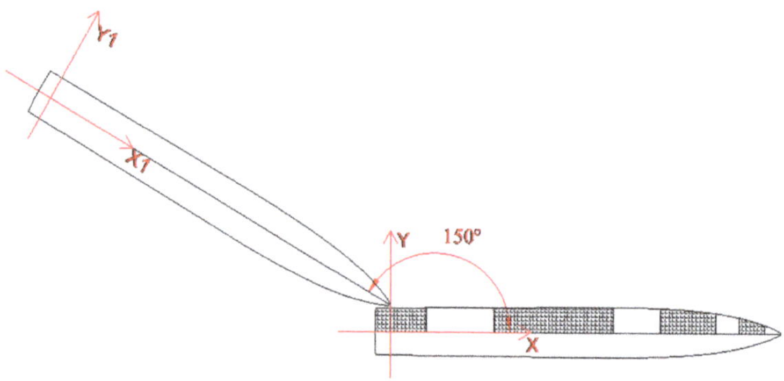

Figure 4. Diagram of ship collision.

To simplify the calculation process and ensure the validity of the results, the relative impact velocity method is used when setting the speed, where the struck ship is assumed to be stationary, and the relative impact velocity of the striking ship is 5 knots. This approach streamlines the calculation process while ensuring the accuracy of the results.

Figure 5 shows the finite element model of the ship, which was established using the ABAQUS finite element software. The collision location is the stern side of the struck ship. Since this paper mainly focuses on investigating the structural response of the collision zone of the struck ship, the striking ship was set as a rigid body, whereas the struck ship was set as a deformable body, with the collision zone of the struck ship meshed. To optimize simulation computation time, the collision region of the struck ship is enhanced with a finer mesh, whereas coarser meshes with a mesh size of 0.25 m are employed for the remaining areas of both the struck and striking ships. Both striking ship and struck ship were modeled with the four-node shell element S4R from the ABAQUS element library. A mesh sensitivity analysis is performed on the collision region of the struck ship to ascertain an appropriate mesh size. Figure 6 demonstrates the collision region models of the struck ship at varying mesh sizes of 0.1 m, 0.05 m, and 0.025 m, respectively.

The collision force curves obtained from the three different mesh sizes mentioned above are shown in Figure 7. It can be observed that these collision force curves exhibit good convergence, with minimal differences in the computed results when using mesh sizes of 0.05 m and 0.025 m. Therefore, for the subsequent analysis, a uniform mesh size of 0.05 m is adopted for the collision region of the struck ship.

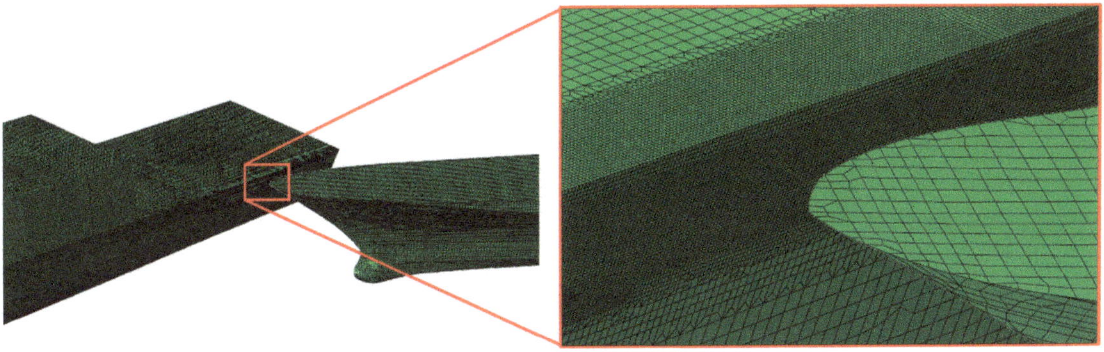

Figure 5. FEM of ship collision.

(a) (b) (c)

Figure 6. (**a**) Mesh size of 0.1 m, (**b**) mesh size of 0.05 m, (**c**) mesh size of 0.025 m.

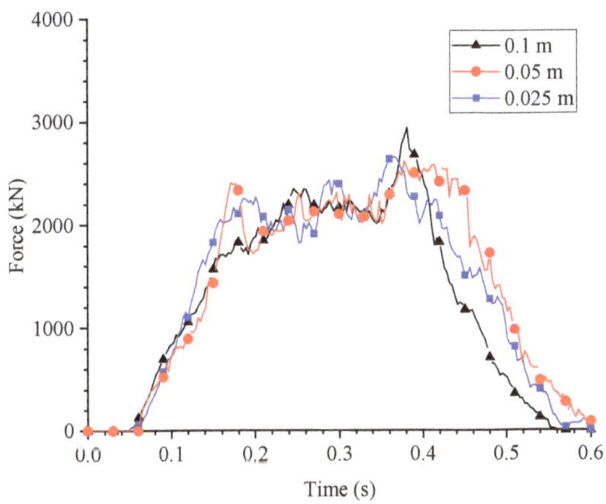

Figure 7. Collision force curves at different mesh sizes.

The primary parameters of both vessels are presented in Table 1, indicating that the striking ship has a total mass of 1000 T and the struck ship has a total mass of 3850 T.

Table 1. Principle parameters of collision ships.

Collision Ships	Total Length/m	Beam Molded/m	Molded Depth/m	Total Mass/T
Struck ship	132.00	16.00	8.60	3850
Striking ship	83.00	9.77	6.68	1000

For the collision simulation, the model uses general contact with both normal and tangential contact attributes. The normal contact attribute is set to hard contact while the tangential contact attribute uses the penalty function method, with a friction coefficient of 0.3. The Cowper-Symonds model, which considers the material's strain rate sensitivity, is employed as the constitutive model for the hull material [34,35]:

$$\frac{\sigma_d}{\sigma_y} = 1 + \left(\frac{\dot{\varepsilon}}{D}\right)^{\frac{1}{q}} \tag{24}$$

where σ_d is dynamic yield stress, σ_y is static yield stress, $\dot{\varepsilon}$ is equivalent strain rate, and D and q are constants set to 40.4 and 5, respectively.

To simulate ship collision in a water medium, the impact of water on the collision effect is taken into account by including the attached water mass [36]. In the case of ship–ship collision, the transverse motion mainly affects the struck ship, while the forward and backward motion mainly affects the striking ship. Therefore, this paper only considers the attached water mass of the transverse and the forward–backward motion. Equations (25) and (26) represent the empirical formulas for the added water mass of the struck ship and the striking ship, respectively:

$$m_1 = (0.4 \sim 1.3)m \tag{25}$$

$$m_2 = (0.02 \sim 0.07)m \tag{26}$$

where m is the hull mass. The attached water mass of the struck ship is 0.4 m and that of the striking ship is 0.03 m.

Figure 8 shows the model structure diagram of the collision area of the struck ship. The shell plating and the frame act as the primary energy-absorbing components in this region during an impact load. Therefore, this paper aims to adjust the thickness of the shell plating and the size of the frame web while retaining the original structure of the struck ship.

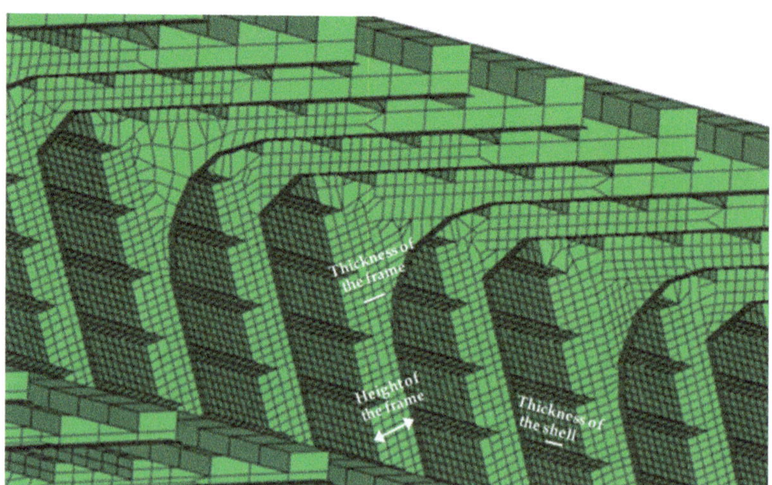

Figure 8. Model structure diagram of collision area.

The thickness of the shell plating and the thickness and height of the frame web are considered design variables, with the original shell plating of the struck ship having a thickness of 13 mm, a frame size of 240 mm × 6 mm/40 mm × 8 mm, and a longitudinal bone size of 100 mm × 7.5 mm. To ensure the rationality of the structure, the design domain of the shell plating thickness ranges from 13 mm to 16 mm, the design domain of the frame web thickness ranges from 4 mm to 7 mm, and the design domain of the frame web height ranges from 200 mm to 260 mm. The structural weight is considered a constraint condition, and the weight calculated from the 11th sample point in Tables 2 and 3 serves as the threshold that should not be exceeded in a lightweight design. Therefore, its value should be less than or equal to 4860.96 kg to incorporate lightweight design principles.

Table 2. Orthogonal design and finite element simulation results for working condition 1.

Sample Number	Thickness of the Shell Plating/(mm)	Thickness of the Frame Web/(mm)	Height of the Frame Web/(mm)	Weight/(kg)	Peak Collision Force/(N)	Collision Depth/(m)	Plastic Energy Absorption/(J)	Objective Function Value
1	13	4	200	4161.92	2,508,050	0.082	118,796.65	0.357
2	13	5	220	4295.25	2,651,900	0.079	123,713.37	0.368
3	13	6	240	4437.30	2,640,630	0.075	128,236.35	0.416
4	13	7	260	4580.71	2,849,940	0.069	132,640.73	0.365
5	14	4	220	4415.32	2,681,830	0.078	116,599.61	0.378
6	14	5	200	4514.94	2,675,860	0.076	118,022.23	0.411
7	14	6	260	4693.83	2,746,050	0.071	123,799.87	0.441
8	14	7	240	4798.00	2,677,790	0.069	121,210.58	0.480
9	15	4	240	4669.56	2,292,020	0.071	102,084.79	0.486
10	15	5	260	4806.95	2,365,960	0.071	105,583.03	0.537
11	15	6	200	4860.96	2,869,170	0.073	116,529.87	0.469
12	15	7	220	5008.76	2,826,550	0.070	113,533.07	0.507
13	16	4	260	4920.06	2,231,700	0.069	99,321.75	0.583
14	16	5	240	5030.26	2,332,510	0.069	100,145.03	0.610
15	16	6	220	5128.84	2,251,270	0.069	100,168.92	0.677
16	16	7	200	5220.97	2,450,980	0.068	96,076.65	0.631

Table 3. Orthogonal design and finite element simulation results for working condition 2.

Sample Number	Thickness of the Shell Plating/(mm)	Thickness of the Frame Web/(mm)	Height of the Frame Web/(mm)	Weight/(kg)	Peak Collision Force/(N)	Collision Depth/(m)	Plastic Energy Absorption/(J)	Objective Function Value
1	13	4	200	4161.92	2,120,530	0.073	95,290.00	0.381
2	13	5	220	4295.25	3,644,950	0.069	120,226.30	0.292
3	13	6	240	4437.30	3,533,160	0.066	118,995.62	0.336
4	13	7	260	4580.71	3,208,110	0.067	128,030.34	0.490
5	14	4	220	4415.32	2,202,090	0.070	92,556.30	0.448
6	14	5	200	4514.94	2,354,290	0.067	94,286.20	0.445
7	14	6	260	4693.83	3,495,660	0.063	116,734.40	0.420
8	14	7	240	4798.00	3,447,750	0.061	110,576.50	0.426
9	15	4	240	4669.56	2,267,900	0.067	90,899.90	0.521
10	15	5	260	4806.95	3,369,870	0.065	115,508.20	0.504
11	15	6	200	4860.96	3,626,310	0.063	112,863.40	0.471
12	15	7	220	5008.76	3,656,210	0.061	109,058.00	0.493
13	16	4	260	4920.06	2,435,650	0.065	89,656.30	0.592
14	16	5	240	5030.26	2,395,300	0.062	91,386.60	0.624
15	16	6	220	5128.84	2,418,970	0.060	91,136.70	0.643
16	16	7	200	5220.97	2,402,330	0.059	86,735.70	0.663

The design variables, constraints, and objective functions presented above have been consolidated, and the mathematical model for optimizing the crashworthiness of the hull structure under a single working condition is formulated in Equation (27):

$$\begin{aligned} \min \quad & F = \alpha_M \cdot f(M) + \alpha_D \cdot f(D) - \alpha_P \cdot f(P) + \alpha_E \cdot f(E) \\ s.t. \quad & 13 \text{ mm} \leq t_w \leq 16 \text{ mm} \\ & 4 \text{ mm} \leq t_f \leq 7 \text{ mm} \\ & 200 \text{ mm} \leq h \leq 260 \text{ mm} \\ & m \leq 4860.96 \text{ kg} \end{aligned} \tag{27}$$

Taking into account the crashworthiness optimization problem under multi-working conditions with different collision positions, the mathematical model can be represented by Equation (28):

$$\begin{aligned} \min \quad & F(\boldsymbol{X}) = \sum_{i=1}^{n} \omega_i \cdot \left[W_1 \cdot \left| f_{i1}(\boldsymbol{X}) - f_{i1}^*(\boldsymbol{X}) \right| + W_2 \cdot \left| f_{i2}(\boldsymbol{X}) - f_{i2}^*(\boldsymbol{X}) \right| \right] \\ s.t. \quad & 13 \text{ mm} \leq t_w \leq 16 \text{ mm} \\ & 4 \text{ mm} \leq t_f \leq 7 \text{ mm} \\ & 200 \text{ mm} \leq h \leq 260 \text{ mm} \\ & m \leq 4860.96 \text{ kg} \end{aligned} \tag{28}$$

3.2. Orthogonal Design Samples

To address the multi-working condition crashworthiness optimization problem examined in this paper while considering the rationality of the orthogonal table design and structural manufacturability, the L_{16} (4^5) orthogonal table was adopted for orthogonal design. The factors considered in the design were the thickness of the shell plating, the thickness of the frame web, and the height of the frame web, with each factor having four levels in the design domain. Specifically, the thickness of the shell plating had levels of 13 mm, 14 mm, 15 mm, and 16 mm; the thickness of the frame web had levels of 4 mm, 5 mm, 6 mm, and 7 mm; and the frame web height had levels of 200 mm, 220 mm, 240 mm, and 260 mm.

The response results (weight, peak collision force, collision depth, plastic energy absorption) of all sample points within the orthogonal table were calculated sequentially using the ABAQUS finite element software, with the results presented in Tables 2 and 3.

3.3. Training of BPNN Surrogate Model

Based on the predetermined parameters of the BPNN, BPNN surrogate models were constructed for two collision working conditions using Tables 2 and 3, respectively. In these models, the input layer has three nodes, the hidden layer has seven nodes, and the output layer has four nodes. The genetic algorithm iteration plot for Condition 1 is shown in Figure 9a, and the corresponding loss function curve for the BPNN is shown in Figure 9b. Similarly, the genetic algorithm iteration plot for Condition 2 is shown in Figure 10a, and the associated loss function curve for the BP neural network is shown in Figure 10b.

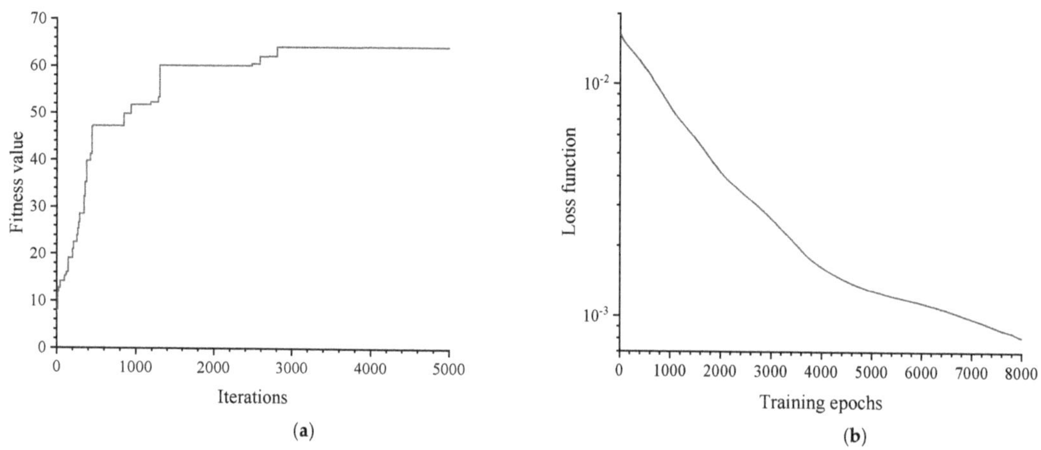

Figure 9. (**a**) Genetic algorithm iteration plot (Condition 1); (**b**) loss function curve for the BPNN (Condition 1).

From the GA fitness curve, it can be observed that the optimal fitness value changes dramatically in the early stage of algorithm iteration and does not have significant changes in the later stage. This is because the optimization objective of GA is the numerical minimal BP neural network error, which theoretically can only approach the optimal value of 0 indefinitely, and there is no truly minimal point. Therefore, when the iteration number is not set, GA will continue to optimize indefinitely, but the fitness value will become increasingly flat with the increase of the iteration number. From the BP loss function curve, it can be seen that the network converges to an accuracy of nearly 10^{-3} after 8000 training iterations, and it only takes 2 s, indicating that the BP neural network in this paper is excellent in both training accuracy and efficiency.

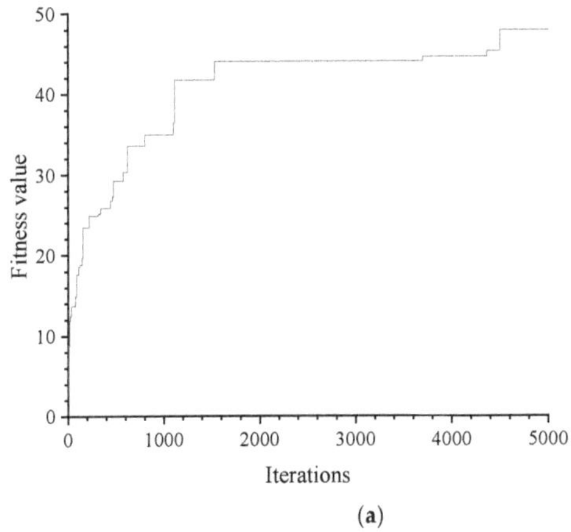

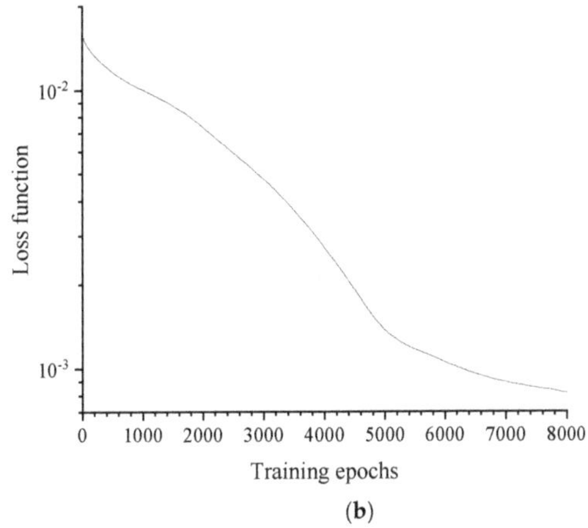

Figure 10. (**a**) Genetic algorithm iteration plot (Condition 2); (**b**) loss function curve for the BPNN (Condition 2).

3.4. Optimal Results

The two BPNN surrogate models established in the previous section are employed in optimization to provide effective predicted values for the objective function (Equation (28)) in the multi-working condition optimization. The ideal points for each optimization objective in Equation (28) are obtained through single-objective optimization using SSA based on the optimization objectives in Equation (27). By employing SSA, each crashworthiness index for individual working conditions is optimized separately, enabling the attainment of the optimal value for each index. In the sparrow search algorithm configuration, the population size is set at 100, with 20% dedicated to explorers and 10% to investigators. The security value is 0.8, and the number of iterations is 500. Subsequently, the obtained ideal points are substituted into Equation (28), and the sparrow search algorithm is employed once again to optimize the calculation, resulting in optimized outcomes for multiple working conditions. The convergence curve of the optimization process is shown in Figure 11.

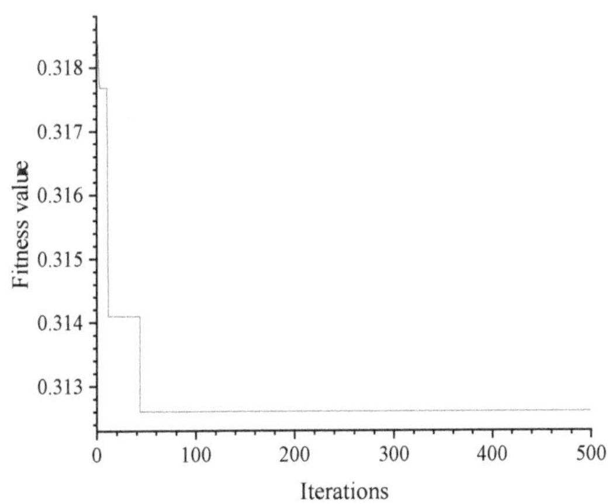

Figure 11. SSA iteration plot.

The output of the optimization yields the following results: the shell plating thickness is 13 mm, the frame web thickness is 5.36 mm, and the frame web height is 200 mm.

Compared to the predictive results of BPNN, the simulated values obtained through finite element calculations are more realistic. Therefore, this paper further utilizes ABAQUS finite element software to simulate and calculate the optimized results, while also validating the generalization ability of the BPNN. Tables 4 and 5 present the comparative results between the BPNN predicted values and the finite element simulated values under two collision conditions.

Table 4. Comparison between predicted and simulated values for Condition 1.

	Weight/kg	Peak Collision Force/(N)	Collision Depth/(m)	Plastic Energy Absorption/(J)
Predicted values	4339.19	2,535,273	0.075	121,148.98
Simulated values	4317.80	2,689,010	0.072	122,891.17
Relative error (%)	0.50	5.72	4.17	1.42

Table 5. Comparison between predicted and simulated values for Condition 2.

	Weight/kg	Peak Collision Force/(N)	Collision Depth/(m)	Plastic Energy Absorption/(J)
Predicted values	4351.93	3,386,878	0.069	117,932.80
Simulated values	4317.80	3,619,960	0.069	118,838.70
Relative error (%)	0.79	6.44	0	0.76

Based on the comparative results, the modified BPNN model showcased remarkable predictive accuracy in this paper, as evidenced by the minimal relative errors observed for various performance indicators under two collision conditions. For Condition 1, the relative errors ranged from 0.5% to 5.72% for different indicators, while for Condition 2, the relative errors ranged from 0.79% to 6.44% for different indicators. Notably, a perfect prediction (0% error) was achieved for one indicator in Condition 2. These findings unequivocally demonstrate the superior predictive efficacy and strong generalization capability of the modified BPNN model proposed in this paper. Such remarkable performance underscores the model's potential for reliable predictions and highlights its suitability for practical applications in the field of ship collision. It can effectively replace laborious and time-consuming finite element calculations during the optimization process of the sparrow search algorithm.

To highlight the effectiveness of multi-working condition optimization, a comparison is made between the optimization results of multi-working condition optimization and the separate optimization results of two collision conditions, as shown in Table 6. In this table, "Optimization Condition" refers to the working condition currently used for optimization, and "Calculation Condition" refers to the working condition for which the optimization results are calculated using the optimized conditions.

Table 6. Comparing optimization results.

Optimization Condition	Calculation Condition	Thickness of the Shell Plating/(mm)	Thickness ofthe Frame Web/(mm)	Height of the Frame Web/(mm)	Weight/kg	Peak Collision Force/(N)	Collision Depth/(m)	Plastic Energy Absorption/(J)	Objective Function Value
Original design	Condition 1	13	6	240	4437.30	2,640,630	0.075	128,236.35	0.416
	Condition 2	13	6	240	4437.30	3,533,160	0.066	118,995.62	0.336
Condition 1	Condition 1	13	4	260	4204.80	2,681,300	0.079	123,571.30	0.313
	Condition 2	13	4	260	4204.80	3,231,880	0.073	118,193.30	0.340
Condition 2	Condition 1	13	5	220	4295.25	2,651,900	0.079	123,713.37	0.368
	Condition 2	13	5	220	4295.25	3,644,950	0.069	120,226.3	0.292
Multi-conditions	Condition 1	13	5.36	200	4317.80	2,689,010	0.072	122,891.17	0.287
	Condition 2	13	5.36	200	4317.80	3,619,960	0.069	118,838.70	0.301

From the calculated results of different optimization conditions in Table 6, it can be observed that compared to Condition 1 (collision area at the center of the plate), Condition 2 (collision area at the intersection of the plate and frame) experiences significantly higher collision force, but with less collision depth and plastic energy absorption. This indicates that the structural strength is higher at the intersection of the plate and frame compared to the center of the plate, highlighting the need to focus on the collision performance of Condition 1. However, in actual ship collision accidents, the collision area exhibits a certain level of randomness, so the collision performance of other areas (such as Condition 2) should not be overlooked.

The table presents the structural response when optimizing Condition 1 independently and calculating the results for Condition 2 based on the optimization results of Condition 1, as well as the structural response when optimizing Condition 2 independently and calculating the results for Condition 1 based on the optimization results of Condition 2. From these comparative results, it can be observed that when only pursuing the collision performance of Condition 1, the collision performance of Condition 2 is inferior to that achieved through the separate optimization of Condition 2, with objective function values of 0.34 and 0.292, respectively. Similarly, when only pursuing the collision performance of Condition 2, the collision performance of Condition 1 is also inferior to that achieved through the separate optimization of Condition 1, with objective function values of 0.368 and 0.313, respectively.

Therefore, in order to find a relative balance between the two conditions, this paper performed joint optimization considering multi-working conditions. By comparing the objective function of multi-working condition optimization with the objective functions of separate optimization for Condition 1 and Condition 2, it can be observed that the objective function values for Condition 1 in the multi-working condition optimization are smaller than the objective functions obtained from separate optimizations of Condition 1 and Condition 2: $0.287 < 0.313 < 0.368$. On the other hand, the objective function values for Condition 2 in the multi-working condition optimization fall between the objective function values obtained from separate optimizations of Condition 1 and Condition 2: $0.292 < 0.301 < 0.34$. This indicates that in the multi-working condition optimization approach presented in this paper, by considering both collision conditions simultaneously, the optimal solution can meet the relative optimality under both collision conditions. Moreover, the objective functions of Condition 1 and Condition 2 are significantly lower than the original design, with a reduction of 31% for Condition 1 and 10.4% for Condition 2, demonstrating excellent optimization results.

In terms of changes in structural dimensions, it can be concluded that maintaining the same shell plating thickness while appropriately increasing the thickness of the frame web and reducing its height contributes to improving the structural collision performance.

4. Discussion

4.1. Analysis of Improvement in BPNN

In order to analyze the learning ability of the modified BPNN on samples more intuitively, the modified BPNN is trained and compared with the original BPNN using sample data from two collision conditions. The comparative results are shown in Figure 12.

From Figure 12, it can be observed that when the preset accuracy is 10^{-4}, regardless of the collision conditions, the original BPNN fails to reach the preset accuracy within the specified number of training iterations. This indicates that the training efficiency of the BPNN based on the traditional gradient descent method is low, particularly when the learning rate is small, leading to slow convergence. On the other hand, the modified BPNN, which incorporates adaptive learning rate adjustment, is able to converge to the preset accuracy in fewer training iterations. Furthermore, compared to the original BPNN, the modified BPNN achieves a smaller loss value, indicating that the genetic algorithm (GA) can help the traditional BP neural network model escape local optima traps and enhance the model's learning ability. Consequently, after the modification, the BPNN exhibits significant improvements in both training accuracy and convergence speed.

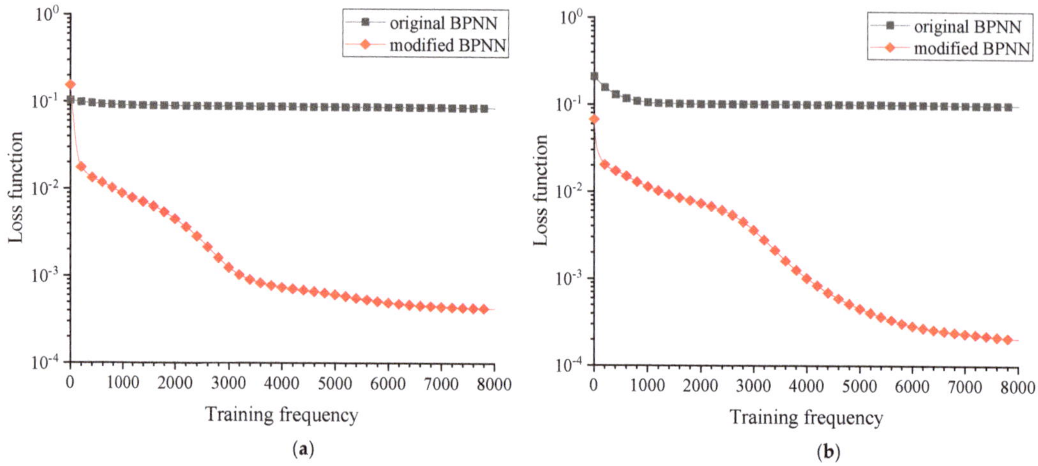

Figure 12. (**a**) Comparative results of BPNN before and after modification (Condition 1); (**b**) comparative results of BPNN before and after modification (Condition 2).

4.2. Analysis of Ship Crashworthiness Performance before and after Optimization

This paper aims to achieve relative optimality among various collision resistance indicators by considering the influence of multiple structural responses on structural crashworthiness during multi-objective optimization.

From the optimization results in Table 6, it can be observed that the total plastic energy absorption of the optimized structure has been reduced in both collision conditions, with percentage reductions of 4.16% and 0.14%, respectively. This indicates a decrease in the plastic deformation occurring in the collided region of the struck ship's structure. Regarding the peak collision forces, the optimized structure shows improvements compared to the original structure in both collision conditions, with increases of 1.80% and 2.40%, respectively. This suggests an enhanced ability of the optimized structure to withstand collision loads.

As for the parameter of maximum collision depth in the struck ship's hull structure, the optimized structure exhibits a 4.00% reduction under Condition 1. Although there is a 4.55% increase in the maximum collision depth of the optimized structure under Condition 2, when considering multiple weighted collision resistance indicators, the final objective function shows a decrease.

Figure 13 presents the damage deformation contour maps of the struck ship before and after optimization in two collision conditions. For the purpose of comparison, in this paper, a critical value of 150 MPa is set in the post-processing module of ABAQUS. The structural regions with stress levels below 150 MPa are represented in gray, while the regions with stress levels above 150 MPa are displayed with a continuous color gradient ranging from red to blue, corresponding to the magnitude of stress. By comparing the areas of the colored regions, it can be observed that the optimized structure has a smaller area of high-stress regions compared to the structure before optimization.

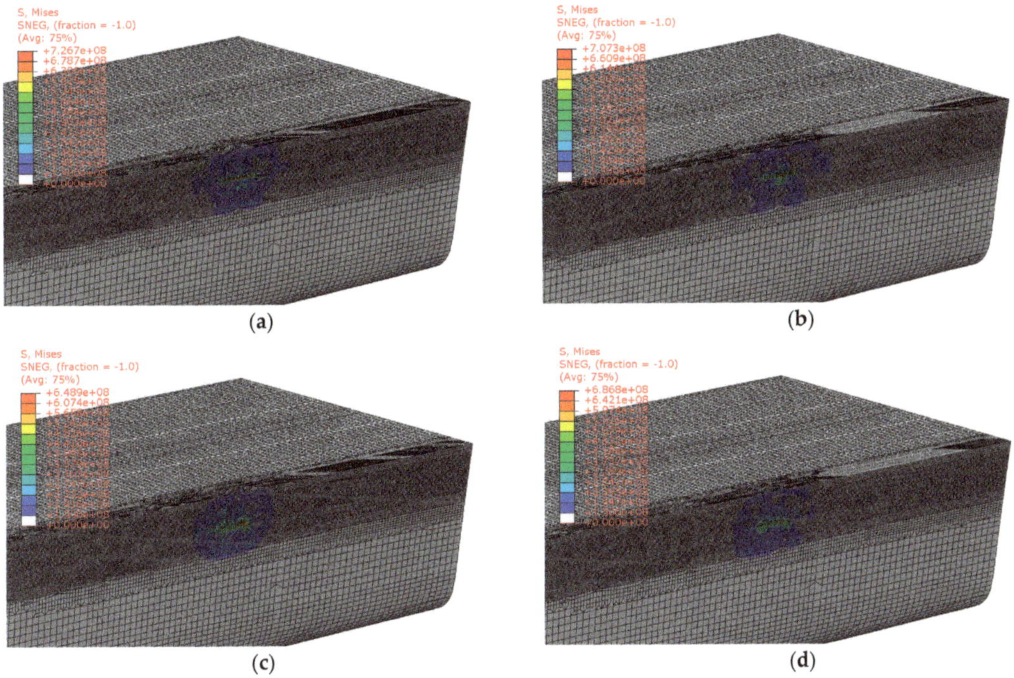

Figure 13. (**a**) Before optimization (Condition 1); (**b**) after optimization (Condition 1); (**c**) before optimization (Condition 2); (**d**) after optimization (Condition 2).

5. Conclusions

This paper presents a multi-working condition structural optimization design method, called BP-TSSA, based on BPNN, TOPSIS, and SSA, with the objective of enhancing the collision resistance performance of ship structures. The proposed method achieves significant improvements in the structural collision resistance performance compared to the original structure, and it also enables the attainment of relatively optimal structural designs for two different collision conditions.

The main results are as follows:

1. For the multi-working condition collision optimization problem in two different collision conditions, the optimized objective functions using the BP-TSSA algorithm proposed in this paper show a reduction of 31% and 10.4%, respectively, compared to the original designs. This indicates that the BP-TSSA algorithm proposed in this paper achieves good results in handling multi-working condition, multi-objective, and multi-parameter collision optimization problems for ship structures, demonstrating its excellent optimization capability;
2. The optimized results of the BPNN prediction align well with the finite element simulation results. This demonstrates that the BPNN surrogate model proposed in this paper possesses sufficient generalization ability;
3. The objective function values obtained for both collision conditions in the multi-working condition optimization are very small. This indicates that the multi-working condition optimization method proposed in this paper achieves relative optimality between the two collision conditions by considering them simultaneously.

In conclusion, the optimization method proposed in this paper can effectively enhance the collision resistance performance and optimization efficiency of ship structures. It can serve as an optimization tool for further design of ship structures and contribute to improving maritime safety.

Author Contributions: Conceptualization, W.Q. and K.L.; methodology, W.Q.; software, W.Q.; formal analysis, H.L.; data curation, S.Z.; visualization, H.L.; writing—original draft, W.Q.; writing—review and editing, Z.G. and K.L.; supervision, K.L. and J.W. All authors have read and agreed to the published version of the manuscript.

Funding: This research was funded by the National Natural Science Foundation of China (Grant No. 52171311; Grant 52271279) and the Jiangsu Province Graduate Research and Practice Innovation Program Project under Grant SJCX22_1958.

Institutional Review Board Statement: Not applicable.

Informed Consent Statement: Not applicable.

Data Availability Statement: Not applicable.

Conflicts of Interest: The authors declare no conflict of interest.

References

1. Fang, J.; Sun, G.; Qiu, N.; Kim, N.H.; Li, Q. On design optimization for structural crashworthiness and its state of the art. *Struct. Multidiscip. Optim.* **2017**, *55*, 1091–1119. [CrossRef]
2. Jiang, P.; Zhou, Q.; Shao, X. *Surrogate Model-Based Engineering Design and Optimization*; Springer: Berlin/Heidelberg, Germany, 2020.
3. Angione, C.; Silverman, E.; Yaneske, E. Using machine learning as a surrogate model for agent-based simulations. *PLoS ONE* **2022**, *17*, e0263150. [CrossRef] [PubMed]
4. Cho, S.; Kim, M.; Lyu, B.; Moon, I. Optimization of an explosive waste incinerator via an artificial neural network surrogate model. *Chem. Eng. J.* **2021**, *407*, 126659. [CrossRef]
5. Karsh, P.; Kumar, R.; Vaishali; Dey, S. Radial Basis Function-Based Uncertain Low-Velocity Impact Behavior Analysis of Functionally Graded Plates. In *Machine Learning Applied to Composite Materials*; Springer: Berlin/Heidelberg, Germany, 2022; pp. 77–97.
6. Mao, L.; Zhu, X.; Huang, Z.; Li, Y. Shock response of composite lattice sandwich structure subjected to underwater explosion. *Chin. J. Ship Res.* **2022**, *17*, 253–263.
7. Wei, Y.; Zhong, Q.; Wang, D. Ultimate strength prediction of I-core sandwich plate based on BP neural network. *Chin. J. Ship Res.* **2022**, *17*, 125–134.
8. Peng, W.; Zhang, J.; Shi, M.; Li, J.; Guo, S. Low-frequency sound insulation optimisation design of membrane-type acoustic metamaterials based on Kriging surrogate model. *Mater. Des.* **2023**, *225*, 111491. [CrossRef]
9. Lin, Y.; He, J.; Li, K. Hull form design optimization of twin-skeg fishing vessel for minimum resistance based on surrogate model. *Adv. Eng. Softw.* **2018**, *123*, 38–50. [CrossRef]
10. Prebeg, P.; Andric, J.; Rudan, S.; Jambrecic, L. Multiobjective ship structural optimization using surrogate models of an oil tanker crashworthiness. In *Marine Design XIII*; CRC Press: Boca Raton, FL, USA, 2018; pp. 459–470.
11. Chen, D.; Jing, L.; Yang, F. Optimal design of sandwich panels with layered-gradient aluminum foam cores under air-blast loading. *Compos. Part B Eng.* **2019**, *166*, 169–186. [CrossRef]
12. Herranen, H.; Pabut, O.; Eerme, M.; Majak, J.; Pohlak, M.; Jaan, K.; Saarna, M.; Allikas, G.; Aruniit, A. Design and testing of sandwich structures with different core materials. *Mater. Sci.* **2012**, *18*, 45–50. [CrossRef]
13. Jiang, Z.; Gu, M. Optimization of a fender structure for the crashworthiness design. *Mater. Des.* **2010**, *31*, 1085–1095. [CrossRef]
14. Ali, T.; Peng, Y.; Jinhao, Z.; Kun, L.; Renchuan, Y. Crashworthiness optimization method for sandwich plate structure under impact loading. *Ocean Eng.* **2022**, *250*, 110870. [CrossRef]
15. Yan, J.; Li, L. Multi-objective optimization of milling parameters–the trade-offs between energy, production rate and cutting quality. *J. Clean. Prod.* **2013**, *52*, 462–471. [CrossRef]
16. Ke, L.; Liu, K.; Wu, G.; Wang, Z.; Wang, P. Multi-objective optimization design of corrugated steel sandwich panel for impact resistance. *Metals* **2021**, *11*, 1378. [CrossRef]
17. Liu, Z.; Lu, J.; Zhu, P. Lightweight design of automotive composite bumper system using modified particle swarm optimizer. *Compos. Struct.* **2016**, *140*, 630–643. [CrossRef]
18. Li, Y.; Yang, Q.; Chang, T.; Qin, T.; Wu, F. Multi-load cases topological optimization by weighted sum method based on load case severity degree and ideality. *Adv. Mech. Eng.* **2020**, *12*, 1687814020947510. [CrossRef]
19. Bo, L.; Zhang, M.; Hanyu, L.; Hong, N. Optimization design containing dimension and buffer parameters of landing legs for reusable landing vehicle. *Chin. J. Aeronaut.* **2022**, *35*, 234–249.
20. Deb, K.; Deb, K. Multi-objective optimization. In *Search Methodologies: Introductory Tutorials in Optimization and Decision Support Techniques*; Springer: Berlin/Heidelberg, Germany, 2013; pp. 403–449.
21. Stanimirovic, I.P.; Zlatanovic, M.L.; Petkovic, M.D. On the linear weighted sum method for multi-objective optimization. *Facta Acta Univ.* **2011**, *26*, 49–63.

22. Chiley, V.; Sharapov, I.; Kosson, A.; Koster, U.; Reece, R.; Samaniego de la Fuente, S.; Subbiah, V.; James, M. Online normalization for training neural networks. *Adv. Neural Inf. Process. Syst.* **2019**, 32, 8433–8443.
23. Smith, A.E.; Coit, D.W.; Baeck, T.; Fogel, D.; Michalewicz, Z. Penalty functions. *Handb. Evol. Comput.* **1997**, *97*, C5.
24. Yoon, K.P.; Hwang, C.-L. *Multiple Attribute Decision Making: An Introduction*; Sage Publications: New Delhi, India; Thousand Oaks, CA, USA; London, UK, 1995.
25. Su, L.; Zhang, J.; Wang, C.; Zhang, Y.; Li, Z.; Song, Y.; Jin, T.; Ma, Z. Identifying main factors of capacity fading in lithium ion cells using orthogonal design of experiments. *Appl. Energy* **2016**, *163*, 201–210. [CrossRef]
26. Rumelhart, D.E.; Hinton, G.E.; Williams, R.J. *Learning Internal Representations by Error Propagation*; California University San Diego La Jolla Institute for Cognitive Science: La Jolla, CA, USA, 1985.
27. Sampson, J.R. Adaptation in natural and artificial systems (John H. Holland). *SIAM Rev.* **1976**, *18*, 529–530. [CrossRef]
28. Kingma, D.P.; Ba, J. Adam: A method for stochastic optimization. *arXiv* **2014**, arXiv:1412.6980.
29. Xue, J.; Shen, B. A novel swarm intelligence optimization approach: Sparrow search algorithm. *Syst. Sci. Control Eng.* **2020**, *8*, 22–34. [CrossRef]
30. Yan, S.; Yang, P.; Zhu, D.; Zheng, W.; Wu, F. Improved sparrow search algorithm based on iterative local search. *Comput. Intell. Neurosci.* **2021**, *2021*, 6860503. [CrossRef] [PubMed]
31. Chen, H.; Li, W.; Yang, X. A whale optimization algorithm with chaos mechanism based on quasi-opposition for global optimization problems. *Expert Syst. Appl.* **2020**, *158*, 113612. [CrossRef]
32. Meng, X.-B.; Gao, X.Z.; Lu, L.; Liu, Y.; Zhang, H. A new bio-inspired optimisation algorithm: Bird Swarm Algorithm. *J. Exp. Theor. Artif. Intell.* **2016**, *28*, 673–687. [CrossRef]
33. Reynolds, A. Liberating Lévy walk research from the shackles of optimal foraging. *Phys. Life Rev.* **2015**, *14*, 59–83. [CrossRef]
34. Jones, N. Some comments on the modelling of material properties for dynamic structural plasticity. *Mech. Prop. Mater. High Rates Strain* **1989**, *102*, 435–445.
35. Jones, N. *Structural Impact*; Cambridge University Press: Cambridge, UK, 2011.
36. Minorsky, V. *An Analysis of Ship Collisions with Reference to Protection of Nuclear Power Plants*; Sharp (George G.) Inc.: New York, NY, USA, 1958.

Journal of
*Marine Science
and Engineering*

MDPI

Article

Experimental Research on the Stiffness Step between the Main Hull and Superstructure of Cruise Ships

Zhiyong Pei [1,2], Bin Yang [3], Guangwu Liu [1,2,*] and Weiguo Wu [1,2]

1 Green & Smart River-Sea-Going Ship, Cruise and Yacht Research Center, Wuhan University of Technology, Wuhan 430063, China; zhiyong_pei@whut.edu.cn (Z.P.); mailjt@163.com (W.W.)
2 Hubei Province Engineering Research Center on Green & Smart River-Sea-Going Ship, Wuhan 430063, China
3 School of Naval Architecture, Ocean and Energy Power Engineering, Wuhan University of Technology, Wuhan 430063, China; 290315@whut.edu.cn
* Correspondence: gliu@whut.edu.cn

Abstract: The demand for larger passenger capacity and more entertainment facilities has led to the rapid growth of the cruise tourism market. The superstructure of cruise ships is designed to be plumper, with numerous decks and complex structural forms. To control the weight and the center of gravity, the bending stiffness of the superstructure is always designed to be weaker than that of the main hull, resulting in a stiffness step. Currently, there is no satisfactory method to accurately estimate the influence of the stiffness step between the main hull and superstructure on the structural response of cruise ships. In the present research, an experimental analysis is conducted to investigate the stiffness step between the main hull and superstructure of a typical cruise ship. By comparing the longitudinal stress distribution characteristics with and without the stiffness step with the theoretical results, the influence of the stiffness step on the longitudinal strength is investigated. Furthermore, the maximum stress and the bending efficiency of the superstructure are also discussed. The present research is of reference significance for the structural safety and reliability design of cruise ships.

Keywords: cruise ships; superstructure; stiffness step; experimental analysis; structural response

Citation: Pei, Z.; Yang, B.; Liu, G.; Wu, W. Experimental Research on the Stiffness Step between the Main Hull and Superstructure of Cruise Ships. *J. Mar. Sci. Eng.* **2023**, *11*, 1264. https://doi.org/10.3390/jmse11071264

Academic Editor: Erkan Oterkus

Received: 3 June 2023
Revised: 14 June 2023
Accepted: 18 June 2023
Published: 21 June 2023

1. Introduction

For high technology and high value-added entertainment ships, generous living space and a comfortable environment are the most important aspects of cruise ships, which results in plump superstructure arrangements. The lightweight design and the control of the center of gravity bring a series of challenges to the structural design of large superstructures. Due to the arrangement of life-saving passageways and lifeboats, the connection between the main hull and superstructure will be retracted inwards to a certain extent. In addition, the bending stiffness of the superstructure is always designed to be weaker than that of the main hull due to the purpose of the lightweight design. Therefore, there is a stiffness step between the main hull and superstructure, which affects the force transformation.

The cruise ship is typically designed with a large superstructure consisting of multiple decks, as shown in Figure 1. To meet the entertainment requirement, complex structural forms such as atrium and theater are also included, which sets them apart from other merchant ships.

For cruise ships, the interaction mechanism between the main hull and superstructure is more complex. The superstructure adopts a lightweight design for the control of the center of gravity, which results in the weaker bending stiffness of the superstructure. Furthermore, the inward structure between the main hull and superstructure aggravates the stiffness difference, as shown in Figure 1. Due to the existence of different stiffness, the deformation between the main hull and superstructure is inconsistent. Therefore, it is necessary to conduct a model experiment to investigate the influence of the stiffness step between the main hull and superstructure on the structural response of cruise ships.

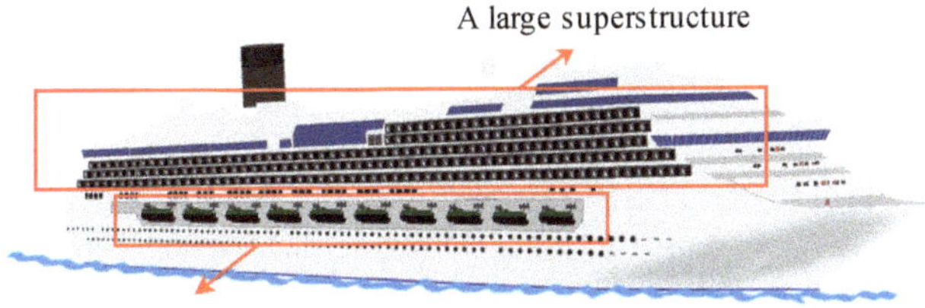

Figure 1. The cruise ship with a large superstructure.

The theoretical methods that consider the stiffness characteristics of cruise ships include the simple beam theory, two-beam theory, coupled beam theory, and multi-beam method. The longitudinal stress is generally estimated using simple beam theory in the initial design [1]. The simple beam theory is derived from the thin-walled beam theory, based on the assumption of plane section. The longitudinal stress σ on cross-section along height can be calculated according to Equation (1).

$$\sigma = -\frac{M}{I_y} \cdot z \tag{1}$$

M is the bending moment of the cross-section. I_y is the moment of inertia on the neutral axis. z is the distance from the calculation position to the neutral axis. Simple beam theory is widely used in the initial design due to its advantages of simplicity. However, the deformation of the main hull and superstructure cannot be consistent when the longitudinal bending occurs, and the longitudinal stress distribution along height does not conform to the plane section assumption. When this theory is used to calculate longitudinal stress, errors may occur [2].

Crawford and Ruby [3] proposed the two-beam theory based on the traditional simple beam theory. The fundamental idea is to consider the main hull and superstructure as two independent beams, taking into account the longitudinal shear and vertical forces generated by the interaction between them. Many researchers subsequently investigated and developed the theory on simple beam theory with certain assumptions. Bleich [4] discussed in detail the inconsistency in the curvature of the main hull and superstructure. Terazawa and Yagi [5] applied the energy method to estimate the longitudinal stress in the main hull and superstructure using assumed structural stress distribution modes. However, the beam equilibrium formulas must be modified to accommodate different interactions between the main hull and superstructure when the two-beam theory is adopted.

Naar [6] proposed the couple beam theory for the multi-deck superstructure of cruise ships. It is assumed that the cruise ship can be composed of multiple thin-walled beams coupling with longitudinal and vertical springs. Morshedsolouk [7] modified the couple-beam equilibrium equations proposed by Naar. Toming [8] assumed that the main deck and each deck are thin-walled beams with bending and axial stiffness so that the ship structure can be divided into spring-distributed beams based on vertical stiffness and shear stiffness. When using the couple beam theory for structural response analysis, the longitudinal stress distribution will no longer be continuous. In addition, the couple beam theory needs to use FEM to solve the beam equilibrium equations, which makes it difficult to popularize the application in the initial design.

Yang et al. [9] proposed the multiple-beam method to examine the stress-distribution characteristics and the bending efficiency of the superstructure. The method assumes that the main hull and each deck of the superstructure can be regarded as thin-walled beams, considering the vertical forces and horizontal shear forces interacting between adjacent decks. The multiple-beam method is suitable in initial design without finite element analysis results.

Reasonable theoretical analysis can make preliminary judgments on longitudinal stress distribution and provide a reference basis for the structural design of superstructures. However, this approach may no longer be effective for the plump superstructure of cruise ships. With the development of computational technologies, numerical simulation methods have become powerful tools.

The interaction between the main hull and superstructure is so complex that it has become one hotspot of research for a long time. Pauling and Payer [10] investigated the interaction between the main hull and superstructure using the finite element method. Mitchell [11] further researched the difference in stiffness between the main hull and superstructure and discussed how to build a reasonable finite element model. Fransman et al. [12] proposed a simplified modeling principle for the numerical simulation in detail and put forward a simplified calculation method for the longitudinal stress of cruise ships.

Furthermore, the issue of reducing the weight of the superstructure of cruise ships to a minimum while ensuring that it possesses sufficient strength has become a challenge that every designer must consider. Mackney et al. [13] designed a series of hull-superstructure interaction models to investigate the parameters that might affect the bending efficiency of the superstructure, such as superstructure length, number of decks, and deck spacing. As pointed out by Fricke and Gerlach [14], the superstructure undergoes a different bending curvature from the main hull. During the early stages of ship construction, the design of the superstructure was typically based on the strength of the main hull. [15]. Pei et al. [16] investigated the efficiency of superstructure participating in the longitudinal bending and analyzed the effects of different connection types between the main hull and superstructure on structural response.

In addition to the numerical method, the model experiment is an effective way to reveal the mechanism. Wu et al. [17] designed a large-scale model using the similarity theory and carried out experimental research on the bending efficiency of the superstructure. The research indicates that the main hull participating in the longitudinal bending is the middle deck, not the uppermost deck. Zhu et al. [18] conducted bending experiments on one ship with a long superstructure and concluded that the opening of the side of the superstructure could reduce the degree to which the superstructure participates in longitudinal bending. Shi et al. [19,20] carried out a collapse experiment to research the superstructure's effectiveness. Then, the increment–iterative relationship is revised for the ultimate strength evaluation of the cruise ship. For the time being, the experimental research on the stiffness step between the main hull and plump superstructure of cruise ships has kept its virginity.

In the present paper, a typical cruise ship is considered as the research object. The structural features, including the multiple decks and the stiffness step between the main hull and superstructure, are analyzed first. A similar model considering the stiffness step between the main hull and superstructure is designed, and a model experiment is then carried out. The structural response and stiffness step are discussed and analyzed. The research has meaningful for the structural safety and reliability design of superstructures for cruise ships.

2. Experimental Details

The experimental model is designed based on the scantling of actual cruise ship. In order to consider the stiffness step, an inward retracted structure is designed in the connection area between the main hull and superstructure.

2.1. Description of the Model

The main parameters of the model are shown in Table 1. The characteristics of the cross-section are shown in Table 2. The dimensions of the longitudinal components are shown in Table 3. The transverse bulkheads adopt a thicker plate with 16 mm to ensure the loading. The model uses a pipe with $\Phi 39 \times 3/45 \times 3$ to simulate the pillar of the cruise ship. The cross-section of the model is shown in Figure 2. As shown in Figure 2, Inner deck/250 mm represents the structure at this position is the inner deck, and the distance from the inner deck to the baseline is 250 mm. And the annotations in the other figure all represent this meaning.

Table 1. The main parameters of the experiment model.

	Width (mm)	Height of Main Hull (mm)	Height of Deck (mm)
Model	2720	1410	250

Table 2. The section characteristics of the experiment model.

Parameter	Model
Height of neutral axis (mm)	697.507
Moment of inertia ($mm^2 \cdot mm^2$)	5.43×10^{10}

Table 3. The dimensions of longitudinal components.

Components	Dimension(mm)
Bottom plate	6
Inner bottom plate	6
Bottom girder	5
Bottom longitudinal	-36×5
Inner longitudinal	-36×5
Side longitudinal	-25×6
Side plate	4
Deck plate	2
Deck longitudinal	-12×5

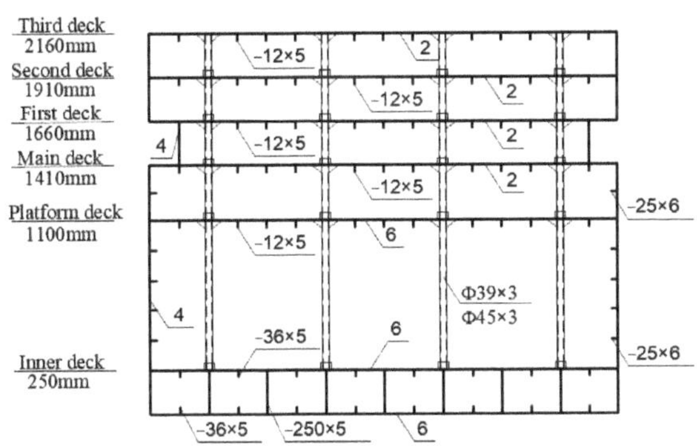

Figure 2. The sectional characteristics of the model.

Based on objective conditions such as the experiment site and the loading capacity, the geometric similarity ratio between the cruise ship and the experimental model has been determined as C_L = 12:1. The experimental model was designed as a thin-walled structure. Suppose the thickness similarity ratio is taken to be the same as the geometric similarity ratio, which will result in a very small plate thickness of the model, which is obviously impossible to achieve [21]. Therefore, the thickness is regarded as an independent parameter, and the thickness similarity ratio between the prototype and the model is finally determined as C_t = 4:1 by comprehensively considering the experimental cost, laboratory condition, and processing technology.

The mid-ship section of the prototype is selected for a similar design to simplify the model structure. The prototype and model need to meet the linear bending similarity criterion, which means that the prototype and model are guaranteed to have similar moments of inertia in the cross-section and similar height of the neutral axis [22]. The similarity condition of the moment of inertia can be expressed as

$$C_I = C_L^3 C_t \tag{2}$$

C_I is the ratio of moments of inertia between the prototype and the model. To ensure that the strain of members at corresponding positions in the prototype and model meet the similarity relationship, the height of the neutral axis must be similar, which can be expressed as

$$C_e = C_L \tag{3}$$

C_e is the height similarity ratio of the neutral axis between the prototype and model. After the model design has satisfied the bending similarity criterion, it is necessary to compare the sectional characteristics of the prototype and model to verify the similarity and ensure the reasonableness of the model design [23]. The comparison of sectional characteristics between the prototype and model is shown in Table 4. The height error of the neutral axis and the section of inertia of the prototype and model are all within 1%. Therefore, it can be determined that the section characteristics between the prototype and model are similar when bending occurs.

Table 4. The comparison of sectional characteristics between the prototype and model.

Parameter	Symbol	Value
Moment of inertia of prototype ($mm^2 \cdot mm^2$)	M_s	3.79×10^{14}
Moment of inertia after being scaled ($mm^2 \cdot mm^2$)	M_s/C_I	5.48×10^{10}
Moment of inertia of model ($mm^2 \cdot mm^2$)	M_m	5.43×10^{10}
Error of the moment of inertia (%)	-	0.887
Height of neutral axis of prototype (mm)	H_s	8408.6
Height of neutral axis after being scaled (mm)	H_s/C_e	700.7
Height of neutral axis of model (mm)	H_m	697.5
Error of the height of neutral axis (%)	-	0.458

The model has a total length of 13,700 mm, a width of 2720 mm, a main hull height of 1410 mm, a superstructure layers spacing of 250 mm, and a total height of 2160 mm. Here, the model experiment under hogging bending is simulated with a four-point bending mechanism, as shown in Figure 3, including two loading blocks used to produce bending moments, two transition blocks connected to loading blocks for the force transmission, and a test block.

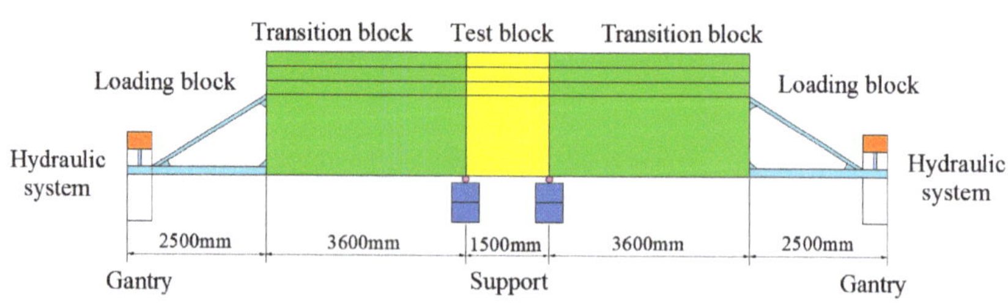

Figure 3. Schematic diagram of experimental model.

2.2. Experimental Setup

In the experimental procedure, the hydraulic system and gantry are included in the loading block to provide a downward load to the model, which is used to produce the bending moment, as shown in Figure 4.

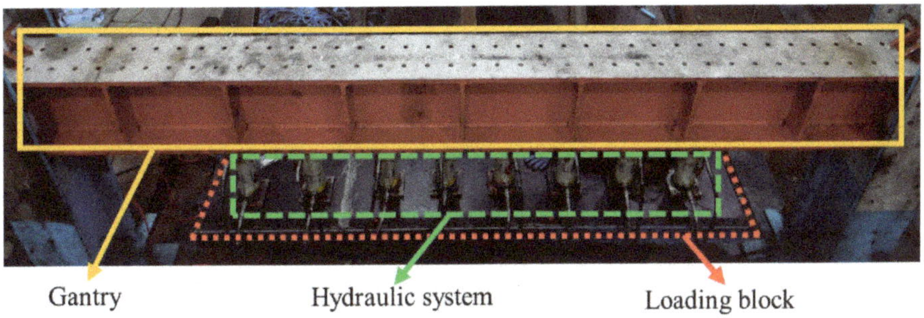

Figure 4. The loading system of the experimental model.

At the bottom of the transverse bulkheads of the test block, a combination of round steel and supports is used to provide reaction force for the experimental model, which can produce uniformly distributed bending moments in the test block, as shown in Figure 5.

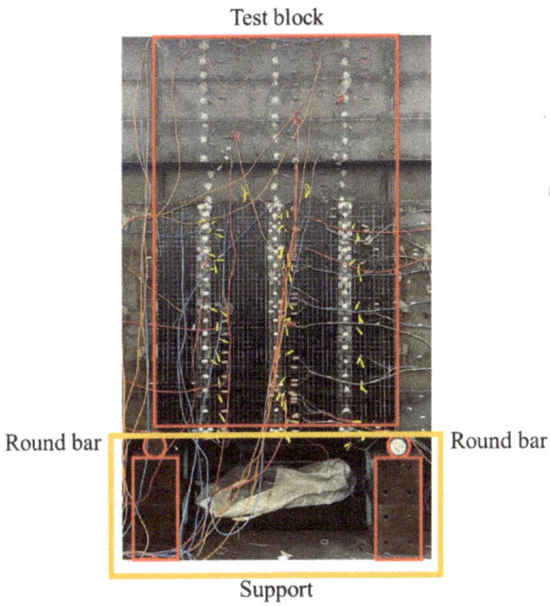

Figure 5. The simple support at the bottom of the test block.

In order to monitor the changes of the longitudinal stress in critical areas during the test, single strain gauges are adopted to measure the longitudinal strain variation. Then, the main focus of the research is on the middle of the ship, including the decks, bottom plates, and side shells. Therefore, single strain gauges are arranged along the ship length at three sections near the midship, as shown in Figure 6.

Loading is performed in the elastic range of the experiment model, and the loading order is shown in Table 5. In order to analyze the stiffness step between the main hull and superstructure, five groups of models with different deck forms are designed, as shown in Figure 7. The details are as follows.

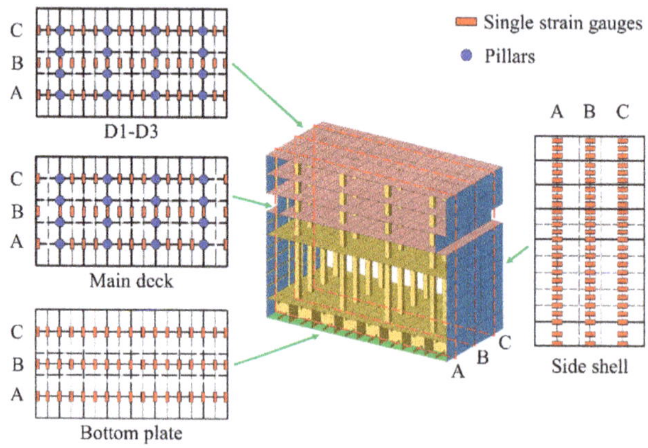

Figure 6. Arrangement of gauges in the test block.

Table 5. The loading order of the experimental model.

Order	Load (t)
1	4
2	8
3	12
4	16
5	20
6	24
7	28
8	32
9	36
10	40

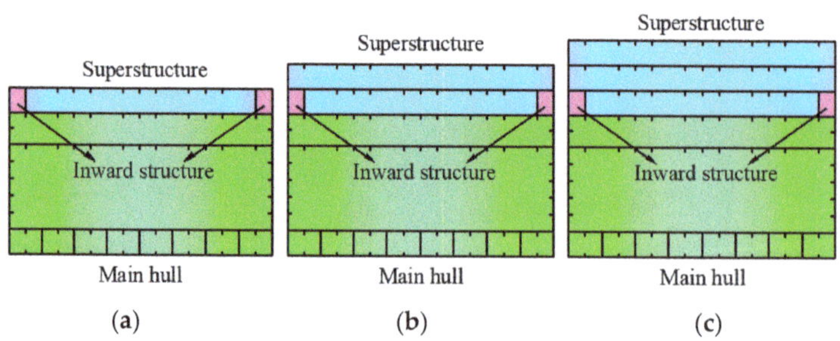

Figure 7. Different layers of deck and inward structure for stiffness step. (**a**) M1. (**b**) M2 and M3. (**c**) M4 and M5.

M1 is the model of welding the first deck onto the main hull. There is a certain degree of indent inward at the side shell between the main deck and the first deck, as shown in Figure 7a.

M2 is the model of welding the second deck onto M1. There is a certain degree of indent inward at the side shell between the main deck and the first deck. M3 is the model that encloses the inward structure of M2, as shown in Figure 7b.

M4 is the model of welding the third deck onto M3, and there is no inward structure at the side shell between the main deck and the first deck. M5 is the model that exists the inward structure, as shown in Figure 7c.

The main process of the experiment is as follows.

Install the first superstructure to the main hull and then arrange the strain gauges at the corresponding positions, as shown in Figure 7a. Since the experiment model will produce residual stress during the manufacturing process, it is necessary to preload and unload within the elastic range (3–5 times), and the loading value is slightly greater than the maximum value of the experimental loading. Then, review the experimental measurement data to eliminate and repair the existence of bad points. Further, debug the loading system to prepare for the experiment. Graded loading is adopted in the experiment to observe the linear law of external load and strain. Each group is repeated more than twice to ensure the accuracy of the experiment data.

After completing the experiment of M1, the second superstructure was welded to the model, as shown in Figure 7b. Repeat the experiment steps and carry out the experiment of M2 and M3.

Similarly, the third superstructure was welded to the model after completing the experiment of M2 and M3, as shown in Figure 7c. Repeat the experiment steps and carry out the experiment of M4 and M5.

2.3. Experimental Results

In M3 and M4 conditions, the sides of the main hull extend directly to the superstructure, and the force from the main hull can be transferred to the superstructure. In M1, M2, and M5 conditions, the side at the connection between the main hull and superstructure is indented, so the force from the main hull cannot be effectively transferred to the superstructure. When there is a stiffness step between the main hull and superstructure, the transmission of forces at the connection between the main hull and superstructure will change, which will affect the longitudinal stress distribution along the height. Therefore, the influence of the stiffness step can be studied by analyzing the results of the longitudinal stress distribution along the height.

During the model experiment, all the data can be recorded automatically at each loading step. As the experiment model belongs to a symmetrical model, the strains measured on port and starboard are basically the same, and the strains measured on cross-section A and cross-section C are basically the same. This paper selects the experiment data of cross-section A and cross-section B (midship) on starboard as the research object with a series of data collations and analyses. As the trend of the longitudinal stress distribution along height is consistent with the increase of load, the experiment results of 20 t are selected to analyze the influence of the stiffness step.

2.3.1. Experiment Results of M1

M1 is the model experiment with one superstructure deck, and there is the stiffness step between the main hull and superstructure. The distribution of longitudinal stress along the height under 10 groups of loading is shown in Figure A1, and the distribution of longitudinal stress along the height of 20 t is shown in Figure 8. The longitudinal stress in cross-sections A and B are roughly linear along height in the main hull (0–1410 mm), with a step change in the magnitude of the stress above the main hull (1410–1660 mm). Furthermore, the longitudinal stress of cross-section B is slightly greater than that of cross-section A above the main hull (1410–1660 mm), which indicates that the longitudinal stress increases as the section gets closer to midship.

2.3.2. Experiment Results of M2 and M3

M2 is the model experiment with two superstructure decks, and there is the stiffness step between the main hull and superstructure. M3 is the model experiment with two superstructure decks, and there is no stiffness step between the main hull and the superstructure. The experimental results of M2 and M3 under 10 groups of loading are shown in Figure A2, and the experimental results of 20 t for cross-section B are shown in Figure 9.

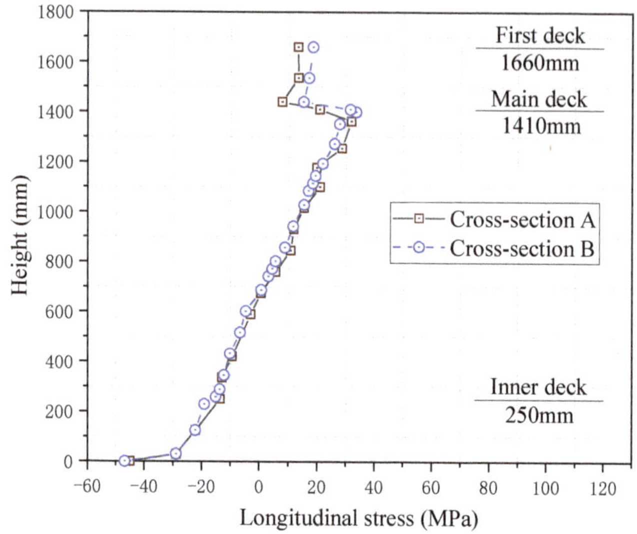

Figure 8. Longitudinal stress distribution along height of M1.

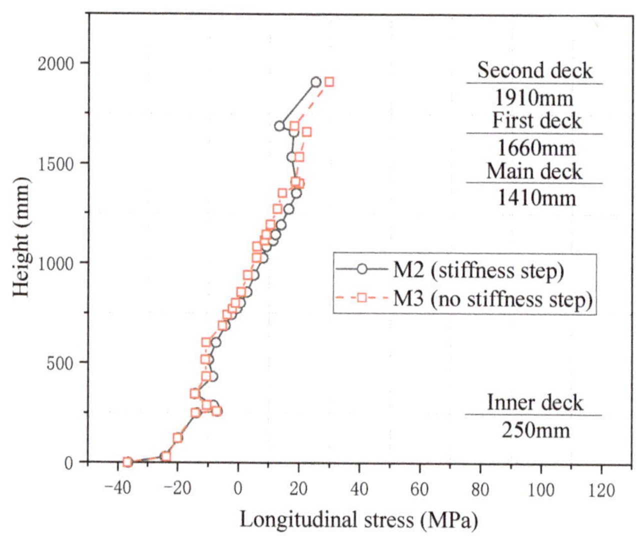

Figure 9. Longitudinal stress distribution along height of M2 and M3 for cross-section B.

The following conclusions can be drawn from Figure 9. The longitudinal stresses of M2 (stiffness step) and M3 (no stiffness step) are almost consistent from bottom to inner bottom (0–250 mm). The longitudinal stresses of M3 (no stiffness step) are lower than that of M2 (stiffness step) from the inner bottom to the main deck (250–1410 mm), while the longitudinal stresses of M3 (no stiffness step) become higher than that of M2 (stiffness step) from main deck to second deck (1410–1910 mm). The results indicate that the participation of the superstructure in the longitudinal bending will decrease when there is a stiffness step.

2.3.3. Experiment Results of M4 and M5

M4 is the model experiment with three superstructure decks, and there is no stiffness step between the main hull and the superstructure. M5 is the model experiment with three superstructure decks, and there is the stiffness step between the main hull and superstructure. The experimental results of M4 and M5 under 10 groups of loading are shown in Figure A3, and the experimental results of 20 t are shown in Figure 10.

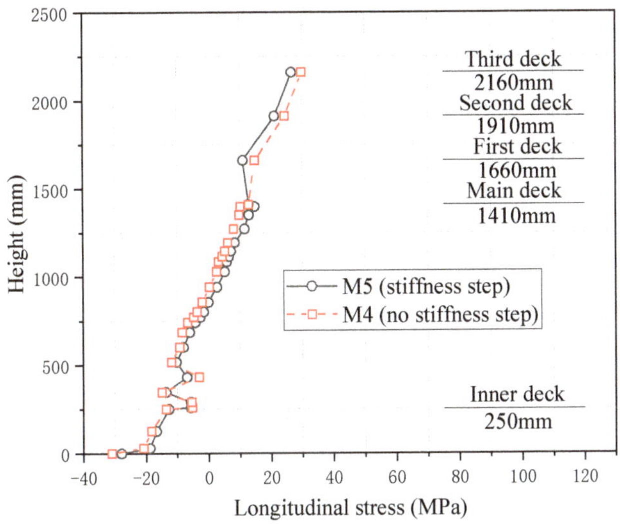

Figure 10. Longitudinal stress distribution along height of M4 and M5 for cross-section B.

3. Analysis and Discussion

As the influence of the stiffness step on the longitudinal stress in the midship position (cross-section B) is greater than that in the other two cross-sections, the superstructure of the midship cross-section is taken as the research object.

3.1. Longitudinal Stress Distribution Characteristics

Simple beam theory is based on the assumption of planar cross-sections. The results obtained using this theory can be considered as the results when the superstructure fully participates in the longitudinal bending. Moreover, these results can serve as the basis for determining the effective degree of participation of the superstructure in longitudinal bending. The experiment results of 20 t for the cross-section B (midship) of the superstructure with two decks and three decks are compared with the results of simple beam theory, as shown in Figures 11 and 12. The experimental results show that, different from ordinary ships, the plump superstructure of cruise ships does not participate in the longitudinal bending. Furthermore, the existence of the stiffness step between the main hull and superstructure further reduces the degree to which the superstructure participated in the longitudinal bending.

3.2. Bending Efficiency of Superstructures

While the analysis of longitudinal stress distribution characteristics in Section 3.1 highlights the difference between the results with and without the stiffness step, it is challenging to further analyze the influence of the stiffness step on superstructure effectively. In order to investigate the influence of the stiffness step on the longitudinal stress intuitively, the stiffness step on the structural response is analyzed in terms of the bending efficiency of the superstructure.

The extent of the superstructure participating in the longitudinal bending can be measured by the bending efficiency. The bending efficiency is commonly defined as follows [9]:

$$v = \frac{\sigma_{sv}}{\sigma_s} \tag{4}$$

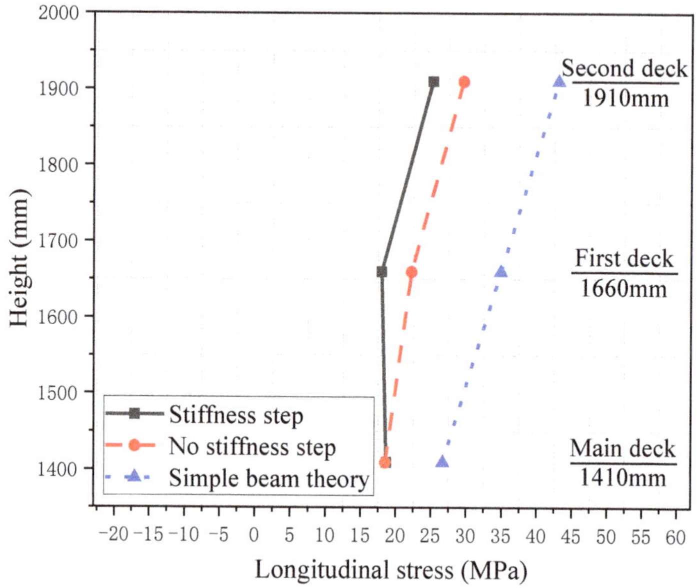

Figure 11. The longitudinal stress distribution of two superstructure decks.

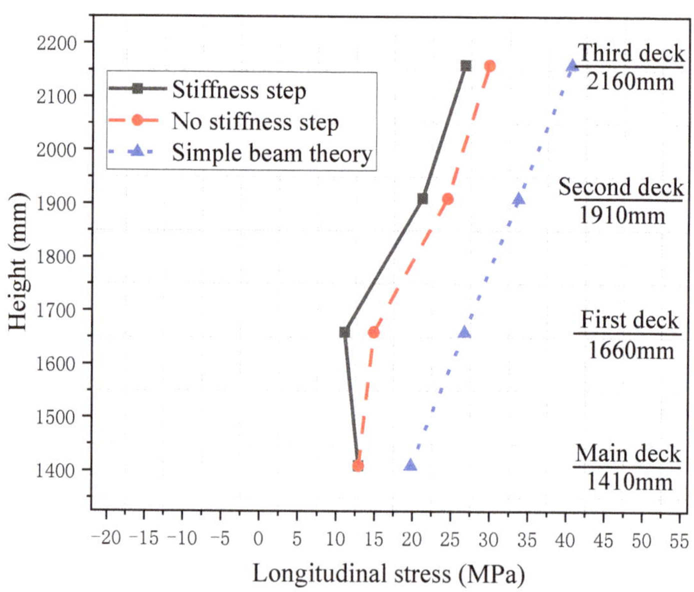

Figure 12. The longitudinal stress distribution of three superstructure decks.

where v is the bending efficiency of the superstructure, σ_{sv} is the actual longitudinal stress, σ_s is the longitudinal stress when the superstructure fully participates in the longitudinal bending. In the research, σ_s is calculated using the simple beam theory. Furthermore, the bending efficiency of the superstructure with two decks is shown in Figure 13, while the bending efficiency of the superstructure with three decks is shown in Figure 14.

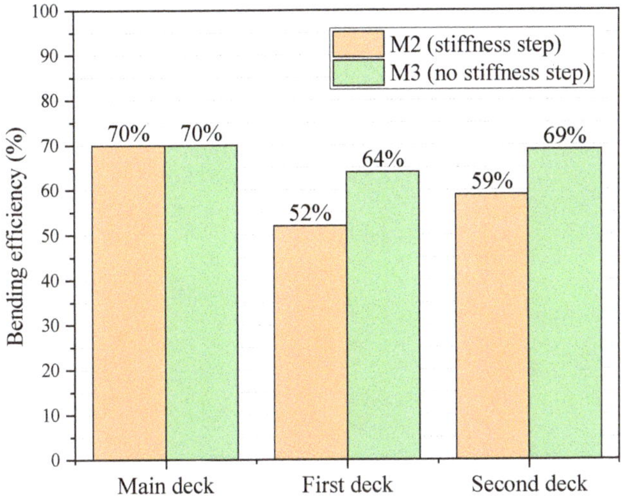

Figure 13. The bending efficiency of the superstructure with two decks (20 t).

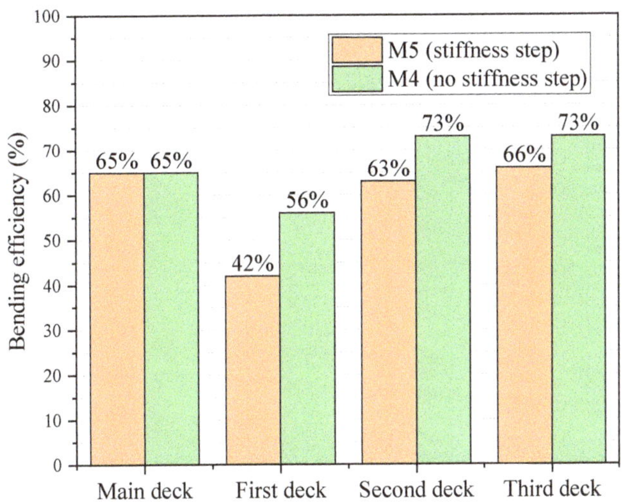

Figure 14. The bending efficiency of the superstructure with three decks (20 t).

The following conclusions can be drawn from Figures 13 and 14. At the main deck, the bending efficiency of the superstructure is essentially the same with and without the stiffness step, which indicates that the stiffness step has little effect on the main deck.

For superstructures with two decks, the presence or absence of the stiffness step at the first and second decks has a large effect on the bending efficiency of the superstructure, with the difference remaining at around 10%.

For superstructures with three decks, the difference in bending efficiency remains about 14% at the first deck with and without the stiffness step. At the second deck, the difference in bending efficiency remains about 10% at the first deck with and without the stiffness step. While for the third deck, the difference in bending efficiency remains around 8% with and without the stiffness step.

3.3. *Maximum Longitudinal Stress and Deformation*

According to the results of traditional beam theory, the maximum stress location will occur at the uppermost deck of the model. The results indicate that the maximum stresses for all five sets of conditions under hogging bending moment occur at the bottom deck of the model, as shown in Figures 8–10. The stresses in the uppermost deck are significantly less than the theoretically calculated stresses, as shown in Figures 11 and 12.

The stresses in the bottom plate and the second deck for M2 and M3 are shown in Figure 15a. The stresses in the bottom plate and the third deck for M4 and M5 are shown in Figure 15b. The stresses in the bottom plate are greater than the stresses in the uppermost deck. Furthermore, the stresses with the stiffness step are less than the stresses without the stiffness step in the uppermost deck. Due to the stiffness step between the main hull and the superstructure, the main hull cannot transfer the force to the superstructure completely and effectively. As a result, thin plates are commonly used in the superstructure design of cruise ships.

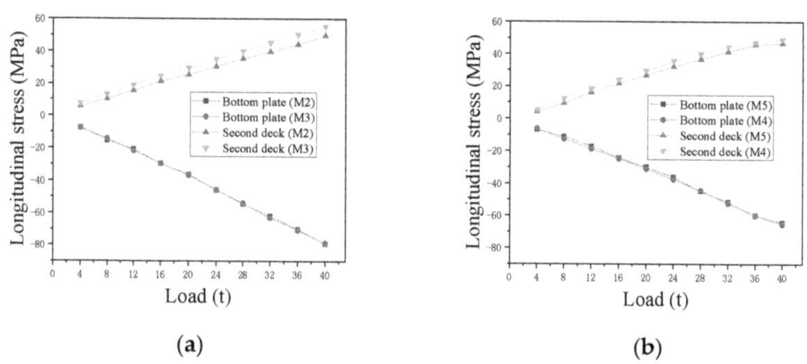

Figure 15. The stresses in the bottom plate and the uppermost deck. (**a**) Longitudinal stress for M2 and M3 (**b**) Longitudinal stress for M4 and M5.

4. Conclusions

There are differences in stiffness between the main hull and superstructure of cruise ships. Currently, there is a lack of a simple evaluation method to analyze the influence of the stiffness step on the structure response. In this paper, the stiffness step experiment of cruise ships is designed and carried out. Based on the experimental results, the influence of the stiffness step on force transformation and structural response is analyzed and discussed. The research is meaningful for the structural safety and reliability of superstructure. The main conclusions are as follows.

Due to the influence of the stiffness step between the main hull and superstructure, the traditional beam theory cannot correctly estimate longitudinal stress distribution along the height of large cruise ships. Therefore, the superstructure and the main hull should be calculated as two beams with different stiffness in the initial design.

Based on a series of experimental results, the longitudinal stress in the model with the stiffness step is less than that without the stiffness step. Additionally, the bending efficiency of the superstructure with the stiffness step is also less than that without the stiffness step. The above analysis indicates that the presence of the stiffness step prevents the superstructure from effectively participating in the longitudinal bending of the main hull.

The experimental results demonstrate that the maximum stress did not occur at the uppermost deck, indicating that the presence of the stiffness step is advantageous for the lightweight and safe design of the superstructure on cruise ships.

Author Contributions: Conceptualization, B.Y. and Z.P.; methodology, B.Y.; software, G.L.; validation, W.W.; writing—original draft preparation, B.Y.; writing—review and editing, G.L., Z.P. and B.Y.; supervision, Z.P. and G.L. All authors have read and agreed to the published version of the manuscript.

Funding: This research received no external funding.

Institutional Review Board Statement: Not applicable.

Informed Consent Statement: Not applicable.

Data Availability Statement: Not applicable.

Acknowledgments: This work was supported by Hubei Provincial Engineering Research Center on Green & Smart River-sea-going Ship, China.

Conflicts of Interest: The authors declare no conflict of interest.

Nomenclature

The following symbols are used in this manuscript:

Symbol	Description
σ	Longitudinal stress
M	Bending moment of cross-section
I_y	Moment of inertia on the neutral axis
z	Distance from the calculation position to the neutral axis
C_L	Geometric similarity ratio between the prototype and the model
C_t	Thickness similarity ratio between the prototype and the model
C_I	Ratio of moments of inertia between the prototype and the model
C_e	Height similarity ratio of the neutral axis between the prototype and the model
M_s	Moment of inertia of the prototype
M_m	Moment of inertia of the model
H_s	Height of neutral axis of the prototype
H_m	Height of neutral axis of the model
v	Bending efficiency of the superstructure
σ_{sv}	Actual longitudinal stress
σ_s	Longitudinal stress when the superstructure fully participates in the longitudinal bending

Appendix A

The experimental results of M1 under 10 groups of loading are shown in Figure A1.

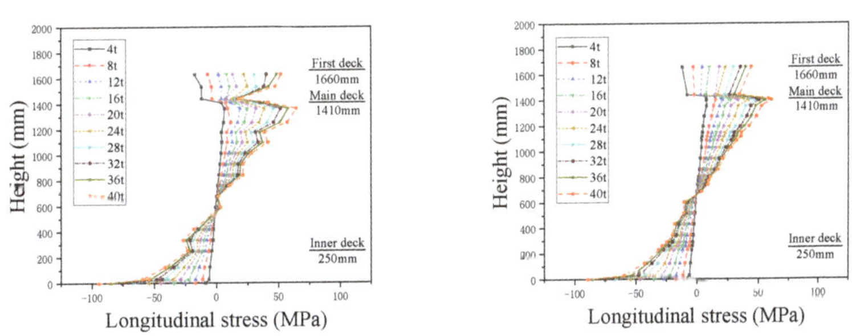

(**a**) (**b**)

Figure A1. Longitudinal stress distribution along height of M1 under 10 groups of loading. (**a**) Cross-section A (**b**) Cross-section B.

The experimental results of M2 and M3 under 10 groups of loading are shown in Figure A2.

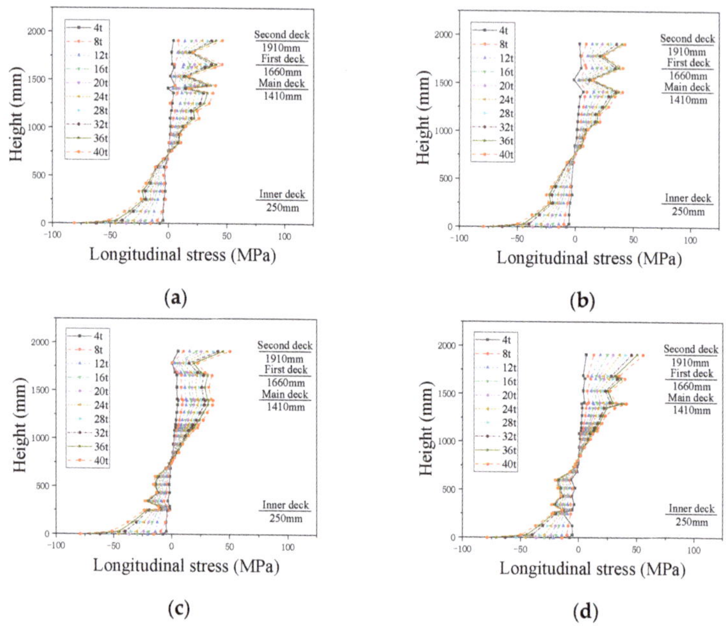

Figure A2. Longitudinal stress distribution along height of M2 and M3 under 10 groups of loading. (**a**) Cross-section A of M2 (**b**) Cross-section A of M3 (**c**) Cross-section B of M2 (**d**) Cross-section B of M3.

The experimental results of M4 and M5 under 10 groups of loading are shown in Figure A3.

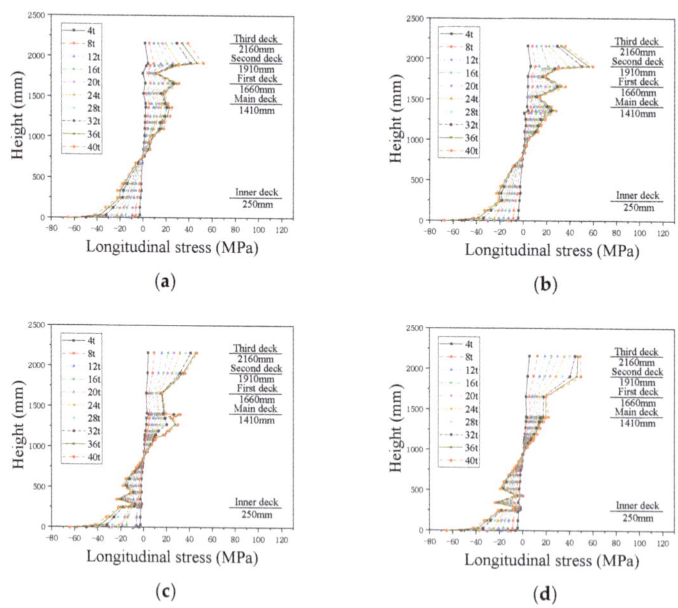

Figure A3. Longitudinal stress distribution along height of M4 and M5 under 10 groups of loading. (**a**) Cross-section A of M4 (**b**) Cross-section A of M5 (**c**) Cross-section B of M4 (**d**) Cross-section B of M5.

References

1. Wu, F.; Zhu, X.; Mei, Z. *Ship Structural Mechanics*; National Defense Industry Press: Beijing, China, 2010; pp. 18–34.
2. Yang, B.; Pei, Z.; Wu, W. Research on Stress Distribution Characteristics of Cruise Ship Based on Modified Two-beam Theory. *J. Ship Mech.* **2022**, *26*, 1503–1513.
3. Crawford, L.; Ruby, W. *Model Tests on Hull-Deckhouse Interaction*; Ship Structure Committee: Washington, DC, USA, 1955; Volume 4, pp. 693–699.
4. Bleich, H. Nonlinear distribution of bending stresses due to distortion of the cross section. *J. Appl. Mech.* **1952**, *29*, 95–104. [CrossRef]
5. Terazava, K.; Yagi, J. Stress distribution in deckhouse and superstructure. *Soc. Nav. Archit. Jpn.* **1964**, *9*, 51–150.
6. Naar, H.; Varsta, P.; Kujala, P. A theory of coupled beams for strength assessment of passenger ships. *Mar. Struct.* **2004**, *17*, 590–611. [CrossRef]
7. Morshedsolouk, F.; Khedmati, M. An extension of coupled beam method and its application to study ship's hull-superstructure interaction problems. *Lat. Am. J. Solids Struct.* **2011**, *8*, 265–290. [CrossRef]
8. Toming, R.; Kerge, E.; Naar, H. Hull and superstructure interaction using coupled beam method. In *International Symposium on Practical Design of Ships and Other Floating Structures*; Technical University of Denmark: Lyngby, Denmark, 2016; pp. 851–859.
9. Yang, B.; Pei, Z.; Wu, W. Stress-distribution characteristics of cruise ship based on multiple-beam method. *Ocean Eng.* **2022**, *266*, 112646. [CrossRef]
10. Paulling, J.; Payer, H. Hull-deckhouse interaction by finite-element calculations. *Trans. Soc. Nav. Archit. Mar. Eng.* **1968**, *76*, 281–296.
11. Mitchell, G. Analysis of structural interaction between a ship's hull and deckhouse. *Trans. R. Inst. Nav. Archit.* **1978**, *120*, 121–136.
12. Fransman, J. The influence of passenger ship superstructures on the response of the hull girder. *Trans. R. Inst. Nav. Archit.* **1988**, *131*, 55–83.
13. Mackney, M. Superstructure Effectiveness in the Preliminary Assessment of the Hull Behavior. *Mar. Technol.* **1999**, *36*, 29–43. [CrossRef]
14. Fricke, W.; Gerlach, B. Effect of large openings without and with windows on the shear stiffness of side walls in passenger ships. *Ships Offshore Struct.* **2015**, *10*, 256–271. [CrossRef]
15. Pei, Z.; Shen, W.; Yang, P.; Wu, W. *Hull Strength and Structure Design*; Science Press: Beijing, China, 2017; pp. 202–210.
16. Pei, Z.; Ma, Z.; Zhu, B. Research on the bending efficiency of superstructure to hull girder strength of inland passenger ship. *Ocean Eng.* **2019**, *195*, 106762.
17. Wu, W.; Duan, H.; Pei, Z.; Wong, C. Ship superstructure participation in the longitudinal bending strength test study. *J. Ship Mech.* **2001**, *3*, 62–68.
18. Zhu, X. Experimental study on effect of cut-apart side wall plate on the longitudinal bending. *Shipbuild. China* **2003**, *3*, 67–74.
19. Shi, G.; Gao, D. Analysis of hull girder ultimate strength for cruise ship with multi-layer superstructures. *Ships Offshore Struct.* **2019**, *14*, 698–708. [CrossRef]
20. Shi, G.; Gao, D. Model experiment of large superstructures' influence on hull girder ultimate strength for cruise ships. *Ocean Eng.* **2021**, *222*, 108626. [CrossRef]
21. Araneda, J. Dimensional-directional analysis by a quaternionic representation of physical quantities. *J. Frankl. Inst.* **1996**, *333*, 113–126. [CrossRef]
22. Pei, Z.; Xie, S.; Yang, B.; Ao, L. Research on model design of structural collapse test of large cruise ship. *J. Wuhan Univ. Technol.* **2022**, *46*, 242–246.
23. Zhang, S.; Khan, I. Buckling and ultimate capability of plates and stiffened panels in axial compression. *Mar. Struct.* **2009**, *22*, 791–808. [CrossRef]

Journal of
Marine Science and Engineering

MDPI

Article

Numerical and Experimental Analysis of Fire Resistance for Bulkhead and Deck Structures of Ships and Offshore Installations

Shuai Zong, Kun Liu *, Weijian Qiu, Zhenguo Gao and Jiaxia Wang

School of Naval Architecture and Ocean Engineering, Jiangsu University of Science and Technology, Zhenjiang 212003, China; 220110101103@stu.just.edu.cn (S.Z.); 15751774907@163.com (W.Q.); zhenguo.gao@just.edu.cn (Z.G.); jxwang66@yeah.net (J.W.)

* Correspondence: kunliu@just.edu.cn; Tel.: +86-135-1169-2085; Fax: +86-0511-8444-6543

Abstract: Investigating the loss of integrity (E) in cabin walls and decks, as well as the role of insulation capabilities, holds significant implications for preventing serious human, economic and environmental damage caused by the ignition of cabins in ships and ocean platforms due to fires and explosions. In this study, the fire resistance of A-60 class ship bulkheads and decks was evaluated through two groups of standard fire resistance tests. In the first test, the steel structure side of the bulkhead was exposed to the fire, while in the second test, the mineral wool and L-shaped stiffeners side of the deck was exposed to the fire. Numerical material models for steel and mineral wool were established based on standards, and the temperature distribution and structural deformation were simulated using Abaqus. The results showed a good correlation with the experimental data. The maximum and average temperature increases on the unheated surface of the bulkhead during the standard fire resistance test were 158 °C and 136 °C, respectively. The corresponding values for the deck were 176 °C and 138 °C. Upon the conclusion of the experiment, the maximum displacement deformation in the direction towards the furnace from the center of the cabin wall was 54 mm, and from the center of the deck, the maximum displacement deformation towards the furnace was 28 mm. This research can provide guidance for the design of fire-resistant ship compartment structures.

Keywords: fire resistance test; temperature distribution; structural deformation

Citation: Zong, S.; Liu, K.; Qiu, W.; Gao, Z.; Wang, J. Numerical and Experimental Analysis of Fire Resistance for Bulkhead and Deck Structures of Ships and Offshore Installations. *J. Mar. Sci. Eng.* **2023**, *11*, 1200. https://doi.org/10.3390/jmse11061200

Academic Editor: Saad Saad-Eldeen

Received: 14 May 2023
Revised: 30 May 2023
Accepted: 1 June 2023
Published: 9 June 2023

1. Introduction

Ships and offshore structures carrying oil and gas are often subjected to high-temperature impact due to the large number of accidents including fires and explosions. Their structural collapse during fires causes not only significant casualties but also environmental catastrophes such as oil spills [1–3]. HSE (2007) reported that fire and explosion were considered the main hazards in more than 60 offshore accidents that occurred in the last 40 years [4], such as the Sanchi oil tanker accident on 6 January 2018 in the East China Sea and the U.S.S. Bonhomme Richard accident on 12 July 2020 in San Diego shown in Figure 1 [5,6].

The heat generated by the fire and explosion is transmitted through steel structures of the bulkheads and deck, which can ignite fires in other cabins, resulting in severe loss of personnel and economy and damage to the environment. In order to prevent the occurrence of this kind of accident risk, the structural members must meet the fire resistance grade specification determined by the standard fire resistance test and the tests of fire resistance properties for marine structures, such as decks and bulkheads in ships and offshore installations. These tests must be rigorously performed in adherence to the stipulations delineated in the Safety of Life at Sea (SOLAS) Regulation II-2/17 [7]. These evaluations are further guided by resolutions and directives issued by the International Maritime Organization (IMO) as well as recommendations from member nations. Additionally, pertinent standards established by classification societies, including the American Bureau of Shipping

(ABS) [8] and the Russian Maritime Register of Shipping (RS) [9], are integrated into the evaluation process. The selection of an appropriate fire temperature model is crucial for accurately representing fire behavior and ensuring that the simulated conditions closely resemble real-world scenarios. Typically, fire resistance performance of structures is assessed using two distinct testing methodologies: the standard temperature regime and the hydrocarbon compound fire conditions. In Europe and the United States, the combustion of hydrocarbons and fire development are primarily examined in accordance with the hydrocarbon fire curve, wherein initial fire temperatures may reach 1000 °C or even exceed this threshold [9,10]. Conversely, in the Russian oil and gas complex (O&G) fire resistance evaluations, standard curves are employed to simulate fire environments for decks and bulkheads, as specified by the ISO-834 standard developed by ISO Technical Committee 92 [11]. However, it is important to note that analogous requirements for ship decks and bulkheads are exclusively regulated by stipulations pertaining to standard fire conditions.

(a) (b)

Figure 1. (**a**) Sanchi oil tanker and (**b**) U.S.S. Bonhomme Richard (right) accidents.

In the 1950s, the international community commenced efforts to address the impact of fires on ships and offshore installations, conducting extensive research on temperature distribution and their fire resistance capabilities within structures. The European Steel Construction Association and the British Standard Institution have conducted numerous experimental studies using single component as an analysis object (such as steel beams and steel columns). The physical properties and mechanical properties of steel structures at high temperatures were investigated, and specifications for fire-resistant design of steel structures were written based on the test data [12,13]. Since then, many countries such as the United States and Europe have also carried out a large amount of theoretical and test research on material performance [14,15], nodes and components [16,17], and corresponding steel structure fire-resistant design standards have been produced. Based on the European standard, Hanus carried out a comprehensive high-temperature test of S500 M steel and studied the dynamic response process of a steel beam under rapid and slow heating. It was found that the failure temperature range of the steel beam was between 550 °C and 650 °C [18]. Based on a large number of experimental data and theoretical research results, many countries, regions and organizations have proposed the applicable constitutive models for fire resistance of steel structures. Zhang [19], Agarwal [20], Yao and Tan [21,22], Buchaan [23], etc. have studied the deformation damage mechanism of different structures, components and material performance of steel at high temperature based on relevant specifications and derived the analytical expression of damage deformation and bearing capacity analysis of specific structural injury in the event of a fire, which promotes the development of structural damage deformation prediction technology for fire accidents.

Nassiraei H. [24] investigated the behavior of T/Y-joints reinforced with collar plates under compressive loading at elevated temperatures. Finite element modeling and experimental validation were conducted, revealing significant increases in initial stiffness and ultimate strength with increased collar plate length and thickness. A design formula was developed to determine the ultimate strength of these joints at elevated temperatures.

Due to the inability to recreate the real accident scenario well, the current research on the temperature distribution of offshore platform fire accidents mainly adopts numerical simulation methods. Kim [25] studied the fire resistance of FPSO's upper deck based on the temperature distribution data obtained from the jet fire test of steel tubes and compared the test results using a numerical method, which provided valuable data support for the fire resistance design of FPSO's superstructure. Based on this, Seo [26] proposed a method based on a risk-based optimization design procedure, utilizing probabilistic sampling and numerical simulation of a fire to determine fire accident loads. Computational fluid dynamics (CFD) software was used to conduct an A60 test based on thermal response analysis results and compute the temperature distribution, thereby establishing an optimized scheme for a thin-walled structure in a living area. In addition, the Paik team [27,28] of Busan National University in South Korea has used the method of combining experiment, theory and simulation: the structural response mechanism affected by fire was deeply explored, and the research process of structural response of offshore platform structure in complex accidents was promoted, which provides the basis for the safety design of offshore platform structures and the emergency response mechanism of accidents.

In order to ensure the safety of ships and offshore installation structures, most classification societies use insulation materials, which are usually non-combustible materials or materials with low combustion spread to strengthen the fire protection design of key cabins (such as cargo tanks and cabins) when a fire occurs. According to SOLAS Regulation II-2/17 [29] and the parameters of the fire-resistance limits and temperature exposure modes, the ships and offshore installation structures of the deck and bulkheads that separate industrial rooms are made of non-flammable materials and designed with certain fire-resistance classes: A, B, C and H (standard regime—A, B, C classes, and hydrocarbon—H class) [30]. Seo J.K. simulated the fire resistance of the A-60-level steel structure and calculated the temperature distribution for a certain road according to the simulation results. The analysis results show that the safety design of the sea cabin on the sea compartment complies with the specification requirements [31]. LeMoyne Boyer conducted full-scale tests on 21 steel bulkheads of different thicknesses and densities in the A-60, A-30, A-15 and A-0 classes according to IMO Resolution A 517 [32]. In addition, Zhou et al. conducted a numerical simulation of heat transfer to analyze the thermodynamic behavior associated with the generation of the cutting access openings in hull structures using the flame cutting method [33]. Park and Song also carried out fire resistance tests according to the test procedures specified in the MSC specification and analyzed fire resistance according to the material type of the bulkhead penetration [34].

The purpose of this study is to validate the numerical analysis and experimental results of transient heat transfer in A-60 grade steel structures. We utilized the fire test procedure (FTP) to verify the A-60 fire resistance specified in the Marine Safety Committee (MSC), including the design of the specimens, furnace temperature conditions and the setting of heating time. We conducted two sets of fire resistance tests on A-60 grade ship cabin walls and decks under standard fire conditions. Experimental and numerical analysis of the integrity loss and insulation capability of the cabin walls and decks were performed. This research can provide guidance for the design of fire-resistant ship cabin structures.

2. A-60 Class Decks and Bulkheads Fire Test Procedure (FTP)

2.1. Bulkheads and Decks Structure Specimen Design

The external dimensions of the A-60 class bulkhead core are 3020 mm in width and 2480 mm in height, with a steel plate thickness of 4.5 mm. For structural enhancement, six pieces of angle steel, each measuring 65 mm by 65 mm by 6 mm, have been incorporated as reinforcement, arranged at intervals of 600 mm. The A-60 class deck core, on the other hand, has external dimensions of 3020 mm in length and 2420 mm in width, with an identical steel plate thickness of 4.5 mm. To bolster its structural integrity, five pieces of angle steel, each sized 100 mm by 75 mm by 7 mm, have been employed as reinforcing elements, also arranged at intervals of 600 mm.

The insulation material for the test sample is ABM-FR6O marine fire-resistant mineral wool board. The specifications and dimensions of the insulation materials for the bulkheads are 1200 mm × 600 mm × 75 mm (90 kg/m^3), 1200 mm × 600 mm × 25 mm (90 kg/m^3) and 1200 mm × 600 mm × 50 mm (90 kg/m^3) (Figure 2a). The insulation materials for the deck have dimensions of 1200 mm × 600 mm × 50 mm and 1200 mm × 600 mm × 25 mm (Figure 3a). On the steel plate surface (bulkhead/deck) with the supporting material, a layer of 75 mm/50 mm-thick rock wool board is laid, and on the surface of the supporting material, a layer of 25 mm/50 mm-thick, 115 mm/175 mm-wide (on the middle 4 supporting materials) and 90 mm/125 mm-wide (on the two vertical edge supporting materials) rock wool board is laid. They are all laid using the stud and fastening ring process. On both sides of the supporting material angle steel, a layer of 25 mm/50 mm-thick rock wool board that is level with the angle steel is placed vertically, and inside the supporting material angle steel, a 25 mm/50 mm-thick rock wool board is filled; on the top and bottom ends of each row of the supporting material in the width direction of the bulkhead core, a 50 mm-thick, 90 mm/150 mm-high rock wool board is also placed vertically. The maximum distance between the studs and the rock wool board seam on the bulkhead core is 100 mm; the distance between the studs in the width direction of the bulkhead core is 223 mm, 285 mm and 288 mm and in the height direction is 238.5 mm, 200 mm and 400 mm. The maximum distance between the studs and the rock wool board seam on the supporting material is 50 mm; the distance between the studs on the supporting material is 283 mm, 300 mm and 100 mm. There are about 20 studs per meter on the bulkhead (including those on the supporting material) (Figures 2b and 3b).

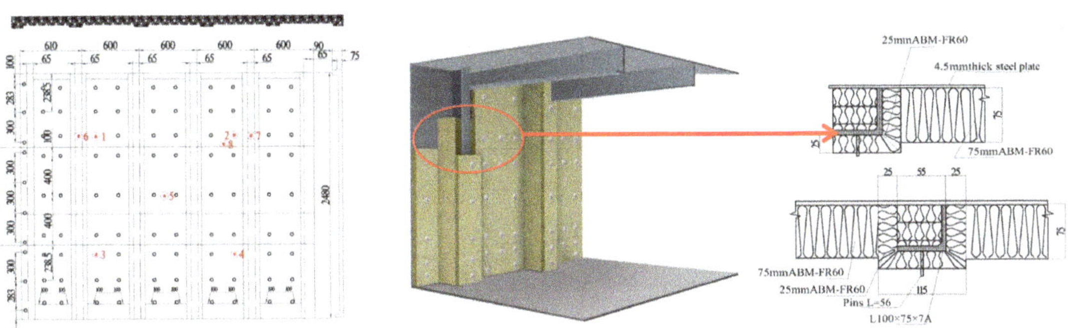

(a) (b)

Figure 2. (**a**) Location of thermocouples on bulkhead A-60; (**b**) section of A-60 bulkhead specimen. (unit: mm).

In the standard fire resistance test, the bulkhead was tested in the vertical position with the steel structure side exposed to the fire, and the side with mineral wool and L-shaped stiffeners of the deck is exposed to the fire. The bulkhead and deck were mounted within a steel restraint frame with a 50 mm-thick refractory concrete lining. The temperature of the sample was measured by thermocouples. A total of 8 thermocouples were arranged on the unheated surface of the specimen. Thermocouples No. 1 to No. 5 were located at the centers of the four 1/4 regions of the bulkhead specimen and the center of the whole specimen, respectively. No. 6 and 7 were located at the height of 3/4 of the two supporting materials, and No. 8 was located at the corresponding position of the heat insulation material seam. All the thermocouples mentioned above were used to measure the temperature rise at each corresponding point, where No. 1 to No. 5 were also used to calculate the average temperature rise (Figures 2a and 3a).

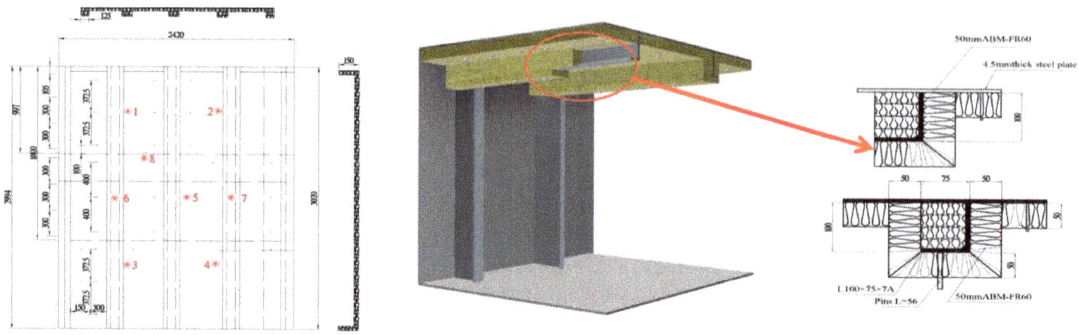

(a) (b)

Figure 3. (**a**) Location of thermocouples on deck A-60; (**b**) section of A-60 deck specimen. (unit: mm).

2.2. Experimental Apparatus and Experimental Procedure

The actual fire resistance test was carried out according to the FTP specification specified in MSC.307(88):

(1) The initial laboratory ambient temperature was maintained at 20 °C, accompanied by a relative humidity of 68%. The average temperature of the unheated surface of the specimen was consistently measured at 20 °C as well.
(2) The specimens underwent testing in a vertical furnace featuring two distinct furnace opening dimensions: 3000 mm × 3000 mm and 3000 mm × 2400 mm. Within the furnace, one thermocouple was installed per square meter, resulting in a total of nine plate thermocouples. The temperature measurement points were situated at a distance of 100 mm from the surface of the specimen facing the fire. The average inner furnace temperature was determined according to the standard temperature curve.
(3) During the heating test, the furnace pressure was maintained at a slight positive pressure relative to the pressure in the laboratory. The pressure setting and control at the bottom of the bulkhead sample's upward 2000 mm height was 12 Pa.
(4) A total of eight thermocouples were arranged on the unexposed surface of the specimen. Thermocouples 1 through 5 were located at the center of the four quarter sections of the cabin wall specimen and the center of the entire specimen. Thermocouples 6 and 7 were positioned at the 3/4 height of the two stiffening members, while thermocouple 8 was situated at the corresponding location of the insulation material seam. All of these thermocouples were employed to measure the temperature rise at their respective points, with thermocouples 1 through 5 also being utilized for calculating the average temperature rise. Refer to Figures 2 and 3 for further details.
(5) Temperature readings were continuously monitored and documented during the entire testing duration. The specimen was deemed to comply with the regulations if it withstood the specified flame temperature for 60 min without smoke or flames penetrating the gap, and the structural integrity remained uncompromised. Furthermore, the average temperature increase of the unheated specimen surface was not to exceed 140 °C, and the maximum temperature of the unheated specimen surface was not to surpass 180 °C at the conclusion of the 60 min interval.

2.3. The Furnace Control Temperature Curve

Experimental A-60 class bulkhead and deck specimens were tested to determine their limit state during fire exposure according to IMO FTP Code Part 3 [35]. There are several heating curves for the temperature field analysis of structures in general fire accidents, such as the ASTM-E119 standard temperature rise curve developed in North America and elsewhere and the ISO-834 standard temperature rise curve developed by the International Organization for Standardization [36].

The heating furnace was configured to create a standard heating state according to the ISO-834 standard warming curve, as shown in Equation (1).

$$T = T_0 + 345\log(8t + 1) \tag{1}$$

where T is the temperature inside the furnace in °C, and T_0 is the initial temperature inside the furnace; t is the time in minutes from the start of the test.

The comparison between the actual temperature inside the furnace (Actual) and the standard temperature curve (ISO-834) specified by ISO is shown in Figure 4. As shown in the figure, the actual temperature curve inside the furnace is in good agreement with the ISO standard temperature curve, and the temperature change specified by MSC.307(88) during the fire resistance test meets the internal test temperature requirements.

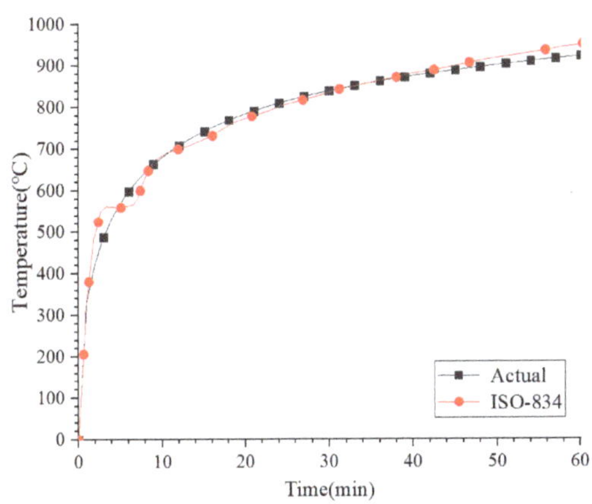

Figure 4. Verification of the time–temperature curve for the furnace.

3. Fire Resistance Test Results

The actual fire resistance tests were conducted in accordance with the FTP Code specified in MSC.307(88). Both the deck and the bulkhead were installed in a vertical fire testing furnace, which was designed for testing A-class reinforced concrete frames for ships and offshore facilities. The furnace temperature was continuously controlled to comply with the standard combustion curve prescribed in ISO-834-1.

3.1. The Structural Integrity of Specimen

The A-60 deck and bulkhead fire resistance test was conducted at the Far East Fire Test Center. Figures 5–8 depict the structural damage state and temperature distribution of the fire-resistant Class A-60 decks and bulkheads, both prior to and following the fire test.

Figure 6 presents the unheated surface of the A-60 deck following a 60 min fire test. Figure 6a illustrates the preparations prior to the fire test on the deck, while Figure 6b displays the state of the deck after being heated for 60 min. The annotations within the figure denote the respective locations where thermocouples were employed to measure temperature.

(a) (b)

Figure 5. Heated surface of the A-60 deck before (**a**) and after (**b**) fire test (60 min) (The enlarged area in (**b**) is the damage area of mineral wool and marked.)

(a) (b)

Figure 6. Unheated surface of the A-60 deck before and after fire test (60 min).

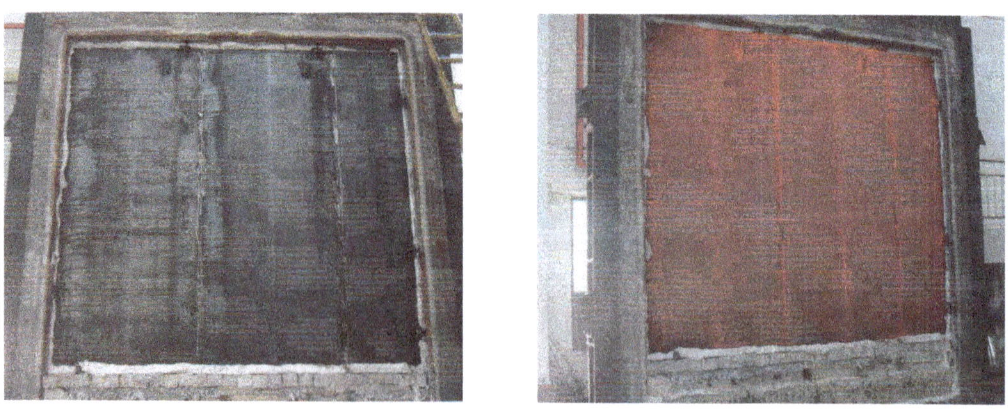

(a) (b)

Figure 7. Heated surface of the A-60 bulkhead before (**a**) and after (**b**) fire test (60 min).

(a) (b)

Figure 8. Unheated surface of the A-60 bulkhead before and after fire test (60 min). (**a**) shows the preparation of the bulkhead before the fire test, and (**b**) shows the state of the bulkhead after heating for 60 min.

Figure 7 shows the heating surface of A-60 bulkhead before and after 60 min of fire test. Based on the observed test results: (1) No flames crossed the unheated surface of the test specimen during the entire testing time. (2) No cracks, holes or other visible changes were observed on the sample, and no deflection values were measured. Gap measurements were conducted using 6 mm and 25 mm gauges. The 6 mm gauge could not penetrate the specimen. (3) The cotton pad ignition test was carried out on the unheated surface of the sample, and the results showed that the cotton pad was not ignited and there was no smoldering phenomenon. The heated surface remained intact without visible damage.

3.2. The Structural Thermal Insulation

According to the FTP specification specified in MSC.307(88), thermocouples are arranged on the unheated surface of the A-60 deck and bulkhead for temperature measurement. As shown in Tables 1 and 2, at the end of 60 min, the maximum temperature rise for any thermocouple on the unheated surface of the A-60 deck was 177 °C, with an average temperature rise of 138 °C for thermocouples 1 through 5. At the conclusion of the 60 min period, the maximum temperature rise for any thermocouple on the unheated surface of the A-60 bulkhead was 158 °C, with an average temperature rise of 137 °C for thermocouples 1 through 5. The structure met the regulatory requirements.

Table 1. Experimental temperature change of different thermocouples on unheated surface of A-60 deck.

Te (°C)	**Time (min)**						
	0	**10**	**20**	**30**	**40**	**50**	**60**
No. 1	0	10	42	66	95	120	140
No. 2	0	13	36	59	85	108	124
No. 3	0	10	35	58	82	106	123
No. 4	0	8	35	62	94	125	150
No. 5	0	8	39	65	97	128	154
Mean	0	9	37	62	90	117	138
No. 6	0	9	28	57	92	138	177
No. 7	0	5	31	59	87	123	155
No. 8	0	6	46	74	106	132	154

Table 2. Experimental temperature change of different thermocouples on unheated surface of A-60 bulkhead.

Te (°C)	Time (min)						
	0	**10**	**20**	**30**	**40**	**50**	**60**
No. 1	0	3	16	44	74	141	156
No. 2	0	1	9	32	46	66	92
No. 3	0	2	16	47	94	146	158
No. 4	0	1	14	47	75	142	158
No. 5	0	1	12	31	49	112	122
Mean	0	1	13	40	67	121	137
No. 6	0	2	9	16	34	53	68
No. 7	0	1	10	23	42	61	84
No. 8	0	2	12	44	64	117	156

According to the test reports, the temperature measured by thermocouples in the furnace was determined as an absolute value, and the temperature on the unheated surface was recorded and displayed as the difference between the ambient temperature and the temperature on the unheated surface.

3.3. The Structural Thermal Insulation

The deformation records of the A-60 deck and bulkhead centers during the test are presented in Table 3.

Table 3. A-60 deck center deformation displacement table after the test.

	0 min	**15 min**	**30 min**	**45 min**	**60 min**
Deck center (mm)	0	0	5	16	28
Bulkhead center (mm)	0	103	83	55	54

Based on the deformation table, it can be observed that the center of the A-60 deck did not undergo any deformation during the first 15 min. After 30 min, as the temperature increased, the deck center began to exhibit concave deformation towards the furnace interior, with the deformation gradually increasing. At the end of the 60 min test, the maximum deformation at the center was 28 mm. During the fire resistance test for the A-60 bulkhead, the bulkhead center began to show concave deformation towards the furnace interior within the first 15 min, with a maximum value of 103 mm. Subsequently, as the temperature increased, the deformation gradually decreased, and at the end of the 60 min test, the bulkhead center deformation was 54 mm.

4. Simulation of Bulkhead and Deck Section Heating

4.1. The Finite Element Model

Finite element simulation analysis of the heating process for the bulkhead and deck sections was conducted using the finite element software Abaqus2020. The temperature distribution and structural deformation of the deck and bulkhead were analyzed through sequential thermomechanical coupling analysis. The time step and the total loading time were set to 60 s and 3600 s, respectively. The results of each sub-step were saved to facilitate the post-processing and observation of the temperature field over time.

To investigate the influence of mesh size on the computational results, a mesh sensitivity analysis was performed for the numerical analysis of the deck and bulkhead. The models were divided into meshes with different sizes, considering mesh sizes of 10 mm, 12 mm, 15 mm, 20 mm, 25 mm and 30 mm, using a uniform meshing approach. Thermal-mechanical coupling analysis was carried out for the models with different mesh sizes to obtain the temperature distribution and structural deformation results of the deck and bulkhead. The results are shown in Figures 9 and 10.

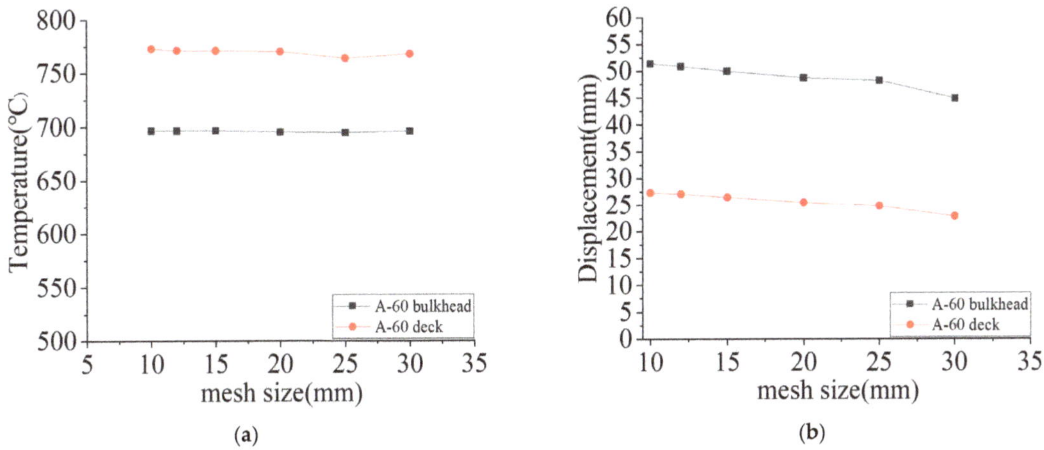

Figure 9. (**a**) The variation of structural temperature with mesh size. (**b**) The variation of displacement deformation with mesh size.

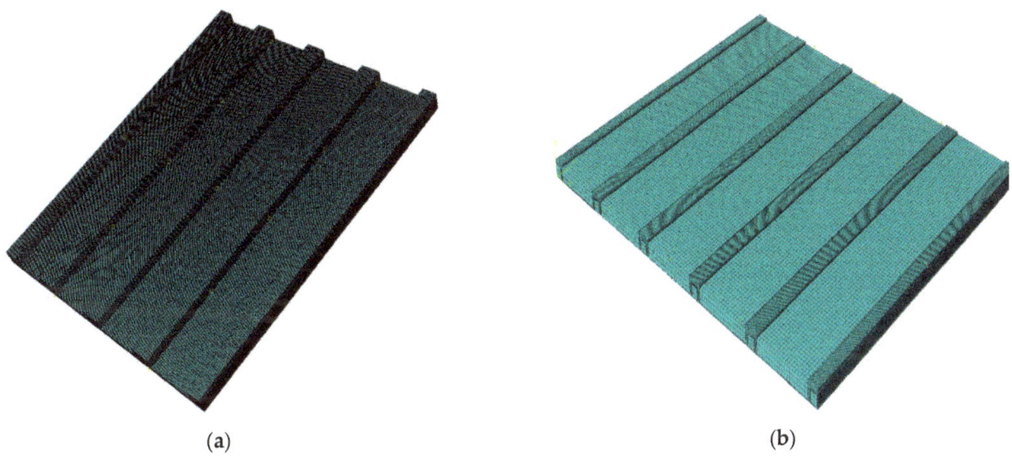

Figure 10. (**a**) Finite element model of A-60 deck; (**b**) finite element model of A-60 bulkhead.

It can be seen from the calculation results of Figure 9a that the temperature distribution of deck and bulkhead does not change significantly with the change of grid size. This shows that in the numerical simulation analysis, the mesh size has little effect on the temperature distribution during the structural fire resistance test.

According to the calculation results of Figure 9b, with the increase of mesh size, the displacement at the center point of deck and bulkhead structure decreases gradually. The deformation trends of deck and bulkhead structures with different mesh sizes are consistent with the experimental results. Considering both calculation accuracy and calculation efficiency, the structural grid size of 10 mm is selected for finite element numerical calculation.

As shown in Figure 10, the element types used in the finite element model of the A-60 deck and bulkhead for a heat transfer analysis are eight-node linear heat transfer tetrahedron solids. In order to calculate the accuracy, the finite element mesh size of the mineral wool is 10 mm, and the mesh size of the steel plate is also 10 mm. In order to transfer heat between the mineral wool and steel sheets, binding coupling constraints are used between the mineral wool and the steel plates.

4.2. Materials and Methods

In the study of structural responses of steel constructions and mineral wool under high-temperature conditions, it is crucial to account for the impacts of temperature elevations on the performance of both steel structures and mineral wool. Determining the material parameters of steel and mineral wool at varying temperatures is necessary, including specific heat capacity, thermal conductivity, convective heat transfer coefficient and thermal expansion coefficient, among others. This lays the foundation for subsequent numerical simulation analyses of steel structures and mineral wool under the influence of high temperatures.

In the simulation of heating, the thermal conductivity equation is used in the three-dimensional temperature field [37], as shown in Equation (2).

$$\rho C \frac{\partial T}{\partial t} = k\left(\frac{\partial^2 T}{\partial x^2} + \frac{\partial^2 T}{\partial y^2} + \frac{\partial^2 T}{\partial z^2}\right) \quad (2)$$

where T is the temperature inside the furnace in °C; k is the thermal conductivity in W/(m·K); t is the time in minutes from the start of the test in °C; C is the specific heat capacity in J/(kg·K) and ρ is the density in kg/m^3. During numerical analysis, both thermal conductivity and specific heat capacity vary with temperature. Therefore, in subsequent calculations, it is imperative to consider the effects induced by temperature changes. However, the influence of increasing temperature on density is not pronounced and is thus disregarded in this context.

Before the standard fire test occurs, the specimen is at ambient temperature, and it is assumed that the entire structure section temperature is uniform and equal to the ambient temperature T_0. The ambient temperature is averaged according to the test report and is assumed to be 20 °C.

4.2.1. Convective Heat Transfer Coefficient of Steel and Mineral Wool

There are four types of heat transfer relations on the boundary of the specimen: convective heat exchange between the surface of the specimen and the high-temperature gas inside the furnace, radiant heat from the high-temperature gas to the specimen, radiant heat from the furnace wall to the specimen and radiant heat from the specimen to the outside. The boundary condition of the specimen under the condition of the fire resistance test is the third type of boundary, as shown in Equation (3) [38].

$$-k\frac{\partial T}{\partial n} = \alpha(T_0 - T) \quad (3)$$

where α is the convective heat transfer coefficient in W/(m^2·K), mean 25 W/(m^2·K); k is the thermal conductivity in W/(m·K); T_0 is the initial temperature inside the furnace in °C.

4.2.2. Radiative Heat Transfer Coefficient of Steel and Mineral Wool

The radiation heat transfer condition is set at the boundary of the model. The radiation heat transfer is determined according to Equation (4) [39].

$$F_n = k_{SB} \times \beta \times \left(T^4 - T_0^4\right) \quad (4)$$

where k_{SB} is the Stefan–Boltzmann constant in W/(m^2·K^4), mean 5.67 × 10^{-8} W/(m^2·K^4); β is the surface absorption coefficient, and the surface absorption coefficient is set at 0.5 [37].

The material characteristics of mineral wool differ from those of bulkhead and deck. Its thermal conductivity and specific heat capacity increase with the increase in temperature. Previous research results have shown this, as indicated in Table 4 [40,41], "K_{wool}" is employed to denote the thermal conductivity of mineral wool.

Table 4. The main characteristics of mineral wool for structures.

Te (°C)	K_{wool} W/(m·K)			C_{wool} J/(kg·K)		
	10	**100**	**300**	**10**	**100**	**300**
	0.035	0.046	0.085	840	860	900

4.2.3. Physical Properties of Steel

In investigating the response of steel structures under high-temperature conditions, the initial step is to ascertain the temperature distribution within the structure. Unlike the mechanical properties of steel under normal temperature static loads, the strength and stiffness of steel significantly decline when it exceeds a certain temperature, a phenomenon referred to as "steel softening". At this point, the steel structure undergoes substantial plastic deformation, losing its original load-bearing capacity. This section collates and summarizes the physical and mechanical properties of Grade A steel material models ranging from 20 to 1000 °C, including specific heat capacity, thermal conductivity, convective heat transfer coefficient and thermal expansion coefficient. Notably, the impact of increased temperature on steel density and Poisson's ratio is negligible, and thus this analysis disregards the temperature effects on these factors (with density assumed as 7850 kg/m^3 and Poisson's ratio as 0.3). This provides a foundation for subsequent numerical simulation analyses.

4.2.4. Specific Heat Capacity of Steel

Specific heat capacity refers to the amount of heat absorbed or released by a unit mass of steel for a unit temperature rise or drop, with units of J/(kg·K) or J/(kg·°C).

The European Standard (EC3) [13] stipulates:

$$
\begin{array}{ll}
C_{\text{steel}} = 425 + 7.73 \times 10^{-1}T - 1.69 \times 10^{-3}T^2 & \\
+2.22 \times 10^{-6}T^3 & 20 \le T < 600\ ^\circ\text{C} \\
C_{\text{steel}} = 666 + 13002/(738 - T) & 600 \le T < 735\ ^\circ\text{C} \\
C_{\text{steel}} = 545 + 17820/(T - 731) & 735 \le T < 900\ ^\circ\text{C} \\
C_{\text{steel}} = 650 & 900 \le T < 1200\ ^\circ\text{C}
\end{array} \quad (5)
$$

The specific heat capacity of 20–700 °C steel calculated with reference to EC3 is shown in Table 5.

Table 5. Specific heat capacity of steel at elevated temperatures.

Te (°C)	20	100	200	300	400	500	600	700	800	900	1000
C_{steel} J/(kg·K)	439	487	529	564	605	666	760	1008	803	650	650

4.2.5. Thermal Conductivity of Steel

Thermal conductivity k refers to the amount of heat transferred per unit area per unit time through a unit thickness of steel, under steady heat transfer conditions and with a temperature difference of 1 °C between the upper and lower surfaces. Its units are either A or B. According to the European Standard (EC3), thermal conductivity decreases with increasing temperature, and the rate of change slows when the temperature exceeds 800 °C.

$$
\begin{array}{ll}
k_{\text{steel}} = 54 - 3.33 \times 10^{-2}T & 20 \le T < 800\ ^\circ\text{C} \\
k_{\text{steel}} = 27.3 & 800 \le T < 1200\ ^\circ\text{C}
\end{array} \quad (6)
$$

Thermal conductivity of steel at 20–1000 °C calculated according to European Standard (EC3) is shown in Table 6.

Table 6. The heat conductivity coefficient of steel at elevated temperatures.

Te (°C)	20	100	200	300	400	500	600	700	800	900	1000
k_{steel} W/(m·K)	53.3	50.7	47.3	44	40.7	37.4	34	30.7	27.3	27.3	27.3

4.2.6. Stress–Strain Relationship of Steel

As the temperature rises to a certain level (generally above 300 °C), the strength and stiffness of steel gradually decrease and rapidly decline after surpassing a certain temperature (typically 600 °C). Consequently, the structure undergoes substantial plastic deformation, losing its load-bearing capacity.

Different standards provide diverse theoretical stipulations for the reduction coefficients (TRCs) of yield strength and modulus of elasticity of steel materials under high-temperature conditions. According to the theoretical formulas of different standards [12,13,42,43], TRCs for yield strength and modulus of elasticity were calculated, as shown in Figure 11.

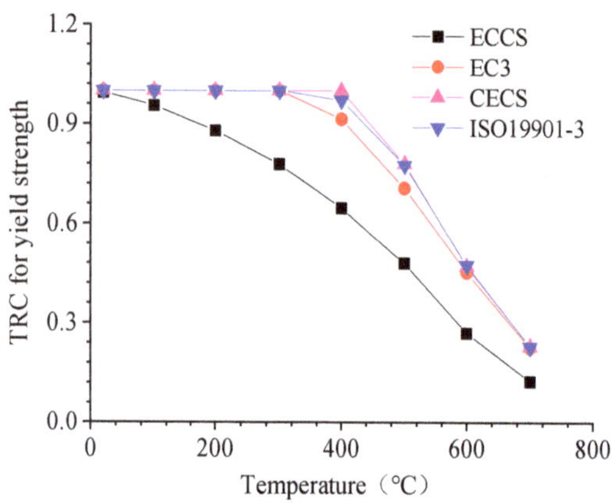

(a)

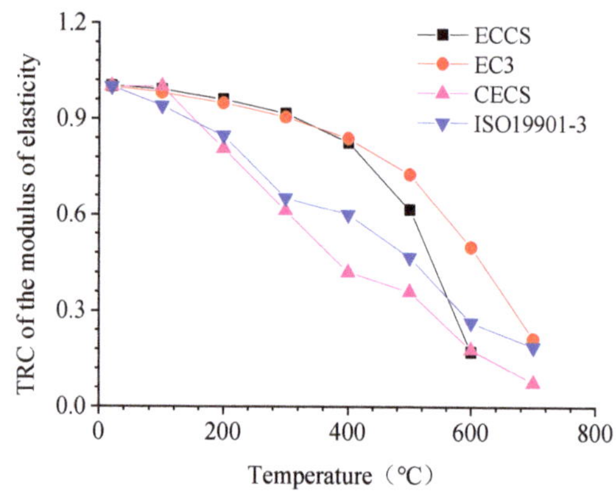

(b)

Figure 11. (**a**) The yield strength–temperature relationship and (**b**) the elastic modulus–temperature relationship.

From the Figure 11a, it can be observed that several standards show the same trend for the relationship between yield strength and temperature. The yield strength of steel gradually decreases before reaching 300 °C. When the temperature exceeds 300 °C, the yield strength rapidly decreases with increasing temperature. When the temperature reaches 700 °C, the yield strength stipulated by these standards is only 10% to 20% of that at room temperature.

From the Figure 11b, the relationship between the modulus of elasticity and temperature stipulated by the standards also follows the same trend. The modulus of elasticity as defined by ECCS and EC3 decreases slowly before 300 °C and rapidly declines after 300 °C. In comparison, the modulus of elasticity as defined by CECS and ISO19901-3 continually decreases in the range of 20 °C to 700 °C, without the apparent "slow-then-rapid" phenomenon.

In this study, the reduction coefficients for the modulus of elasticity and yield strength of steel materials at high temperatures, as defined by ISO19901-3, were taken as the standard. The tensile test data of marine grade A steel [44] were adjusted accordingly, and the elastic modulus of the material at different temperatures and the engineering stress-strain relationship are shown in Figures 12 and 13.

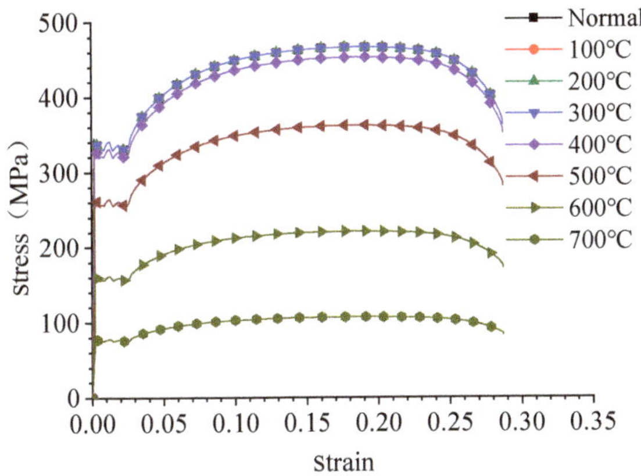

Figure 12. The stress–strain relationship of materials at different temperatures.

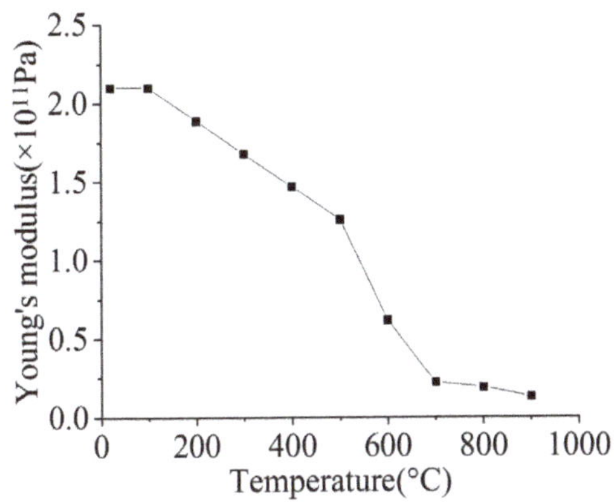

Figure 13. The Young's modulus–temperature relationship.

In finite element numerical simulation calculations, it is necessary to correct the engineering stress–strain relationship and convert it into the true stress–strain relationship. In this study, the "combined material" relationship curve [45] was adopted to address the conversion from engineering to true stress–strain relationship:

The true stress and strain before necking can be calculated from the engineering stress and strain using the following equation:

$$\sigma_t = \sigma_e(\varepsilon_e + 1) \tag{7}$$

$$\varepsilon_t = \ln(\varepsilon_e + 1) \tag{8}$$

The true stress–strain curve after necking:

$$\sigma_t = C\varepsilon_t^n \tag{9}$$

where $n = \ln(1 + A_g)$; $C = R_m(e/n)^n$; $A_g = 1/(0.24 + 0.01395R_m)$; R_m is the ultimate tensile stress from the tensile test, e is the base of the natural logarithm and A_g is the maximum uniform strain related to the ultimate tensile stress.

By connecting the above two segments with a straight line, the converted true stress–strain curve can be directly used for the material input in the simulation, as shown in Figure 14.

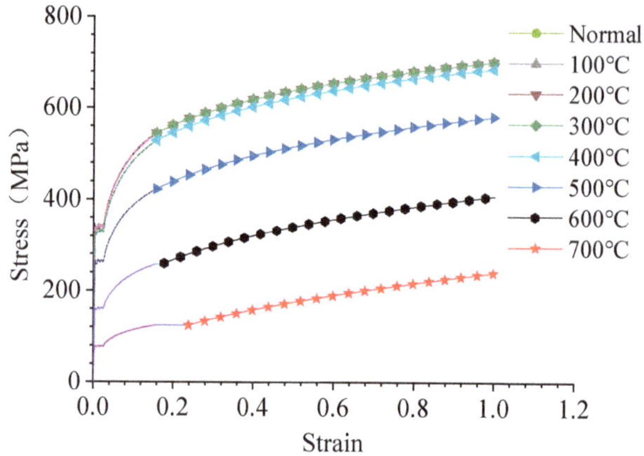

Figure 14. True stress–strain curves at different temperatures.

5. Results and Discussion

5.1. Structure Temperature Distribution

Based on the thermomechanical coupling analysis, the temperature distribution of the structure was calculated and mapped onto the finite element models (a) and (b). Figures 15 and 16 show the contour maps of the overall temperature distribution on the deck and cabin walls after 60 min of heating.

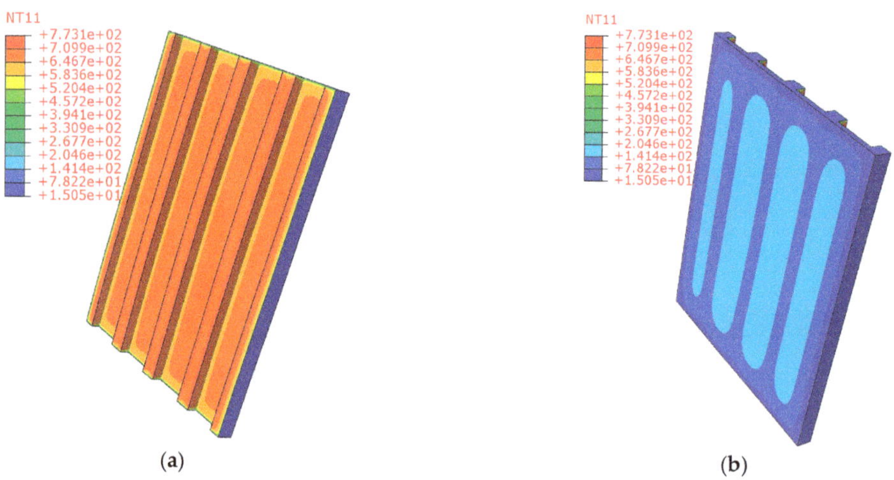

Figure 15. (**a**) Surface exposed to the inside furnace of the A-60 deck; (**b**) surface unexposed to the inside furnace of the A-60 deck.

As Table 7 shows, the temperature of each thermocouple area on the unheated surface of the A-60 deck changes evenly over time. In the first 30 min, due to the protection provided by the mineral wool, the temperature of most regions in the sample changes little. As the temperature continues to rise, the temperature change gradient in the center of the sample increases. The temperature change around the reinforcing rib in the middle gradually becomes higher than in other areas.

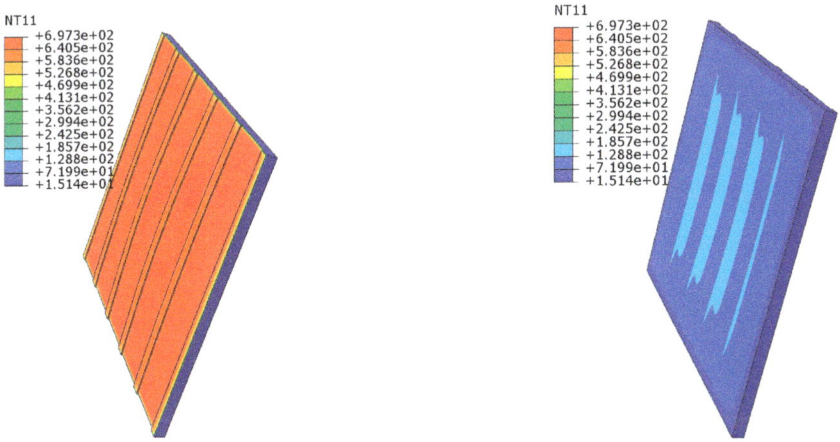

(**a**) (**b**)

Figure 16. (**a**) Surface exposed to furnace of A-60 bulkhead; (**b**) surface unexposed to furnace of A-60 bulkhead.

Table 7. Simulated temperature change of different thermocouples on unheated surface of A-60 deck.

Te (°C)	**Time (min)**						
	0	**10**	**20**	**30**	**40**	**50**	**60**
No. 1	0	14	39	67	95	119	140
No. 2	0	9	33	56	83	105	127
No. 3	0	6	34	55	84	106	125
No. 4	0	6	33	63	95	128	153
No. 5	0	11	42	64	98	131	155
Mean	0	9.2	36.2	61	91	117.8	140
No. 6	0	6	27	57	91	139	179
No. 7	0	4	28	60	86	124	152
No. 8	0	8	47	77	107	132	152

The results of the finite element analysis indicate that at the end of 60 min, the maximum temperature rise of any thermocouple is 179 °C. The average temperature rise of thermocouples 1 to 5 is 140 °C. The structure meets the FTP specification specified in MSC.307(88).

The results of the finite element analysis, shown in Table 8, indicate that the outer edge of the sample maintains its shape, while the interior of the sample begins to degenerate. Since the outer edge of the sample is close to high-temperature airflow, the temperature is high, and the gradient is large. On the other hand, the deepening of the sample causes both temperature and gradient to decrease. Due to the protection provided by the mineral wool in the first 30 min, the temperature in most areas inside the sample remains low. As the temperature inside the furnace continues to rise, the gradients at the outer edge continue to increase and gradually move toward the middle of the sample. The temperature of the tendon gradually decreases, forming five low-temperature zones with uniform temperatures.

The results of the finite element analysis indicate that at the end of 60 min, the maximum temperature rise of any thermocouple is 158 °C. The average temperature rise of thermocouples 1 to 5 is 137.2 °C. The structure meets the FTP specification specified in MSC.307(88).

Table 8. Simulated temperature change of different thermocouples on unheated surface of A-60 bulkhead.

Te (°C)	Time (min)						
	0	**10**	**20**	**30**	**40**	**50**	**60**
No. 1	0	3	14	47	75	140	156
No. 2	0	1	8	34	45	67	94
No. 3	0	5	14	50	96	145	158
No. 4	0	1	17	44	78	113	158
No. 5	0	1	13	29	47	118	120
Mean	0	2.2	13.2	40.8	68.2	116.6	137.2
No. 6	0	5	10	15	32	56	66
No. 7	0	2	11	24	42	64	86
No. 8	0	2	12	41	61	116	158

5.2. Structural Deformation

In the nonlinear finite element analysis, a thermomechanical coupled analysis was employed to calculate the central displacement of the structure under thermal loads. The deformation records at the center of the A-60 deck and cabin walls during the simulation process are shown in Table 3. At the end of the simulation, the deformation cloud diagrams of the deck and cabin walls are presented in Figure 17.

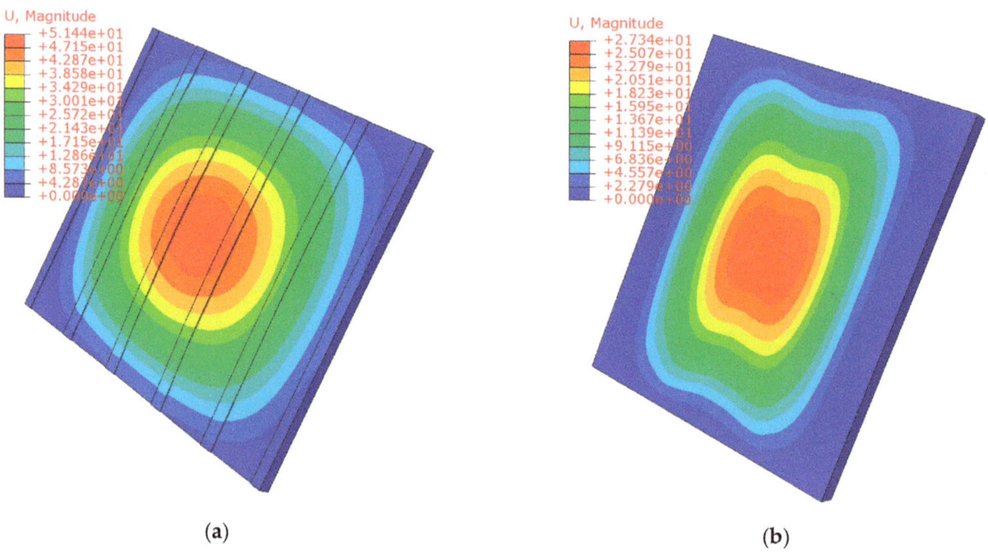

Figure 17. (**a**) Deformation of A-60 bulkhead; (**b**) deformation of A-60 deck.

Figure 17a,b respectively show the structural deformation of the deck and bulkhead at 60 min during the thermomechanical coupling analysis. From the figures, it can be observed that the maximum thermal deformation of the bulkhead specimen is 51.44 mm, while the maximum deformation of the deck specimen is 27.34 mm. Both maximum deformations occur at the center of the specimens.

According to the deformation Table 9, it can be observed that the deformation at the center of the A-60 deck is relatively slow within the first 15 min. After 30 min, as the temperature rises, the deformation at the deck center begins to accelerate, and the amount of deformation gradually increases. At the end of the 60 min test, the maximum deformation at the center is 27.3 mm. During the fire resistance test on the A-60 bulkhead, a larger deformation occurs at the bulkhead center within the first 15 min, with a maximum value of 105 mm. Subsequently, as the temperature rises, the amount of deformation gradually decreases, and at the end of the 60 min test, the deformation at the bulkhead center is 51.44 mm.

Table 9. A-60 deck center deformation displacement table.

	0 min	15 min	30 min	45 min	60 min
Deck center (mm)	0	1.8	6.7	18	27.3
Bulkhead center (mm)	0	105	84.5	57.8	51.44

5.3. Comparative Analysis of Test and Simulation

Experimental and simulated average and maximum temperature rise record curves of the unheated surface of the A-60 deck and bulkhead are shown in Figures 18 and 19.

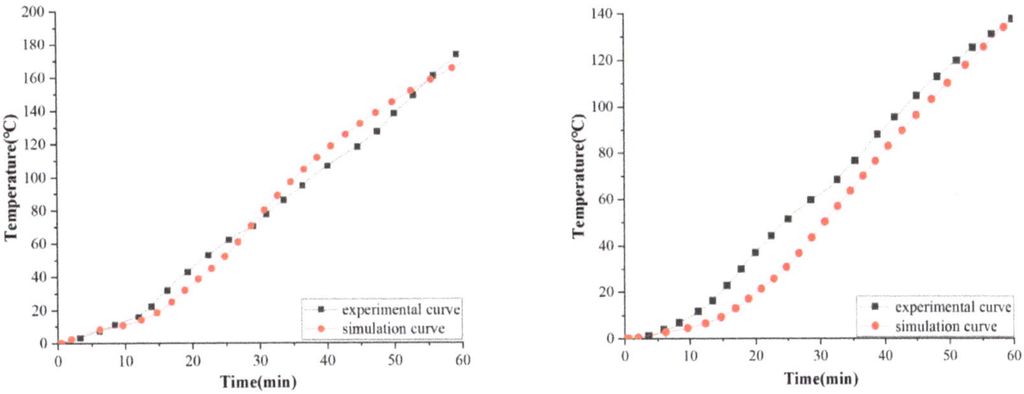

(a) (b)

Figure 18. (**a**) Experimental and simulated maximum temperature rise record curves of unheated surface of A-60 deck; (**b**) experimental and simulated average temperature rise record curves of unheated surface of A-60 deck.

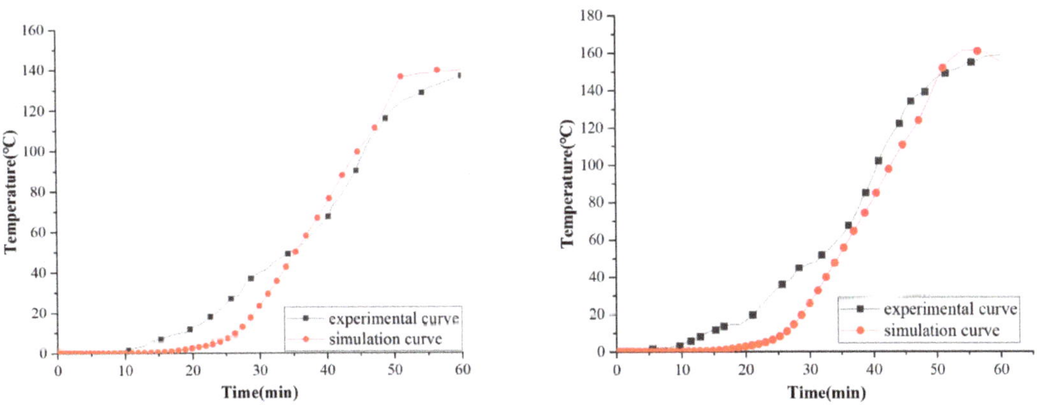

(a) (b)

Figure 19. (**a**) Experimental and simulated maximum temperature rise record curves of unheated surface of A-60 bulkhead. (**b**) Experimental and simulated average temperature rise record curves of unheated surface of A-60 bulkhead.

The simulation results using Abaqus on the heating of the A-60 cabin wall and deck show that the temperature of the deck and cabin wall increases slowly between 20 and 30 min and increases quickly between 30 and 50 min. During the 50–60 min period, the temperature of the test specimen increases gradually. The simulation results for the cabin wall have an error of within 25% during the 20–30 min period, but the simulation results for the deck and cabin wall match the experimental results well (with a difference of no more than 5%). The main reason for this error is that the test report does not provide information about the presence of additional fire retardants (including water-based adhesives) in the partition structure, and mineral wool is a dry fire retardant containing organic matter and water, which cannot be simulated in the simulation.

As shown in Figure 20, during the heating process, the deck structure undergoes deformation rather slowly at the initial stage. With the increase in temperature, thermal expansion begins at the central point of the deck structure, causing the structure to shift towards the furnace. In the heating process of the bulkhead structure, given that the heating surface is on one side of the steel plate and considering that the thermal conductivity of the steel plate is significantly higher than that of the mineral wool, the steel plate undergoes rapid thermal expansion and deformation at the onset of heating under the effect of gravity. Once the mineral wool absorbs enough heat, thermal expansion begins in the mineral wool structure, leading to a slow shift of the structural deformation in the opposite direction. The displacement changes measured at the centers of the deck and bulkhead during the test align well with the results of the simulation calculations, with the error value within 10 percent. These simulation calculations can provide a reference for the subsequent fire resistance tests.

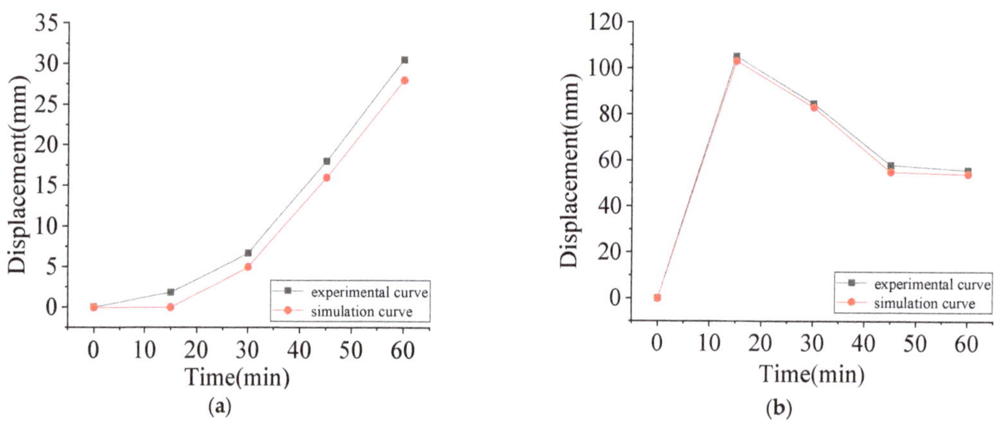

Figure 20. (**a**) Experimental and simulated displacement curves of unheated surface of A-60 deck. (**b**) Experimental and simulated displacement curves of unheated surface of A-60 bulkhead.

6. Conclusions

This study employed both experimental and numerical simulation techniques to investigate the temperature distribution and structural deformation of the A-60 deck and bulkhead during fire resistance tests.

The primary conclusions of this research include the following points:

- In this study, the fire resistance performance of the A-60 deck and bulkhead was numerically analyzed using Abaqus software. The simulation analysis of the standard fire resistance test for A-60 deck and bulkhead was conducted by selecting appropriate finite element units and material parameters. The simulation results showed good agreement with the experimental results (with an error within 5%), indicating the effectiveness of this study in predicting the performance of the samples and saving time and costs.

- In the standard fire resistance test of the A-60 ship deck and bulkhead, the average temperature rise on the unexposed surface at 60 min did not exceed 140 °C compared to the initial temperature, and the temperature rise at any point on the unexposed surface did not exceed 180 °C compared to the initial temperature. Throughout the entire testing process, no smoke or flame penetration occurred, and the structure remained intact. The structure met the specification requirements. Therefore, this study provides guidance for the design of fire-resistant ship compartment structures.
- There is a certain discrepancy between the finite element numerical analysis results and the fire resistance test. The main reason for this discrepancy is the lack of information in the test report regarding the presence of additional fire retardants (including water-based adhesives) in the partition structure. Mineral wool, being a dry fire retardant containing organic matter and water, cannot be accurately reproduced in the simulation. Further research is needed to strengthen the study of the thermal performance of mineral wool at high temperatures.

Author Contributions: Conceptualization, S.Z. and W.Q.; Writing—original draft, S.Z.; Writing—review and editing, Z.G. and K.L.; Visualization, K.L. and J.W. All authors have read and agreed to the published version of the manuscript.

Funding: This research was funded by the National Natural Science Foundation of China (Grant No. 52171311; Grant No. 52271279).

Institutional Review Board Statement: Not applicable.

Informed Consent Statement: Not applicable.

Data Availability Statement: Not applicable.

Conflicts of Interest: The authors declare no conflict of interest.

References

1. Paik, J.K.; Czujko, J.; Kim, B.J.; Seo, J.K.; Ryu, H.S.; Ha, Y.C.; Janiszewski, P.; Musial, B. Quantitative assessment of hydrocarbon explosion and fire risks in offshore installations. *Mar. Struct.* **2011**, 24, 73–96. [CrossRef]
2. Andryushkin, A.Y.; Kirshina, A.A.; Kadochnikova, E.N. The evaluation of the fire-retardant efficiency of intumescent coatings of steel structures exposed to high-temperature gas flows. *Fire Explos.* **2021**, 30, 14–26. [CrossRef]
3. Gravit, M.; Zimin, S.; Lazarev, Y.; Dmitriev, I.; Golub, E. Fire simulation of bearing structures for natural gas module plant. In *International Scientific Siberian Transport Forum*; Springer: Cham, Switzerland, 2019; pp. 365–376.
4. Veritas D, N. Accident statistics for floating offshore units on the UK continental shelf 1980–2005. Health and Safety Executive (HSE) Report. [Online] (Updated 2007). Available online: http://www.hse.gov.uk/research/-rrpdf/rr567.pdf (accessed on 1 October 2013).
5. Vinnem, J.E.; Røed, W. Offshore Risk Assessment Vol. 1. In *Principles, Modelling and Applications of QRA Studies*; Springer: Berlin/Heidelberg, Germany, 2014.
6. Woo, D.; Seo, J.K. Numerical validation of the two-way fluid-structure interaction method for non-linear structural analysis under fire conditions. *J. Mar. Sci. Eng.* **2021**, 9, 400. [CrossRef]
7. Zisimopoulos D, A. Use of Fiber Reinforced Plastics in Ship Construction: A Study of SOLAS regulation II-2/17 on Alternative Design and Arrangementsfor Fire Safety. 2016. Available online: https://dspace.lib.ntua.gr/xmlui/handle/123456789/42464?show=full (accessed on 10 May 2023).
8. Russian Maritime Register of Shipping. Available online: https://rs-class.org/en/ (accessed on 22 November 2021).
9. Imran, M.; Liew, M.S.; Nasif, M.S.; Gracia, E.M.; Danyaro, K.U.; Niazi, M.U. Thermal and Mechanical Response of Partially Protected Steel I-Beam under Fire. *MATEC Web Conf. EDP Sci.* **2018**, 203, 06009. [CrossRef]
10. Gravit, M.; Gumerova, E.; Bardin, A.; Lukinov, V. Increase of fire resistance limits of building structures of oil-and-gas complex under hydrocarbon fire. In *Energy Management of Municipal Transportation Facilities and Transport*; Springer: Cham, Switzerland, 2017; pp. 818–829.
11. *ISO 834-75*; Fire-Resistance Tests—Elements of Building Construction. ISO: Geneva, Switzerland, 1975.
12. ECCS. European recommendations for the fire safety of steel structures. In *Calculation of the Fire Resistant*; Elsevier Scientific Pub. Co.: Amsterdam, The Netherlands, 1983.
13. British Standard Institution. *Structural Use of Steelwork in Building*; British Standard Institution: Chiswick, UK, 2000.
14. Kirby, B.R.; Preston, R.R. High temperature properties of hot-rolled, structural steels for use in fire engineering design studies. *Fire Saf. J.* **1988**, 13, 27–37. [CrossRef]

15. Twilt, L. Strength and deformation properties of steel at elevated temperatures: Some practical implications. *Fire Saf. J.* **1988**, *13*, 9–15. [CrossRef]
16. Anderberg, Y.; Forsen, N.; Aasen, B. *Measured and Predicted Behaviour of Steel Beams and Columns in Fire*; Department of Fire Safety Engineering and Systems Safety, Lund University: Lund, Sweden, 1986; Volume 1, pp. 259–269.
17. SkowrońSki, W. Buckling fire endurance of steel columns. *J. Struct. Eng.* **1993**, *119*, 1712–1732. [CrossRef]
18. Hanus, F.; Vassart, O.; Caillet, N.; Franssen, J.M. High temperature full-scale tests performed on S500M steel grade beams. *J. Constr. Steel Res.* **2017**, *133*, 448–458. [CrossRef]
19. Zhang, C.; Li, G.Q.; Wang, Y.C. Predictability of buckling temperature of axially loaded steel columns in fire. *J. Constr. Steel Res.* **2012**, *75*, 32–37. [CrossRef]
20. Agarwal, A.; Choe, L.; Varma, A.H. Fire design of steel columns: Effects of thermal gradients. *J. Constr. Steel Res.* **2014**, *93*, 107–118. [CrossRef]
21. Tan, K.H.; Yao, Y. Fire resistance of reinforced concrete columns subjected to 1-, 2-, and 3-face heating. *J. Struct. Eng.* **2004**, *130*, 1820–1828. [CrossRef]
22. Yao, Y.; Tan, K.H. Fire resistance of four-face heated reinforced concrete columns. *J. Struct. Eng.* **2003**, *129*, 1220–1229.
23. Buchanan, A.H.; Abu, A.K. *Structural Design for Fire Safety*; John Wiley & Sons: Hoboken, NJ, USA, 2017.
24. Nassiraei, H. Static strength of tubular T/Y-joints reinforced with collar plates at fire induced elevated temperature. *Mar. Struct.* **2019**, *67*, 102635. [CrossRef]
25. Kim, B.J.; Seo, J.K.; Park, J.H.; Jeong, J.S.; Oh, B.K.; Kim, S.H.; Park, C.H.; Paik, J.K. Load characteristics of steel and concrete tubular members under jet fire: An experimental and numerical study. *Ocean. Eng.* **2010**, *37*, 1159–1168. [CrossRef]
26. Seo, J.K.; Lee, S.E.; Park, J.S. A method for determining fire accidental loads and its application to thermal response analysis for optimal design of offshore thin-walled structures. *Fire Saf. J.* **2017**, *92*, 107–121. [CrossRef]
27. Kim, S.J.; Lee, J.; Kim, S.H.; Seo, J.K.; Kim, B.J.; Ha, Y.C.; Paik, J.K.; Lee, K.S.; Park, B.; Ki, M.S.; et al. Nonlinear structural response in jet fire in association with the interaction between fire loads and time-variant geometry and material properties. *Ocean. Eng.* **2017**, *144*, 118–134. [CrossRef]
28. Darbra, R.M.; Palacios, A.; Casal, J. Domino effect in chemical accidents: Main features and accident sequences. *J. Hazard. Mater.* **2010**, *183*, 565–573. [CrossRef] [PubMed]
29. Flis, L. Application of the rules of the American Bureau of Shipping (ABS) to design mobile offshore structures. *Marit. Tech. J.* **2016**, *205*, 17–28. [CrossRef]
30. ABS Rules for Building and Classing Facilities on Offshore Installations. Rules for Building and Classing Facilities on Offshore Installations 2021. Available online: https://infostore.saiglobal.com/en-au/?productID=1787173 (accessed on 10 May 2023).
31. Kim, S.J.; Lee, J.; Paik, J.K.; Seo, J.K.; Shin, W.H.; Park, J.S. A study on fire design accidental loads for aluminum safety helidecks. *Int. J. Nav. Archit. Ocean. Eng.* **2016**, *8*, 519–529. [CrossRef]
32. Boyer, L.M. *Thermal Radiation from Marine Fire Boundaries: Evaluation and Analysis of A-60, A-30, A-15, and A-0 Bulkhead Assemblies*; Wyman-Gordon Co.: Worcester, MA, USA, 1993.
33. Zhou, B.; Han, X.; Guo, W.; Liu, Y.; Tan, S.K. Numerical and experimental study on cutting access opening in ship structure. *J. Ship Prod. Des.* **2017**, *33*, 12–23. [CrossRef]
34. Park, W.C.; Song, C.Y. Heat transfer characteristics of bulkhead penetration piece for A60 class compartment II: Fire resistance test for piece material and insulation types. *J. Ocean. Eng. Tech.* **2019**, *33*, 340–349. [CrossRef]
35. Res I M O. MSC 307 (88)-International Code for Application of Fire Test Procedures. Annex1, Part3–Part4, 2010. Available online: https://wwwcdn.imo.org/localresources/en/KnowledgeCentre/IndexofIMOResolutions/MSCResolutions/MSC.307(88).pdf (accessed on 10 May 2023).
36. Maraveas, C.; Tsavdaridis, K.D.; Nadjai, A. Fire Resistance of Unprotected Ultra Shallow Floor Beams (USFB): A Numerical Investigation. *Fire Technol.* **2017**, *53*, 609–627. [CrossRef]
37. Markus, E.S.; Snegirev, A.Y.; Kuznetsov, E.A. *Numerical Simulation of a Fire Using Fire Dynamics*; St. Petersburg Polytech-Press: St. Petersburg, Russia, 2021; p. 175.
38. *ENBS 1-2:2002*; Eurocode 1: Actions on Structures—Part 1-2: General Actions—Actions on Structures Exposed to Fire. British Standards: Chiswick, UK, 1991.
39. ELCUT. Modeling of Two-Dimensional Fields by the Finite Element Method. Available online: https://elcut.ru/free_doc_r.htm (accessed on 6 January 2022).
40. Code of Practice. SP 50.13330.2012. *Thermal Performance of the Buildings.* Available online: https://docs.cntd.ru/document/1200095525 (accessed on 11 January 2022).
41. Paudel, D.; Rinta-Paavola, A.; Mattila, H.P.; Hostikka, S. Multiphysics modelling of stone wool fire resistance. *Fire Technol.* **2021**, *57*, 1283–1312. [CrossRef]
42. *CECS 200:2006*; China Association for Engineering Construction Standardization—Technical Code for Fire Safety of Steel Structures in Buildings. China Planning Publishing House: Beijing, China, 2006.
43. Neumann, N. Study of ISO 19901-3 Building Code Correspondence Factor for Eurocode 3. *Am. Soc. Mech. Eng.* **2013**, *55324*, V02AT02A023.

44. Liu, K.; Wang, Z.; Tang, W.; Zhang, Y.; Wang, G. Experimental and numerical analysis of laterally impacted stiffened plates considering the effect of strain rate. *Ocean. Eng.* **2015**, *99*, 44–54. [CrossRef]
45. Richard, A.; Villavicencio, F. *Response of Ship structural components to Impact Loading*; Lisbon University of Technology: Lisbon, Portugal, 2012.

Journal of
Marine Science and Engineering

MDPI

Article

The Oscillating Behavior of Trawl Codends Including Various Geometric Configurations of Simulated Catch

Feng Zhang [1], Hao Tang [1,2,3,4,5,*], Nyatchouba Nsangue Bruno Thierry [1,*], Wei Liu [1], Qiuyang Sun [1], Meixi Zhu [1], Can Zhang [1], Xuhao Guo [1], Chenxu Shan [1], Liuxiong Xu [1,2,3,4,5] and Fuxiang Hu [6]

1 College of Marine Sciences, Shanghai Ocean University, 999 Huchenghuan Road, Lingang New District, Shanghai 201306, China
2 National Engineering Research Center for Oceanic Fisheries, Shanghai 201306, China
3 Key Laboratory of Oceanic Fisheries Exploration, Ministry of Agriculture and Rural Affairs, Shanghai 201306, China
4 The Key Laboratory of Sustainable Exploitation of Oceanic Fisheries Resources, Shanghai Ocean University, Ministry of Education, Shanghai 201306, China
5 Scientific Observing and Experimental Station of Oceanic Fishery Resources, Ministry of Agriculture and Rural Affairs, Shanghai 201306, China
6 Faculty of Marine Science, Tokyo University of Marine Science and Technology, Minato, Tokyo 108-8477, Japan
* Correspondence: htang@shou.edu.cn (H.T.); nsanguet@yahoo.fr (N.N.B.T.); Tel.: +86-21-6190-0309 (H.T.); +86-186-2182-9173 (N.N.B.T.); Fax: +86-21-6190-0304 (H.T.); +86-186-2182-9173 (N.N.B.T.)

Abstract: Codends are the posterior components of trawl nets that collect the catch and play a crucial role in the selectivity process. Due to the accumulation of catch and the variety of catch types, the quality of catch and trawl selectivity can be negatively impacted. Therefore, this study aims to investigate the effects of various catch configurations on the hydrodynamic characteristics, geometrical profile, and fluttering motions of the codend in a flume tank. A codend structure was designed and tested using various catch configurations, including grooved-type configurations (canvas, green canvas, basketballs) and spherical configurations (table tennis balls filled with water, balloons filled with water, and balls made of twine) in the flume tank. The sea trial data were compared with the flume tank data. The results indicate that there were no significant differences in the codend profiles between the different catch configurations. The drag of the codend with a grooved-type configuration was 13.63% greater than that obtained using a spherical configuration as the catch. The wavelet coefficient obtained from the codend drag revealed that the oscillations of the codend with a grooved-type catch configuration began at a periodicity of 0.07 s and were more intense than that of the codend with the spherical catch configuration. Moreover, these amplitudes increased as the codend flow velocity increased. The wavelet analysis results showed that the dominant frequency of the periodic high-energy coherent structures for the codend drag and codend displacements was detected at a low-frequency. In terms of displacement oscillation characteristics, the table tennis ball filled with water was an approximate substitute for real catch during the sea trial because the difference in wavelet coefficients for the codend displacements in amplitude and the period between the model codend with the table tennis ball filled with water and the full-scale codend was 91% and 89%, respectively. The findings of this study confirm the feasibility of replacing real catch with simulated catch configurations with similar shapes in model testing. They can provide basic scientific data for improving the hydrodynamic characteristics and selectivity of the codend structure.

Keywords: codend; simulated catch; resistance characteristics; oscillation characteristics

Citation: Zhang, F.; Tang, H.; Thierry, N.N.B.; Liu, W.; Sun, Q.; Zhu, M.; Zhang, C.; Guo, X.; Shan, C.; Xu, L.; et al. The Oscillating Behavior of Trawl Codends Including Various Geometric Configurations of Simulated Catch. *J. Mar. Sci. Eng.* **2023**, *11*, 1026. https://doi.org/10.3390/jmse11051026

Academic Editor: Abdellatif Ouahsine

Received: 31 March 2023
Revised: 4 May 2023
Accepted: 9 May 2023
Published: 11 May 2023

1. Introduction

The codend structure, located at the rear of the trawl net, is a critical component of the trawl net that is used to collect the catch during the trawling process, with a drag that accounts for 10% of the total drag of the trawl net [1,2]. Thus, understanding its hydrodynamic behavior and fluttering motions is crucial for determining trawl stability, improving catch performance, and comprehending the juvenile fish escape process in codends. The underwater shape, selectivity, and hydrodynamic characteristics of the codend are highly influenced by catch configurations inside the codend and dependent on the structure's flexible characteristics, mesh size, twine diameter, and fishing operation parameters [3–5].

Indeed, depending on the catch type, catch accumulations, and catch shape, the codend structure bulges and oscillates, resulting in a variation in the codend's front mesh opening, which affects the drag, shape, and stability of the codend [6]. Consequently, the scientific community has been increasingly focused on addressing one of the primary challenges, which is the development and innovation of codend design based on simulated catch configurations that accurately emulate the shape of actual catches within a full-scale codend. This development aims to enhance the selective properties of the codend and promote greater sustainability within the fisheries sector [7,8].

Several studies have reported on experimental studies on the analysis of codend behavior over the last few decades. O'Neill et al. [9] discovered that as catch accumulates, the front mesh of the codend closes and the codend shape bulges, increasing codend motions and codend drag. Therefore, when various types of fish enter in the codend, they are gradually pressed and blocked at the end-part of the codend, causing a change in codend volume, which then affects the trawl's drag and water filtering performance [10]. Priour et al. [11] compared the shape of the T90 mesh and traditional T0 mesh on the codend and discovered that the T90 codend is more conducive to juvenile fish escape, while the T90 codend with catch has a larger geometric configuration bulge and greater drag than the T0 codend [12]. Indeed, the presence of catch inside the trawl net will reduce the mesh opening of the codend, which will limit the flow passage through the codend and lead to the creation of vortex shedding behind the codend and turbulent flow due to the codend wake. Theses turbulent flows will create a pressure gradient on the codend surface that will create intense codend motions and may affect selectivity via the fish response, such as swimming speed and maneuverability [13–15]. Thus, keeping the trawl system in a relatively stable state is required to reduce codend oscillation, which causes skin damage to the fish and a decline in quality [16,17]. As a result of flume tank experiments, some progress has been made in the analysis of the complex interaction between a flexible fluttering trawl net and turbulent flow. According to Priour and Prada [18], the codend oscillations and configurations are primarily determined by the shape of catch and mesh, and as the catch obstructs the water flow passage directly through the mesh, codend oscillations are produced [19]. Druault and Germain [20] investigated the flow in the unsteady wake formed behind the fluttering codend structure and discovered that the codend oscillations can be caused by the local hydrodynamic effects that occur in the wake zone as well as the motion of the fluttering codend. Thierry et al. [21,22] used an electromagnetic current meter to investigate the distribution of the flow field inside and around the bottom trawl net model, revealing that changes in the surrounding unsteady flow produced oscillations in different parts of the trawl net, affecting catch quality and the overall trawl net performance.

Furthermore, various catch configurations have recently been used to simulate catches inside the codend rather than actual catches such as fish to avoid adverse effects on precision instruments and water quality. O'Neill and O'Donoghue [16] used small polystyrene nettings of sufficient size as the simulated catch to measure the geometric shape of the trawl codend and analyze its shape and pressure distribution. In the codend selectivity experiments, O'Neill et al. [23] used water-filled balloons instead of real catch, indicating that the codend oscillations were rather obvious. Bouhoubeiny et al. [13] used Particle Image Velocimetry (PIV) technology to measure the wake effect and flow field around the codend with a hemispherical cap and water-filled balloon instead of the actual catch, and discovered that the oscillating motion of the codend model was very significant. Madsen et al. [24] used the small plastic bags filled with water in the flume tank to analyze the behavior of different codends, revealing that the drag of two-panel T90 codends was lower than that of other codends, while indicating that the T0 codend oscillation effects were greater than those of other codends. Thierry et al. [2] used an elliptical-shaped piece of polyvinyl chloride canvas attached to the posterior of the codend and positioned slightly over the top of the codend, similar to Kim [14], to analyze the motions of the shaking codend in the flume tank. Thierry et al. [2] and Liu et al. [7] used the table tennis balls filled with, on average, 0.045 kg of water each to simulate the fish catch inside the gear and demonstrated that the codend with catch created more instabilities of the codend motion, which can influence both the catch quality and the whole trawl performance compared to the codend without catch.

However, despite the existing work on the subject and the available data, there is no relevant conclusion from the experimental studies evaluating the effect of geometric catch configuration on the hydrodynamic characteristics of codend linked to its fluttering motion and comparing them to real fish in the sea trial in order to solve the problem of selectivity and better understand fish behavior. Furthermore, in previous studies, the different types of catches in the study of codend oscillation were not taken into account in order to reliably improve the drag and understand the catch process and fish selectivity.

This study aims to evaluate the effects of the geometric catch configuration on codend oscillation motions, drag, and the geometrical shape of the codend. Thus, a model codend was designed and assembled based on the full-scale codend used in midwater trawl fisheries and was tested in the flume tank using six kinds of simulated catch configurations, such as grooved and spherical types. The measurements were conducted under five flow velocities, and the measured data were compared with those obtained using the sea trials in order to discuss the relevant information that could be used to improve codend selectivity, reduce energy consumption, and understand fish behavior. The continuous wavelet transform was used to analyze the time-frequency characteristics of the temporal codend drag and fluttering codend motions. The findings of this study will assist researchers in selecting the most appropriate simulated catch types and will provide basic scientific data for optimizing the codend structure.

2. Materials and Methods

2.1. Evaluation of the Full-Scale Prototype and Data Collection

The Antarctic krill trawl net selected and designed for this study was a four-panel midwater trawl net commonly used by Chinese fishing vessel "Long Teng" of China National Fisheries Corp. The main specifications and characteristic parameters of the vessel are listed in Table 1. The circumference of the trawl net at the mouth was 300 m, the trawl length was 132.8 m, the headline length was 55.38 m, and the fishing line length was 54.88 m. The trawl was made of 6.0 mm and 4.0 mm polyethylene (PE) twine diameters, with mesh sizes ranging from 400 mm in the wing (total length of 20 m) and the 1st trawl body section to 200 mm in 2nd to 7th trawl body sections, and 144 mm in the remaining trawl body (total length of 88.8 m) sections and the codend (30 m). The liner was made of polyamide (PA) with a mesh size of 16 mm and twine diameter of 2.0 mm. The vertical opening was maintained by floats distributed on the headline and providing a total buoyancy of 21.462 kN as well as 300 kg chains on the fishing line and two 300 kg heavy bobs attached to each wing-end (Figure 1).

Table 1. Main dimension of Antarctic krill trawler "Long Teng".

Total Length (m)	Main Engine Power (kW)	Total Tonnage (t)	Max Speed (kn)	Moulded Depth (m)	Moudeld Breadth (m)
120.7	5296	7765	13	12	19

Full-scale observation of the Antarctic krill trawl in action was carried out in the South Orkney Islands, the Antarctic Peninsula, and the South Shetland Islands (60° S; 45° W and 64° S; 59° W, respectively) during the fishing operation conducted between February and March of 2020. The towing speeds were recorded based on the data displayed by the vessel's GPS radar and ranged from 2 to 3 knots. The data related to the codend depths (the distance from the sea surface) were obtained using DR (Dr-1050 bathymetric meter, Canada) hydroacoustic trawl monitoring sensors attached to the fishing gear on the upper middle point of the codend. According to the manufacturer's specifications, this instrument had a measuring range of 0–750 m and a specified accuracy of ±0.05%. The total operating time was longer and varied between 254 and 7254 s according to the Antarctic krill school size, which was convenient for data recording and analysis accuracy setting. The data were evaluated every 6 s to obtain information about the entire process, such as towing speed and codend depth.

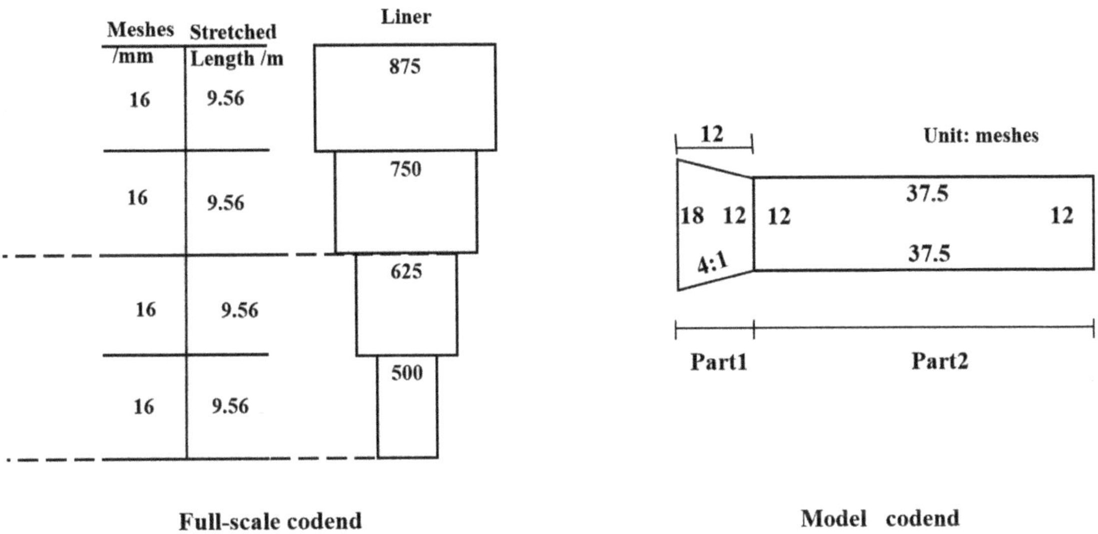

Figure 1. The schematic diagram of the full-scale codend and the model codend structure.

2.2. Codend Design

The model trawl codends were selected and designed in model scale based on modified Tauti's law, with particular attention to the Antarctic fisheries. This codend model was designed based on a 1:20 scale ($\lambda = 10$) in the length scale of the codend model, a 1:3.6 small-scale ($\lambda' = 3.6$) in mesh size and twine diameter, and a 1:10.54 time ratio ($\tau = 10.54$) using Tauti's law [25–27]. The codend was designed by assembling four pieces of netting using a mesh size of 40 mm and a twine diameter of 1.11 mm using polyethylene (PE) twine materials with diamond-shaped meshes. The codend's overall length was 1.5 m. Each piece of netting was joined by a codend with a circumference of 72 meshes and an extension with a circumference of 48 meshes (Table 2 and Figure 1).

Table 2. Specifications and parameters of experiment codend.

	Twine Materials	Mesh Size (mm)	Knot Direction	Codend Length (mm)	Cutting Ratio
Extension	PE	40	TO	480	4:1(/NBNBN/3)
codend	PE	40	TO	1500	AN

2.3. Design of the Simulated Catch Configurations

The physical fishing net can accommodate catch with a diameter of approximately 2.4 m. For the experiment, simulated catch with a diameter of D = 24 cm was used to maintain consistency in size. The size of the catch determines the geometric configuration of the codend, which can be divided into different shapes such as grooves, semicircles, and spheres. Based on the simulated catch configurations, this study divided them into two categories: grooved simulated catch (canvas, green canvas, and basketball) and spherical simulated catch (table tennis ball filled with water, balloons filled with water, and balls made of twine). The simulated catches had different water filtration properties. The weights of the two types of simulated catch were quite different in air, but only slightly different in water. Table 3 shows the specific weights. While the upstream surface of the grooved simulated catch was concave and that of the spherical simulated catch was convex, both types ensured that the simulated catch in water had the same upstream surface area.

Table 3. Simulated catch weights in air and water.

The Type of The Simulated Catch	Grooved Type Simulated Catch			Spheroidal Simulated Catch		
	Canvas	Green Canvas	Basketball	Table Tennis Ball	A Ball Made of Twine	Balloon
Material (cotton content)	Cotton (85–95%)	Cotton (81–90%)	polyurethane	Carbon nitride	Polyethylene	Latex
Thickness (mm)	0.4	0.9	1.5			
Weight in air (N)	0.25	0.51	1.04	10.6	6.03	14.5
Weight in water(N)	0.16	0.30	0.57	2.7	0.92	2.1

2.4. Experimental Setup and Procedures

The physical experiments were carried out in the flume tank at the National Center for Ocean Fishing Engineering Technology Research of the Shanghai Ocean University (NERCOF). The tank's test section was 15.0 m in length, 3.5 m wide, and 2.3 m deep, holding 530 tons of freshwater. The flume tank was a horizontal and circular water channel, and the flow was driven by four contra-rotating impellers using constant-speed hydraulic delivery pumps with a rated power of 132 kW, a maximum water depth of 2.0 m, and a flow velocity range of 0–1.5 m/s. A side-viewing window on one side of the flume tank allowed users to observe and record a video of the codend's behavior during testing. The codend behavior was recorded using a camera with a frequency of 59 Hz per frame image and a resolution of 1920 × 1080 pixels^2. During the experiments, the water temperature in the flume tank was kept constant at 17.6–18.4 °C. Figure 2 depicts the specific settings of each test's instruments and equipment. To accurately measure hydrodynamic forces, the codend model was attached to a circular rigid frame with a diameter of 36 cm and submerged in water to a depth of 10 cm. A current meter was installed approximately 2.0 m upstream of the codend model to detect the flow velocity. The hydrodynamic force signals were measured using a six-component load cell with a capacity of 5 kgf each and a specified accuracy of 2%. As measured using the load cell, these hydrodynamic force signals were amplified using a dynamic strain amplifier (DPM-6H). Afterwards, these signals were sent to an A/D converter and thereafter to a computer. During the experiment, we recorded the mean of 240 data points obtained at a frequency of 4 Hz over a period of 60 s.

The drag force of the rigid frame was first measured before conducting experimental measurements on different codends. Firstly, the drag force of the rigid frame was measured at five different flow velocities. Secondly, the codend opening was wrapped around the rigid frame, and the drag force of the codend and frame was measured in the empty stage using the same method. Finally, the combined drag force of the codend, frame, and simulated catch was measured after adding six different catch configurations. The measurements were taken using flow velocities of 0.5, 0.6, 0.7, 0.8, and 0.9 m/s.

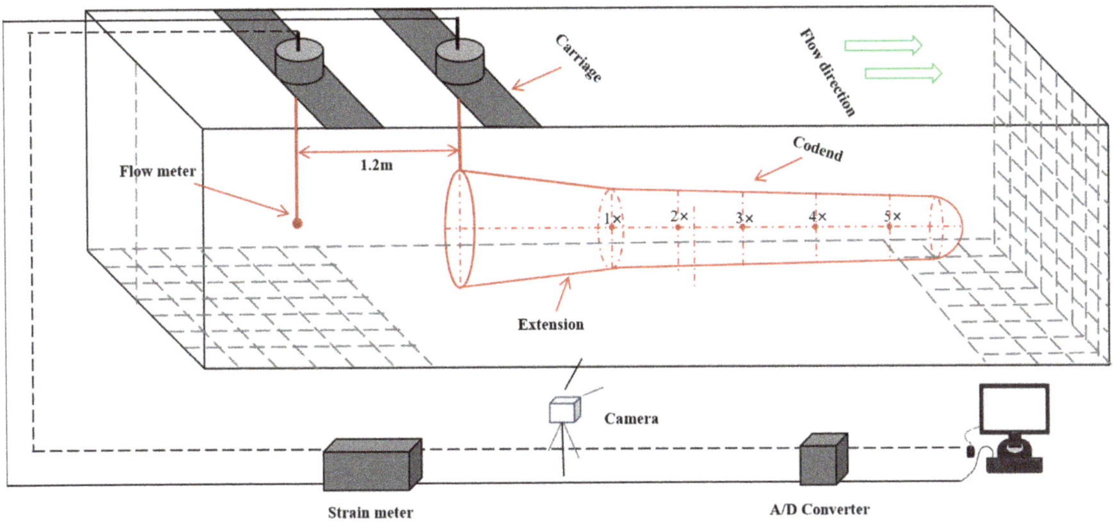

Figure 2. Flume tank and equipment.

2.5. Data Extraction and Methods

2.5.1. Data Processing of the Shape of Codend

During the experiments, a series of videos describing the codend fluttering motions were recorded from the front view of the flume tank observation port using a video camera that was fixed in position and zoomed in and out (Figure 3). Video cameras were used to capture the trawl actions and net shape geometry. From the recorded video footage, a series of images separated by 0.25 s were selected and imported into graph digitizing software to extract the coordinates of the model net's characteristic points based on a plane-coordinate system.

Following that, the coordinates representing the codend behaviors were interpolated, allowing the temporal motions of the different codends to be determined. Initially, a series of 240 images were acquired from the recorded video footage at a frame rate of 4 Hz. These images were used to extract the coordinates of the distinctive points of the codend model with different catch configurations based on a plane-coordinate system. A reference bar was used to calibrate the measurements and reduce the impact of camera lens stresses, water refraction, and parallax on the extraction of these coordinates.

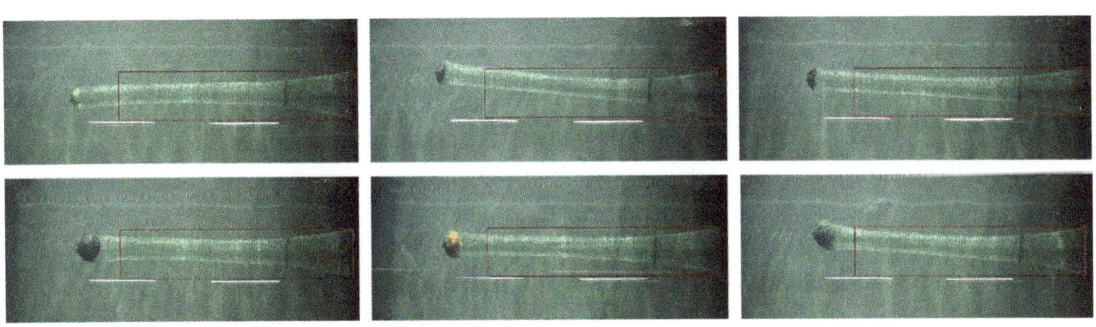

Figure 3. The camera footages of codend (**lower**: Spheroidal; **upper**: Grooved type).

2.5.2. Pearson Correlation Test

In this study, we utilized the Pearson correlation test to investigate the relationship between the oscillation period and amplitude of the actual codend and those of the model codend containing varied simulated catch. The Pearson product-moment correlation coefficient can be calculated using Equation (1):

$$\rho_{XY} = \frac{Cov(X,Y)}{\sqrt{D(X)}\sqrt{D(Y)}} = \frac{E\{[X-E(X)][Y-E(Y)]\}}{\sqrt{D(X)}\sqrt{D(Y)}} \quad (1)$$

where E is the mean, D is the variance, and are the standard deviation of variables X and Y, and $E\{[X\text{-}E(X)]\,[Y-E(Y)]\}$ is the covariance of variables X and Y, denoted by $Cov(X, Y)$, i.e., $Cov(X, Y) = E\{[X-E(X)]\,[Y-E(Y)]\}$.

2.5.3. Wavelet Analysis

The wavelet transform is an effective tool for analyzing non-periodic signals. It is critical in the context of codend motion analysis to detect coherent structures emerging from fluttering motions and to associate unsteady turbulent flow. In this study, the wavelet analysis method was used to determine the time-periodicity of the codend oscillation and codend drag. The wavelet analysis was carried out using MathWorks' Matlab 2019B software package and its wavelet toolbox. The following is the analysis procedure:

The wavelet function is the foundation of wavelet analysis. It refers to a class of oscillating functions that can quickly decay to zero, such as the wavelet function $\psi(t)\in L^2(R)$, and satisfy:

$$\int_{-\infty}^{+\infty} \psi(t)dt = 0 \quad (2)$$

where $\psi(t)$ is the basic wavelet function, which can be expanded and translated on the time axis to form a family of function systems:

$$\psi_{a,b}(t) = |a|^{-1/2}\psi\left(\frac{t-b}{a}\right) \text{ in } a,b \in R,\ a \neq 0 \quad (3)$$

where a and b are the scale and the position parameter, respectively. In Morlet mother wavelet, a is roughly equal to $1/f$ or period in numerical value.

In the sub-wavelet given by Equation (3), for a given energy-limited signal $f(t) \in L^2(R)$, its wavelet transform is:

$$W_f(a,b) = |a|^{-1/2}\int_R (t)\overline{\psi}\left(\frac{t-b}{a}\right)dt \quad (4)$$

where $W_f\,(a,b)$ is the wavelet transform coefficient of a signal or square integrable function f(t) at the scale a and the location b, and $\overline{\psi}(\frac{x-b}{a})$ is the complex conjugate function of $\psi(\frac{x-b}{a})$.

The Morlet wavelet was chosen as the complex wavelet in this study, and the phase difference between its real and imaginary parts is $\pi/2$. The modulus represents the number of a certain scale component, and the phase can be used to study the singularity and real-time periodicity of the signal.

3. Results

3.1. Morphological Changes of Codend

As illustrated in Figure 4, the shape of the codend structure tends to tighten as flow velocity increases. The results showed that the outlines profiles of codends with different simulated catch were essentially similar, and the horizontal length of the codend structure increased as flow velocity increased. Under the same flow velocity conditions, the horizontal length increase in the codend containing the canvas is 0.013% greater than the original codend length, while the horizontal length increase in the codend containing the ball made of twine is 0.009% greater than the original codend length. There are some differences between the codend filled with grooves and the spherical simulated catch as flow velocity increases. The end bulge of the codend with grooved catch configuration was approximately 23.54 cm, and the end bulge of the codend with spherical catch configuration was approximately 25.13 cm.

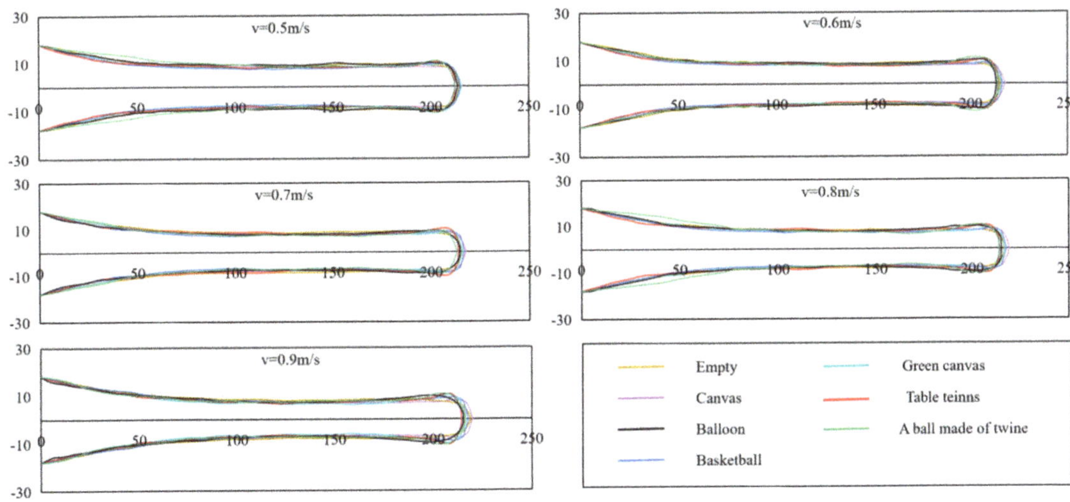

Figure 4. The shape transformation for codend with simulated catch at different flow velocity.

3.2. *Morlet Wavelet Transform of Codend Resistance of Different Types of Simulated Catch*

In this study, a flow velocity of 0.7 m/s (corresponding to the actual towing speed of 2.4 knots at sea) was chosen as the basic flow velocity of the codend model. According to Figure 5 the drag force of the codend with a grooved catch configuration was about 22.17 N on average, while that of a spherical catch configuration was about 19.20 N. Moreover, the drag force of the codend filled with green canvas was the greatest (22.7 N), while the drag force of codend with balloons filled with water was the lowest (17.9 N). Wavelet coefficients of codend drag with the grooved catch configuration mainly oscillated intensively between 30 and 60 s and were observed at the frequency of 1.3–2.2 Hz for canvas and green canvas, except for the codend with the basketball, in which the high wavelet spectrum coefficient were observed at a high frequency of 2.9–4 Hz at t = 38–42 s. For the case of codends with a spherical catch configuration, the drag oscillation was slower, and the larger wavelet spectrum was obtained at a frequency varying from 1.84 to 2.87 Hz at t = 10–40 s, 2.14 to 2.87 Hz at t = 35–50 s, and 1.04 to 2.87 Hz at t = 0–30 s for the oscillation of a table tennis ball filled with water, balloons filled with water, and balls made of twine, respectively. It was observed that the codend containing the basketball oscillated the most violently, while the codend containing the table tennis ball filled with water oscillated evenly throughout the process. The wavelet coefficient of the codend drag with grooved catch configuration began to oscillate at the frequency of 0.87 Hz first, then oscillated more intensively than the codend containing the spherical catch configuration.

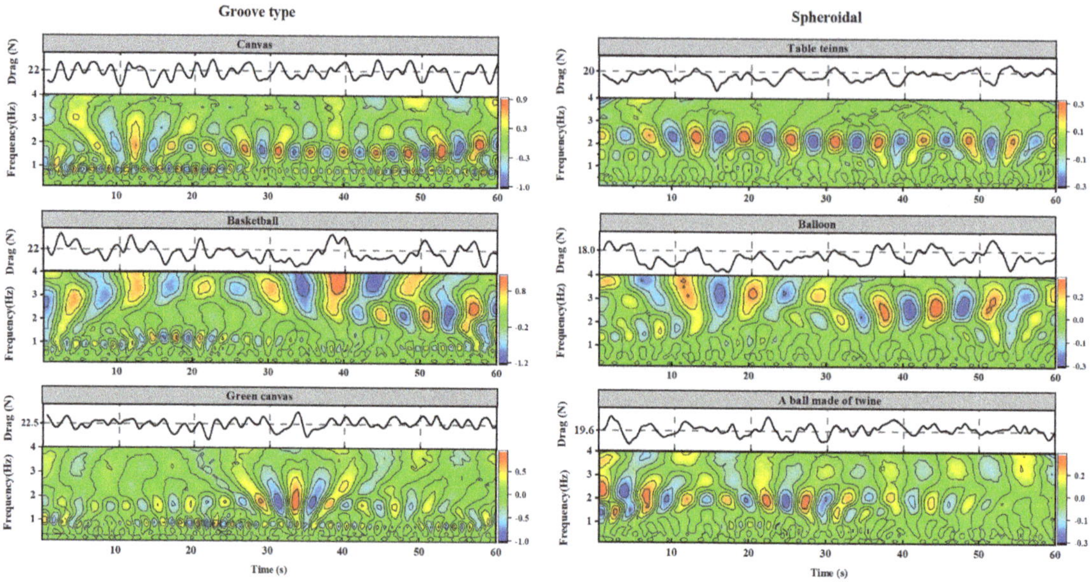

Figure 5. Morlet wavelet transform of codend drag with different simulated catch.

3.3. The Oscillation Characteristic of the Codend Drag Force Based on Main Period Morlet Wavelet Transform

Figure 6 depicts the oscillation of the main periodicity wavelet coefficients of codend drag with various catch configurations plotted according to the time scale of the codend drag oscillation. On average, the oscillation periodicity of wavelet coefficients of the codend drag using grooved catch configuration was about 2.49 s, while that of the spherical catch configuration was about 2.83 s, indicating that the oscillation frequency of the codend with grooved catch configuration was faster. It was observed that the codend with green canvas has the fastest oscillation frequency (0.51 Hz) compared to other catch types. The average amplitude of the wavelet coefficient of the codend drag with grooved catch configuration was approximately 0.98, while the average amplitude of the wavelet coefficient of the codend drag with spherical catch configuration was approximately 0.4. The average amplitude of the wavelet coefficient of the codend drag filled with the grooved catch configuration was approximately 2.4 times greater than that of the codend drag with spherical catch configuration, indicating that the codend filled with the grooved catch configuration is more oscillatory. In addition, the largest amplitude of the wavelet coefficient (1.3) was observed on the codend drag-filled basketball (Table 4).

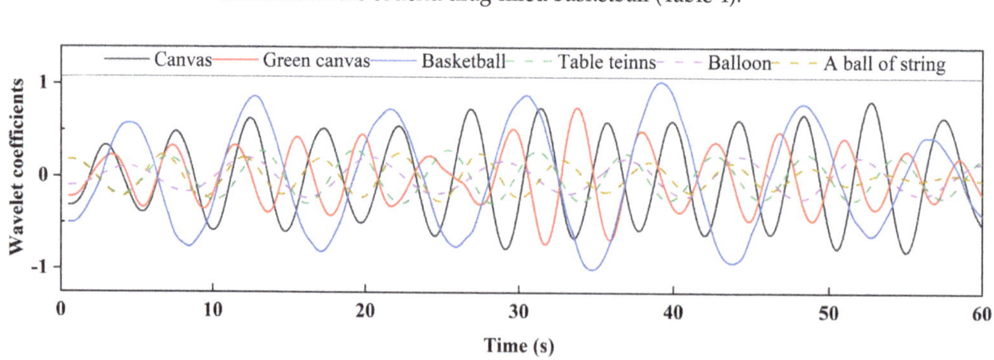

Figure 6. Principal period wavelet coefficient oscillation of codend drag with simulated catch.

Table 4. Oscillation period amplitude of wavelet coefficient of codend drag with simulating catch.

The Type of The Simulated Catch	Grooved Type Simulated Catch			Spheroidal Simulated Catch		
	Canvas	Basketball	Green Canvas	Table Tennis Ball	Balloon	A Ball Made of Twine
Frequency (Hz)	0.48	0.29	0.51	0.37	0.29	0.42
Amplitude	0.81	1.3	0.85	0.51	0.37	0.34

3.4. Morlet Wavelet Transform Characteristics of Spatial Displacement of Codend with Different Types of Simulated Catch

Figure 7 depicts the Morlet wavelet transform performed on the transverse (x-direction) and longitudinal (z-direction) displacements of the codend with various simulated catch configurations. According to the findings, the average transverse length of the codend with grooved catch configuration was approximately 217.60 cm, which was greater than that of the codend with spherical catch configuration (~213.67 cm). The amplitude of transverse displacement for the codend with all the simulated catch configurations varied little, ranging from −2 to 2 cm. The wavelet spectrum coefficient of all the codend oscillations in the transverse direction varied slightly. During time intervals of 0–15 s and 45–60 s, dominant periodic positive and negative peaks were observed at frequencies ranging from 2.06 to 4 Hz, indicating the presence of strong periodic oscillations. This indicated that the simulated catch configurations had a low effect on the transverse displacement of the codend.

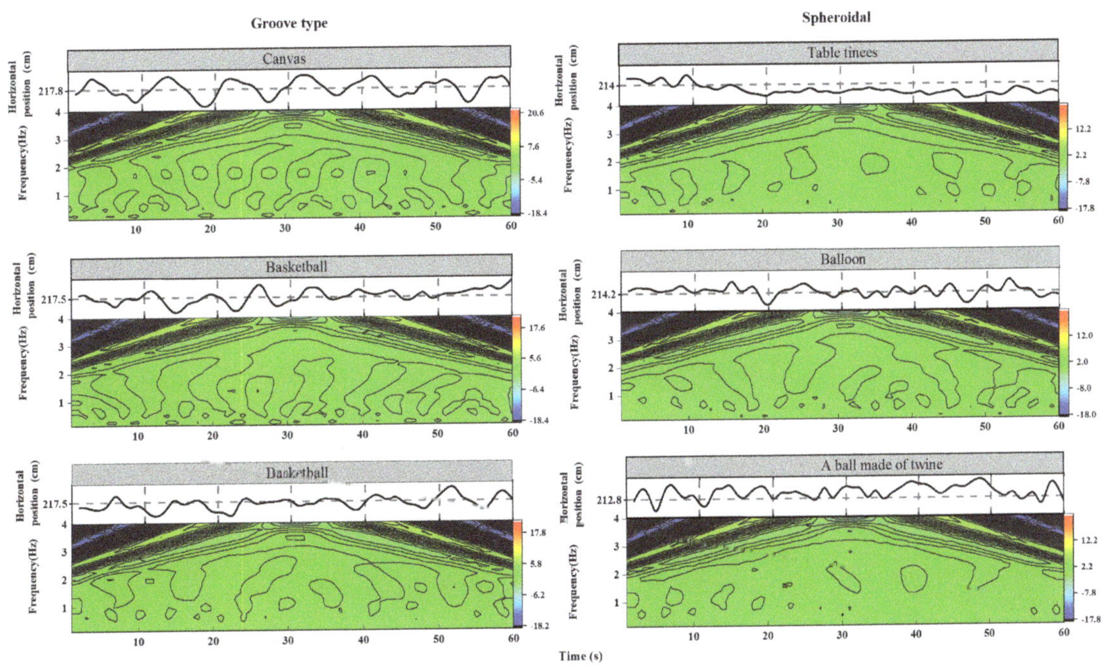

Figure 7. Morlet wavelet transform of different simulated catch codend displacements in x-direction.

Figure 8 displays the time evolution of the codend oscillations in the z-direction. The codend motions in the longitudinal direction were influenced by the simulated catch configuration and characterized by quasi-periodic oscillations. The oscillation amplitude in the z-direction of the codend with a grooved catch configuration was greater than that of the codend with a spherical catch configuration. Additionally, the amplitude of the longitudinal oscillations of the codend with green canvas was much higher compared to those of the codend with other catch configurations.

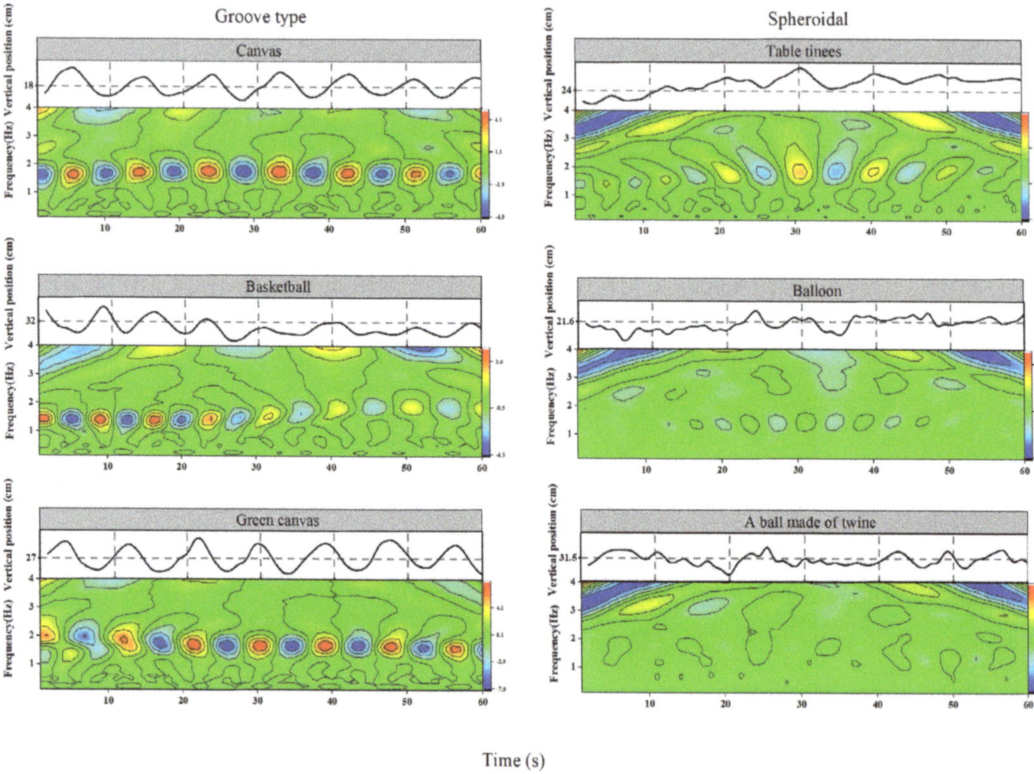

Figure 8. Morlet wavelet transform of different simulated catch codend displacements in z-direction.

The wavelet coefficient of the longitudinal displacement of the codend with a grooved catch configuration began to oscillate at approximately 0.1 s, and its oscillations were relatively intense. However, the wavelet spectrum energy of the codends with spherical catch configuration in the z-direction was lower compared to those obtained on the codend with grooved catch configuration. A higher wavelet spectrum coefficient of the codend oscillations in the z-direction was obtained at a frequency of 1.4–2 Hz for the catch configurations. Furthermore, it was observed that the codend containing the table tennis ball filled with water oscillated more violently, while the codend containing the filled-twine ball oscillated more smoothly.

3.5. The Oscillation Characteristic of the Codend Spatial Displacement Based on Main Period Morlet Wavelet Transform

Figure 9 depicts the oscillation of the main periodicity wavelet coefficients for codend displacement in both directions for all the simulated catch configurations. The amplitudes of the wavelet coefficients for codend displacement in the x-direction were relatively small and basically remained in the range of −1–1. Among them, the codend with the green canvas had the greatest amplitude (1.29), while the codend with the table tennis ball had the least amplitude (0.35). The average amplitude of the wavelet coefficients of the codend with a grooved catch configuration was about 0.91, which is about 2.86 times that of the codends with a spherical catch (~0.32). The average amplitudes of the wavelet coefficients for the codend with grooved catch configuration was ~8.79 in the z-direction, which was 7.57 times greater than that of the codends with spherical catch configuration (Table 5).

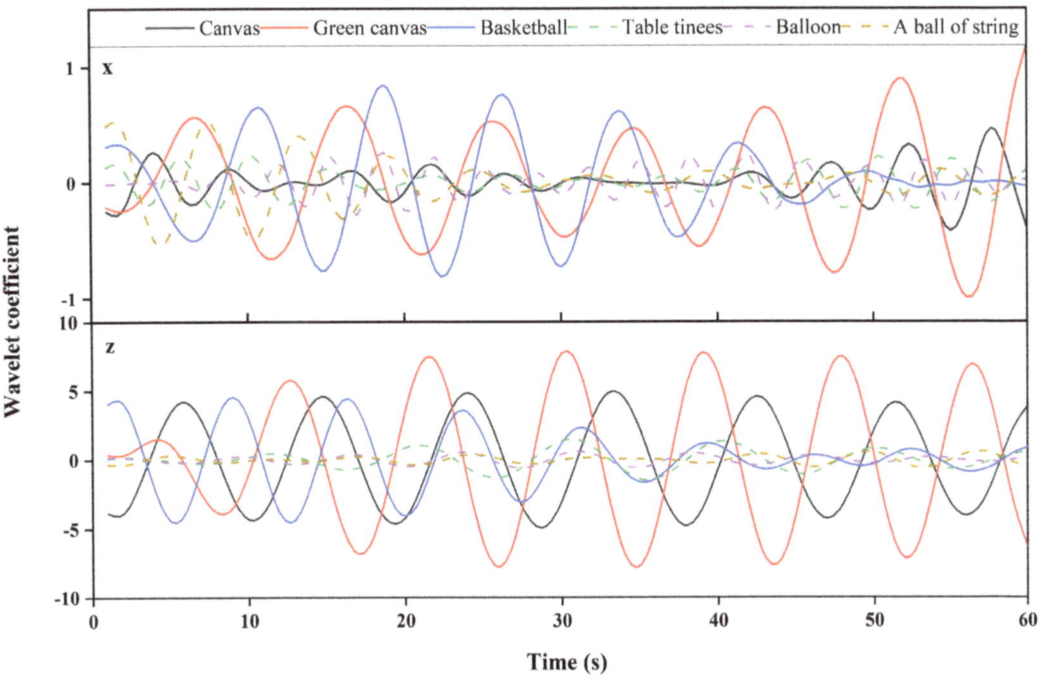

Figure 9. Oscillations of principal period wavelet coefficients of simulated catch displacement of codend (**upper**: x-direction; **lower**: z-direction).

Table 5. Oscillation period amplitude of wavelet coefficient of simulated catch codend displacement.

The Type of The Simulated Catch		Grooved Type Simulated Catch			Spheroidal Simulated Catch		
		Canvas	Basketball	Green Canvas	Table Tennis Ball	Balloon	A Ball Made of Twine
X	Frequency (Hz)	0.74	0.61	0.53	0.9	1.25	0.67
	Amplitude	0.35	1.08	1.29	0.26	0.32	0.37
Z	Frequency (Hz)	0.52	0.54	0.61	0.5	0.63	0.61
	Amplitude	8.91	4.54	13.72	1.94	0.78	0.69

3.6. Longitudinal Displacement of Codend in Sea Trials

During fishing operations, the codend depth decreased as flow velocity increased (Figure 10). Indeed, at the lower flow velocity (0.5 m/s), the codend depth was the greatest (147.59 m), while at the higher flow velocity (0.9 m/s), it was the smallest (38.58 m). At $0.5 < V \leq 0.7$ m/s, the depth of the codend changed greatly; this depth decreased by 35.5% each time the flow velocity increased by 0.1 m/s.

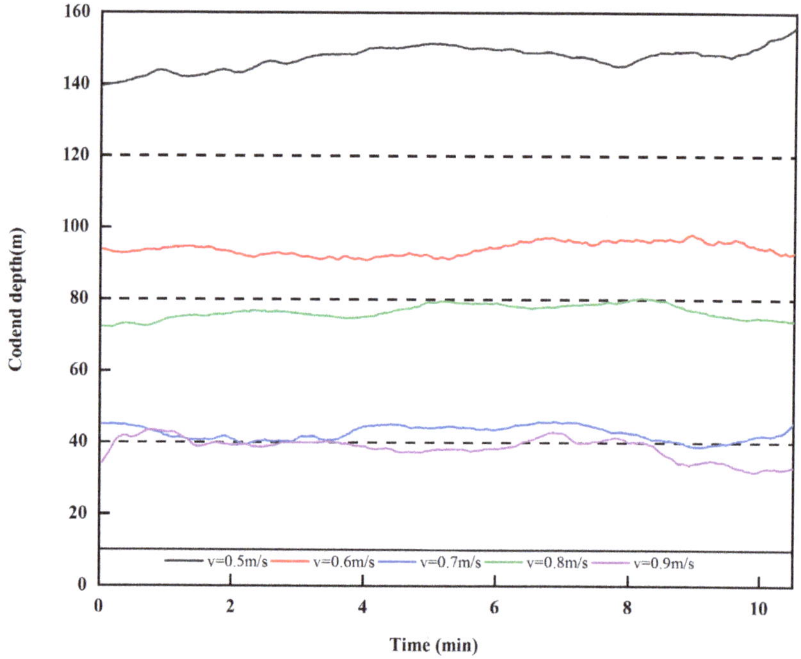

Figure 10. Longitudinal displacement of codend measured at sea.

3.7. The Oscillation Characteristic of the Codend Longitudinal (z-Direction) Displacement Based on Morlet Wavelet Transform in Sea Trials

The higher wavelet spectrum coefficient of the codend oscillations in the depth-direction was obtained at the frequency of 0.13–0.17 Hz at t = 25–30 s and 5–10 s for v = 0.5, 0.6, and 0.9 m/s. However, when v = 0.7 and 0.8 m/s, two large events were visible on the codend depth oscillation. For the first even, a higher wavelet coefficient was observed at a frequency of 0.08–0.09 Hz at the times of 0.1 s, 5 s, and 13 s during the fishing operation. In addition, for the second even, a higher wavelet coefficient was observed at the frequency of 0.13–0.17 Hz at t = 23–28 s (Figure 11). These findings imply that flow velocity has an impact on the codend depth oscillation and that it has evolved into a higher periodicity structure with much more energy than small-scale structures. The wavelet analysis results revealed that alternate strong positive and negative peaks of wavelet coefficients appeared. The wavelet coefficients of the depth displacement of the codend oscillated most intensely when the flow velocity was v = 0.7 m/s, while at v = 0.8 m/s, the wavelet coefficients of the depth displacement of the codend oscillated slowly.

As shown in Figure 12, at the flow velocity of 0.5 m/s, the oscillation frequency of wavelet coefficients of the longitudinal displacement of the codend was the slowest, and it increased as the flow velocity increased. The oscillation frequency was the fastest at v = 0.8 and 0.9 m/s. When the flow velocity was $v \leq 0.7$ m/s, the amplitude of the measured longitudinal wavelet coefficients increased as the flow velocity increased, and when $v > 0.7$ m/s, it decreased as the flow velocity increased (Table 6).

Table 6. Oscillation period amplitude of wavelet coefficient measured longitudinal (z-direction) displacement of codend.

Flow Velocity (m/s)		0.5	0.6	0.7	0.8	0.9
Measured	Frequency (Hz)	0.03	0.04	0.04	0.05	0.05
	Amplitude	1.72	1.85	2.33	0.81	2.02

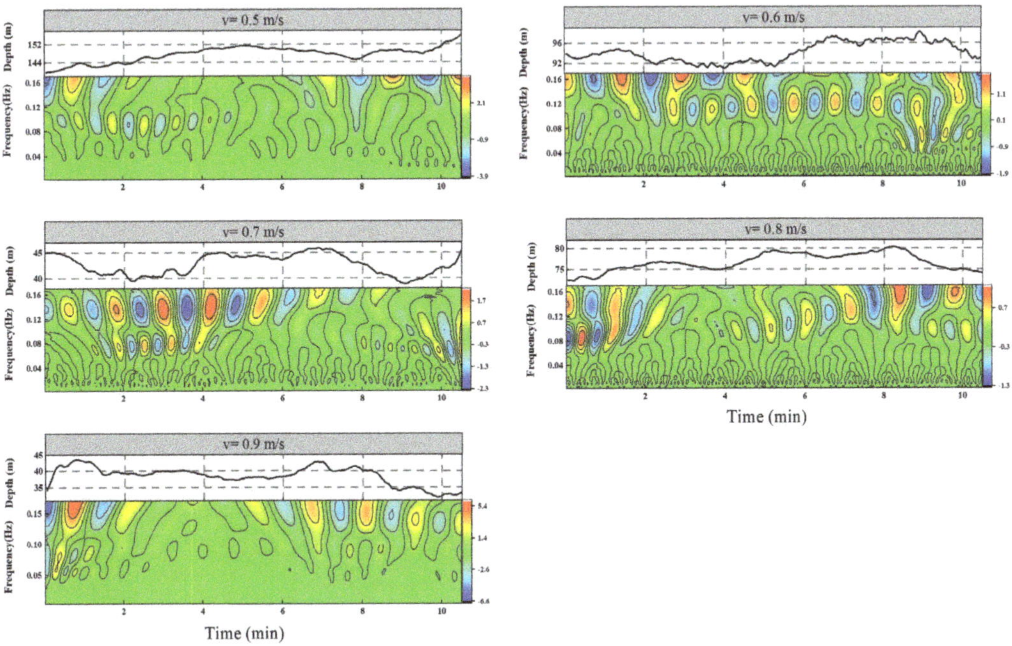

Figure 11. Morlet wavelet transform of longitudinal (z-direction) displacement of codend measured at sea.

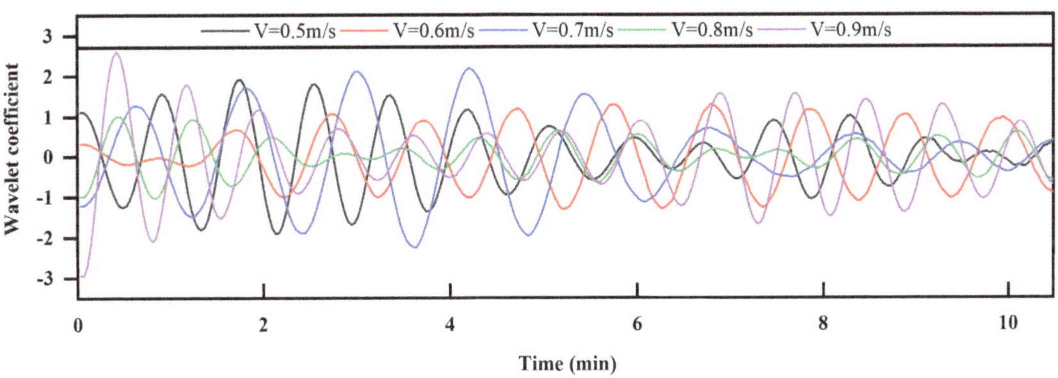

Figure 12. Oscillations of principal period wavelet coefficients of measured longitudinal (z-direction) displacement of codend.

3.8. Contrast and Analysis of Measured and Test Results at Sea

Both the sea measurement and flume test flow velocities range from 0.5 to 0.9 m/s, and there is an obvious up-and-down vibration in the longitudinal position at the end of various types of simulated catch codend structures (Table 7). Figure 13 shows the analysis of displacement fluctuation using the wavelet transform, and the Pearson correlation coefficient is used to test the correlation between the analysis results and the actual codend oscillation. Based on the test results, the oscillation periods of the codend wavelet coefficients for both measured and simulated catches are highly correlated (50%), strongly correlated (16.7%), and extremely weak. Furthermore, the oscillation periods of the codends with canvas, green canvas, and table tennis balls are highly correlated, with correlation coefficients of 90%, 90%, and 89%, respectively.

However, the amplitudes of the wavelet coefficients of the codend measured at sea and simulated catches measured in the flume tank were extremely strongly correlated in 33.3%, and extremely weakly correlated in 50%. The amplitudes of the codends containing the table tennis ball filled with water and balloons filled with water are highly correlated among them, with correlation coefficients of 91% and 89%, respectively.

Table 7. Oscillation period amplitudes of measured and tested codend wave coefficients.

	Flow Velocity (m/s)	Measured	Canvas	Green Canvas	Basketball	Tennis Ball	Balloon	A Ball Made of Twine
Frequency (Hz)	0.5	0.03	0.43	0.42	0.45	0.43	0.48	0.56
	0.6	0.04	0.45	0.43	0.47	0.48	0.56	0.60
	0.7	0.04	0.52	0.54	0.61	0.50	0.63	0.61
	0.8	0.05	0.57	0.63	0.41	0.60	0.33	0.63
	0.9	0.05	0.63	0.63	0.32	0.58	0.49	0.72
Amplitude	0.5	1.72	13.03	15.41	5.04	1.32	0.9	0.53
	0.6	1.85	10.59	7.58	6.84	1.4	0.71	1.13
	0.7	2.33	8.91	13.72	4.54	1.94	0.78	0.69
	0.8	0.81	8.67	11.44	1.78	0.79	2.69	0.77
	0.9	2.02	9.44	11.36	4.26	2.06	1.21	0.93

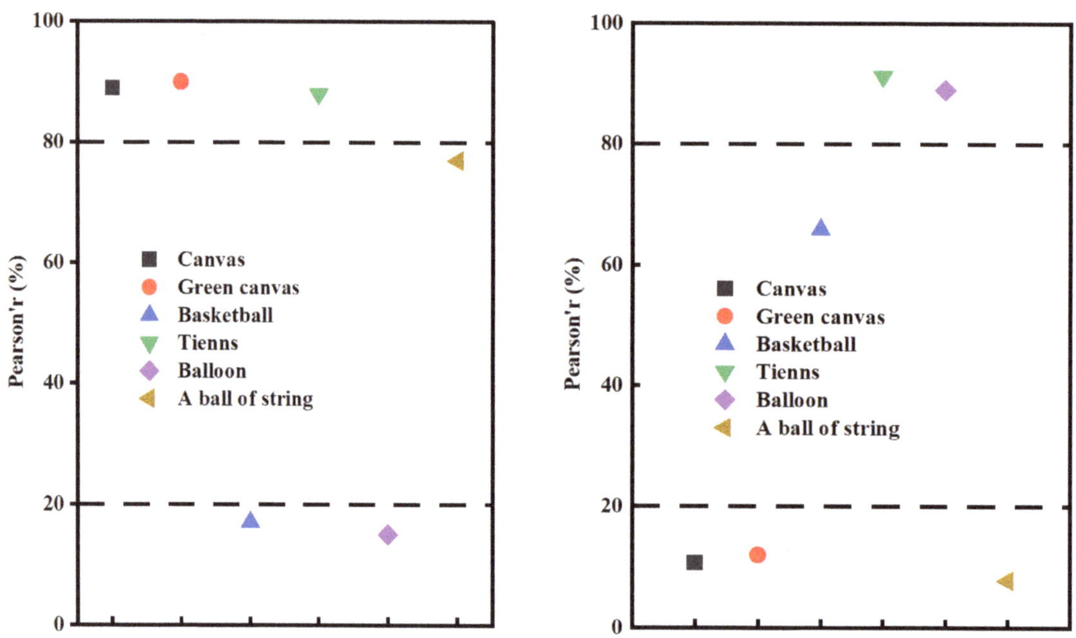

Figure 13. Correlation test between measured and tested codend oscillation frequency and amplitude.

4. Discussions

4.1. Effects of Simulated Catches on the Morphology of Codend

The main factors influencing the motion of the flexible codend and its shape were the towing speed, ship motion caused by wave or wind, and the catch inside. As a result, it was discovered in this study and that of O'Neill et al. [23] that due to catch accumulation inside the codend and under flow velocity, the codend front meshes closed and the tail swelled. This was due to the fact that the more the catches accumulated in the codend, the more the length of the codend increased and the width narrowed, which was consistent with the finding of Meyler et al. [28].

However, Cui et al. [29] observed changes in codend morphology using underwater cameras, and the results revealed that as towing speed increased, the oncoming surface curved into an arc in the horizontal direction. The bending degree of the panel increased as the towing speed increased. This is consistent with the findings of this study, which showed that the horizontal length of codends with each catching configuration increased as flow velocity increased. Furthermore, the findings of this study showed that the contour of the codend with different simulated catch configurations were essentially the same, which contradicts the conclusion by Hermann et al. [30] that a change in catch types would result in a change in the codend shape, providing a spherical bottom with a continuous accumulation of catches. However, due to the different materials used in the catch configurations as mentioned above, and the water filterability through the codend, as flow velocity increased, the channel between the codend narrowed and the horizontal length increased, with no discernible length difference, indicating that the simulated catch configuration had little influence on the length of the codend.

4.2. Effect of Simulated Catch Configuration on the Drag Force and Codend Motions

It was demonstrated in this study that the main factors that influenced the codend drag were the flow velocity and catch factors such as catch configurations, in addition to the wave and wing during fishing operations at sea. However, the change in codend hydrodynamic characteristics, as an important component of the trawl system, determines the hydrodynamic characteristics and stability of the trawl system [31]. Nevertheless, it was demonstrated in this study that codend drag with different simulated catch configurations increased as flow velocity increased. This was because the presence of the catch in the codend caused reflux, resulting in a pressure difference, and the greater the flow velocity, the greater the drag force. It was found in this study that drag forces obtained on the codend with grooved catch configurations was greater compared to those obtained using the spherical catch configuration. This difference in drag was attributed to the material density of the catch types. Another plausible explanation is that the grooved catch configuration made the codend more unstable, unlike the spherical catch configuration, during the flume tank experiment. This instability led to a decrease in water flow velocity and an increase in pressure on the codend surface. This increase in pressure allows the codend with grooved catch configurations to oscillate more than codend with spherical catch configuration, which decreases its mesh opening, limits the flow passage, and thus increases its drag. The wavelet results on the codend drag showed that the wallet coefficient spectrum peak shifted from low frequency (large-scale structures) to high frequency (small-scale structures). This was more pronounced on the codend with grooved catch and was in agreement with finding of Liu et al. [7] and Thierry et al. [2].

The codend oscillated in both the longitudinal and transverse directions during the experiments, which is because the codend attached to a trawl oscillated in both the longitudinal and vertical directions in response to the dynamic interaction of a fishing vessel under trawling conditions [9,23]. These codend oscillations occurred because the presence of the catch inside the codend reduced the mesh opening of the front part of the codend, limiting the flow through the codend and resulting in vortex shedding. This vortex shedding created vertical pressure on the codend, resulting in codend oscillations. However, the results of this study revealed that the longitudinal displacement of the codend filled with various simulated catch configurations changed little, and the spatial wavelet positive and negative oscillations were not visible in the time-frequency domain. This is because of the catch and the effect of water flow; the mesh closure may have been close to the maximum. Thus, it was found that the codend oscillations and codend morphology changed with the catch size and mesh size. However, it was found that the codend with grooved catch configurations oscillated with greater amplitude compared to the codend with spherical catch configuration. The reason for this phenomenon can be explained by the fact that the grooved catch configurations weighed less in comparison to the spherical catch configuration. After comparing the wavelet results of the codend oscillations obtained during the sea trial and the motions of the model codend with the spherical catch configuration, it was observed that there was a similarity between the two. However, the codend oscillations during the sea trial were influenced by factors such as warp length and towing speed. Therefore, it is reasonable to believe that the spherical catch configuration would be the better choice to replace the real catches during the model experiment.

The morphological features of the catch inside the codend can impact its selectivity. Due to the limited space and the limited stamina of the fish inside the codend, it is commonly believed that most of the fish entering the codend are already exhausted. During trawling operations, the codend often undergoes oscillation, inducing panic among fish when they collide with the netting and disturbing the balance of the school, leading to collisions between individuals and ultimately compromising the quality of the catch. Moreover, the oscillation of the codend can obstruct the escape of smaller fish through the mesh, causing them to accumulate in the back of the codend and thus affecting the selectivity of the gear. During fishing operations, vessels require a significant amount of energy, and most modern fishing vessels rely on fossil fuels. In general, resistance in fishing gear can lead to increased energy consumption. Optimizing fishing gear can reduce gear resistance, leading to increased profitability for the fishery and promoting maximum ecological sustainability.

4.3. Simulated Catch Selection

Due to difficulties in controlling the actual catch during model tests, which can significantly affect the precision of flume instruments and water quality, Meyler et al. [28], O'Neill et al. [23], Bouhoubeiny et al. [11], and other researchers utilized simulated catch. However, it is important to acknowledge that there were significant differences between the flume model test and the sea test. While the sea test was conducted on the trawl system under varying and complex sea and operation conditions, the model test had its limitations. The complexity of the sea conditions and operation environment also had a significant impact on the codend, and the actual catch varied from the simulated catch used in the model tests. The experimental model used in this study is based on the way the actual catch accumulates in the sea. When the real catch is smaller, it resembles a grooved shape attached to the inner wall of the codend meshes, but as the catch continues to accumulate, it eventually assumes a hemispherical or spherical shape. The simulated catch in the codend model can be categorized into two types based on its contour: the grooved catch configuration (canvas, green canvas, and basketball) and the spherical catch configuration (table tennis ball filled with water, balloons filled with water, and balls made of twine). During the experiment, a simulated catch configuration was used in the form of a ball made of polyethylene twine instead of real catch. The results showed that the polyethylene twine ball had no significant effect on the drag and displacement oscillation of the codend. This finding is consistent with the study conducted by O'Neill and Donoghue [16], in which polystyrene small nettings of sufficient size were used as a simulated catch configuration to measure the geometric shape of the trawl codend and the pressure distribution on it. However, differences were observed in the material and shape of the simulated catch. The reason why the catch configuration did not affect the drag and motion of the codend was that the ball made of polyethylene twine had a porous structure that allowed water to flow freely through the codend, resulting in a decrease in vertical pressure due to the reduction in vortex shedding effects. Another simulated catch configuration used in this study was water-filled balloons, and the results showed that the codend oscillations obtained with these water-filled balloons, rather than real catch, were less significant than those of other catch configurations. This could be due to two reasons: firstly, the water flow was able to easily pass through the smooth surface of the balloon, causing a small vortex to form around it, and secondly, the weight of the water-filled balloons in the water reduced the oscillation amplitude. However, these results were different from the findings of O'Neill et al. [23], who also used water-filled balloons instead of real catch and found that there was an obvious codend oscillation effect ($p < 0.05$). This study showed that using a grooved catch configuration allowed greater and easier oscillation of the codend compared to a spherical configuration ($p < 0.05$). The reason for this finding is that the grooved catch configuration had poor water filtration, which significantly modified the mesh opening and resulted in a decrease in flow passage. This decrease led to the development of greater vortex shedding around the codend, which increased transverse motions and codend drag [8,32]. In contrast to the grooved catch configuration, the smooth surface of the spherical catch configuration and the small spacing between the catches reduced vortex shedding around the codend, resulting in the weaker oscillation amplitude of the codend.

The wavelet transform was utilized in this study to demonstrate that the codend drag and motions with different simulated catch configurations were out of equilibrium, indicating the presence of coherent structures that were identified. The Pearson correlation coefficient was then used to test the correlation between codend oscillations obtained on the codend during the physical model test and during the sea trial. The higher the correlation coefficient was, the closer the measurement results were to each other. The results showed that the oscillation period of the wavelet coefficients of the codend at sea had a strong correlation (50%) and the amplitude had a strong correlation (33.3%), indicating that the codend of the simulated catch was close to the measured period, but the amplitude was not necessarily close to the measured amplitude.

5. Conclusions

This study was carried out to experimentally evaluate the impact of simulated catch configurations on hydrodynamic characteristics, codend shape, and codend motions, with the aim of choosing the better simulated catch to replace the real catch during flume tank experiments. To this end, a model codend was designed and tested in the flume tank using six different simulated catch types, including grooved and spherical types. The results were compared to sea trial measurements. The wavelet transform method was used to analyze the non-stationary time series of the oscillatory phenomena of the drag force and codend motions. The main conclusions are as follows:

(1) The horizontal length of the codends with different simulated catch configurations increased with increasing flow velocity, but the range of the increase was not obvious, and the simulated catch configuration had little effect on the overall longitudinal displacement (codend motion in x-direction) of the codends ($p > 0.05$).

(2) The drag and displacement oscillations of the codend with grooved catch configuration were obvious, and the oscillations were more severe than those of the codend with spherical catch configuration. Additionally, the longitudinal displacement amplitude of the codend with grooved catch configurations was approximately 8.79 times greater than that of the codend with spherical catch configuration.

(3) The findings of the wavelet transform analysis on the codend drag and codend motions showed that the wavelet coefficients of the codends with grooved catch configurations were greater than those of codends with spherical catch configurations. Additionally, intense oscillations were observed in the low frequencies for all simulated catch configurations.

(4) The correlation coefficient of the codend period with the water-filled table tennis ball was 89%, with an amplitude of 91%, which was closer to the actual measurement. The simulated catch used in the flume test is an approximation within a certain range rather than a catch set that fully matches the actual law of change. Therefore, this study suggests using the tennis ball as the simulated catch because it provided the oscillations that were consistent with those obtained during the sea trial.

Author Contributions: Methodology, F.Z.; Investigation, F.Z., H.T., W.L., Q.S., M.Z., C.Z., X.G. and C.S.; Resources, F.Z; Data curation, F.Z; Writing—original draft, F.Z and H.T.; Writing—review & editing, N.N.B.T., W.L, L.X. and F.H. All authors have read and agreed to the published version of the manuscript.

Funding: This study is partially supported by the Natural Science Foundation of Shanghai, grant number 23ZR1427000 and the National Natural Science Foundation of China, grant number Grand No. 31902426.

Institutional Review Board Statement: Not applicable.

Informed Consent Statement: Not applicable.

Data Availability Statement: The data that support the fundings of this study are available from the corresponding author upon reasonable request.

Acknowledgments: The authors thank the scientific observers who collected the scientific data onboard commercial trawlers Long Teng.

Conflicts of Interest: The authors declare no conflict of interest. The funders had no role in the design of the study; in the collection, analyses, or interpretation of data; in the writing of the manuscript; or in the decision to publish the results.

References

1. Stewart, P.A.M.; Ferro, R.S.T. Four experiments investigating codend drag. *Fish. Res.* **1987**, *5*, 349–358. [CrossRef]
2. Thierry, B.N.N.; Tang, H.; Zhang, J.; Liu, W.; Xu, L.X.; Hu, F.X. Experimental analysis of the influence of gear design and catch weight on the fluid–structure interaction of a flexible codend structure used in trawl fisheries. *Appl. Sci.* **2023**, *13*, 2505. [CrossRef]
3. Broadhurst, M.K.; Larsen, R.B.; Kennelly, S.J.; McShane, P.E. Use and success of composite square-mesh codends in reducing bycatch and in improving size-selectivity of prawns in Gulf St. Vincent, South Australia. *Fish. Bull.* **1999**, *97*, 434–448.
4. Broadhurst, M.K.; Millar, R.B.; Wooden, M.E.L.; Macbeth, W.G. Fish. *Manag. Ecol.* **2006**, *13*, 81–92. [CrossRef]
5. Priour, D. Numerical optimisation of trawls design to improve their energy efficiency. *Fish. Res.* **2009**, *98*, 40–50. [CrossRef]
6. Wileman, D.A.; Ferro, R.S.T.; Fonteyne, R.; Millar, R.B. Manual of methods of measuring the selectivity of towed fishing gears. *ICES Coop. Res. Rep. No.* **1996**, *126*, 215. Available online: https://core.ac.uk/reader/45439858 (accessed on 3 February 2022).
7. Liu, W.; Tang, H.; Xu, L.X.; Dong, S.H.; Xu, L.X.; Hu, F.X. Effect of cutting ratio and catch on drag characteristics and fluttering motions of midwater trawl codend. *J. Mar. Sci. Eng.* **2021**, *9*, 256. [CrossRef]
8. Thierry, B.N.N.; Tang, H.; Xu, L.X.; You, X.X.; Hu, F.X.; Achile, N.P. Hydrodynamic performance of bottom trawls with different materials, mesh sizes, and twine thicknesses. *Fish. Res.* **2020**, *221*, 105403. [CrossRef]
9. O'Neill, F.G.; Knudsen, L.H.; Wileman, D.A.; McKay, S.J. Cod-end drag as a function of catch size and towing speed. *Fish. Res.* **2005**, *72*, 163–171. [CrossRef]
10. Pichot, G.; Germain, G.; Priour, D. On the experimental study of the flow around a fishing net. *Eur. J. Mech.* **2009**, *28*, 103–116. [CrossRef]
11. Priour, D. Modelling axisymmetric codends made of hexagonal mesh types. *Ocean Eng.* **2014**, *92*, 1–11. [CrossRef]
12. Cheng, Z.H.; Paul, D.W.; David, K. Hydrodynamic performance of full−scale T0 and T90 codends with and without a codend cover. *Ocean Eng.* **2022**, *10*, 440. [CrossRef]
13. Bouhoubeiny, E.; Germain, G.; Druault, P. Time-resolved PIV investigations of the flow field around rigid codend net structure. *Fish. Res.* **2011**, *108*, 344–355. [CrossRef]
14. Kim, H.Y. Analysis of the turbulent flow and tilt in the codend of a bottom trawl during fishing operations. *Ocean Eng.* **2013**, *64*, 100–108. [CrossRef]
15. Blevins, R.D.; Saunders, H. Flow-induced vibration. *J. Mech. Des.* **1977**, *101*, 6. [CrossRef]
16. O'Neill, F.G.; O'Donoghue, T. The fluid dynamic loading on catch and the geometry of trawl codends. *Proc. R. Soc. Lond. A* **1997**, *453*, 1631–1648. [CrossRef]
17. Wan, R.; Jia, M.X.; Guan, Q.L.; Huang, L.Y.; Cheng, H.; Zhao, F.F.; He, P.G.; Hu, F.X. Hydrodynamic performance of a newly-designed Antarctic krill trawl using numerical simulation and physical modeling methods. *Ocean Eng.* **2019**, *179*, 173–179. [CrossRef]
18. Priour, D.; Prada, A. An experimental numerical study of the catch weight influence on trawl behavior. *Ocean Eng.* **2015**, *94*, 94–102. [CrossRef]
19. Jones, E.G.; Summerbell, K.; O'Neill, F. The influence of towing speed and fish density on the behaviour of haddock in a trawl codend. *Fish. Res.* **2008**, *94*, 166–174. [CrossRef]
20. Druault, P.; Germain, G. Analysis of hydrodynamics of a moving trawl codend and its fluttering motions in flume tank. *Eur. J. Mech. B Fluid* **2016**, *60*, 219–229. [CrossRef]
21. Thierry, N.N.B.; Tang, H.; Xu, L.X.; Hu, F.X.; You, X.X.; David, M.A.; Achille, N.P. Identifying the turbulent flow developing inside and around the bottom trawl by electromagnetic current velocity meter approach in the flume tank. *J. Hydrodyn.* **2021**, *33*, 636–656. [CrossRef]
22. Thierry, B.N.N.; Tang, H.; Achile, N.P.; Xu, L.X.; Zhou, C.; Hu, F.X. Unsteady turbulent flow developing inside and around different parts of fluttering trawl net in flume tank. *J. Fluid Struct.* **2022**, *108*, 103451. [CrossRef]

23. O'Neill, F.G.; McKay, S.J.; Ward, J.N.; Strickland, A.; Kynoch, R.J.; Zuur, A.F. An investigation of the relationship between sea state induced vessel motion and codend selection. *Fish. Res.* **2003**, *60*, 107–130. [CrossRef]
24. Madsen, N.; Hansen, K.; Madsen, N.A.H. Behavior of different trawl codend concepts. *Ocean Eng.* **2015**, *108*, 571–577. [CrossRef]
25. Tauti, M. A relation between experiments on model and on full-scale of fishing net. *Nippon Suisan Gakkai Shi* **1934**, *3*, 171–177. [CrossRef]
26. Hu, F.X.; Matuda, K.; Tokai, T. Effects of drag coefficient of netting for dynamic similarity on model testing of trawl nets. *Fish. Sci.* **2001**, *67*, 84−89. [CrossRef]
27. Tang, H.; Hu, F.X.; Xu, L.X.; Dong, S.C.; Zhou, C.; Wang, X.F. Variations in hydrodynamic characteristics of netting panels with various twine materials, knot types and weave patterns at small attack angles. *Sci. Rep.* **2019**, *9*, 1923. [CrossRef] [PubMed]
28. Meyler, L.; Petrone, G.; Cammarata, G. Simulation of net structures hydrodynamic fields. In *Modelling and Simulation*; Intech Open: London, UK, 2008; pp. 261–282. [CrossRef]
29. Cui, Y.; Guan, C.Y.; Wan, R.; Huang, B.; Li, J. Numerical simulation of a flatfish cage system in waves and currents. *Aquac. Eng.* **2013**, *56*, 26–33. [CrossRef]
30. Herrmann, B.; Priour, D.; Krag, L.A. Simulation-based study of the combined effect on codend size selection of turning meshes by 90° and reducing the number of meshes in the circumference for round fish. *Fish. Res.* **2007**, *84*, 222–232. [CrossRef]
31. Balash, C.; David, S.; Jonathan, B.; Giles, T.; Neil, B. Drag characterisation of prawn-trawl bodies. *Ocean Eng.* **2016**, *113*, 18–23. [CrossRef]
32. Bouhoubeiny, E.; Druault, P.; Germain, G. Phase-averaged mean properties of turbulent flow developing around a fluttering sheet of net. *Ocean Eng.* **2014**, *82*, 160–168. [CrossRef]

Journal of
Marine Science and Engineering

MDPI

Article

Experimental and Numerical Studies on Bending and Failure Behaviour of Inflated Composite Fabric Membranes for Marine Applications

Yunling Ye [1,2], Jin Gan [1,3], Huabing Liu [2,*], Qingfei Guan [1,3], Zhongyi Zheng [1,3], Xiaolin Ran [4] and Zi'ang Gao [1,2]

1 Key Laboratory of High Performance Ship Technology, Wuhan University of Technology, Ministry of Education, Wuhan 430063, China
2 Green & Smart River-Sea-Going Ship, Cruise and Yacht Research Center, Wuhan University of Technology, Wuhan 430063, China
3 School of Naval Architecture, Ocean and Energy Power Engineering, Wuhan University of Technology, Wuhan 430063, China
4 China Automotive Engineering Research Institute Co., Ltd., Chongqing 401122, China
* Correspondence: huabingliu@whut.edu.cn

Abstract: Owing to their excellent physical characteristics of lightweightiness, compactness and rapid deployment, the inflated membrane structures satisfy the demands of maritime salvage and military transportation for long-distance delivery and rapid response. Exploring the failure behaviour of inflated membrane structures can greatly contribute to their widespread applications in ocean engineering. In this research, the main objective is to comprehensively investigate the bending and failure behaviour of inflated membrane structures. Thus, the Surface-Based Fluid Cavity method is employed to set up the finite element model (*FEM*) which is compared to the experimental results to verify its reliability. In parallel, the effects of internal pressure and wrinkles are discussed. An empirical expression of the ultimate bending loading was fitted by face-centred composite designs of the Response Surface Methodology. The results of experiments and *FEM* show that the bearing capacity of the inflated membrane structure is positively correlated with the internal pressure but decreased obviously with the occurrence and propagation of wrinkles. The deformation behaviour and the stress distribution are similar to those of the traditional four-point bending beam, and the local instability induced by wrinkles will cause structural failure. In addition, the numerical model and the proposed expression showed deviations below 5% in relation to the experimental measures. Therefore, the *FEM* and proposed expression are high of reliability and have important engineering guiding significance for the application of inflated membrane structures in ocean engineering.

Keywords: inflated structure; composite fabric membrane; fluid cavity model; model test; bending stiffness; ultimate strength

Citation: Ye, Y.; Gan, J.; Liu, H.; Guan, Q.; Zheng, Z.; Ran, X.; Gao, Z. Experimental and Numerical Studies on Bending and Failure Behaviour of Inflated Composite Fabric Membranes for Marine Applications. *J. Mar. Sci. Eng.* **2023**, *11*, 800. https://doi.org/10.3390/jmse11040800

Academic Editor: Md Jahir Rizvi

Received: 5 March 2023
Revised: 28 March 2023
Accepted: 4 April 2023
Published: 7 April 2023

1. Introduction

Inflated membrane structures are commonly fabricated using continuously woven or braided composite fabrics, which are inflated with a sufficient quantity of gas to maintain their shape and structural rigidity [1]. The excellent physical characteristics including lightweightiness, compact transportation volume and rapid deployment widen their application in the civilian and aerospace fields [2]. Additionally, inflated membrane structures also exhibit fail-safe tendencies (reversible wrinkling) which endow them to recover the original shape and loading capacity once the load is removed [3].

In recent years, due to the requirements of maritime rescue and military logistics for long-distance delivery and rapid response, as shown in Figure 1, the U.S. military took the lead in proposing floating inflated structures including the Lightweight Modular Causeway System (LMCS) [4,5] and the Rapidly Installed Breakwater (RIB) [6]. The inflated structures provide unique solutions for temporary structures in marine engineering. These new-type structures possess significant load-carrying capacities per unit weight and the efficiency of storage, transportation and deployability is improved by the application of inflated bladders. However, nonlinear material characteristics and fluid-structure interactions meet significant challenges in designing these structures. In particular, the difficulty would be enlarged when analyzing the failure behaviour of inflated structures because of their pressure-dependent natural characteristic. Therefore, research on the bearing capacity of the inflated structures is one of the key factors to guide the design, and it also has great significance for promoting the application of the inflated composite fabric membrane structure in marine engineering.

(**a**) (**b**)

Figure 1. New-type floating inflated structures: (**a**) LMCS prototype; (**b**) RIB.

Inflated membrane structures play the role of flexural members in engineering applications; thus, researchers always simplify them as "beams". Up to now, some researchers have studied inflated beams' bending behaviour. For instance, Wielgosz and Thomas [7,8] proposed a finite element model of the inflated beam based on the Timoshenko beam theory and simplified the inflated gas pressure as a following force. Apedo et al. [9] proposed a 3D *FEM* utilizing the virtual work principle to predict the bending deflection of orthotropic braided beams. The accuracy of the method was verified by comparing it with experiments. Subsequently, Nguyen et al. [10] improved Apedo's model [9] by associating the equation with the state of inflated prestress. Furthermore, Elsabbagh [11] improved the model of Apedo [9] to analyse the effect of the internal pressure and the diameter on the wrinkling load of the inflated beam, and compared the result with the calculation example of Apedo, founding that the deviation in the linear state was satisfied. However, the above methods have not presented satisfying accuracy in the post-wrinkling stage.

Apart from the global bending behaviour, the structural failure induced by local wrinkling has also gained particular interest. Wrinkling refers to the formation of small, localized folds or creases in the fabric of the inflated membrane structure. It could occur as a result of changes in the tension or shape of the fabric [12]. Stein and Hedgepeth [13] proposed the membrane model to predict the wrinkling load, assuming that fabric compressive stresses were not admissible. The wrinkling and failure loads were only dependent on the internal pressure and the mean radius [14]. Furthermore, Veldman [15] developed a thin-shell model and accounted for the material properties to analyse wrinkling loads. Wang et al. [16] improved the thin-shell model of Veldman and proposed the concept of the wrinkling factor to estimate the critical wrinkling load. Then, Ji et al. [17] established a multi-scale wrinkling analysis model to investigate the bending stiffness and load–deflection behaviour of inflated beams subjected to bending loads and validated the accuracy of the prediction model by comparing it with experimental results. Wang et al. [18] conducted a series of experiments to investigate the interaction between gas and membrane, and they found the wrinkling load of the inflated beam was sensitive to the gas-membrane coupling relationship. In the previous works, the coupling effect of internal gas and structures due to the external load of the inflated membranes is ignored. In fact, most of the above-mentioned studies simplified the inflated gas pressure by considering it to be a uniform surface pressure. However, the mechanical characteristics of inflated composite fabric membranes, unlike the traditional composite structure, exhibit high nonlinearity due to internal pressure dependence [19]. Moreover, the internal pressure increases significantly with an increase in the load [20]. In this study, thus, the fluid cavity method is adopted to simulate the coupling effect between internal gas and structure, allowing for an accurate prediction of the structural response under external loads.

The purpose of this work is to investigate the bending and failure behaviour of inflated composite fabric membranes and to characterize the effect of pressure, membrane material and the coupling effect of internal gas and structures. Thus, four-point bending experiments were conducted to reveal the structural response behaviour of the inflated membranes subjected to bending load. In parallel, the fluid cavity method was adopted to establish a *FEM* with the material properties determined from the tensile tests. Subsequently, the bending and failure behaviour was analysed by both experiments and *FEM*, and the effects of the internal pressure and wrinkles were also discussed. Once the *FEM* reliability has been verified by comparing it with the experimental result, a simulation plan would be designed by design-of-experiment (DoE) to fit an empirical expression of the ultimate bending loading. Finally, the empirical expression was adjusted and compared with numerical predictions to verify its reliability.

2. Materials and Methods

2.1. Methods for Experimental Tests

2.1.1. Experimental Samples

The inflated samples were fabricated from coated composite fabric membranes that exhibit desirable characteristics such as low elastic modulus, exceptional extensibility and high tensile strength. As depicted in Figure 2a, the membrane's typical configuration consists of substrate and coating [21], where the former serves as the structural layer with commendable mechanical characteristics which is composed of two mutually orthogonal yarn families. The yarn is manufactured from polyester threads. The longitudinal direction of weaving is referred to as the warp (axial) direction, while the transverse direction is termed the weft direction. The warp yarns are straight, while the weft yarns interlace around the warp yarns in an alternating fashion [22]. The coating, made of polyvinyl chloride (PVC), imparts self-cleaning, durability and impermeability to the membrane, and is also referred to as the functional layer. In order to manufacture the PVC membrane, the coating material is sprayed onto the substrate using specialized equipment.

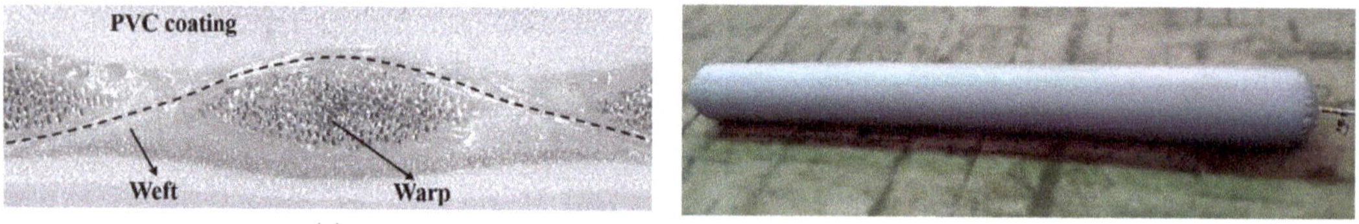

(a) (b)

Figure 2. (**a**) Typical configuration of the fabric membrane; (**b**) Photo of the inflated sample.

A photograph of the inflated sample was presented in Figure 2b, which was in the form of a straight thin-walled beam. The nominal dimensions of the sample were a length of 4 m and a diameter of 400 mm. However, it would be noteworthy that the actual diameter of the sample underwent variations with changes in internal pressure across all cases. For instance, at an internal pressure of 10 kPa, the real diameter of the sample was measured to be 413.81 mm, while it was 415.75 mm at 20 kPa.

2.1.2. Setup

A four-points bending experimental configuration was designed under simply supported boundary conditions to investigate the bending behaviour of inflated samples. The diagram of the test setup was shown in Figure 3.

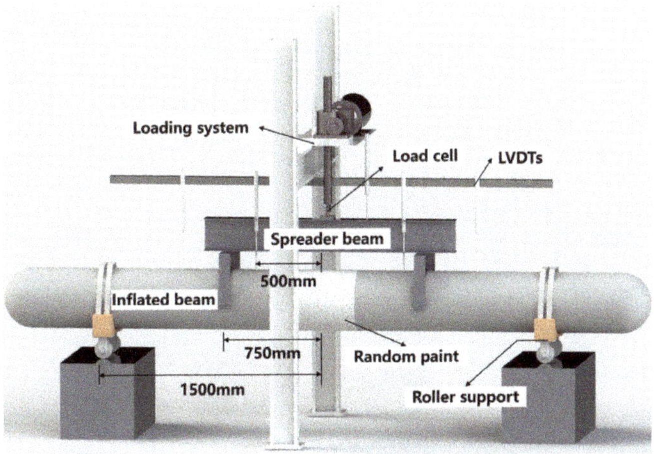

Figure 3. Four-points bend test setup diagram.

The loads were applied using an electric jackscrew and distributed evenly to a pair of load frames through the I-beam. As illustrated in Figure 4a, the load frames, which were 50 mm wide and separated by a distance of 1.5 m, transferred the forces to the samples for testing. The inelastic strap was hinged on the frame to eliminate potential knife-edge effects on the load process and permit follower-type reactions whereby the force vectors were allowed to rotate as the inflated samples deflected. Furthermore, the support consisted of a wooden saddle equipped with two wheels that were designed to match the diameter of the tested samples and provide end support (refer to Figure 4b). To prevent the saddles from tipping, straps were tightly fastened over the top of the sample at each end. This simple support system enabled the sample ends to rotate freely and move horizontally during the test. The span from center to center of rollers was 3 m.

Furthermore, the concentrated force, the internal pressure and the deflection were measured by the pressure transducer, the precise barometer and the linear variable differential transformers (LVDTs), respectively. In addition, to pay attention to examining the response of the midspan region, the Digital Image Correlation system was employed to capture the deflection and strain exactly in this region.

2.2. Numerical Modeling Verification

2.2.1. Calculation Method of Inflated Membrane Structure Based on Fluid Cavity Model

The load–deflection response of the inflated membrane is highly dependent on internal pressure; nevertheless, the volume variation caused by structural response will change the magnitude of internal pressure [1,23]. Thus, one crucial aspect of modeling inflated membrane structures is accurately simulating the variation in internal gas pressure throughout the loading process within the numerical model.

The conventional approach involves approximating the internal gas as uniformly distributed pressure on the surface of the membrane, which fails to describe the complex coupling effect between the gas and the structure. Surface-based Fluid Cavity Modeling [24] can effectively solve the problem of fluid-structure coupling relationships under hydrostatic conditions. For this method, the fluid pressure is calculated by the volume of the fluid cavity, so that the variation of the cavity volume caused by the external load can be considered. It is widely used in the airbag inflation process [25], the airship structural response analysis, and the mechanical property analysis of the inflated string dome structure [26]. The method relies on the principles of the ideal gas theory, and its underlying state equation is presented in Equation (1).

$$PV = nRT, \tag{1}$$

where P is the absolute pressure, V is the gas volume, n is the number of internal gas substances, R is the gas constant and T is the thermodynamic temperature.

A typical fluid cavity element F3D4 is demonstrated in Figure 5, which is a 5-node pyramid element. The fluid cavity wall node is located at the bottom of the element, and the top node is the fluid cavity reference node which should be located in the interior of the cavity.

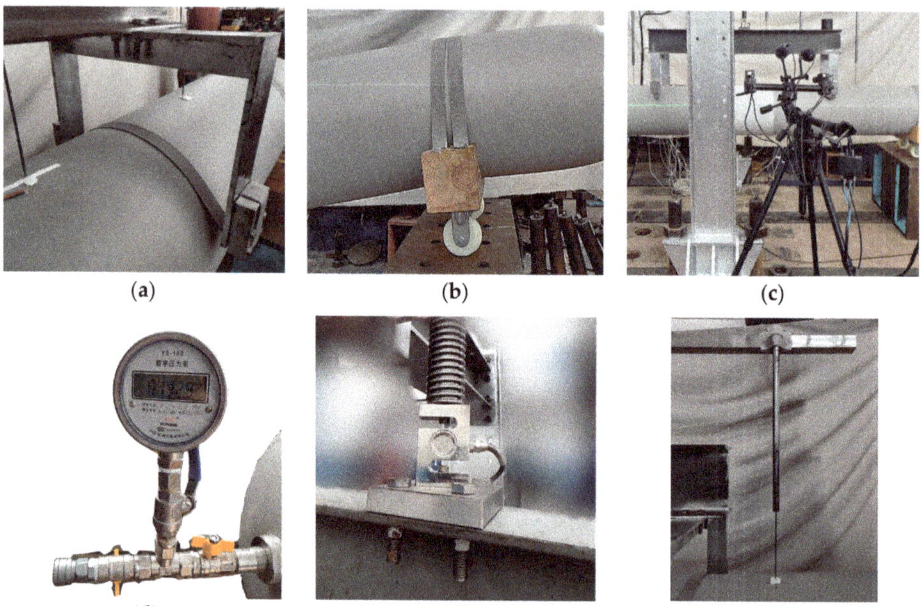

Figure 4. Four-points bend test setup configuration: (**a**) loading frame; (**b**) roller support; (**c**) Digital Image Correlation measure system; (**d**) air pressure gauge; (**e**) s-type load cell; (**f**) LVDT.

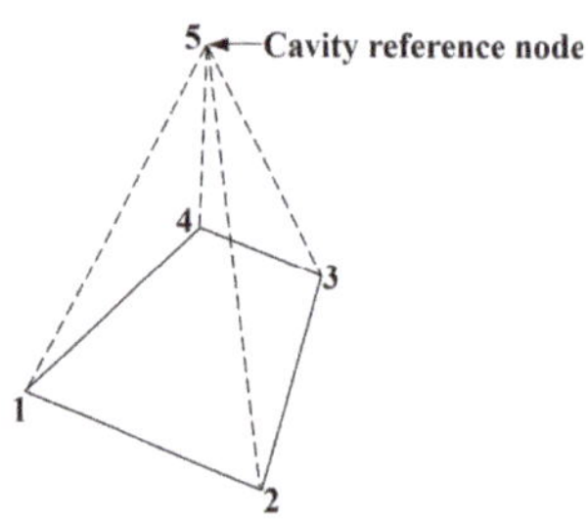

Figure 5. F3D4 element.

2.2.2. Numerical Model

A nonlinear finite element model was developed by the commercial finite element software Abaqus. The total length and the diameter of the model were 4000 mm and 400 mm respectively. As shown in Figure 6, the bladder was modelled by the 4-node membrane element. Two symmetrical concentrated loads were applied to this model, the loading point was 750 mm away from the midspan and the 4-node shell element was employed to model the loading straps. Moreover, the support saddles were modelled by the 8-node solid element. In order to prevent relative sliding, the contacts between each component were all restrained by bonding.

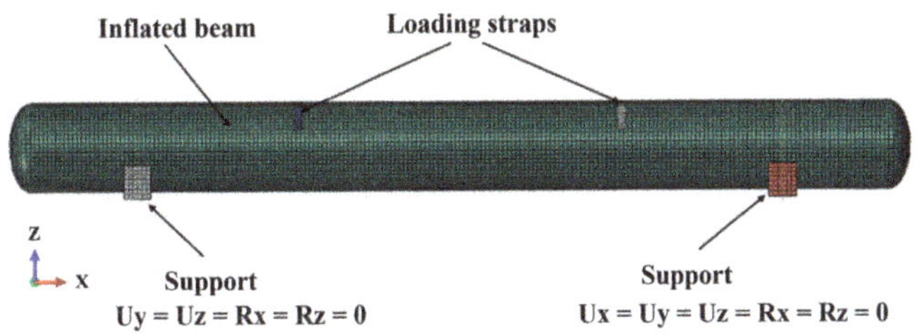

Figure 6. Numerical modeling.

To simulate simply supported boundary conditions, the X-axial displacement of the right support and Y-axis rotation of the two support ends were released, while other degrees of freedom were constrained.

2.2.3. Material Property

The membrane material was considered as the orthotropic material; thus, the four material parameters E_{12}, E_{21}, υ_{21}, and G_{12} should be ascertained to define the material properties of the numerical model. The stress-strain relationship is as follows:

$$\begin{Bmatrix} \varepsilon_{12} \\ \varepsilon_{21} \\ \gamma_{12} \end{Bmatrix} = \begin{bmatrix} 1/E_{12} & -\upsilon_{12}/E_{12} & 0 \\ -\upsilon_{12}/E_{12} & 1/E_{21} & 0 \\ 0 & 0 & 1/G_{12} \end{bmatrix} \begin{Bmatrix} \sigma_{11} \\ \sigma_{22} \\ \tau_{12} \end{Bmatrix} \tag{2}$$

The mechanical parameters of the membrane material were obtained by the three-orientation uniaxial tensile test (see Figure 7). The tensile test configuration was devised in accordance with the ASTM D4851 standard [27].

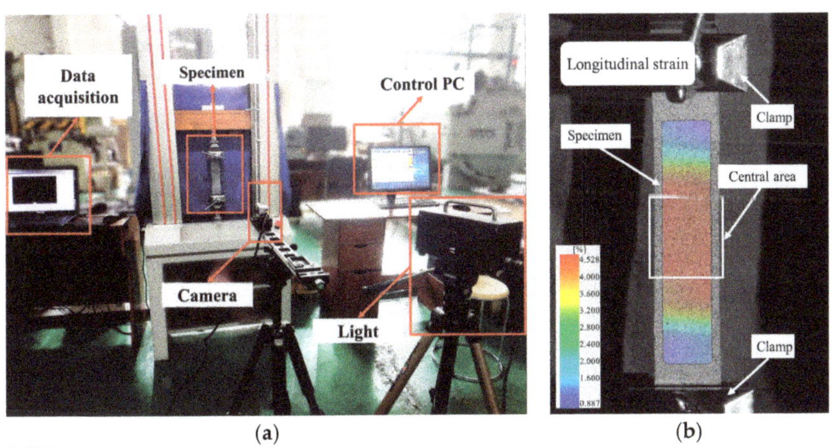

Figure 7. Uniaxial tensile test of the fabric membrane: (**a**) schematic diagram of samples (**b**) longitudinal strain contours of the samples.

2.3. Methods for Parametric Analysis and Fitting of Design Expression

2.3.1. Simulation Plan

The Response Surface Methodology (RSM) is an analysis method integrating the experiment and FE model, in which the regression analysis is used to establish the mathematical model of design variables and target response values [28–30]. Figure 8 showed the simulation scheme that was designed by employing the combination of two face-centred composite designs of experiments (CCF-DoE). Each group of designs included a center point, 2^k "corner points" and 2k "star points". The distances from the project center to the "corner points" were equal to ±2 and ±1 coded units respectively.

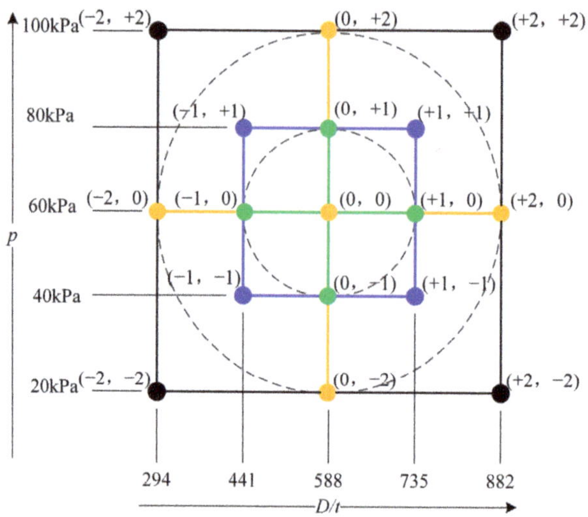

Figure 8. Coded scheme of the design of the experiment for the plan of simulations.

In this study, two regression variables ($k = 2$) were considered, $X1$: D/t and $X2$: p, with ranges of $294 \leq D/t \leq 882$ and 20 kPa $\leq p \leq$ 100 kPa, respectively (D was the diameter of the model, t was the thickness of the membrane and p was the internal pressure). Thus, the uncoded and coded values from each regressor level were summarized in Table 1.

Table 1. Design of experiments for the parametric analysis plan of simulations.

Block	Type	X1	X2 (kPa)	D/t	p (kPa)
1	2k	−2	−2	294	20
1	2k	+2	+2	882	100
1	2k	−2	+2	294	100
1	2k	+2	−2	882	20
2	(*)	−2	0	294	60
2	(*)	0	−2	588	20
2	(*)	+2	0	882	60
2	(*)	0	+2	588	100
2	(*)	0	0	588	60
3	2k	−1	−1	441	40
3	2k	+1	+1	735	80
3	2k	−1	+1	441	80
3	2k	+1	−1	735	40
4	(*)	−1	0	441	60
4	(*)	0	−1	588	40
4	(*)	+1	0	735	60
4	(*)	0	+1	588	80

Nomenclature type: 2k—Corner point; (*)—Star point.

2.3.2. Parametric Analysis

To evaluate the contribution of each regressor variable and their interactions, a parametric analysis was conducted utilizing a multiple linear model analysis of variance (ANOVA) which was further optimized using the least-squares method (LSM).

2.3.3. Design Expression

The empirical response surface model of the ultimate loading was established by multiple linear regression which considered $k = 2$ regressor variables, $X1$: D/t and $X2$: p, by the regression coefficients βj ($j = 0, 1, \cdots, k$).

$$F(X_j) = \beta_0 + \sum_{j=1}^{k} \beta_j X_j + \sum_{j=1}^{k} \beta_{jj} X_j^2 + \sum \sum_{i<j=2}^{k} \beta_{ij} X_i X_j + \sum_{j=1}^{k} \beta_{jjj} X_j^3 + \sum \sum_{i<j=2}^{k} \beta_{iij} X_i^2 X_j + \sum \sum_{i<j=2}^{k} \beta_{ijj} X_i X_j^2, \quad (3)$$

where β_j, β_{jj}, β_{jjj} were the expected first-order, second-order and third-order effects in the ultimate loading $F(X_j)$, respectively, from the changes of regression variables X_j when the rest of the variables X_i were constrained. B_{ij}, β_{iij}, β_{ijj} represented the interaction between regression variables X_i and X_j.

Substituting the regression variables in Equation (3), the cubical model of ultimate loading $F(X_j)$ was conducted.

$$F(D/t, p) = \beta_0 + \beta_1(D/t) + \beta_2 p + \beta_3(D/t)p + \beta_4\left(D/t\right)^2 + \beta_5 p^2 + \beta_6\left(D/t\right)^2 p + \beta_7(D/t)p^2 + \beta_8\left(D/t\right)^3 + \beta_9 p^3, \quad (4)$$

The final expression would depend on the result of the parametric analysis, and part terms of Equation (4) would be disregarded as the p-value > 0.01.

3. Results and Discussions

3.1. Experimental Results and Numerical Model Verification

3.1.1. Four-Points Bending Experimental Results

Five measured points' load–deflection curves for the inflated sample with 30 kPa internal pressure were plotted in Figure 9. The load–deflection curves of the symmetric measured points (e.g., 1/6 spam and 5/6 span or 1/3 spam and 2/3 span) were almost coincident, which proved that the bending load system was symmetric and there was no torsion load during the test process.

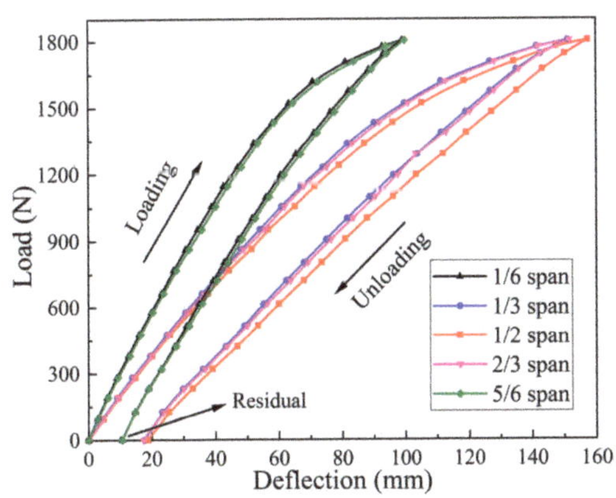

Figure 9. Load–deflection curves for the inflated sample with 30 kPa internal pressure.

The load–deflection curves exhibited an initial linear response, which was succeeded by a softening behaviour. Likewise, the inflated samples presented two typical deflected states with the increase of load. Under the initial load (as shown in Figure 10a), the deflected curve transits smoothly along the axis of the samples which just seemed like a common beam. With the increase of load, the deflection curve developed into the trilinear model (as shown in Figure 10b), loading points were inflection points of the trilinear and the loading section of the inflated sample (between the two loading points) was almost straight. Theoretically, the bending moments in the loading section are equal and gradually decrease from the loading point to the adjacent boundary, which leads mainly to this typical deformation behaviour of the thin-walled inflated beam.

(**a**) (**b**)

Figure 10. Deflected process of the inflated sample: (**a**) initial loading; (**b**) close to ultimate load.

Besides, the inflated samples would not recover when unloaded completely, and the residual deformation increased with the increase of the internal pressure. The reason for the phenomenon was related to the crimp interchange of the woven substrate and the viscoelasticity of the polyester fabrics [31]. The residual deformation would fade away with time or the occurrence of inflating and deflating alternately.

In addition, when the load reached the limit value, the compressive stress caused by the bending moment would counteract the tensile stress induced by internal pressure, which would result in significant local deformation (shown in Figure 11). Especially, it would compress the membrane sheet along the axial direction on the compressive part and trigger the local membrane to generate strangulation or wrinkles. This wrinkling behaviour would lead to buckling which was similar to that of the circular tube subjected to pure bending load [32]. Nevertheless, as a result of the support provided by the internal pressure, the formation of wrinkles did not lead to immediate structural buckling failure but instead degraded the structural stiffness of the inflated beam, which was also validated by Malm's experiments [33]. This was also reflected in the load–deflection curve, which exhibited a softening effect in the post-wrinkle stage (see Figure 9). The weaker local carrying capacity in the wrinkling region would also enhance the trilinear deflected pattern.

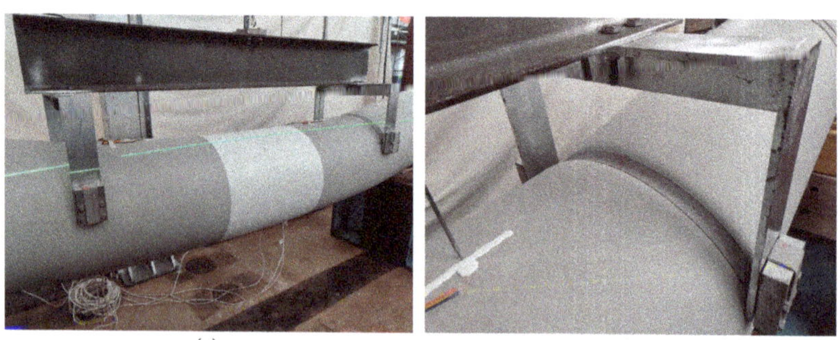

(**a**) (**b**)

Figure 11. The local deformed pattern of the inflated sample close to ultimate load: (**a**) bending region; (**b**) loading point.

3.1.2. Material Level Test Results

The stress–strain curves for fabric membrane straps were illustrated in Figure 12. The stress–strain curves represented high nonlinearity during the loading process [34]. Nevertheless, stage 1 of the stress–strain curves could cover the range of this study and this part of the curves could be considered linearity; thus, the elastic moduli of warp and weft were obtained by fitting the slopes. The details of the main material parameters were summarized in Table 2.

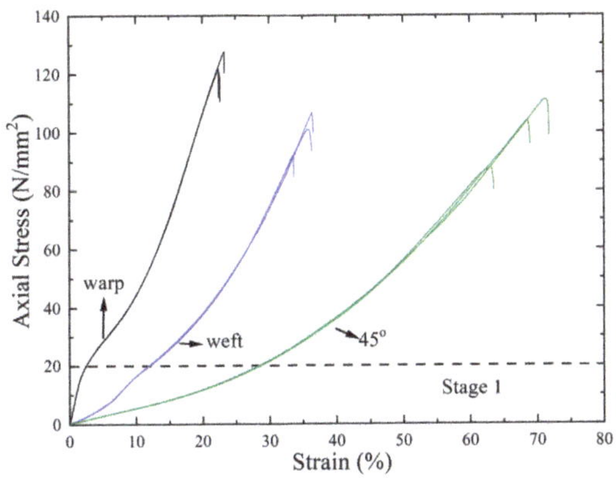

Figure 12. Stress–strain curves for fabric membrane strap.

Table 2. Main material parameters of fabric membranes.

Properties	Values
Base fabric material	Polyester
Type of coating	PVC
Thickness (mm)	0.68
Density (kg/m^3)	1650
Tensile strength (warp/weft/45°) (N/50 mm)	4116/3453/3126
Breaking elongation (warp/weft/45°) (%)	22.27/35.80/70.11
Elastic modulus (warp E_{12}/ weft E_{21}) (MPa)	1019/232
Shear modulus G_{12} (MPa)	19.29
Poisson's ratio υ_{12}	0.81

Where ρ is the density, E_{12} and E_{21} are the elastic moduli in the longitudinal and latitudinal directions, respectively, and υ_{12} is Poisson's ratio. G_{12} is the shear modulus obtained from Equation (5) [35]. E_{45} and υ_{45} are the elastic moduli and Poisson's ratio in the 45° direction, respectively.

$$G_{12} = \frac{E_{45}}{2(1+\upsilon_{45})}, \tag{5}$$

3.2. *Numerical Modeling Verification*

3.2.1. Mesh Properties

Figure 13a showed the midspan load–displacement curves of the 60 kPa internal pressure model under the mesh sizes of 40 mm, 30 mm, 20 mm and 15 mm, respectively. It could be seen that the slopes of curves in the linear stage were almost the same under different mesh sizes, which meant that the mesh size within the selected range did not affect the calculation accuracy of the structural bending stiffness. However, the variation in mesh size would change the ultimate bearing capacity. Moreover, a reduction in mesh size to 20 mm resulted in an ultimate loading value that was almost indistinguishable from that obtained with a mesh size of 15 mm. Meanwhile, as shown in Figure 13b, the calculation time of the model with the mesh size of 15 mm had reached 43 h which was the maximum one. Considering the calculation accuracy and efficiency, the mesh size of 20 mm was finally selected.

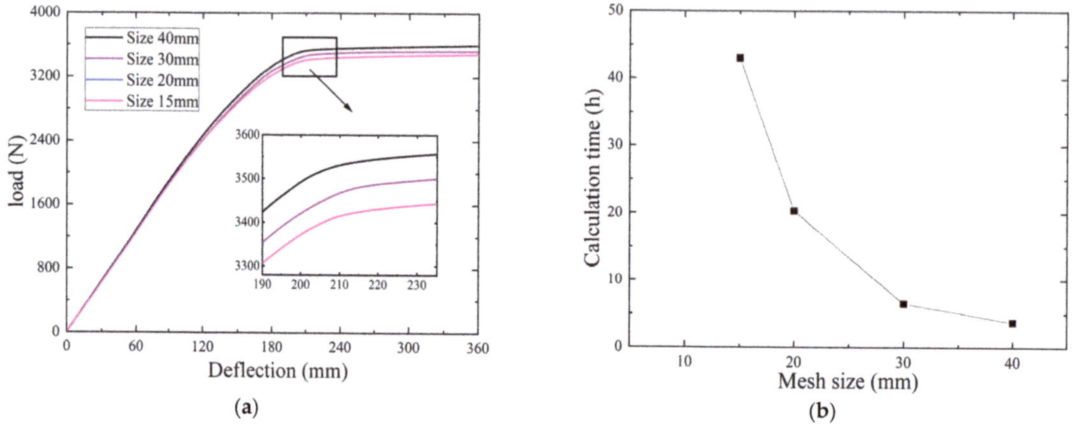

Figure 13. The effect of mesh size on the accuracy and efficiency: (**a**) Load–deflection curves with different mesh sizes (**b**) Calculation vs mesh size.

3.2.2. Deformation Behaviour and Failure Mode Analysis

The typical deformed patterns of the FE model of the inflated membrane structures were shown in Figure 14. In comparison with Figure 10, it was apparent that the overall deformed patterns based on the *FEM* were consistent with the experimental results.

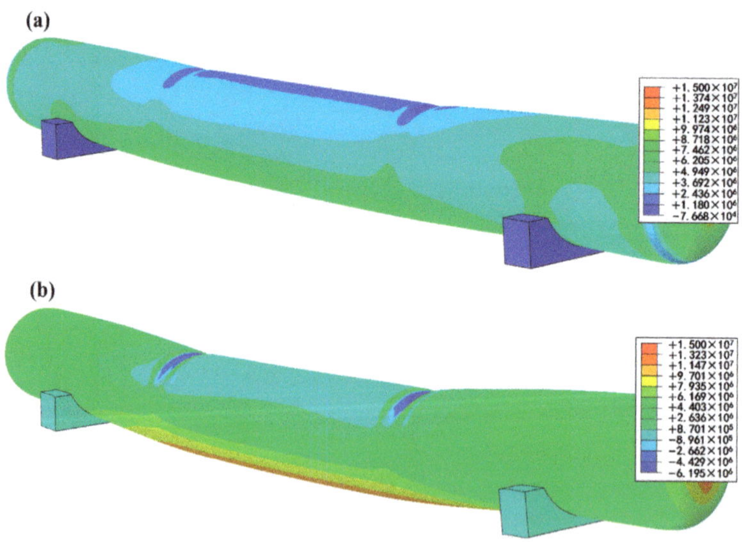

Figure 14. The deformed patterns of the FE model: (**a**) initial loading; (**b**) close to ultimate load.

Besides, it was found that the bearing capacity of inflated structures decreased significantly with the appearance of wrinkles in the experiment. Then, the development of wrinkles would induce a special failure mode of inflated structures which was the structural instability caused by local buckling. This phenomenon had also been verified in the numerical simulation. When the load reached the inflexion point where the load–displacement curve changed from straight to curved, wrinkles occurred on the upper surface of the model in the bending region. As the applied load intensified, the wrinkles underwent a progressive transformation from multiple creases to a singular groove, culminating in the creation of the wrinkling hinge, as shown in Figure 15, the structure lost its bearing capacity at this moment. Therefore, the appearance of wrinkles would weaken the bearing capacity of the inflated membranes, and the further evolution of wrinkle hinges would lead to the failure of this structure [36].

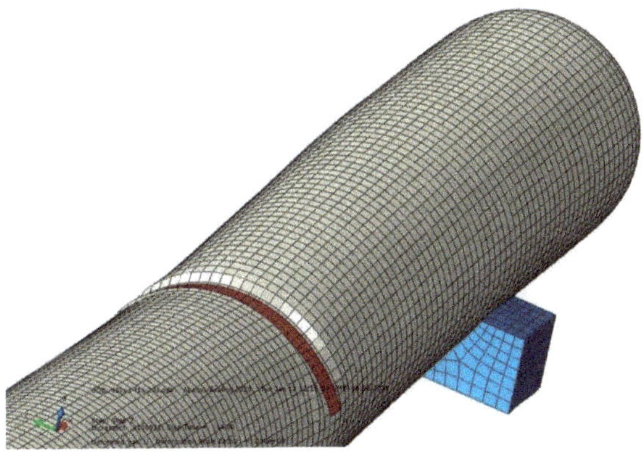

Figure 15. The local wrinkle under ultimate loading.

As shown in Figure 16, the FE model results could match the patterns of the experimental load–deflection curves under the internal pressure conditions of 10 kPa, 20 kPa, 30 kPa, 40 kPa and 50 kPa, and they were nearly consistent under the initial stages with the same internal pressure especially. In addition, with the increase of internal pressure, the failure load and the structural stiffness were significantly improved. Therefore, increasing the internal pressure was an effective way to strengthen the bearing capacity of inflated membrane structures.

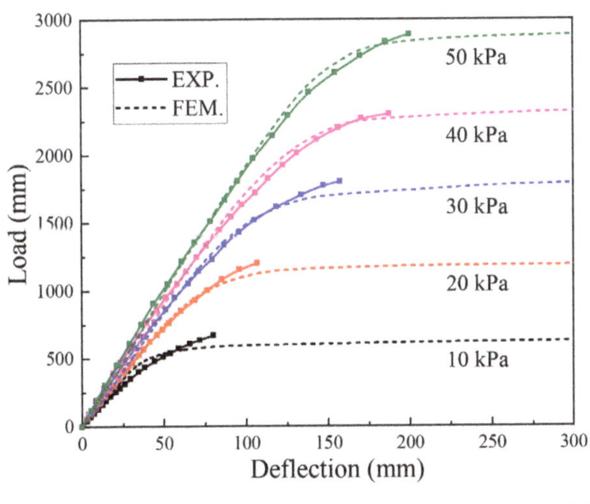

Figure 16. Numerical and experimental comparison for the load–deflection response of midspan of inflated structures.

For quantitative analysis, the ultimate load and the structural stiffness were two important parameters to characterize the load-bearing performance of inflated membrane structures. To quantify the differences between numerical predictions and experimental measurements, the Expression (6) was utilized to determine the deviations.

$$\begin{cases} F_{dev} = \left| \frac{F_{FEM}}{F_{EXP}} - 1 \right| \\ K_{dev} = \left| \frac{K_{FEM}}{K_{EXP}} - 1 \right| \end{cases} \quad (6)$$

As shown in Table 3, the ultimate load increased with the increase of internal pressure. Nevertheless, the deviations between the *FEM* and experimental values of the ultimate load in all cases were less than 5%. Furthermore, the initial response of the inflated membrane structures, especially in the pre-wrinkle stage, is of interest because the inflated membrane structures would likely be designed to stay within the linear portion of the load–displacement response on a mission. Thus, the linear stage of the load–deflection curve was fitted and the slope of the fitted line was calculated to present the equivalent structural stiffness in the linear stage. The comparative results of the *FEM* and the experiment showed that the minimum deviation was less than 3% while the maximum value was not more than 5%. Therefore, the *FEM* has high reliability for the prediction of the ultimate load and the equivalent stiffness for the inflated membrane structure.

Table 3. Values and deviations for numerical and experimental results of the ultimate load and the equivalent structural stiffness.

Internal Pressure (kPa)	Ultimate Load (N)			Equivalent Structural Stiffness (N/mm)		
	Exp.	*FEM*	Deviation	*Exp.*	*FEM*	Deviation
10	651.99	622.22	4.57%	12.00	12.58	4.83%
20	1201.73	1184.44	1.44%	14.53	15.27	4.75%
30	1803.55	1748.00	3.08%	16.21	16.56	2.16%
40	2320.45	2282.22	1.65%	17.32	17.95	3.64%
50	2887.52	2808.89	2.72%	18.19	18.52	1.81%

3.2.3. Comparative Analysis of Stress or Strain Response

The stress state of the inflated structure under the bending load was extracted from *FEM*. The axial tensile stress was basically equal everywhere under the effect of internal pressure load, which was in good agreement with the stress distribution law obtained from net theory [37] (due to geometric variation, the stress had a gradient distribution at both ends of the membranes, which was not within the scope of this study). As the load increased, the tensile stress in the upper part of the model gradually decreased (see Figure 14). Especially, the stress concentration would occur in the local loading region, which would eventually lead to model failure with further development.

In order to verify the strain response of the inflated model under the bending load, the Digital Image Correlation (DIC) system was employed to measure the strain field in the midspan region of the experimental sample. Figure 17a showed the typical strain nephogram of the midspan region under 30 kPa internal pressure. The upper part was compressed with the lower part tensioned, which was proved by the *FEM* result. The strain data of midspan measuring points t and b were probed to quantify the loading–strain response in the loading process, which were plotted in Figure 17b. It was found that the curves of the *FEM* results were in good agreement with the experimental ones which meant that the *FEM* results of the strain response were responsible. Nevertheless, the strain of the measuring point t was followed with the measured values through most of the loading process but then deviated after wrinkles occurred. The reason could be that the membrane element M3D4R without bending stiffness was used for describing the inflated structure behaviour in the *FEM*, which would lose its bearing capacity immediately when encountering axial compressive stress [19], whereas the actual membrane material had the compressive stiffness which could withstand a certain axial compressive stress. Thus, the bearing capacity of the inflated structure in the *FEM* was smaller than that of the actual structure in the post-wrinkling stage; it also led to a smaller *FEM* compressive strain value than the test result after the wrinkle occurred.

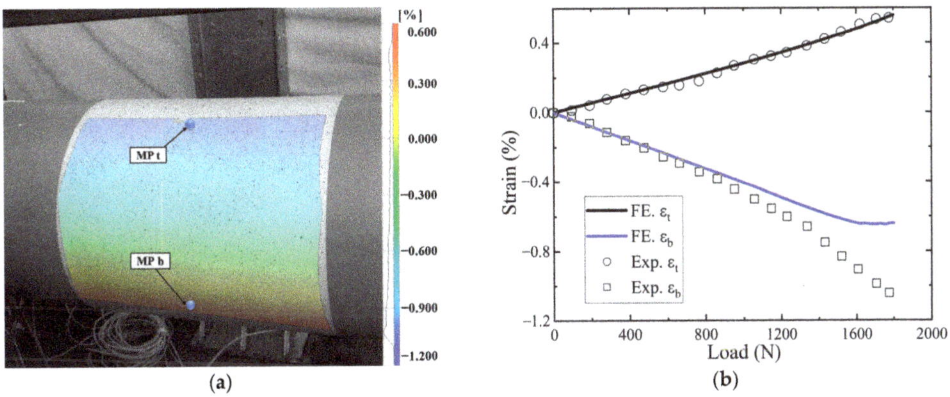

Figure 17. The axial strain–loading response of the midspan region: (**a**) the strain nephogram from the DIC; (**b**) the loading–strain curves.

3.3. Parametric Analysis and Fitting of Design Expression Results

Based on the CCF-DoE design method, 17 groups of experiments were designed, and the ultimate load was calculated using the fluid cavity simulation method. Some results were shown in Table 4.

Table 4. Results of the ultimate load from RSM and *FEM*.

Order	X1	X2 (kPa)	D/t	p (kPa)	F_{FEM} (N)	F_{RSM} (N/m)	Error (%)
1	−2	−2	294	20	230.22	229.75	−0.20
2	+2	+2	882	100	20,800.00	20,784.32	−0.08
3	−2	+2	294	100	730.40	736.82	0.88
4	+2	−2	882	20	3802.00	3778.69	−0.61
…	…	…	…	…	…	…	…
14	−1	0	441	60	1445.67	1464.01	1.27
15	0	−1	588	40	2512.22	2448.88	−2.52
16	+1	0	735	60	7111.11	7363.11	3.54
17	0	+1	588	80	4987.00	5065.68	1.58

The ANOVA results of the cubic model were shown in Table A1. The Model F-value of 4580.94 presented the model was significant and there was only a 0.01% chance that an F-value this large could occur due to noise.

Besides, in this case, X_1, X_2, X_1X_2, $X_1{}^2$ and $X_1{}^2X_2$ were significant model terms due to their *p*-values of less than 0.01. Moreover, *p*-values greater than 0.10 indicated the model terms were not significant; thus, these terms, including $X_2{}^2$, $X_1X_2{}^2$, $X_1{}^3$, and $X_2{}^3$, could be reduced to improve the prediction model. Table A2 showed the results of the reduction cubic (RC) model, the Model F-value of 10921.12 and the *p*-values of all the terms less than 0.0001 presented the model was significant.

Furthermore, the Predicted R^2 of 0.9997 was in reasonable agreement with the Adjusted R^2 of 0.997, and they were all significantly closer to 1. The signal-to-noise ratio of 377.96 far greater than the specified value 4 was desirable. These all indicated that the regression model had an adequate signal. The final expression of ultimate load $F(D/t, p)$ was denoted by the RC model,

$$F(D/t,p) = 339.990 - 0.748(D/t) + 35.325p - 0.2484(D/t)p - 1.96\times10^{-4}\left(D/t\right)^2 + 5.09\times10^{-4}\left(D/t\right)^2 p, \quad (7)$$

where $294 \le D/t \le 882$ and $20\ \text{kPa} \le p \le 100\ \text{kPa}$.

Additionally, cubic and RC model predictions for the F values were plotted in Figure 18. All of the predicted points were close to the actual line and there was no further deviation of the RC results from the cubic ones. Thus, the RC model was applicable to represent the response surface of the ultimate load F(*D/t*, *p*).

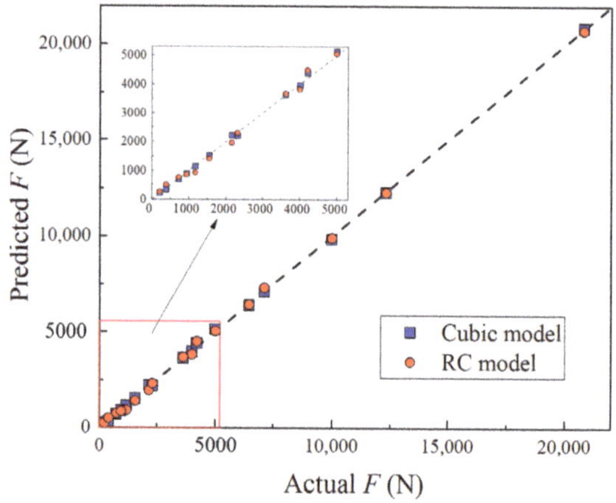

Figure 18. Cubic and RC models predictions for the *F* values.

Figure 19 presented the response surface of the ultimate load for the RC model. The oval isoheights from the contour plot (Figure 19a) represented that the interaction between the regression variables D/t and p were significant. Additionally, from the three-dimensional response surface hyperplane (Figure 19b), it was found that the design points sat basically on the response surface, which indicated that the predicted values of the regression model were in good agreement with the actual values. The deviation values for the RC model and *FEM* of each design point (Table 4) were less than 5%, which indicated that the RC model was highly authentic. Therefore, the proposed design expression had the potential to serve as a feasible alternative to the *FEM* for assessing the ultimate loading of the inflated composite membrane structure within the interesting range.

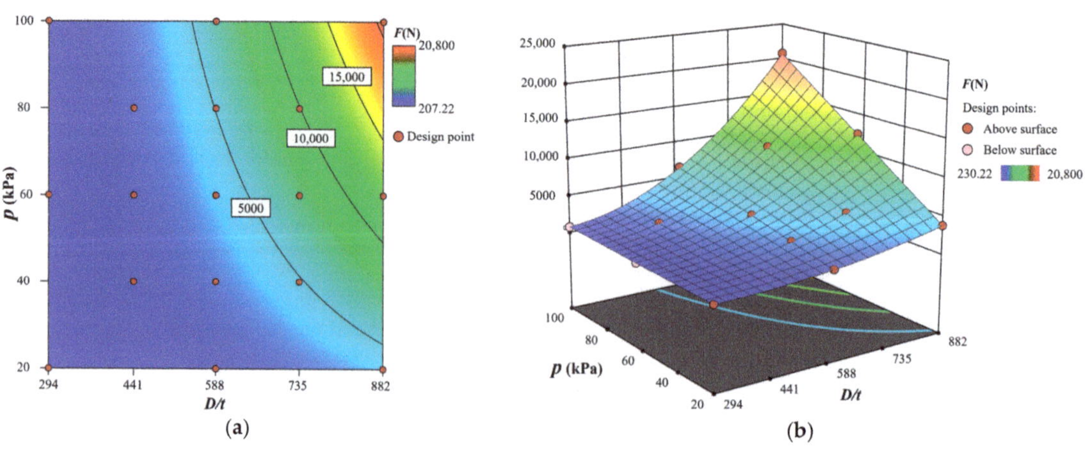

Figure 19. Response surface of the ultimate load for RC model: (**a**) contour plot related to the CCF-DoE design points; (**b**) 3D hyperplane of *F* about the design points within regressor variables space domain.

4. Conclusions

In this research, the bending and failure behaviour that the inflated composite membrane structure represents is investigated by experimental and numerical methods used in marine engineering. The reliability of the numerical model was verified by comparing it with the experimental result, then a simulation plan was designed by design-of-experiment (DoE) to develop an empirical expression for the ultimate bending load. The main conclusions are obtained as follows:

(1) Strong agreement was achieved between numerical results and experimental measurements for both global and local analyses. The *FEM* exhibited remarkable precision in characterizing the response of the inflated composite membrane structure under bending loads, with ultimate loads and equivalent structural stiffness of deviation ranging from 1.65% to 4.57% and 1.81% to 4.83%, respectively, relative to the experimental measurements.

(2) The bending mode was described from different stages observed during the loading process. Under the initial load, the deflected curve transited smoothly along the axis of the inflated sample, which shifted into the trilinear model with the load increasing to the wrinkling load. In parallel, the bearing capacity of inflated structures decreased significantly with the appearance of wrinkles, and the evolution of wrinkles would induce a special failure mode of inflated structures: that was, the instability of the structure caused by local buckling. This behaviour was also verified in the load–deflection curves which exhibited an initial linear response followed by softening behaviour.

(3) The conducted parametric analysis was based on the ANOVA of responses recovered from a CCF-DoE plan of simulations for ultimate load, which adjusted the expected model to a reduction cubic model by eliminating part of insignificant interaction terms. The deviation values between ultimate loads in the RMSE and those obtained from *FEM* were less than 5%. Therefore, the proposed design expression presented a viable alternative to the *FEM* for evaluating the ultimate loading of the inflated composite membrane structure for the interesting range.

In this study, we have solely focused on the ideal four-points bending conditions. However, due to the application of inflated membrane structures in complex marine engineering environments, diverse boundary and loading conditions require further investigation. Therefore, it is important to conduct further research on this matter to address the challenges faced by these structures when subjected to real engineering conditions.

Author Contributions: Conceptualization, Y.Y. and J.G.; methodology, Y.Y.; software, Z.Z. and X.R.; validation, H.L. and Q.G.; formal analysis, H.L. and Z.G.; investigation, H.L. and Z.G.; resources, J.G. and H.L.; data curation, Q.G.; writing—original draft preparation, Y.Y.; writing—review and editing, H.L.; visualization, Y.Y. and Z.G.; supervision, H.L.; project administration, H.L.; funding acquisition, J.G. All authors have read and agreed to the published version of the manuscript.

Funding: This research was funded by the Fundamental Research Funds for the Central Universities (WUT: 2022CG005 and 223373003).

Institutional Review Board Statement: Not applicable.

Informed Consent Statement: Not applicable.

Data Availability Statement: Not applicable.

Acknowledgments: The authors would like to thank funding body for the grant.

Conflicts of Interest: The authors declare no conflict of interest.

Appendix A. ANOVA Results Tables for Regression Models

Table A1. ANOVA results of the cubic model.

Source	Sum of Squares	df	Mean Square	F-Value	p-Value
Regression	4.58×10^8	9	5.08×10^7	4580.94	<0.0001
X_1	2.82×10^7	1	2.82×10^7	2539.59	<0.0001
X_2	5.64×10^6	1	5.64×10^6	507.79	<0.0001
X_1X_2	7.23×10^7	1	7.23×10^7	6514.61	<0.0001
X_1^2	1.89×10^7	1	1.89×10^7	1701.57	<0.0001
X_2^2	8979.02	1	8979.02	0.809	0.3983
$X_1^2X_2$	4.18×10^6	1	4.18×10^6	376.48	<0.0001
$X_1X_2^2$	168.54	1	168.54	0.0152	0.9054
X_1^3	3540.28	1	3540.28	0.319	0.5899
X_2^3	170.54	1	170.54	0.0154	0.9048
Residual	77,690.91	7	11,098.7		
R-squared	0.9998				
Factor	Coefficient Estimate	Standard Error	95% CI Low	95% CI High	VIF
β_0-Intercept	3739.9	41.61	3641.52	3838.28	
β_1–D/t	5824.75	115.58	5551.44	6098.06	9.03
β_2–p	2604.58	115.58	2331.26	2877.89	9.03
β_3–$(D/t)p$	4124.64	51.1	4003.8	4245.48	1
β_4–$(D/t)^2$	2607.66	63.22	2458.18	2757.14	1.1
β_5–p^2	56.86	63.22	−92.62	206.34	1.1
β_6–$(D/t)^2p$	1756.6	90.53	1542.52	1970.67	3
β_7–$(D/t)p^2$	11.16	90.53	−202.92	225.23	3
β_8–$(D/t)^{23}$	80.04	141.72	−255.08	415.16	11.03
β_9–p^3	17.57	141.72	−317.55	352.69	11.03

Table A2. ANOVA results of the reduction-cubic model.

Source	Sum of Squares	df	Mean Square	F-Value	p-Value
Regression	4.58×10^8	5	9.15×10^7	10,921.12	<0.0001
X_1	2.61×10^8	1	2.61×10^8	31,146.93	<0.0001
X_2	2.09×10^7	1	2.09×10^7	2495.57	<0.0001
X_1X_2	7.23×10^7	1	7.23×10^7	8628.63	<0.0001
X_1^2	2.11×10^7	1	2.11×10^7	2521.59	<0.0001
$X_1^2X_2$	5.13×10^6	1	5.13×10^6	612.44	<0.0001
Residual	92,174.63	11	8379.51		
R-squared	0.9998				
Factor	Coefficient Estimate	Standard Error	95% CI Low	95% CI High	VIF
β_0-Intercept	3757.28	32.01	3686.82	3827.74	
β_1–D/t	5899.11	33.43	5825.54	5972.68	1
β_2–p	2616.8	52.38	2501.51	2732.09	2.46
β_3–$(D/t)p$	4124.64	44.4	4026.91	4222.37	1
β_4–$(D/t)^2$	2625.11	52.28	2510.05	2740.17	1
β_5–$(D/t)^2p$	1761.37	71.17	1604.72	1918.03	2.46

References

1. Davids, W.G.; Zhang, H. Beam finite element for nonlinear analysis of pressurized fabric beam–columns. *Eng. Struct.* **2008**, *30*, 1969–1980. [CrossRef]
2. Turner, A.; Kabche, J.; Peterson, M.; Davids, W. Tension/torsion testing of inflatable fabric tubes. *Exp. Tech.* **2008**, *32*, 47–52. [CrossRef]
3. Hulton, A.W.; Cavallaro, P.V.; Hart, C.J. Modal analysis and experimental testing of air-inflated drop-stitch fabric structures used in marine applications. In Proceedings of the ASME International Mechanical Engineering Congress and Exposition, Tampa, FL, USA, 3–9 November 2017; p. V009T012A030.
4. Fowler, J.E.; Resio, D.T.; Pratt, J.N.; Boc, S.J.; Sargent, F.E. *Innovations for Future Gap Crossing Operations*; Engineer Research and Development Center Vicksburg: Vicksburg, MS, USA, 2006.

5. Russell, B.R.; Thrall, A.P.; Padula, J.A.; Fowler, J.E. Reconceptualization and Optimization of a Rapidly Deployable Floating Causeway. *J. Bridge Eng.* **2014**, *19*, 04013013. [CrossRef]
6. Fowler, J.; Pollock, C.; Resio, D.; Turk, G. Development of a rapidly installed breakwater (RIB) system. *Oceanogr. Lit. Rev.* **1998**, *3*, 580.
7. Wielgosz, C.; Thomas, J.-C. Deflections of inflatable fabric panels at high pressure. *Thin-Walled Struct.* **2002**, *40*, 523–536. [CrossRef]
8. Thomas, J.C.; Wielgosz, C. Deflections of highly inflated fabric tubes. *Thin-Walled Struct.* **2004**, *42*, 1049–1066. [CrossRef]
9. Apedo, K.L.; Ronel, S.; Jacquelin, E.; Bennani, A.; Massenzio, M. Nonlinear finite element analysis of inflatable beams made from orthotropic woven fabric. *Int. J. Solids Struct.* **2010**, *47*, 2017–2033. [CrossRef]
10. Nguyen, Q.-T.; Thomas, J.-C.; Le van, A. Inflation and bending of an orthotropic inflatable beam. *Thin-Walled Struct.* **2015**, *88*, 129–144. [CrossRef]
11. Elsabbagh, A. Nonlinear finite element model for the analysis of axisymmetric inflatable beams. *Thin-Walled Struct.* **2015**, *96*, 307–313. [CrossRef]
12. Zhang, L.; Dong, K.; Lu, M.; Zhang, H. A wrinkling model for pneumatic membranes and the complementarity computational framework. *Comput. Mech.* **2020**, *65*, 119–134. [CrossRef]
13. Stein, M.; Hedgepeth, J.M. *Analysis of Partly Wrinkled Membranes*; National Aeronautics and Space Administration: Washington, DC, USA, 1961.
14. Haughton, D.; McKay, B. Wrinkling of inflated elastic cylindrical membranes under flexure. *Int. J. Eng. Sci.* **1996**, *34*, 1531–1550. [CrossRef]
15. Veldman, S. Wrinkling prediction of cylindrical and conical inflated cantilever beams under torsion and bending. *Thin-Walled Struct.* **2006**, *44*, 211–215. [CrossRef]
16. Wang, C.; Du, Z.; Tan, H. Initial wrinkling and its evolution of membrane inflated cone in bending. *Thin-Walled Struct.* **2012**, *59*, 97–102. [CrossRef]
17. Ji, Q.X.; Wang, C.G.; Tan, H.F. Multi-scale wrinkling analysis of the inflated beam under bending. *Int. J. Mech. Sci.* **2017**, *126*, 1–11. [CrossRef]
18. Wang, X.; Fu, H.; Law, S.-s.; Yang, Q.; Yang, N. Experimental study on the interaction between inner air and enveloping membrane of inflated membrane tubes. *Eng. Struct.* **2020**, *219*, 110892. [CrossRef]
19. Cavallaro, P.V.; Johnson, M.E.; Sadegh, A.M. Mechanics of plain-woven fabrics for inflated structures. *Compos. Struct.* **2003**, *61*, 375–393. [CrossRef]
20. Liu, P.; Yue, M.; Feng, S.Z.; Ngamkhanong, C. Structural behaviour of air-inflated beams. *Structures* **2023**, *47*, 1613–1623. [CrossRef]
21. Yingying, Z.; Junhao, X.; Qilin, Z. Advances in mechanical properties of coated fabrics in civil engineering. *J. Ind. Text.* **2016**, *48*, 255–271. [CrossRef]
22. Jekel, C.F.; Venter, G.; Venter, M.P. Modeling PVC-coated polyester as a hypoelastic non-linear orthotropic material. *Compos. Struct.* **2017**, *161*, 51–64. [CrossRef]
23. Kabche, J.P.; Peterson, M.L.; Davids, W.G. Effect of inflation pressure on the constitutive response of coated woven fabrics used in airbeams. *Compos. Part B Eng.* **2011**, *42*, 526–537. [CrossRef]
24. Dassault Systèmes Simulia Corp. SIMULIA User Assistance 2020. 2019. Available online: https://www.3ds.com/support/documentation/users-guides/ (accessed on 28 March 2023).
25. Sosa, E.M.; Wong, J.C.-S.; Adumitroaie, A.; Barbero, E.J.; Thompson, G.J. Finite element simulation of deployment of large-scale confined inflatable structures. *Thin-Walled Struct.* **2016**, *104*, 152–167. [CrossRef]
26. Cao, Z.; Wan, Z.; Yan, J.; Fan, F. Static behaviour and simplified design method of a Tensairity truss with a spindle-shaped airbeam. *J. Constr. Steel Res.* **2018**, *145*, 244–253. [CrossRef]
27. *ASTM D4851-07(2019)e1*; Standard Test Methods for Coated and Laminated Fabrics for Architectural Use. ASTM International: West Conshohocken, PA, USA, 2019. [CrossRef]
28. Myers, R.H.; Montgomery, D.C.; Anderson-Cook, C.M. *Response Surface Methodology: Process and Product Optimization Using Designed Experiments*; John Wiley & Sons: Hoboken, NJ, USA, 2016.
29. Jalani, M.A.; Saad, M.R.; Abdullah, M.F.; Md Naiem, M.A.; Razali, M.N.; Zainal Abidin, N.; Abdul Rahman, M.R. A Statistical Analysis for Optimisation of a Hybrid BBDB-PA in Mantanani Island, Sabah. *J. Mar. Sci. Eng.* **2023**, *11*, 386. [CrossRef]
30. Yin, P.; Wang, K.; Chen, L.; Zhang, Y.; Yang, K.; Wang, J. Horizontal Bearing Capacity and Reliability of Piles in Coastal Soft Soil Considering the Time-Varying Characteristics. *J. Mar. Sci. Eng.* **2023**, *11*, 247. [CrossRef]
31. Chen, J.; Chen, W.; Wang, M.; Ding, Y.; Zhou, H.; Zhao, B.; Fan, J. Mechanical behaviors and elastic parameters of laminated fabric URETEK3216LV subjected to uniaxial and biaxial loading. *Appl. Compos. Mater.* **2017**, *24*, 1107–1136. [CrossRef]
32. Wood, J.D. The flexure of a uniformly pressurized, circular, cylindrical shell. *J. Appl. Mech.* **1958**, *25*, 453–458. [CrossRef]
33. Malm, C.G.; Davids, W.G.; Peterson, M.L.; Turner, A.W. Experimental characterization and finite element analysis of inflated fabric beams. *Constr. Build. Mater.* **2009**, *23*, 2027–2034. [CrossRef]
34. Xu, J.; Zhang, Y.; Wu, M.; Zhao, Y. Experimental analysis of off-axis mechanical behaviors of PVC coated fabrics subjected to cyclic loading. *Polym. Test.* **2019**, *80*, 106090. [CrossRef]
35. Chen, S.; Ding, X.; Yi, H. On the anisotropic tensile behaviors of flexible polyvinyl chloride-coated fabrics. *Text. Res. J.* **2007**, *77*, 369–374. [CrossRef]

36. Wang, C.G.; Liu, M.X.; Kang, J.T.; Xue, Z.M.; Tan, H.F. Bending wrinkling and kink behaviors of inflated beam under local uniform loadings. *Int. J. Mech. Sci.* **2017**, *120*, 136–148. [CrossRef]
37. Clapp, J.D.; Young, A.C.; Davids, W.G.; Goupee, A.J. Bending response of reinforced, inflated, tubular braided fabric structural members. *Thin-Walled Struct.* **2016**, *107*, 415–426. [CrossRef]

Journal of
Marine Science and Engineering

MDPI

Article

A CFD-FEA Method for Hydroelastic Analysis of Floating Structures

Nan Gu [1,2], Deli Liang [3], Xueqian Zhou [1,2,4] and Huilong Ren [1,2,5,*]

1 College of Shipbuilding Engineering, Harbin Engineering University, Harbin 150001, China
2 International Joint Laboratory of Naval Architecture and Offshore Technology between Harbin Engineering University and University of Lisbon, Harbin 150001, China
3 Beijing Institute of Astronautical System Engineering, Beijing 100076, China
4 Yantai Research Institute, Harbin Engineering University, Yantai 264000, China
5 Qingdao Key Laboratory of marine Structure Environmental Adaptability, Qingdao 266071, China
* Correspondence: renhuilong@hrbeu.edu.cn

Abstract: The so-called large multi-body floating offshore structure is a new type of offshore structure with a huge and extremely flat deck area, which has a promising prospect as a floating port and also in applications in the area of marine space exploitation. Due to its unique structural form, the hydrodynamic and structural response characteristics are very complex. Specifically, due to the instantaneous position variation in the body surface, the nonlinearity of the free surface, the interactions between floating bodies, and the elastic deformation of floating bodies, the nonlinear factors are significant and cannot be neglected. For these kinds of problems, methods based on CFD-FEA (computation fluid dynamics–finite element analysis) coupling simulation have significant advantages over traditional methods. In the present paper, the hydrodynamic and structural response characteristics of a large multi-body floating offshore structure are studied using a CFD-FEA method, and the results are compared with those obtained by the potential-flow-based commercial code SESAM/WADAM, and a three-dimensional nonlinear hydroelastic analysis commercial code COMPASS-WALCS-NE. The comparison and the analysis of the results show that the CFD-FEA method presented in this study can well simulate the behavior of the hydroelastic responses of flexible floating structures and has the potential to capture complex nonlinear behaviors.

Keywords: floating structure; CFD-FEA; hydroelasticity; nonlinear wave loads

Citation: Gu, N.; Liang, D.; Zhou, X.; Ren, H. A CFD-FEA Method for Hydroelastic Analysis of Floating Structures. *J. Mar. Sci. Eng.* **2023**, *11*, 737. https://doi.org/10.3390/jmse11040737

Academic Editors: Giuseppe Roberto Tomasicchio and Jens Engström

Received: 12 February 2023
Revised: 12 March 2023
Accepted: 21 March 2023
Published: 28 March 2023

1. Introduction

At present, offshore structures tend to increase in dimension and complexity. The so-called large multi-body floating offshore structure shown in Figures 1 and 2 is a new type of floating offshore structure which has appeared in recent years, which is mainly composed of a substructure, connecting structure (truss or column), upper box structure, and superstructure. The substructure is composed of a row of distributed floating bodies. The length and width of the floating offshore structure are very large, while the height is very small. It is an extremely flat and flexible structure. Its stiffness and vibration natural frequency are low, and the elastic deformation cannot be neglected. The calculation of the floating body as a rigid body may greatly underestimate the bending moment [1]. In addition, under the wave action of multiple floating bodies, the reflected wave of a floating body will affect the motions and loads on other floating bodies, and waves will be amplified or sheltered in some areas. The fluid flow around the floating bodies is very complex, especially when the spacing between dispersed floating bodies is small and the fluid resonates inside the spacing. In addition, the viscous effect is also obvious, which makes it particularly difficult to predict the motions and loads of floating bodies. Since the size of a single floating body in the substructure is relatively small, strong nonlinear phenomena, such as wave overtopping the floating body and the water entry and exit of the floating body, will occur under severe sea conditions. Therefore, it is necessary to analyze the deformation of this large-scale offshore structure under fluid loads and the influence of floating body deformation using nonlinear hydroelastic methods accounting for the fluid viscosity, so as to obtain the wave loads and structural responses. According to the results of the tank model test, the deformation and displacement of each floating body are consistent at each time in head seas. Therefore, in order to reasonably reduce the difficulty and complexity and also to focus on the main problems, the present study chose to analyze the hydroelastic behavior of a single body instead of an entire floating structure and verify the feasibility of the CFD-FEA coupling method. The large multi-body floating offshore structure has many usages. The floating offshore structure of a length of 300 m is an independent and complete module, i.e., a module of a 300 m long structure alone can also be used as a floating port; in this case, a single module can be analyzed. If a larger length is required, such as that used as a floating airport, several modules can be connected by connectors. In the case of a large floating structure comprising of several modules, it is still reasonable and appropriate to start with the hydroelastic analysis of a single module.

Figure 1. Large multi-body floating offshore structure.

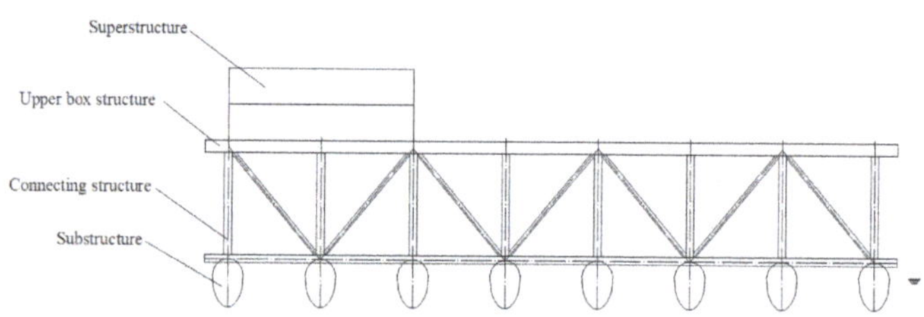

Figure 2. Cross-section of large multi-body floating offshore structure.

Since the late 1970s, the potential-flow-based theory and methods for hydroelasticity analysis have made remarkable progress. Betts et al. [2] proposed a two-dimensional hydroelasticity theory. Wu [3] devised a fully three-dimensional approach using a generalized boundary condition of the fluid–solid interface, and the method has been very successful and widely applied in the community. However, the limitations of the methods based on potential flow theory are also obvious. Compared with potential-flow-based theory, CFD (computational fluid dynamics) methods are more accurate in describing the changes in velocity and pressure in the fluid domain, capturing the nonlinear phenomenon of waves on the free surface, and simulating the fine flow field with hydrodynamic interactions between multiple bodies. Xie et al. [4], Wang and Guedes Soares [5], Jiao and Huang [6], Huang et al. [7], and Jiao et al. [8] applied CFD-based methods to analyze the seakeeping and wave loads of ships and offshore structures. It has been found that, compared with potential-flow-based methods, CFD methods can more realistically simulate the flow field and obtain more accurate numerical results.

In recent years, with the development of CFD technology and the improvement of computer performance, some researchers have begun to study how to couple CFD (computational fluid dynamics) and FEA (finite element analysis) for hydroelasticity analysis. El

Moctar et al. [9] and Oberhagemann [10] combined the CFD method and the Timoshenko beam model for a two-way coupling simulation of ship slamming loads, and validation against the experimental results was also conducted. Because the Timoshenko beam model is relatively simple, only vertical bending and horizontal bending were studied, thus it is not applicable to the torsion problem. Wilson et al. [11] used a CFD method to study the flow of a ship in waves, and then applied the motions and pressure distribution on the ship's hull to the finite element model of the ship's structure for a structural analysis. In order to reduce the amount of calculation, only a one-way coupling method was adopted, i.e., only the influence of fluid on the structure was considered, while the influence of structural deformation on the fluid was neglected. Liu et al. [12] developed a CFD-FEM (computation fluid dynamics–finite element method) coupling simulation method for estimating the hydroelastic responses of the floating elastic plate. For offshore structures with complex forms, the finite element model for the entire hull structure would also be very complex, thus the amount of time required for the fluid–structure coupling calculation would be very large and result in great difficulties in practical applications. Lakshmynarayanana and Hirdaris [13] and Lakshmynarayanana and Temarel [14] studied the symmetrical motions and loads of the S175 container ship in severe waves and under slamming impacts using one-way and two-way CFD-FEA coupling methods. It was shown that the two-way coupling method is able to more accurately estimate the hydroelastic effects of ships in waves. Jiao et al. [15] used the two-way CFD-FEA coupling method to predict the motions, wave loads, and hydroelastic responses of the S175 container ship in regular waves. All the methods adopted in [13–15] are based on the linear Timoshenko beam element to simplify the simulation of the hull structure, neglecting the influence of rotational inertia on the motions and wave loads of offshore structures. This simplification is acceptable for the case of head seas, but not for quartering and beam seas as the effect of rotational inertia is significant and cannot be completely neglected.

To sum up, the existing hydroelastic analysis methods have certain limitations, and there are still some difficulties in calculating the wave loads on the large multi-body floating offshore structure shown in Figure 1. Previous studies focused on traditional ships, such as barges or container ships. Although traditional ship types can also reach 300 m in length, due to the differences in structural forms, the hydrodynamic and structural response characteristics are different. The hydroelastic analysis method for traditional ship types may not be well suited for large multi-body floating offshore structures. For instance, wave overtopping on the structure is a phenomenon that may behave differently between the two types of structures. To capture such behaviors with complex phenomena, the CFD-FEA coupling method, for its inherent natures of methodology, is a possible solution.

In this study, a CFD-FEA coupling simulation approach is devised, where the fluid domain is solved by the RANS (Reynolds-averaged Navier–Stokes)-based commercial code STAR-CCM+ [16], and the structure domain is solved by the commercial FE code ABAQUS [17]. The advantage of this method is that it can fully consider the interaction between elastic structure and fluid through two-way fluid–structure coupling and it can effectively solve strong nonlinear problems such as wave overtopping the floating body and water entry and exit of the floating body. At the same time, it can significantly reduce the computational time of the fluid–structure coupling calculation, and it is applicable to the case of head seas, quartering seas, and beam seas. It is suitable for solving the nonlinear wave loads of large multi-body floating offshore structures accounting for the influence of hydroelasticity. In the present study, a single floating body of the large multi-body floating offshore structure as shown in Figure 1 is studied, and the hydrodynamic and structural response characteristics of a single floating body of the large multi-body floating offshore structure are studied with a CFD-FEA coupling method. By comparing the potential-flow-based numerical results for wave loads with and without accounting for the elasticity, the CFD-FEA-based method for hydroelastic analysis of flexible floating structures is validated. It is found that the CFD-FEA method has the potential to capture the complex phenomena of the wave–structure interaction.

The paper is organized as follows: the problem and method of solution used in this study are described in Section 2. Then, in Section 3, the numerical modeling and mesh sensitivity analysis are carried out with the verification of the numerical method. In Section 4, the numerical method is validated by comparing the potential-flow-based numerical results for wave loads with and without accounting for the elasticity. Finally, the conclusion is drawn in Section 5.

2. Problem Statement and Method of Solution

2.1. Problem Statement

In the wave loads on a single floating body of the large multi-body floating offshore structure, the nonlinear factors of the hydrodynamic problem are significant, such as the instantaneous position variation in the body surface, the nonlinearity of the free surface, the elastic deformation of the floating body. These nonlinear factors have a great influence on the loads and motions of the floating body. In order to account for the nonlinear factors, a two-way fluid–structure simulation approach is adopted by coupling CFD and FEA. The fluid domain is solved using the finite volume method (FVM) using STAR-CCM+, and the structure domain is solved by the finite element method (FEM) using ABAQUS.

For the fluid domain, according to Park et al. [18], the contribution of the compressibility and surface tension of the fluid is small and can be neglected. The mass conservation equation and Navier–Stokes equation are used to describe the flow of viscous fluid:

$$\nabla \cdot v = 0 \tag{1}$$

$$f - \frac{1}{\rho}\nabla p + \mu \nabla^2 v = \frac{\partial v}{\partial t} + (v \cdot \nabla) v \tag{2}$$

where p is the fluid pressure, v is the velocity vector, ρ is the fluid density and μ is the dynamic viscosity, and f is the gravity.

For the structure domain, assuming that the structure is a linear elastic material, the motion equation of the structure is obtained using the finite element method:

$$[M]\{\ddot{u}\} + [C]\{\dot{u}\} + [K]\{u\} = \{F\} \tag{3}$$

where $[M]$ is the mass matrix of the structure, $[C]$ is the damping matrix of structure, $[K]$ is the stiffness matrix of the structure, $\{F\}$ is the external fluid loads from the CFD solution in the fluid domain, including the fluid pressure and shear force, and $\{u\}$ is the motions and deformations of the structure.

2.2. Coupling Approach

The coupling approach is shown in Figure 3, where t_0 is the initial time and Δt is the time step. At the initial time, the finite volume method (FVM)-based code STAR-CCM+ is used to calculate the fluid pressure and shear force on the floating body surface, and then the quantities obtained by the FVM are applied to the floating body surface in the finite element model in ABAQUS by means of data mapping. Under the action of fluid pressure and shear force in the fluid domain, the velocity and acceleration of the finite element nodes of the floating body will vary, which leads to the motions and deformations of the fluid–structure coupling interface. Then, the node displacements are transferred to STAR-CCM+ to update the coupling interface. The pressure field and velocity field obtained by STAR-CCM+ are then transferred to the next time step, as well as the velocity and acceleration of the nodes obtained by ABAQUS.

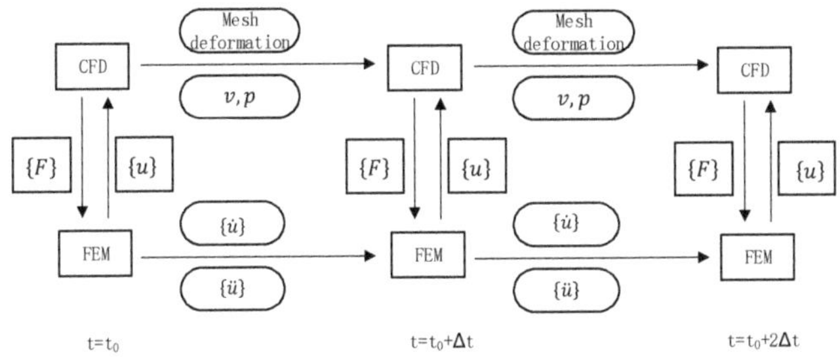

Figure 3. Framework of two-way coupling of CFD-FEA.

2.3. Coordinate Systems

Three right-handed orthogonal coordinate systems are used to describe the floating body motions: earth-fixed coordinate system $O-XYZ$, body-fixed coordinate system $G-x_by_bz_b$, and steadily translating coordinate system $o-xyz$.

The origin of the earth-fixed coordinate system $O-XYZ$ lies on the still water surface. The positive X axis is in the direction of the wave propagation, the positive Z axis is directed upwards, and β is the wave direction, as shown in Figure 4.

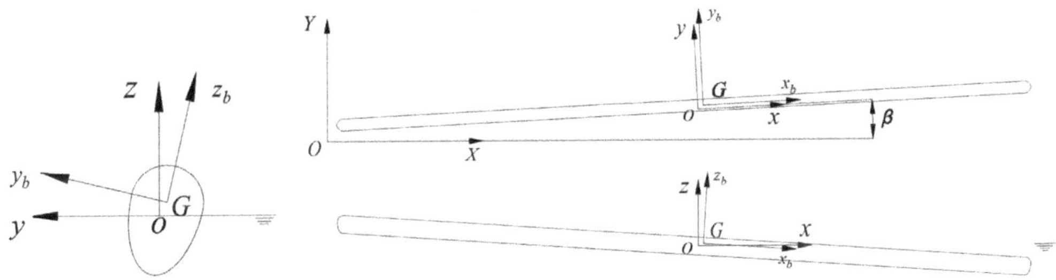

Figure 4. Coordinate systems.

The body-fixed coordinate system $G-x_by_bz_b$ is connected to the ship, with its origin placed at the center of gravity G. The positive x_b axis is in the longitudinal forward direction. The positive y_b axis is pointing in the port side direction. The positive z_b axis is perpendicular to the waterline of the hull and is pointing upward.

The steadily translating coordinate system $o-xyz$ is moving forward at the nominal speed. If the floating body is stationary, the origin of the coordinate system $o-xyz$ is located at the intersection of the still water surface, the midship cross-section and the center plane, and the directions of the $o-xyz$ axes are the same as those of the $G-x_by_bz_b$.

2.4. Backbone Beam

A backbone beam is used to simulate the stiffness of the whole floating body. The stress and deformation of the micro-segment of the backbone beam are analyzed. The micro-segment of the backbone beam is shown in Figure 5 [19]. The beam is located in the $o-xyz$ coordinate system. As shown in Figure 5c, the reference point o' is the intersection of the ox axis and the cross-section of the micro-segment when floating still, i.e., the intersection of Gx_B axis and section of the micro-segment. In Figure 5c, C is the mass center of the micro-segment; S is the shear center of the micro-segment; $Z_C(x)$ is the distance between the mass centroid C and o'; $Z_S(x)$ is the distance between the shear center S and o'; and $\overline{Z}(x)$ is the distance between the center of mass C and the shear center S, $\overline{Z}(x)=Z_S(x)-Z_C(x)$.

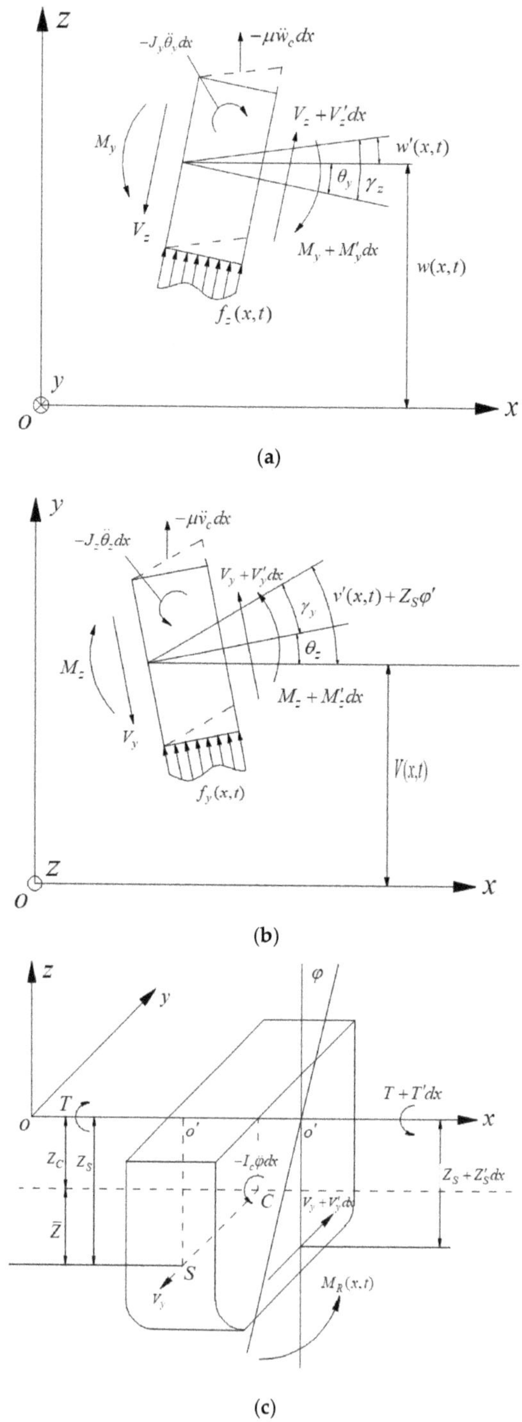

Figure 5. Micro-segment of backbone beam: (**a**) vertical motion; (**b**) horizontal motion; (**c**) torsional motion.

For the vertical bending, as shown in Figure 5a, the differential equations of the motion of the backbone beam are:

$$\begin{cases} \mu(x)\ddot{w}(x,t) = V_z'(x,t) + f_z(x,t) \\ J_y(x)\ddot{\theta}_y(x,t) = M_y'(x,t) - V_z(x,t) \end{cases} \quad (4)$$

For the horizontal bending and torsion, as shown in Figure 5b,c, the differential equations of the motion of the backbone beam are:

$$\begin{cases} \mu(x)\ddot{v}_c(x,t) = V_y'(x,t) + f_y(x,t) \\ J_z(x)\ddot{\theta}_z(x,t) = M_z'(x,t) + V_y(x,t) \\ I_c(x)\ddot{\varphi}(x,t) = T'(x,t) + \overline{Z}(x)V_y'(x,t) + Z_S'(x)V_y(x,t) + M_R(x,t) \end{cases} \quad (5)$$

where $\mu(x)$ is the mass distribution per unit length of the beam; $J(x)$ is the moment of inertia per unit length of the beam; I_c is the torsional moment of inertia per unit length passing through the mass center C and parallel to the GX_b axis, and its relationship with the moment of inertia $I_O(x)$ about the GX_b axis is $I_o(x) = I_C(x) + Z_C^2(x)\mu(x)$; $w(x,t)$ is the vertical bending deformation of the cross-section center; $\theta(x,t)$ is the rotation angle of the section due to bending; $\varphi(x,t)$ is the rotation angle of the cross-section due to torsion; f is the force acting on the unit length of the beam; $M(x,t)$ is the cross-section bending moment; $V(x,t)$ is the cross-section shear force; $T(x,t)$ is the cross-section disturbing force; and $M_R(x,t)$ is the roll moment acting on the unit length of the beam. The subscript y denotes the y direction or rotating about the y axis, and the subscript z stands for the z direction or rotating about the z axis. The dot over the variable denotes the time derivative, and the prime indicates the partial derivative of the length x.

According to the Timoshenko beam theory, the relationship between the force and deformation on the section can be obtained as below. For the vertical bending,

$$\begin{cases} V_z(x,t) = kGA_V(x)\left[\gamma_z(x,t) + \alpha_V(x)\dot{\gamma}_z(x,t)\right] \\ M_y(x,t) = EI_V(x)\left[\theta_y'(x,t) + \beta_V(x)\dot{\theta}_y'(x,t)\right] \end{cases} \quad (6)$$

For the horizontal bending and torsion,

$$\begin{cases} V_y(x,t) = kGA_H(x)\left[\gamma_y(x,t) + \alpha_H(x)\dot{\gamma}_y(x,t)\right] \\ M_z(x,t) = EI_H(x)\left[\theta_z'(x,t) + \beta_H(x)\dot{\theta}_z'(x,t)\right] \\ T(x,t) = C(x)\varphi'(x,t) - [C_l(x)\varphi''(x,t)]' + \Gamma(x)\dot{\varphi}'(x,t) \end{cases} \quad (7)$$

where kGA is the shear stiffness of the section; EI is the bending stiffness of the section; C is the torsional stiffness of the section; α is the shear damping coefficient; β is the bending damping coefficient; Γ is torsional damping coefficient; γ is the shear angle; and C_l is the warping stiffness. Subscripts V and H indicate that the quantities are related to a vertical or horizontal motion, respectively.

The geometric relationship of the deformation of the backbone beam is:

$$w'(x,t) = \gamma_z(x,t) - \theta_y(x,t) \quad (8)$$

$$[v(x,t) + Z_S(x)\varphi(x,t)]' = Z_S'(x)\varphi(x,t) + \gamma_y(x,t) + \theta_z(x,t) \quad (9)$$

In the above Equations (4), (6) and (8) constitute the governing equations of vertical motion, and (5), (7) and (9) constitute the horizontal motion and torsional motion of the backbone beam.

2.5. Section Loads Calculation

Through the above calculation, the pressure distribution, velocity, and acceleration on the surface of the floating body can be obtained, and then the forces and moments at the cross-section of the floating body can be calculated according to the following formula [20].

$$\{\overline{Q}\} = -\iint_{S_x} P\{n\}ds + [M]g - [M][\ddot{\mu}] + \{\Delta\overline{Q}\} \tag{10}$$

where $\{\overline{Q}\} = (N\ SF_Y\ SF_Z\ TM\ BM_Y\ BM_Z)^T$ is the six-component of the section loads, S_x is the wetted surface area of a floating body over a partial length, P is the total fluid pressure and shear force, $\{\Delta\overline{Q}\} = (0\,0\,0\,0\ x\cdot SF_Z - x\cdot SF_Y)^T$ is the additional item, $[M]$ is the mass matrix of the partial length floating body, and $[\ddot{\eta}]$ is the 6-DOF acceleration of the floating body.

3. Numerical Modeling and Simulation

In this section, the numerical modeling is introduced. Then, the extent of the computational domain, boundary condition, time step, and mesh sensitivity analysis are discussed and analyzed, and the numerical method is verified.

3.1. Numerical Modeling

The length of a single floating body of large multi-body floating offshore structure is 300 m, the height is 7.5 m, the width is 5 m, the draft is 5 m, and the mass is evenly distributed along the length of the single floating body. The transverse cross-section of the single floating body is shown in Figure 6, and the thickness of the single floating body plate is 20 mm. The hollow rectangular beam is used as the backbone beam to simulate the stiffness of the whole floating body. The floating body has 21 stations (0–20 stations) in total, and is divided into 20 sections, as shown in Figure 7. The red line in Figure 7 denotes the backbone beam.

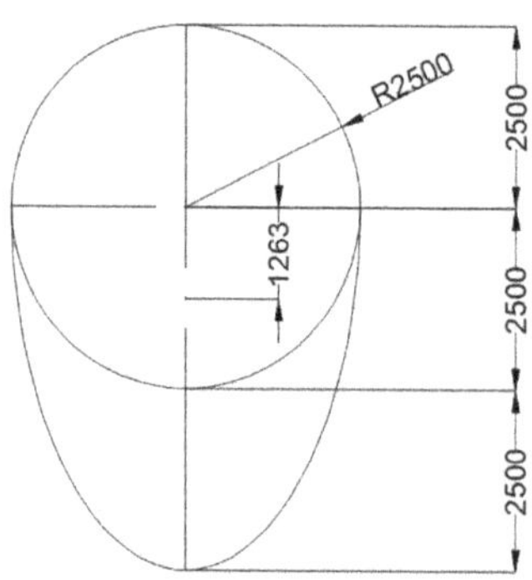

Figure 6. Cross-section of a single floating body.

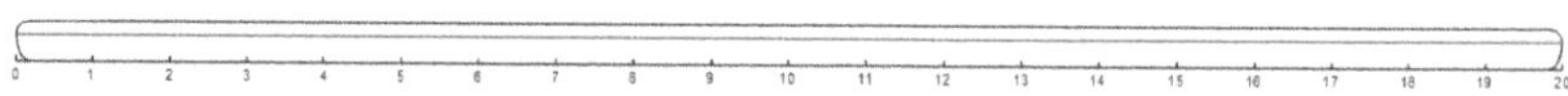

Figure 7. Profile of a single floating body.

The vertical bending stiffness and longitudinal torsional stiffness of the actual floating body cross-section and backbone beam are shown in Table 1.

Table 1. Stiffness parameters.

	Symbols	Units	Actual Value	Backbone Beam	Deviation
Moment of inertia	I_y	m^4	2.61	2.60	−0.30%
Torsional moment of inertia	I_n	m^4	3.74	3.74	0.10%
Vertical bending stiffness	EI_y	$N \cdot m^2$	5.37×10^{11}	5.35×10^{11}	−0.30%
Longitudinal torsional stiffness	GI_n	$N \cdot m^2$	2.96×10^{11}	2.97×10^{11}	0.10%

The single floating body of a large multi-body floating offshore structure is shown in Figure 8. The structural model is composed of two parts, the hull surface and the backbone beam, as shown in Figure 9. The hull surface of the floating body is modeled using shell elements, denoted by the blue surface in Figure 9. The hull surface is the fluid–structure coupling interface, where the external fluid loads are mapped from the fluid domain, which makes no contribution to the overall stiffness and mass of the floating body. The backbone beam is simulated by three-dimensional Timoshenko beam elements, denoted by the red line in Figure 9, which contributes to the stiffness, mass, and moment of the inertia of the floating body. The beam elements are connected to the shell elements by kinematic coupling constraints. The CFD solutions for the external fluid loads in the fluid domain are transferred to the FEA model in the structure domain through the hull surface for structural dynamic analysis. Once the structural dynamic analysis is done, the solutions for the motions and deformations of the backbone beam are obtained. Because of the kinematic coupling constraints between the beam elements and the shell elements, the motions and deformations of the backbone beam will also be fed back to the hull surface for the subsequent hydrodynamic analysis in the fluid domain.

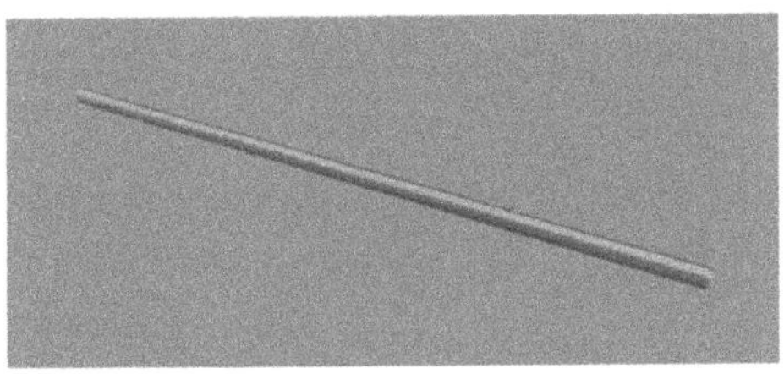

Figure 8. Geometric model of the single floating body.

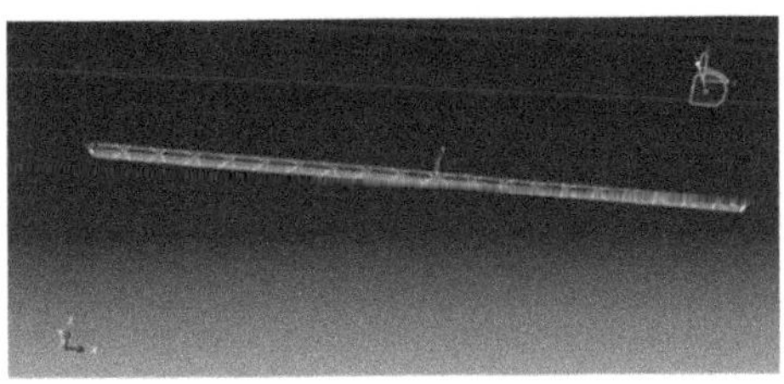

Figure 9. FEA model of the single floating body.

The fluid domain modeling and mesh generation are carried out in STAR-CCM+. The wavelength is set to be L, the length and width of the fluid domain are $4L$, the distance between the static water surface and the bottom of the domain is $2L$, and the distance between the static water surface and the top is $2L$. The boundary conditions are set [8,21] as shown in Figure 10. A pressure outlet is applied at the top boundary. The nonslip wall boundary condition is applied at the floating body surface. Since the floating body is a symmetrical structure, and so is the flow about it in head seas, only half of the flow field needs to be simulated. The symmetry plane boundary condition is applied at the center plane of the structure. Regarding the other four boundary planes, the velocity inlet is applied, where the velocity of the wave is prescribed to avoid the gradient generated from the wall and flow [21]. The volume of fluid (VOF) method is used to track free surfaces at the interface between air and water [22]. The fluid domain is meshed as shown in Figure 11. The extent of the whole computational domain is determined by referring to the recommendations of ITTC [23]. The distance between the bow of the floating body and the inlet boundary is 1.5*L*, and the distance between the stern of the floating body and the outlet boundary is also 1.5*L*. The FVM is used to solve the fluid domain, and the fluid pressure and shear force on the surface of the floating body are obtained. The direction of the force due to the pressure is perpendicular to the surface of the floating body, and the direction of the shear force is tangent to the surface of the floating body.

The Courant number is used to evaluate the time step requirements of a transient simulation for a given mesh size and flow velocity.

$$C = \frac{U\Delta t}{\Delta x} \tag{11}$$

where C is the Courant number, U is the flow velocity, Δt is the representative time step of the simulation, and Δx is the characteristic size of the mesh cell.

In order to ensure the accuracy and stability of the numerical results, the Courant number should be less than one [24]. A constant time step of 0.01 s is chosen for the simulations.

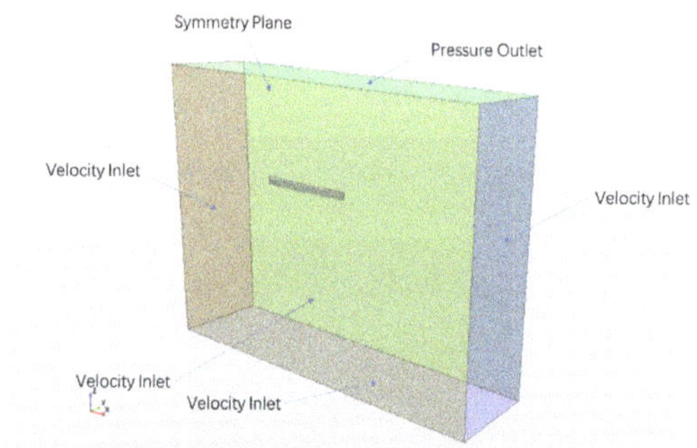

Figure 10. Boundary condition.

Figure 11. *Cont.*

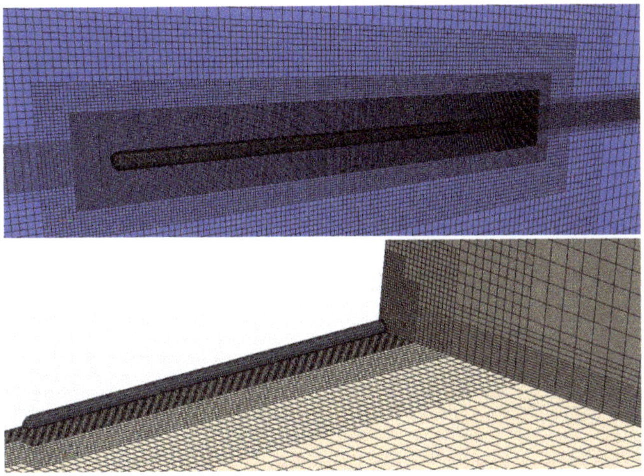

Figure 11. CFD mesh of fluid domain.

The long-crest regular waves are created using linear airy wave theory. Since the offshore structure is in a moored condition during the operation, the forward speed is set to zero in the simulation. In head seas, the heave and pitch motions of the floating body are released, while the surge motion is constrained. Additionally, because of the symmetry, only half of the flow field is simulated, and a symmetrical boundary condition is applied at the center plane, that is, there are no sway, roll, and yaw motions.

3.2. Mesh Sensitivity Analysis

For the mesh convergence analysis, three sets of meshes, as shown in Table 2, are generated for the structure domain and fluid domain mentioned above. A regular wave of 300 m in wave length and 16 m in wave height is used for the mesh sensitivity analysis.

Table 2. Three sets of meshes.

Mesh	Minimum Size of Fluid Domain			Base Size of Structure Domain	Number of Meshes
	x	y	z		
Mesh A	1.56 m	6.00 m	6.00 m	1.50 m	0.8 million
Mesh B	0.78 m	6.00 m	6.00 m	0.75 m	1.5 million
Mesh C	0.24 m	0.24 m	0.24 m	0.50 m	5.2 million

The accuracy of the wave simulation is very important for the accuracy of the simulation of floating body's motions and loads. According to the recommendation by STAR-CCM+, which has also been validated by many researchers [25], a minimum of 40 cells per wavelength and 20 cells per wave height on the free surface is necessary to produce a stable wave with an acceptable dissipation. The mesh created according to the principle is Mesh B, as shown in Table 2, and the mesh which is relatively coarser is Mesh A, while the mesh which is relatively coarser is Mesh C.

The numerical results for the amplitudes of the heave and pitch motion of the single floating body are presented in Figure 12. The values of the heave and pitch motion are only rigid body displacements and do not include the elastic deformations of the floating body. For the heave motion, the results obtained with the three sets of meshes are very close to each other. For the pitch motion, the results obtained with Mesh B are very close to that obtained with Mesh C, with an average difference of 1.7%, but there is a relatively large difference, 21%, between the results obtained with Mesh A and Mesh B.

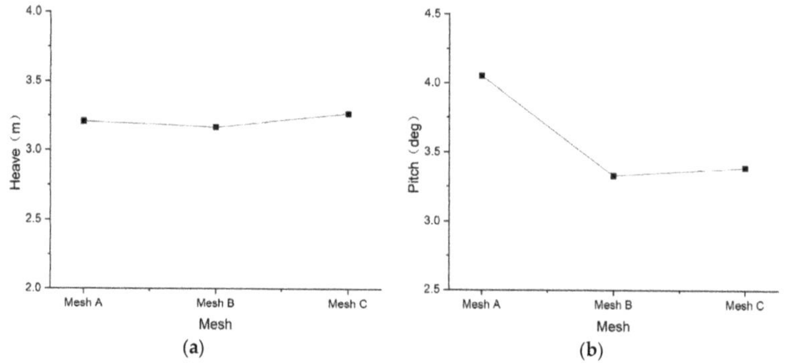

Figure 12. Comparison of the results of three sets of meshes: (**a**) heave of the single floating body; (**b**) pitch of the single floating body.

The numerical results for the amplitudes of the vertical bending moment at midship are presented in Figure 13. Mesh B and Mesh C produce similar results, and the average difference between them is 1.6%. In comparison, there is a relatively large average difference, 35%, between the results obtained with Mesh A and Mesh B.

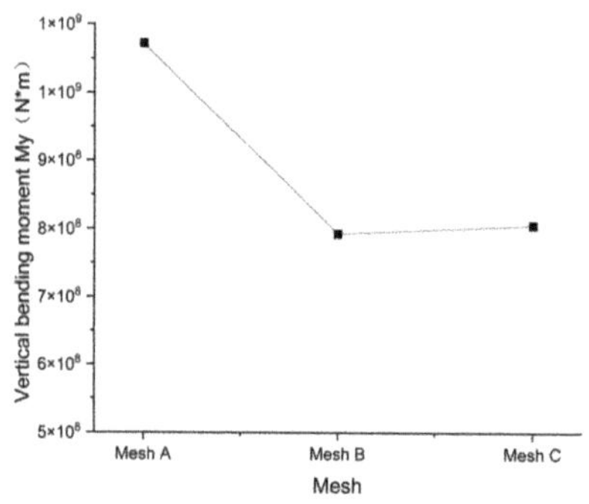

Figure 13. Comparison of numerical results of vertical bending moment at cross-section $x = 0$ m.

From the comparison of the results of three sets of meshes, it is found that the results from Mesh B and Mesh C are very close to each other, which shows the good mesh convergence property of the numerical model. Mesh B, with the meshing principle introduced above, is selected in this paper considering the accuracy and efficiency of the calculation.

3.3. The Influence of Shear Force along the Wetted Surface

Because the fluid is assumed to be inviscid in potential flow theory, the shear force acting on the wetted surface of the floating body is neglected in SESAM/WADAM [26] and COMPASS-WALCS-NE [27]. As described in Section 2.5, the CFD-FEA method can account for the contribution of the shear force to the global solutions. In order to study the influence of shear force on the vertical bending moment of the cross-section, the numerical results obtained with and without shear force by the CFD-FEA method are compared, as shown in Figures 14 and 15. From the comparison, it can be seen that the influence of shear force on the vertical bending moment of the cross-section is very small.

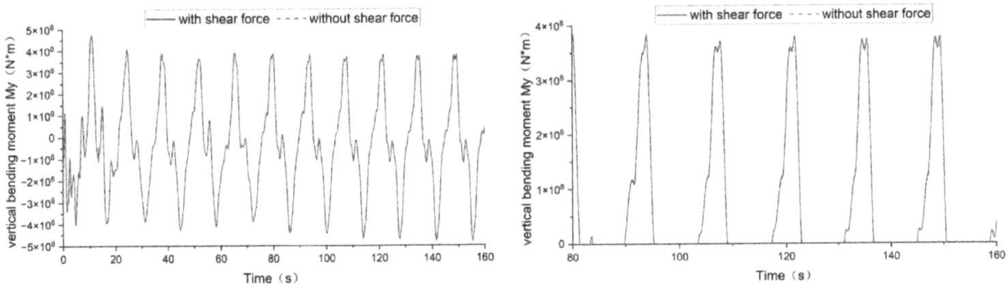

(a) (b)

Figure 14. Comparison of vertical bending moment time history of cross-section $x = -75$ m at wave height 16 m: (**a**) comparison diagram; (**b**) local enlarged diagram.

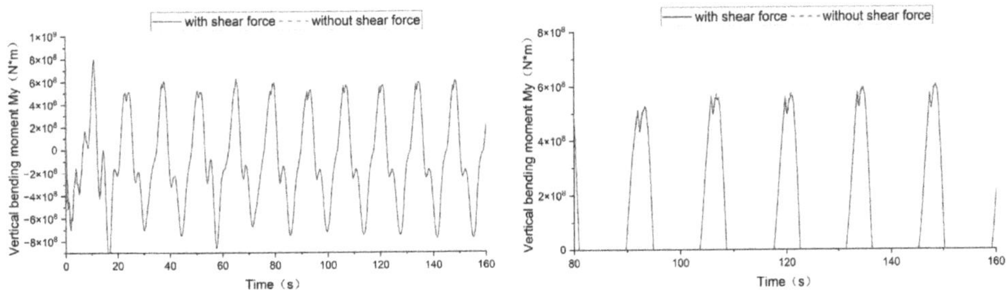

(a) (b)

Figure 15. Comparison of vertical bending moment time history of cross-section $x = 0$ m at wave height 16 m: (**a**) comparison diagram; (**b**) zoom-in of the diagram.

4. Results and Discussion

In this section, the hydrodynamic and structural responses are obtained using the CFD-FEA method. Then, the results are analyzed and compared with those obtained using a potential-flow-based method for wave loads with and without accounting for the elasticity, and the CFD-FEA method for hydroelastic analysis is validated.

4.1. Numerical Results

The simulation is carried out for a range of regular wave heights, as shown in Table 3. The results obtained by the CFD-FEA method are compared with those using the commercial code SESAM. In order to investigate the behavior of the numerical simulation without unnecessary repetitions, one of the regular wave heights in Table 3 is chosen as a typical case for in-depth analysis, namely, head seas with a wave height of 16 m.

Table 3. Regular wave.

Wave Direction	Wave Length	Wave Period	Wave Height
Heading angle 180°	300 m	13.862 s	2.5 m 5.0 m 7.5 m 10.0 m 12.5 m 16.0 m

4.1.1. Wave Simulations

The wave surface elevation of the entire flow field is shown as Figure 16. In Figure 16b, the details of the flow field can be observed clearly, such as water entry and exit of the floating body, elastic deformation of the floating body, and free surface disturbances. Due to the numerical dissipation in the computation and the fluid viscosity, the wave heights are continuously attenuated in the process of wave propagation. In order to verify the accuracy of numerical wave making, several numerical wave height meters are arranged in the flow field to monitor the wave heights at different locations to compare the deviation between the theoretical wave heights and the actual wave heights. The location of numerical wave height meters should be far enough away from the floating body where the influence of the floating body is negligible. Figure 17 shows the actual wave heights at four locations in the flow field. It can be found that the deviations between the actual wave heights and the theoretical wave heights are all within 5%, meeting the accuracy requirements in [25].

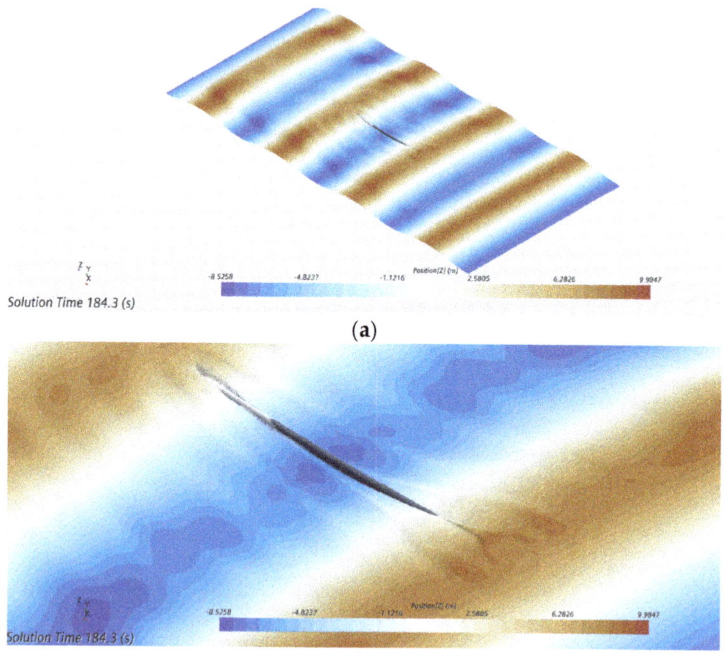

(a)

(b)

Figure 16. (**a**) Wave surface elevation of flow field; (**b**) a close-up of the flow about the structure.

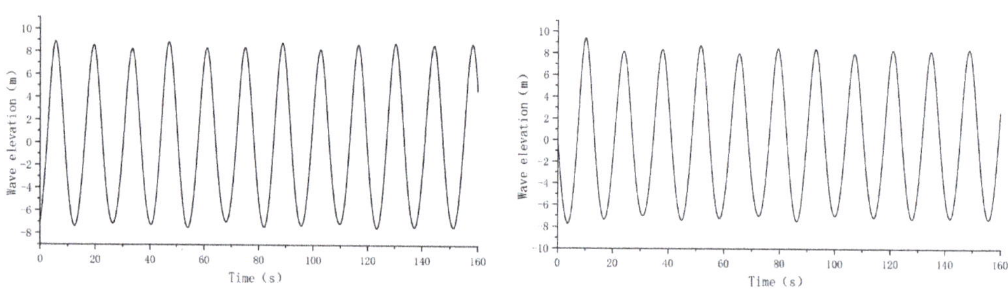

(a)

(b)

Figure 17. *Cont.*

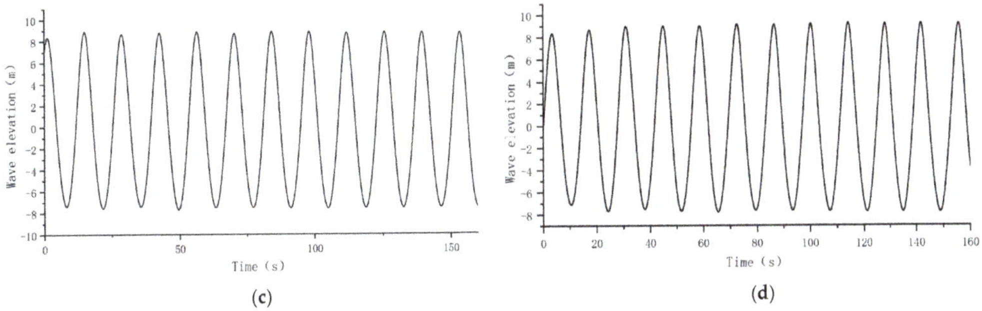

Figure 17. Wave surface elevation at different locations: (**a**) 250 m in front of the bow of the floating body; (**b**) 150 m in front of the bow of the floating body; (**c**) 50 m in front of the bow of the floating body; (**d**) 200 m to the left of the midship section of the floating body.

4.1.2. Motions and Deformations of the Floating Body in Waves

Figure 18a shows the rigid body displacements and the elastic deformations of the floating body in the vertical direction at four time instants within one wave period. The rigid body displacements of the floating body are removed to obtain the vertical elastic deformation of the floating body at four time instants within one wave period, as shown in Figure 18b. It can be seen that at $t = 100$ s, the floating body is in a sagging condition, and at $t = 100\ \text{s} + 1/2T$, the floating body is in a hogging condition, where T is the wave period.

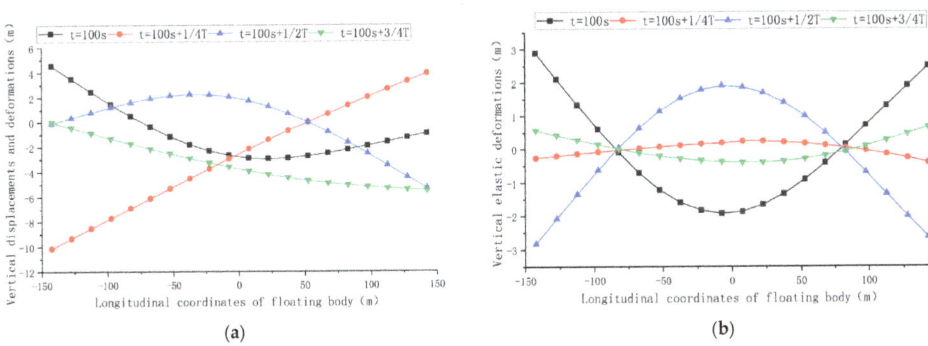

Figure 18. Rigid body displacements and elastic deformations: (**a**) rigid body displacements and elastic deformations in the vertical direction; (**b**) elastic deformations in the vertical direction.

4.1.3. Wave Loads

The pressure on the floating body surface is shown in Figure 19, and the vertical velocity at the center of gravity of each section of the floating body is shown in Figure 20a. Once the pressure, the vertical velocity, and the vertical acceleration are obtained, the vertical bending moment M_y at the cross-sections of the floating body can be obtained from Equation (5), and the time histories are shown in Figure 20b–d.

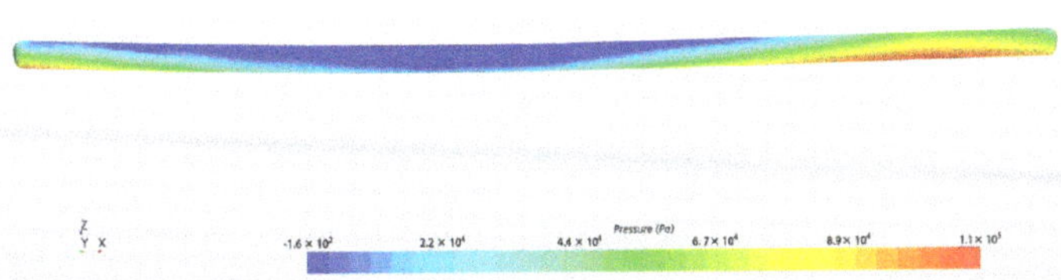

Figure 19. The pressure on the floating body surface.

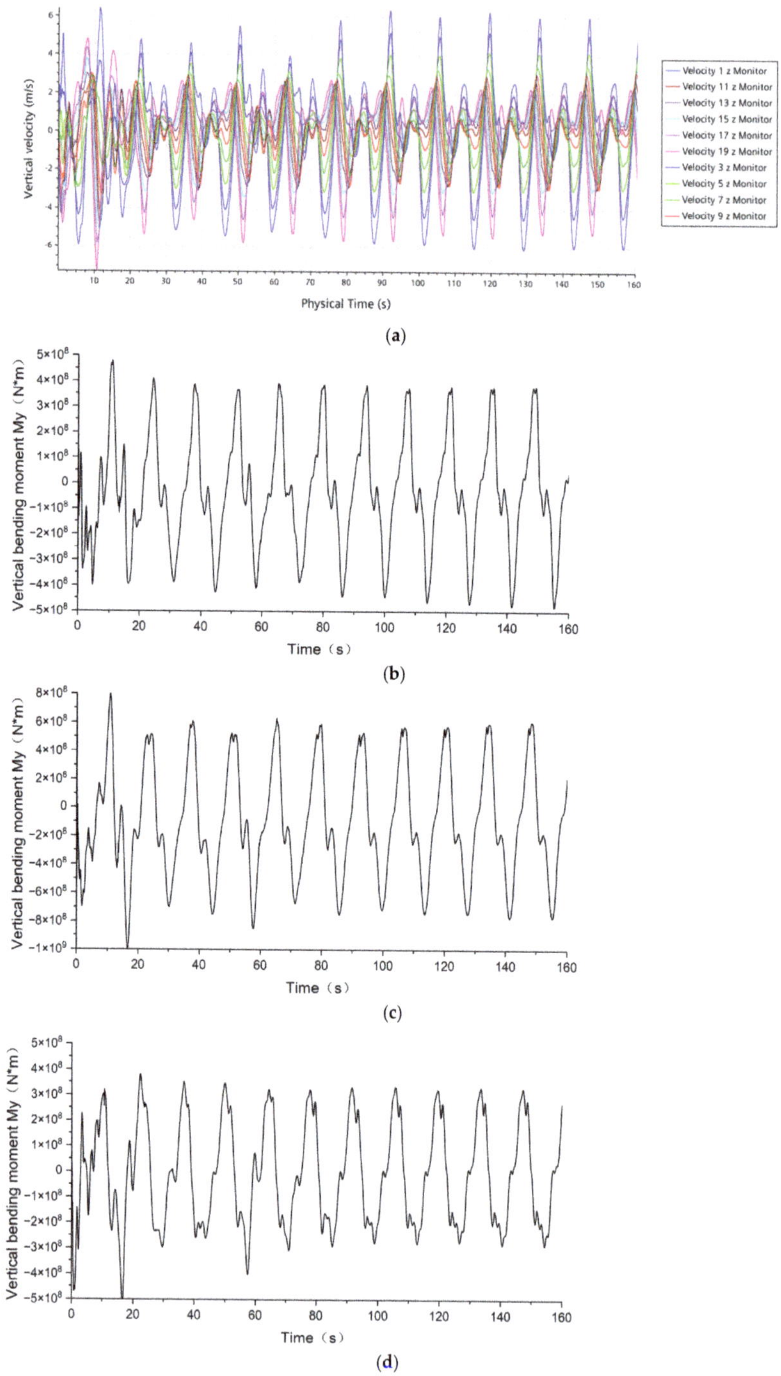

Figure 20. Calculation results of vertical velocity and vertical bending moment: (**a**) vertical velocity at the center of gravity of each section of the floating body; (**b**) vertical bending moment at cross-section $x = -75$ m of the floating body; (**c**) vertical bending moment at cross-section $x = 0$ m of the floating body; (**d**) vertical bending moment at cross-section $x = 75$ m of the floating body.

It can be seen in Figure 20b–d that the results display a periodic pattern of the wave loads on the floating body under the action of regular waves. In order to analyze the change in the wave loads on the floating body, the flow fields at the instant when the vertical bending moment M_y of the cross-section reaches the peak and the trough are compared. For instance, it can be seen in Figure 20c that at $t = 106.84$ s, M_y at the midship cross-section ($x = 0$ m) reaches a peak value; at $t = 113.73$ s, M_y reaches a trough value. Figure 21 shows the side views of the flow field wave surface at $t = 106.84$ s and $t = 113.73$ s, respectively. It can be observed in the figures that when the floating body is in a hogging condition, the vertical bending moment M_y at the middle cross-section reaches a peak value, amd when the floating body is in a sagging condition, the vertical bending moment M_y at the middle cross-section reaches a trough value, that is, the location of the wave crest and trough on the floating body causes a hogging or sagging condition of the floating body. As the wave crest and wave trough move longitudinally along the floating body, the vertical bending moment M_y changes with time.

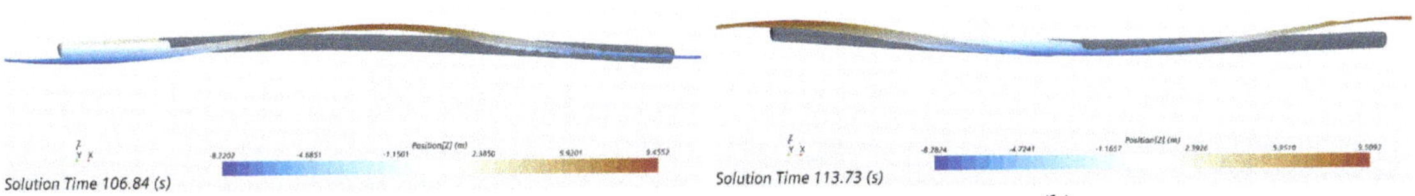

(**a**) (**b**)

Figure 21. Profile of wave surface: (**a**) hogging condition; (**b**) sagging condition.

4.1.4. Fluid Pressure

The locations of the fluid pressure monitoring points on the floating body surface are shown in Table 4, and the time history of the pressure is shown in Figure 22. Although it can be seen in the figure that the pressure extrema occur at the bow, midship, and stern of the floating body at some time instants, but no slamming occurred. In Figure 23, it can be found that, in addition to the periodic extrema of the fluid pressure, at some instants, e.g., 111 s, the fluid pressure at the stern of the floating body also has a local extremum. Since the difference in the fluid force between the bottom and top of the stern is approximately equal to the fluid force at the stern of the floating body, the fluid pressure at the top and the bottom of the stern, and the fluid pressure difference between the bottom and the top of the stern, are compared in Figure 23. In Figure 24, it can be found that at 111 s, the fluid pressures at the top and the bottom of the stern have local extrema, and the difference between the bottom and the top of the stern in the pressure also has local extrema. The instant when the fluid pressure at the stern of the floating body reaches a local extremum, the vertical bending moment at the cross-section of the floating body also has a local extremum, as shown in Figure 20b,c.

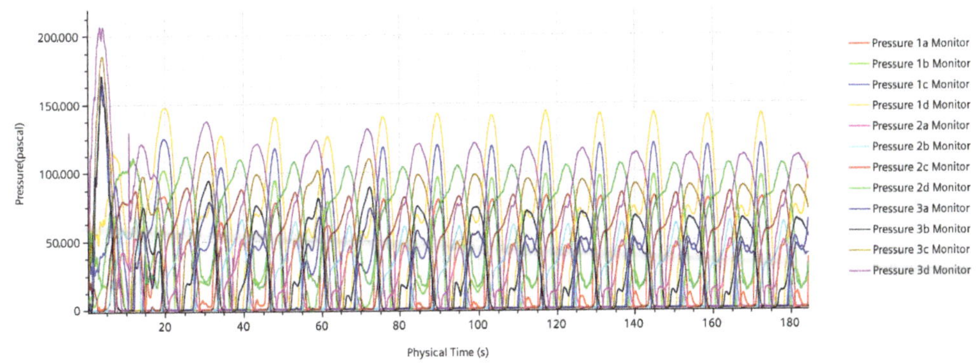

Figure 22. Time history of fluid pressure.

Table 4. Location of fluid pressure monitoring points.

No. of Fluid Pressure Monitoring Point	Location	Coordinates		
		x/m	y/m	z/m
1a	Stern	−147.500	0.000	2.500
1b		−147.500	2.498	0.000
1c		−147.500	2.155	−2.500
1d		−147.500	0.000	−5.000
2a	Midship	0.000	0.000	2.500
2b		0.000	2.498	0.000
2c		0.000	2.155	−2.500
2d		0.000	0.000	−5.000
3a	Bow	147.500	0.000	2.500
3b		147.500	2.498	0.000
3c		147.500	2.155	−2.500
3d		147.500	0.000	−5.000

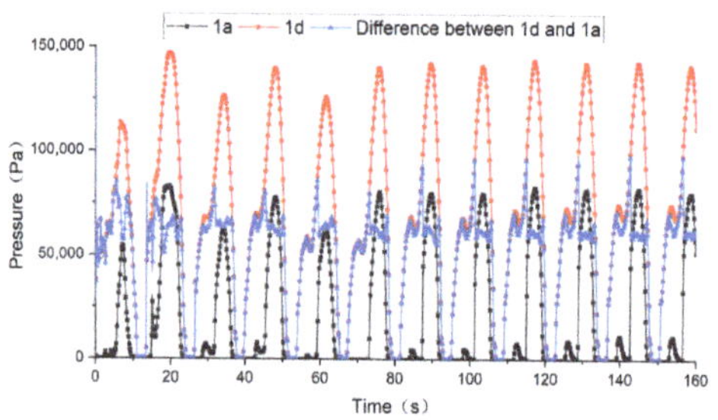

Figure 23. Time history of fluid pressure.

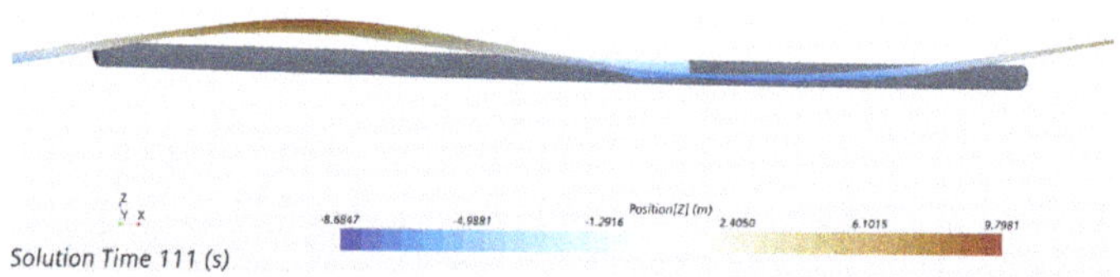

Figure 24. Profile of wave surface at 111 s.

4.2. Validation of the Numerical Method

In order to validate the CFD-FEA coupling method, the wave loads on a single floating body are computed using the WADAM modules of the SESAM Suite and COMPASS-WALCS-NE, and the numerical results are compared with the CFD-FEA results. SESAM is a well-known suite developed by DNV for the hydrodynamic, structural design, and analysis of ships and offshore structures. WADAM is a module of SESAM developed based on the three-dimensional linear frequency domain potential flow theory. Because of the assumption of a rigid body, the elastic deformation of the floating body is ignored. The factors such as surface nonlinearity, fluid viscosity, and slamming of the floating body cannot be accounted for as well. COMPASS-WALCS-NE is a three-dimensional nonlinear hydroelastic analysis application developed by the China Classification Society. This application is based on the three-dimensional potential flow theory and three-dimensional structural dynamics, taking into account the nonlinear wave force caused by the change in the instantaneous wet surface of the hull, the slamming, and other nonlinear loads, and the influence of the structural elastic deformation on the flow field is also accounted for. COMPASS-WALCS-NE has been validated against experimental results in many studies [28–30]; for instance, it has shown that the accuracy of COMPASS-WALCS-NE for nonlinear wave loads is fairly sufficient.

The model used for the simulation with WADAM is shown in Figure 25a, and the one for simulation with COMPASS-WALCS-NE is shown in Figure 25b. During the simulation, the vertical bending moments M_y at the cross-sections of $x = -7.5$ m, 0 m, and 75 m are monitored.

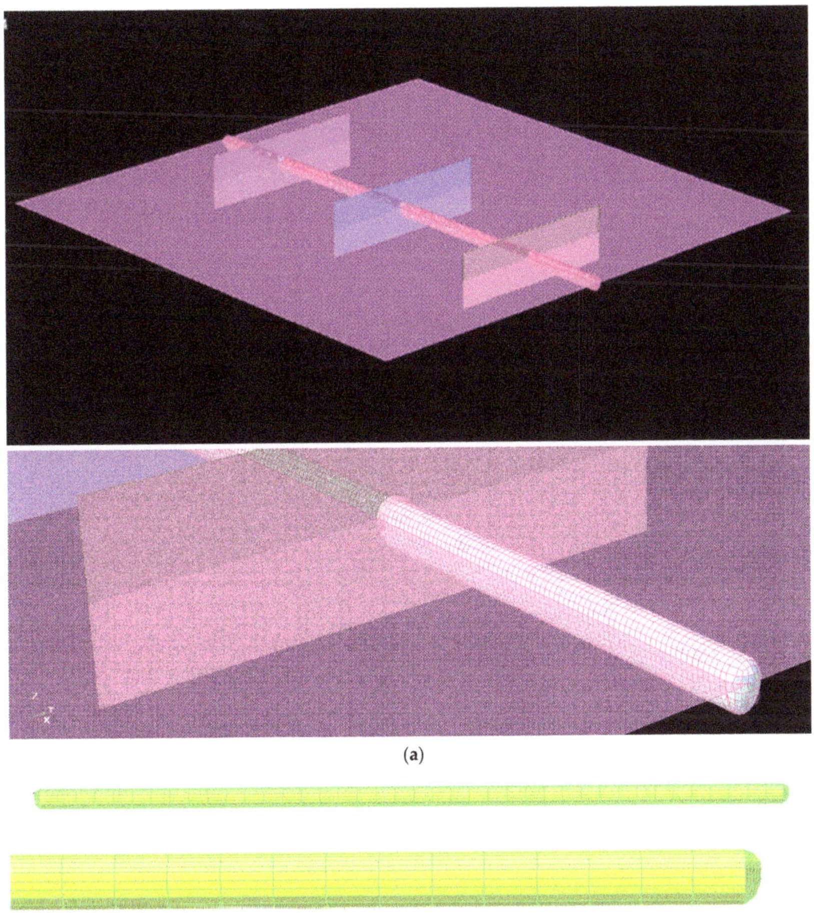

(a)

(b)

Figure 25. Model of a single floating body: (**a**) model for WADAM simulation; (**b**) model for COMPASS-WALCS-NE simulation.

4.2.1. Comparison with WADAM Numerical Results

The numerical results of CFD-FEA and WADAM are compared in Table 5. It can be seen that, for small wave heights, waves do not overtop the floating body, and the wave loads are small, thus the floating body does not have an obvious elastic deformation. In this case, the CFD-FEA method and WADAM produce similar numerical results for the amplitude of the vertical bending moment M_y. However, because the freeboard of the floating body is small, it is easy for the waves to overtop the floating body in case of large wave heights. In this case, the nonlinear characteristics of the instantaneous position variation in the body surface cannot be simply ignored. As the large wave heights result in large wave loads, the floating body will have obvious elastic deformation, and the hydroelastic response becomes significant. The elastic behavior of the body is ignored in WADAM, resulting in a large deviation between the CFD-FEA and WADAM numerical results in the case of large wave heights.

Table 5. Comparison of calculation results of CFD-FEA and WADAM.

Wave Height/m	Method	Amplitude of Vertical Bending Moment M_y/N·m		
		$x = -75$ m	$x = 0$ m	$x = 75$ m
2.5	CFD-FEA	1.052×10^8	2.087×10^8	1.096×10^8
	WADAM	1.219×10^8	2.286×10^8	1.261×10^8
	Deviation	−13.70%	−8.71%	−13.08%
5.0	CFD-FEA	1.741×10^8	3.430×10^8	1.663×10^8
	WADAM	2.438×10^8	4.573×10^8	2.523×10^8
	Deviation	−28.59%	−24.99%	−34.09%
7.5	CFD-FEA	2.355×10^8	5.018×10^8	2.381×10^8
	WADAM	3.657×10^8	6.859×10^8	3.784×10^8
	Deviation	−35.60%	−26.84%	−37.08%
10.0	CFD-FEA	2.373×10^8	5.212×10^8	2.622×10^8
	WADAM	4.877×10^8	9.145×10^8	5.045×10^8
	Deviation	−51.34%	−43.01%	−48.03%
12.5	CFD-FEA	3.326×10^8	6.756×10^8	3.307×10^8
	WADAM	6.096×10^8	1.143×10^9	6.306×10^8
	Deviation	−45.44%	−40.89%	−47.56%
16.0	CFD-FEA	4.742×10^8	7.928×10^8	3.315×10^8
	WADAM	7.802×10^8	1.463×10^9	8.072×10^8
	Deviation	−39.22%	−45.81%	−58.93%

The variation in the numerical results of CFD-FEA and WADAM with the change in wave height is presented in Figure 26a–c. The results obtained with WADAM show that the amplitude of the vertical bending moment M_y increases linearly with the wave height, while the numerical results obtained with CFD-FEA show that the amplitude of the vertical bending moment M_y also increases with the wave height, but at a lower rate. This is because the CFD-FEA method can effectively capture the nonlinear characteristics of an instantaneous position variation in the body surface. Since the freeboard of the floating body is small and large waves can overtop the floating body, the wetted surface of the floating body does not increase "linearly" with the wave height once the wave height exceeds the height of the floating body. It can be seen that the results generated by the CFD-FEA method are more consistent with the reality than WADAM. The numerical method used in WADAM is a three-dimensional linear frequency domain potential flow theory, which assumes that the floating body oscillates slightly about the equilibrium position, and not only is the free-surface boundary condition linearized, but the body boundary condition is approximately satisfied on the average wet surface, which explains that the method is not applicable to solve the wave loads of a small freeboard floating body under severe sea conditions. In addition, due to the small stiffness of the single floating body, the floating body has an obvious elastic deformation in large waves, which is a behavior the method used in WADAM cannot capture.

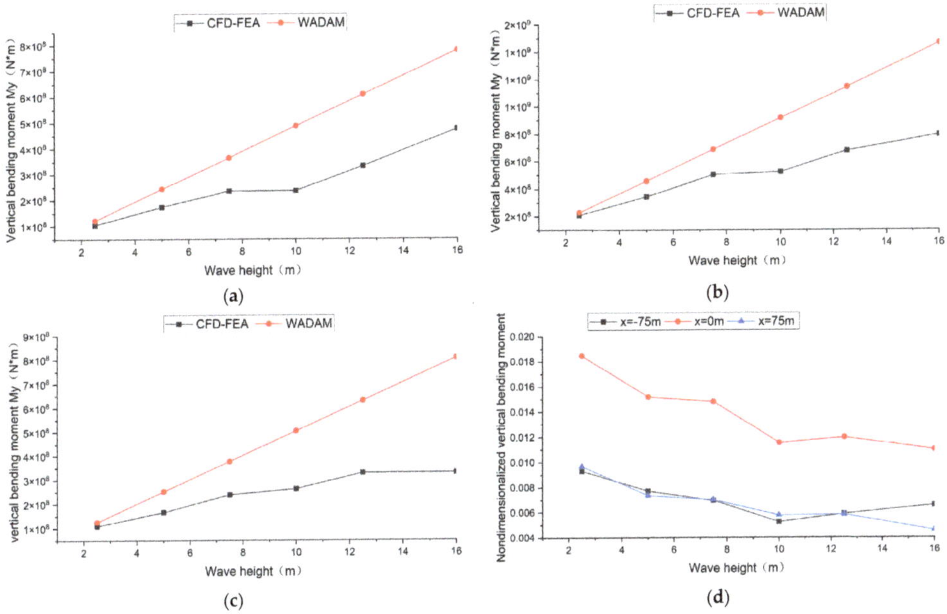

Figure 26. Numerical results of vertical bending moment: (**a**) Comparison of numerical results of the amplitude of vertical bending moment at cross-section $x = -75$ m; (**b**) comparison of numerical results of the amplitude of vertical bending moment at cross-section $x = 0$ m; (**c**) comparison of numerical results of the amplitude of vertical bending moment at cross-section $x = 75$ m; (**d**) nondimensionalized amplitude of vertical bending moment of each cross-section.

The amplitude of the vertical bending moment is nondimensionalized as $M_y/(\rho gL2Bh)$, where M_y is the amplitude of the vertical bending moment, ρ is the water density, L is the floating body length, B is the floating body width, and h is the wave height, as shown in Table 6. It can be observed in Figure 26d that, in general, the nondimensionalized amplitude of the vertical bending moment at the cross-sections decreases with the increase in the wave height. According to the three-dimensional linear frequency domain potential flow theory, the nondimensionalized amplitude of the vertical bending moment at each cross-section does not change with the wave height; however, this is only valid for linear problems. In fact, because of the nonlinearity of the body surface, the nondimensionalized amplitude of the vertical bending moment at the cross-section decreases with the increase in the wave height when the wave overtops the floating body.

Table 6. Nondimensionalized amplitude of vertical bending moment of cross-section by CFD-FEA method.

Wave Height/m	$M_y/(\rho gL2Bh)$		
	Cross-Section $x = -75$ m	Cross-Section $x = 0$ m	Cross-Section $x = 75$ m
2.5	0.0093	0.0185	0.0097
5.0	0.0077	0.0152	0.0074
7.5	0.0069	0.0148	0.0070
10.0	0.0052	0.0115	0.0058
12.5	0.0059	0.0120	0.0059
16.0	0.0066	0.0110	0.0046

4.2.2. Comparison with COMPASS-WALCS-NE Numerical Results

Since the nonlinear characteristics of the hydrodynamic forces of the floating body are obvious in large waves, in order to further investigate the validity of the numerical results of the CFD-FEA method for the case of large wave heights, the time history of a vertical bending moment at each cross-section for h = 12.5 m and h = 16 m obtained with the CFD-FEA method and COMPASS-WALCS-NE are compared in Figures 27 and 28. The curve marked "CFD-FEA" represents the numerical results obtained by the CFD-FEA method accounting for the elastic deformation of the floating body, and the curve marked "COMPASS-WALCS-NE" represents the numerical results obtained by COMPASS-WALCS-NE. It can be seen in Figures 27 and 28 that, about the amplitude of the vertical bending moment M_y, the numerical results of the CFD-FEA method are in a good agreement with the numerical results generated by COMPASS-WALCS-NE. However, the details of the time history curve by the CFD-FEA method are less consistent with those of COMPASS-WALCS-NE. The CFD-FEA method can capture the pressure extrema at the bow, midship, and stern of the floating body at some time instants, even though no slamming occurs, as analyzed in Section 4.1.4. In contrast, the time history curve by COMPASS-WALCS-NE is very smooth, without local extremum. The CFD-FEA method can capture the sudden change in the vertical bending moment M_y during the water entry and exit of the floating body that the potential-flow-based methods cannot.

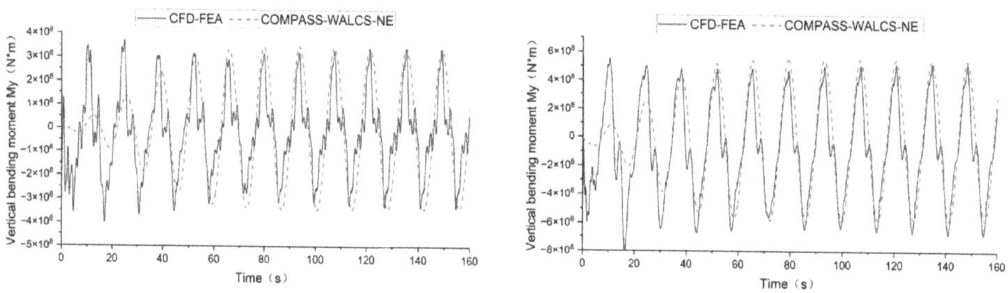

(a) (b)

Figure 27. Comparison of vertical bending moment time history of the cross-section at wave height 12.5 m: (**a**) cross-section $x = -75$ m; (**b**) cross-section $x = 0$ m.

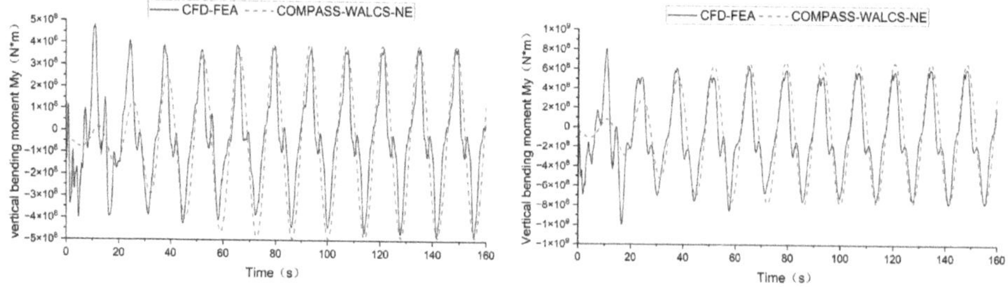

(a) (b)

Figure 28. Comparison of vertical bending moment time history of cross-section at wave height 16 m: (**a**) cross-section $x = -75$ m; (**b**) cross-section $x = 0$ m.

5. Conclusions

In this study, the hydrodynamic and structural responses of an unconventional large floating structure are simulated and analyzed using a CFD-FEA method. The results generated by the CFD-FEA method are compared with those generated by a rigid body linear frequency domain potential flow theory (WADAM) and by a three-dimensional potential flow nonlinear hydroelastic theory (COMPASS-WALCS-NE). Based on the analysis of the numerical results, the following concluding remarks can be drawn:

(1) The proposed CFD-FEA method can effectively exchange the information of the fluid domain and structure domain through the coupling interface, and a two-way coupling computation of the fluid domain and structure domain is realized. The numerical results of the elastic deformation of the floating body are reasonable.

(2) In the case of large wave heights where wave overtopping occurs, the rate of an increase in the amplitude of the vertical bending moment M_y obtained by the CFD-FEA method slowly decreases with the increase in the wave height. The amplitude of the vertical bending moment M_y obtained by the CFD-FEA method is smaller than that by WADAM. Hydroelastic responses of elastic floating structures and nonlinear behaviors are observed in large wave heights, for which case COMPASS-WALCS-NE and the CFD-FEA method are recommended to solve such problems.

(3) For large wave heights, the peak values of the vertical bending moment M_y obtained by the CFD-FEA method are in a good agreement with those by COMPASS-WALCS-NE, which shows that the CFD-FEA method is valid. Moreover, the CFD-FEA method can also capture complex phenomena, such as the details of the change in the vertical bending moment M_y, that the potential-flow-based methods cannot.

In this study, the CFD-FEA method for the hydroelastic response of large floating offshore structures is validated. Not only can the behavior of the hydroelastic responses of elastic floating structures be well estimated by the CFD-FEA method, but also the significant advantages of the CFD-FEA method over the traditional methods in capturing complex nonlinear behaviors are demonstrated.

Author Contributions: Conceptualization, N.G. and H.R.; formal analysis, N.G. and D.L.; investigation, N.G. and X.Z.; methodology, N.G.; project administration, H.R.; resources, H.R.; supervision, X.Z.; validation, N.G., D.L. and X.Z.; writing—original draft, N.G.; writing—review and editing, X.Z. and H.R. All authors have read and agreed to the published version of the manuscript.

Funding: This research received no funding.

Institutional Review Board Statement: Not applicable.

Informed Consent Statement: Not applicable.

Data Availability Statement: Not applicable.

Conflicts of Interest: The authors declare no conflict of interest.

References

1. Andersen, I.M.V. Full Scale Measurements of the Hydro-Elastic Response of Large Container Ships for Decision Support. Ph.D. Thesis, Technical University of Denmark, Copenhagen, Denmark, 2014.
2. Betts, C.V.; Bishop, R.E.D.; Prices, W.G. The symmetric generalized fluid forces applied to a ship in a seaway. *Int. Shipbuild. Prog.* **1977**, *119*, 265–278.
3. Wu, Y.S. Hydroelasticity of Floating Bodies. Ph.D. Thesis, Brunel University, Uxbridge, UK, 1984.
4. Xie, H.; Liu, F.; Liu, X.Y.; Tang, H.Y. Numerical prediction of asymmetrical ship slamming loads based on a hybrid two-step method. *Ocean Eng.* **2020**, *208*, 107331. [CrossRef]
5. Jiao, J.L.; Huang, S.X. CFD simulation of ship seakeeping performance and slamming loads in bi-directional cross wave. *J. Mar. Sci. Eng.* **2020**, *8*, 312. [CrossRef]
6. Wang, S.; Soares, C.G. Slam induced loads on bow-flared sections with various roll angles. *Ocean Eng.* **2013**, *67*, 45–57. [CrossRef]
7. Huang, S.X.; Jiao, J.L.; Chen, C.H. CFD prediction of ship seakeeping behavior in bi-directional cross wave compared with in uni-directional regular wave. *Appl. Ocean Res.* **2021**, *107*, 102426. [CrossRef]
8. Jiao, J.L.; Huang, S.X.; Soares, C.G. Numerical simulation of ship motions in cross waves using CFD. *Ocean Eng.* **2021**, *223*, 108711. [CrossRef]
9. Moctar, O.E.; Oberhagemann, J.; Schellinf, T.E. Free surface RANS method for hull girder springing and whipping. *Trans. Soc. Nav. Archit. Mar. Eng.* **2011**, *119*, 48–66.
10. Oberhagemann, J. On Prediction of Wave-Induced Loads and Vibration of Ship Structures with Finite Volume Fluid Dynamic Methods. Ph.D. Thesis, University of Duisburg-Essen, Duisburg and Essen, Germany, 2016.

11. Wilson, R.V.; Ji, L.; Karman, S.L.; Hyams, D.G.; Sreenivas, K.; Taylor, L.K.; Whitfield, D.L. Simulation of Large Amplitude Ship Motions for Prediction of Fluid-Structure Interaction. In Proceedings of the 27th Symposium on Naval Hydrodynamics, Seoul, Republic of Korea, 5–10 October 2008.
12. Liu, Y.; Zhu, R.Q.; Cheng, Y.; Xie, T.; Li, R.Z. Numerical simulation of hydroelastic responses of floating structure based on CFD-FEM method. *Ocean Eng.* **2020**, *38*, 24–32.
13. Lakshmynarayanana, P.A.; Hirdaris, S. Comparison of nonlinear one- and two-way FFSI methods for the prediction of the symmetric response of a containership in waves. *Ocean Eng.* **2020**, *203*, 107179. [CrossRef]
14. Lakshmynarayanana, P.A.; Temarel, P. Application of a two-way partitioned method for predicting the wave-induced loads of a flexible containership. *Appl. Ocean Res.* **2020**, *96*, 102052. [CrossRef]
15. Jiao, J.L.; Huang, S.X.; Soares, C.G. Viscous fluid-flexible structure interaction analysis on ship springing and whipping responses in regular waves. *J. Fluids Struct.* **2021**, *106*, 103354. [CrossRef]
16. STAR-CCM+ Version 2020.1 Manual. 2020. Available online: https://support.sw.siemens.com/en-US/ (accessed on 1 October 2020).
17. Abaqus 6.14-4 Manual. 2014. Available online: https://www.4realsim.com/abaqus/ (accessed on 1 November 2014).
18. Park, J.C.L.; Kim, M.H.; Miyata, H. Three-dimensional numerical wave tank simulations on fully nonlinear wave-current-body interactions. *J. Mar. Sci. Technol.* **2001**, *6*, 70–82. [CrossRef]
19. Li, H. 3-D hydroelasticity Analysis Method for Wave Load of Ship. Ph.D. Thesis, Harbin Engineering University, Harbin, China, 2009. (In Chinese).
20. Dai, Y.S.; Shen, J.W.; Song, J.Z. *Ship Wave Loads*, 1st ed.; National Defense Industry Press: Beijing, China, 2007; pp. 154–196. (In Chinese)
21. Sun, Z.; Liu, G.J.; Zou, L.; Zheng, H.; Djiddjeli, K. Investigation of Non-Linear Ship Hydroelasticity by CFD-FEM Coupling Method. *J. Mar. Sci. Eng.* **2021**, *9*, 511. [CrossRef]
22. Hirt, C.W.; Nichols, B.D. Volume of Fluid (VOF) Method for the Dynamics of Free Boundaries. *J. Comput. Phys.* **1981**, *39*, 201–225. [CrossRef]
23. ITTC Procedures and Guidelines, 2011. Practical Guidelines for Ship CFD Applications. 7.5-03-02-03. Available online: https://ittc.info/media/1357/75-03-02-03.pdf (accessed on 1 August 2021).
24. Ferziger, J.H.; Peric, M. *Computational Methods for Fluid Dynamics*, 3rd ed.; Springer: Berlin/Heidelberg, Germany, 2002; pp. 135–152.
25. Lakshmynarayanana, P.A.; Temarel, P. Application of CFD and FEA coupling to predict dynamic behaviour of a flexible barge in regular head waves. *Mar. Struct.* **2019**, *65*, 308–325. [CrossRef]
26. SESAM User Manual WADAM. 2017. Available online: https://www.dnv.com/services/frequency-domain-hydrodynamic-analysis-of-stationary-vessels-wadam-2412 (accessed on 10 November 2021).
27. COMPASS-WALCS-N. 2015. Available online: https://www.ccs.org.cn/ccswz/articleDetail?id=201900001000007558&columnId=201900002000000599 (accessed on 12 December 2021).
28. Zhang, K.H.; Ren, H.L.; Li, H.; Yan, L. Nonlinear Hydroelasticity of Large Container Ship. In Proceedings of the 26th International Ocean and Polar Engineering Conference, Rhodes, Greece, 26 June–2 July 2016.
29. Jiao, J.L.; Ren, H.L.; Adenya, C.A. Experimental and Numerical Analysis of Hull Girder Vibrations and Bow Impact of a Large Ship Sailing in Waves. *Shock Vib.* **2015**, *2015*, 10. [CrossRef]
30. Xiao, W.; Wang, H.Y. Wave loads prediction of large scale new type ship. *Ship Boat* **2017**, *167*, 39–46. (In Chinese)

Journal of
Marine Science and Engineering

MDPI

Article

Fracture Prediction of Steel-Plated Structures under Low-Velocity Impact

Burak Can Cerik and Joonmo Choung *

Department of Naval Architecture and Ocean Engineering, Inha University, Incheon 22212, Republic of Korea; bccerik@inha.ac.kr
* Correspondence: heroeswise2@gmail.com

Abstract: In this paper, a validation study of a recently proposed rate-dependent shell element fracture model using quasi-static and dynamic impact tests on square hollow sections (SHS) made from offshore high-tensile strength steel was presented. A rate-dependent forming limit curve was used to predict the membrane loading-dominated failure, while a rate-dependent ductile fracture locus was applied for predicting failure governed by bend loading. The predicted peak force and fracture initiation using the adopted material and fracture model agreed well with the experimental results. The fracture mode was also captured accurately. Further simulations were performed to discuss the importance of the inclusion of dynamic effects and the separate treatment of failure modes. Finally, the shortcomings of the common practice of treatment of rate-effects in low-velocity impact simulations involving fracture were highlighted.

Keywords: ductile fracture; shell element; fracture prediction; strain rate; impact; collision

Citation: Cerik, B.C.; Choung, J. Fracture Prediction of Steel-Plated Structures under Low-Velocity Impact. *J. Mar. Sci. Eng.* **2023**, *11*, 699. https://doi.org/10.3390/jmse11040699

Academic Editors: Bin Liu, Kun Liu, Chenfeng Li and Joško Parunov

Received: 17 February 2023
Revised: 9 March 2023
Accepted: 22 March 2023
Published: 24 March 2023

1. Introduction

The importance of material definitions in simulating the response of steel-plated structures under extreme loads is widely recognized [1–3]. Recently, predicting the fracture of large-scale structures resulting from impact loads using the shell elements in finite element simulations has attracted considerable attention [4–8]. In this context, accurate predictions of fracture are challenging because of the mesh size effects and constitutive assumptions of shell elements, as well as a number of factors that affect the material response under dynamic impact loading conditions.

First of all, under impact loads, the loading speed and resultant strain rates have a substantial impact on material characteristics [9]. In practice, it is often believed that the response to a low-velocity impact is comparable to that under an equivalent quasi-static loading [10]. Experimental studies with scaled models of stiffened panels [11,12] and web girders [13] confirmed this assumption. Yet, a number of comparative experimental and numerical studies have shown that reliable predictions of plastic deformation must take into account the rate-dependency of flow stress [14–16].

The Cowper–Symonds model is the most commonly used model for the strain-rate effect on flow stress [17]. For a given strain rate, the same dynamic hardening factor is applied for all equivalent plastic strains, which is determined solely based on the initial flow stress. This straightforward model is frequently employed in simple rigid-plastic mechanism-based methods to fine-tune the predictions.

On the other hand, when utilizing Cowper–Symonds-type models, caution must be taken as non-linear finite element analysis (NLFEA) results are extremely sensitive to how the rate-dependency of flow stress is catered for [18]. The increase in flow stress depends not only on the rate of deformation but also on the magnitude of strain [19]. This phenomenon is closely associated with thermal softening due to self-heating. A portion of the plastic strain energy is converted to heat, which causes thermal softening to eventually dominate over the increase in flow stress due to rate-hardening. This effect is often neglected in practice because approximate constitutive models often yield acceptable results [20], particularly if plastic deformation is the only concern.

Secondly, fracture prediction with shell elements poses certain challenges. Recent research has indicated that modelling the complex nature of ductile failure with shell elements is not possible using oversimplified constitutive models. The critical failure strain for NLFEA, according to a widely used method, is the elongation at the fracture (%) of specimens for standard tension tests performed at intermediate strain rates [20,21]. According to Cerik et al. [22,23], more complex models must take into account the following in order to simulate fracture failure with finite shell elements: various mechanisms of fracture [24–26], variation in the stress state from the initial conditions to the onset of fracture [27–30], and differentiation between membrane and bending deformation [31].

There are several approaches for dealing with the limitations of shell elements. A straightforward method for simulating fracture in shell elements subjected to quasi-static load that causes membrane stretching, is to adopt a localized necking criterion. Usually such models involve a single constant that can be determined using the flow stress curve of a material [5,32,33]. This approach can be extended for the case of dynamic loading if the rate-dependent flow stress curves are adequately constructed. Cerik and Choung [34,35] recently adopted this approach to define the localized necking locus based on the instantaneous hardening rate. A rate-dependent ductile fracture locus was combined with the localized necking locus (termed as the rate-dependent DSSE-HC model) to predict all possible failure modes, including shear fracture and through-thickness cracking, with shell elements. The mesh size sensitivity of the proposed model was confirmed to be very low.

The model by Cerik and Choung [34] was validated using the dynamic punch indentation tests on metal sheets, given in [36], which were limited only to the failure involving localized necking. In the present study, the new test results presented by Johanessen et al. [37] on square hollow sections (SHS) subjected to quasi-static and dynamic bending were used to validate the proposed model for a more general case of loading that results in different failure modes. These new tests present a challenge for numerically predicting fractures because deformation and failure are more complex than punch-loaded plates, which predominantly undergo membrane stretching. The primary aim is to demonstrate the advantage of the proposed model over single-modelling approaches for fracture. Beside the comparison of predictions with the proposed model with the test results, additional numerical simulations of the test were performed using the simplified approaches used in industrial practice to highlight the benefits of the adopted model. The main contribution of the present work is the validation of the rate-dependent DSSE-HC model by comparing its predictions of the onset of ductile failure for both membrane- and bending-dominated loading conditions.

2. Material Model

For the sake of completeness, the model reported previously in Cerik and Choung [34,35] is summarized.

2.1. Constitutive Model

The constitutive model adopted in the present study follows Roth and Mohr [38]. The logarithmic strain tensor was decomposed into elastic and plastic strain components:

$$\boldsymbol{\varepsilon} = \boldsymbol{\varepsilon}_e + \boldsymbol{\varepsilon}_p \tag{1}$$

The constitutive equation for Cauchy stress, $\boldsymbol{\sigma}$ is as follows:

$$\boldsymbol{\sigma} = \mathbf{C}_{el} : \boldsymbol{\varepsilon}_e \tag{2}$$

where $\mathbf{C}_{el}$ is the isotropic elasticity tensor, which depends on the Young's modulus E and Poisson's ratio ν.

The magnitude of the Cauchy stress tensor is represented by von Mises stress as follows:

$$\bar{\sigma} = \sqrt{3J_2} \tag{3}$$

with J_2 denoting the second invariant of the stress deviator tensor, **s**.

The evolution of the plastic strain tensor, $\boldsymbol{\varepsilon}_p$, is described by the associative flow rule:

$$d\boldsymbol{\varepsilon}_p = d\bar{\varepsilon}_p \frac{\partial \bar{\sigma}}{\partial \boldsymbol{\sigma}} \tag{4}$$

where $d\bar{\varepsilon}_p$ is the increase in the equivalent plastic strain.

The yield surface is expressed as in the given equation:

$$f(\boldsymbol{\sigma}, k) = \bar{\sigma} - k(\bar{\varepsilon}_p, \dot{\bar{\varepsilon}}_p, T) = 0 \tag{5}$$

Here, k represents the deformation resistance. Similar to the Jonhson–Cook model [39], the deformation resistance is assumed as a function of the equivalent plastic strain, $\bar{\varepsilon}_p$, equivalent plastic strain rate, $\dot{\bar{\varepsilon}}_p$, and temperature, T:

$$k(\bar{\varepsilon}_p, \dot{\bar{\varepsilon}}_p, T) = k_\varepsilon k_{\dot{\varepsilon}} k_T \tag{6}$$

The first term, which represents strain hardening, is modelled using a weighted combination of the Swift [40] and Voce [41] laws [42]:

$$k_\varepsilon(\bar{\varepsilon}_p) = \alpha k_S + (1-\alpha) k_V \tag{7}$$

The weighting factor $\alpha \in [0,1]$ requires calibration following a procedure described elsewhere [22,30]. The Swift and Voce laws are expressed as follows:

$$k_S(\bar{\varepsilon}_p) = A(\varepsilon_0 + \bar{\varepsilon}_p)^n \tag{8}$$

$$k_V(\bar{\varepsilon}_p) = k_0 + Q(1 - \exp(-\beta \bar{\varepsilon}_p)) \tag{9}$$

The second term in Equation (6) scales up the flow stress for the strain rates above a certain threshold, $\dot{\bar{\varepsilon}}_0$, as follows [39]:

$$k_{\dot{\varepsilon}}(\dot{\bar{\varepsilon}}_p) = \begin{cases} 1 & \text{if } \dot{\bar{\varepsilon}}_p < \dot{\bar{\varepsilon}}_0 \\ 1 + C\ln\left(\frac{\dot{\bar{\varepsilon}}_p}{\dot{\bar{\varepsilon}}_0}\right) & \text{if } \dot{\bar{\varepsilon}}_p \geq \dot{\bar{\varepsilon}}_0 \end{cases} \tag{10}$$

The last term scales down the flow stress for temperatures above the reference temperature, T_r, as follows [39]:

$$k_T(T) = \begin{cases} 1 & \text{if } T < T_r \\ 1 - \left(\frac{T-T_r}{T_m-T_r}\right)^m & \text{if } T_r \leq T \leq T_m \\ 0 & \text{if } T > T_m \end{cases} \tag{11}$$

where T_m is the melting temperature.

The temperature increase, due to rapid deformation, is related to strain energy as follows:

$$dT = \omega \frac{\eta_k}{\rho C_p} \bar{\sigma} d\bar{\varepsilon}_p \tag{12}$$

In this equation, η_k is a constant, which defines the percentage of the plastic strain energy converted to heat. The other parameters, ρ and C_p, are the density and specific heat of the material, respectively. In the above equation, Roth and Mohr [38] included an extra term

$\omega(\dot{\varepsilon}_p)$. This term is a function of the strain rate and helps to regulate the amount of plastic work converted to heat. It is expressed as follows:

$$\omega(\dot{\varepsilon}_p) = \begin{cases} 0 & \text{if } \dot{\varepsilon}_p < \dot{\varepsilon}_{it} \\ \frac{(\dot{\varepsilon}_p - \dot{\varepsilon}_{it})^2(3\dot{\varepsilon}_a - 2\dot{\varepsilon}_p - \dot{\varepsilon}_{it})}{(\dot{\varepsilon}_a - \dot{\varepsilon}_{it})^3} & \text{if } \dot{\varepsilon}_{it} \le \dot{\varepsilon}_p \le \dot{\varepsilon}_a \\ 1 & \text{if } \dot{\varepsilon}_p > \dot{\varepsilon}_a \end{cases} \tag{13}$$

where $\dot{\varepsilon}_{it}$ and $\dot{\varepsilon}_a$ indicate the two limiting conditions, namely isothermal and adiabatic conditions. It is usually assumed that $\dot{\varepsilon}_{it} = \dot{\varepsilon}_0$ [38]. With this modification of the Johnson–Cook model, thermal softening effects at low strain rates can be eliminated, and thermal softening can be considered only at intermediate and large strain rates, where adiabatic conditions will prevail, and self-heating during large plastic deformation will occur.

2.2. *Fracture Prediction Model for Shell Elements*

The rate-dependent shell element fracture model proposed by Cerik and Choung [34] considers two cases separately: ductile fracture preceded by thinning and without thinning before the initiation of a fracture [26]. The former case was modelled using a fracture indicator, D, as follows:

$$dD = \frac{d\bar{\varepsilon}_p}{\bar{\varepsilon}_f^{pr}} \tag{14}$$

where $\bar{\varepsilon}_f^{pr}(\eta, \bar{\theta}, \dot{\varepsilon}_p)$ is the fracture strain surface of the material. Note that this is valid strictly under proportional strain paths. The stress state is defined using stress triaxiality, η, and the Lode angle parameter, $\bar{\theta}$. However, for the case of plane stress conditions (as in shell elements), the Lode angle parameter can be defined using stress triaxiality. Therefore, stress triaxiality is the sole parameter adequate for defining the stress state in shell elements.

The Hosford–Coulomb (HC) model was used as $\bar{\varepsilon}_f^{pr}(\eta, \bar{\theta}, \dot{\varepsilon}_p)$, a rate-independent version of which is given as:

$$\bar{\varepsilon}_f^{pr}(\eta, \bar{\theta}) = b(1+c)^{\frac{1}{n_f}} \left[\left\{ \frac{1}{2}\left((f_1 - f_2)^a + (f_1 - f_3)^a + (f_2 - f_3)^a \right) \right\}^{\frac{1}{a}} + c(2\eta + f_1 + f_3) \right]^{-\frac{1}{n_f}} \tag{15}$$

$$f_1(\bar{\theta}) = \frac{2}{3}\cos\left(\frac{\pi}{6}(1 - \bar{\theta})\right) \tag{16}$$

$$f_2(\bar{\theta}) = \frac{2}{3}\cos\left(\frac{\pi}{6}(3 + \bar{\theta})\right) \tag{17}$$

$$f_3(\bar{\theta}) = -\frac{2}{3}\cos\left(\frac{\pi}{6}(1 + \bar{\theta})\right) \tag{18}$$

A rate-dependent version of the HC model was introduced by Roth and Mohr [38] by setting the parameter b, in a rate-dependent format:

$$b(\dot{\varepsilon}_p) = \begin{cases} b_0 & \text{if } \dot{\varepsilon}_p < \dot{\varepsilon}_0 \\ b_0\left[1 + \gamma \ln\left(\frac{\dot{\varepsilon}_p}{\dot{\varepsilon}_0}\right)\right] & \text{if } \dot{\varepsilon}_p \ge \dot{\varepsilon}_0 \end{cases} \tag{19}$$

The HC model predicts a fracture if in a shell through-thickness integration point (IP), the fracture indicator, D, exceeds unity:

$$D[IP(i)] \ge 1 \tag{20}$$

Ductile fracture preceded by thinning is predicted using a forming limit curve called DSSE (domain of shell-to-solid equivalence). The localized necking indicator, N, is defined as follows:

$$dN = \frac{d\bar{\varepsilon}_p}{\bar{\varepsilon}^{pr}_{DSSE}} \text{ for } \frac{1}{3} < \eta < \frac{2}{3} \tag{21}$$

The DSSE, $\bar{\varepsilon}^{pr}_{DSSE}[\eta]$, is given by Pack and Mohr [26] as follows:

$$\bar{\varepsilon}^{pr}_{DSSE}(\eta) = b\left(\left\{\frac{1}{2}\left((g_1 - g_2)^d + g_1^d + g_2^d\right)\right\}^{\frac{1}{d}}\right)^{-\frac{1}{p_f}} \text{ for } \frac{1}{3} < \eta < \frac{2}{3} \tag{22}$$

$$g_1(\eta) = \frac{3}{2}\eta + \sqrt{\frac{1}{3} - \frac{3}{4}\eta^2} \tag{23}$$

$$g_2(\eta) = \frac{3}{2}\eta - \sqrt{\frac{1}{3} - \frac{3}{4}\eta^2} \tag{24}$$

In this model, the parameter d is determined using the flow stress curve of the material. For the case of the rate-dependent flow curves, d is not constant and requires to be calculated using the procedure described in [34] at every time increment for the instantaneous values of strain rate and temperature.

Localized necking is assumed to occur only if all integration points through the thickness fulfil the failure criterion:

$$N(z) \geq 1 \; \forall z \in [-t/2, t/2] \tag{25}$$

where t is the shell thickness and z is the through-thickness direction coordinate.

Numerical implementation of the rate-dependent DSSE-HC model and the mesh size sensitivity of the model are described in detail in [34]. Cerik and Choung [34] reported that the fracture predictions using the described model are consistent with the very fine mesh solid element results. It was noted that for predicting the localized necking, the element size to element thickness ratio up to four, yields accurate estimates. Fracture without necking was found to be insensitive to the mesh size except geometrical conditions that require finer meshes. For validation of the model, quasi-static and dynamic impact tests on steel sheets, conducted by Gruben et al. [36], were used. The model used in the present study could predict the response of the test models in both cases and accurately simulate the deviations between the two cases.

3. Description of the Experiments

Johannessen et al. [37] recently reported the material and three-point bending test results used in the present study to validate the described material model. Here, only a summary of the experiments and important test outcomes are given. The SHS test models were made from three different grade high-tensile strength steels. In the present study, only the test results of the models made from offshore grade steel (grade S355NH, in accordance with NS-EN 10225-3) were considered because of the relevance of the offshore steel grade for the application area of the proposed fracture model, marine collision problem, as well as considering the availability of the material test parameters database for the proposed material model that can be assumed to have similar characteristics as the target structure material.

The tests were conducted with 120 × 120 × 5 mm SHS. The test models were 600 mm long. Each model featured a 50 mm notch placed 60 mm off-centre. The purpose of each notch was to provoke failure in a pre-determined location and pose a challenge for the finite element simulations with shell elements in predicting the crack propagation direction. Figure 1 shows the dimensions of the models.

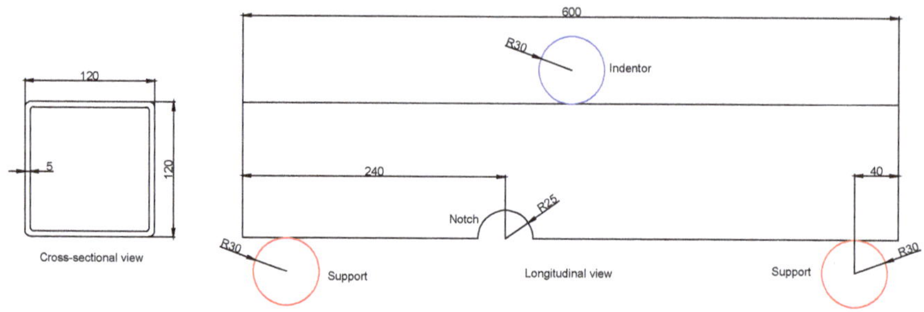

Figure 1. Dimensions of the test models (unit: mm) and the experimental setup in the quasi-static tests.

Standard uniaxial tension tests were performed on dog-bone specimens to characterize the material hardening properties. The test coupons were sampled both in the rolling direction (0°) and transverse direction (90°). Three tests were performed for each direction. The lower yield stress of the material was scattered around 420 MPa. In addition, the test specimens sampled from the transverse direction showed a large spread in hardening and percentage elongation to fracture. The flow stress curves showed considerable scattering, which raises material uncertainties to be considered in the simulation of the tests.

Johannessen et al. [37] performed both quasi-static and low-velocity impact tests. Figure 1 shows a schematic representation of the quasi-static test involving the SHS penetrated by a solid cylinder. The test model rested on two rigid cylinders, 60 mm in diameter. A rigid cylinder was moved with a constant velocity of 10 mm/min towards the model up to 120 mm, before fracture initiated around the notch and propagated diagonally. PEFT sheets were used at the support and contact surfaces between the indentor and model to minimize friction. The tests were repeated three times and showed relatively high scatter. This was associated with the large variability of the material properties.

The low-velocity impact tests were conducted using impact testing equipment, which utilizes a trolley system accelerated with a hydraulic kicking device. In the dynamic tests, the indentor mass was 1470.7 kg, and the velocity just prior to the impact was measured as 4.1 m/s. These yield kinetic energies large enough to cause fracture around the notch of the test models, similar to that observed in the quasi-static tests. In the dynamic tests, the supports and indentor diameter were both 50 mm. A spray lubricant was used to reduce friction between the contact surfaces. The tests were repeated three times, and the repeatability was slightly better than the quasi-static tests. Both the deformation and fracture response of the test models were similar to that in the quasi-static tests. Yet, the peak force was higher because of the strain-rate effects and fracture initiated earlier than in most quasi-static tests.

4. Finite Element Modelling

The tests described in the previous section were simulated using the Abaqus software package (explicit solver), and a user-defined material subroutine (VUMAT) implementing the described material model. The test model was meshed using four-node shell elements (S4R) with five through-thickness integration points. S4R is suitable for most common applications, including crashworthiness analysis. The average element edge length was determined to be 5 mm, which was sufficient to resolve the stress gradients around the notch. The chosen element size also reflects the findings reported in [34] on mesh size sensitivity of the employed fracture model. An element size less than the shell thickness is in general not recommended when using a localized necking criterion for predicting membrane-dominated loading. Rigid elements (R3D4) were used for the supports and indentor. Figure 2 shows a finite element model of the test setup including the boundary conditions.

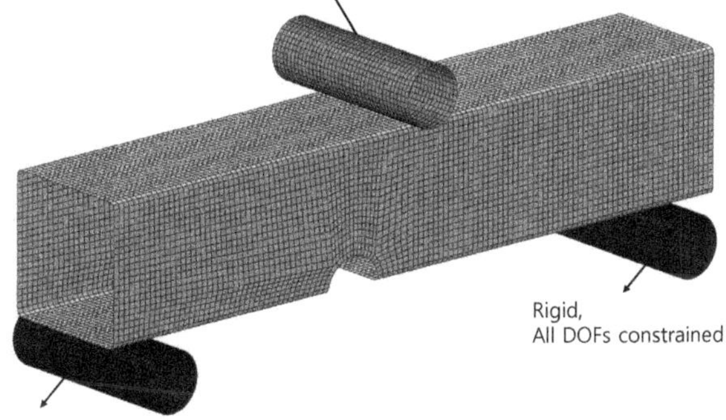

Figure 2. Finite element model of the test model, supports, and indentor.

For the three-point bending simulation of quasi-static and low-velocity impact tests, all the degrees of freedom at the reference node of the rigid supports were constrained except for the displacement in the impact direction. All degrees of freedom at the reference nodes associated with the rigid support were constrained. The quasi-static test was simulated using displacement control, whereas low-velocity impact was simulated by setting an initial velocity to the rigid indentor. Both analyses considered geometric non-linearity and used automatic time incrementing from stable time increment estimates. The duration of the quasi-static simulation was long enough to limit the inertia effects.

The general contact algorithm was used to consider the interaction between the rigid cylinders and the test model. Moreover, self-contact was considered for the surfaces of the impacted structure. According to Johannessen et al. [37], a low-friction coefficient of 0.05 was used for the contact between the surfaces. This value was deemed reasonable considering the applied friction reduction agents applied in the tests.

The material test results in [37] cannot be directly used for calibrating the employed material and fracture models in the present paper as they only consist of standard tension tests. On the other hand, typical test specimens and test conditions for calibrating the rate-dependent material and fracture model, as given in the paper by Roth and Mohr (reference **??**), are many and require fracture specimens with unique geometries to attain the desired stress states. Therefore, it was decided to utilize the existing calibrated material model parameters reported in [30]. In marine structures practice, it is common to classify steel grades based on the yield strength and ultimate tensile strength/yield strength ratio, and steel grades with similar flow stress characteristics (i.e., hardening behaviour) can be assumed to have similar material parameters. The material model parameters were chosen considering the scatter in the test results. An upper bound for the flow stress curve under quasi-static conditions was assumed using the material parameters of the grade DH36 steel, which have been reported in [29,30]. The yield stress of DH36 is 435 MPa, and its hardening rate is comparable to the strain hardening curve of the test coupon exhibiting the largest ultimate tensile strength to yield strength ratio. In addition, a lower bound flow stress curve was defined using the material parameters of the grade AH36 [29,30] similarly. Table 1 lists the hardening law parameters of both grades. Figure 3 shows the hardening curves. The ductility limits of DH36 and AH36 were remarkably different and strongly dependent on the hardening rate. Figure 4 presents the fracture loci of these steel grades. The fracture model parameters were taken from reference [29]. The localized necking loci are plotted with dashed lines, and the fracture loci were plotted with continuous lines.

Table 1. Hardening law parameters of grade AH36 and DH36 steels.

Material	A (MPa)	ε_0 -	n -	k_0 (MPa)	Q (MPa)	β -	α -
AH36	1053	0.005407	0.2194	335	340.2	22.14	0.52
DH36	1058	0.007999	0.1794	444.7	293.1	21.89	0.55

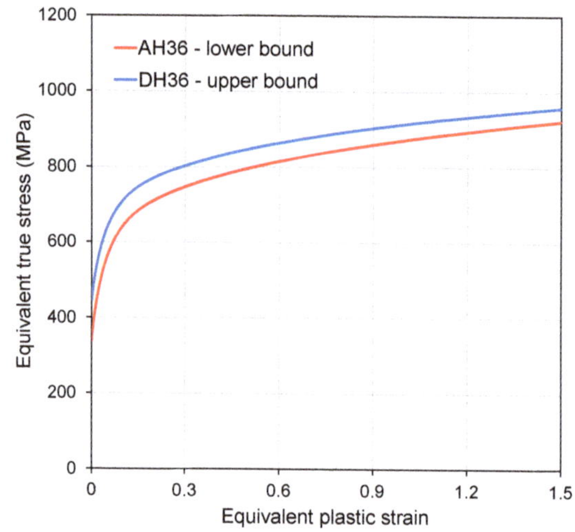

Figure 3. Hardening curves of AH36 and DH36 grade steels [30].

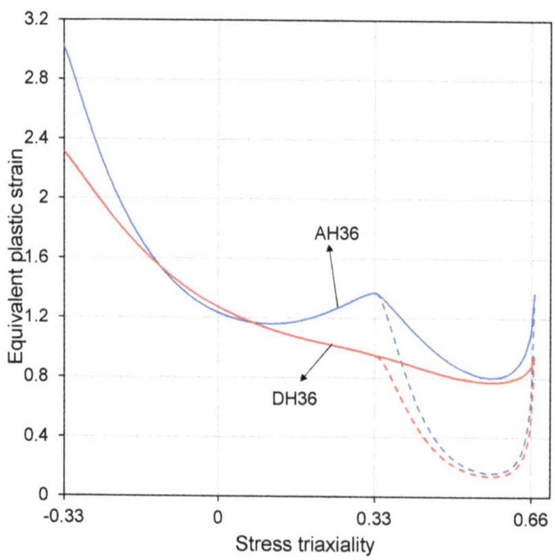

Figure 4. Localized necking (dashed line) and fracture (continuous line) loci of AH36 and DH36 grade steels.

The dynamic flow stress was modelled using the quasi-static flow stress curve of DH36 and dynamic flow stress properties of DP590, which has similar mechanical properties to the steel grades considered in this study [18,34]. We emphasize that in the present study, we do not specifically compare the three materials (moreover, three high-tensile strength structural steels) but use the parameters of these materials as estimates of the subject test material's flow stress and ductile fracture characteristics. Table 2 provides the adopted parameters, and Figure 5 depicts the resulting flow stress at various strain rates. Note that the temperature effect was implicitly considered in these curves. With increasing plastic deformation, the temperature increases and the thermal softening term in the deformation resistance function was calculated accordingly. Figure 6 shows the HC (continuous line) and DSSE (dashed line) loci for the same strain rates. To check the influence of the considered rate effects, the simulations of quasi-static tests were ran ignoring the rate- and temperature-dependent terms in the material model, as reported by [22], i.e., only using the strain-hardening law based on the combined Swift–Voce law and the original (rate-independent) DSSE–HC model. The rate-dependency was included in the low-velocity impact test simulation.

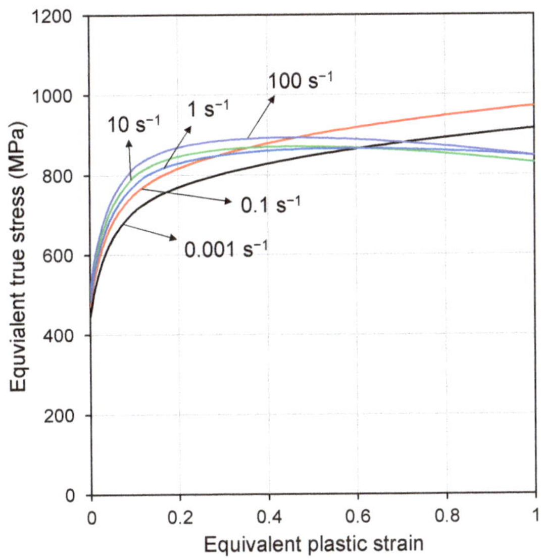

Figure 5. Flow stress curves at various strain rates.

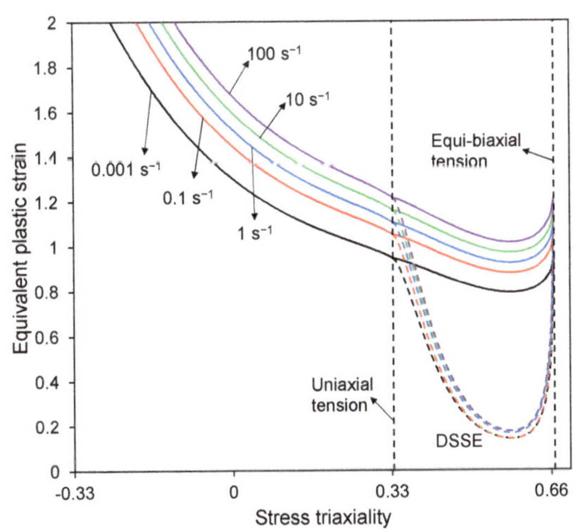

Figure 6. Localized necking and fracture loci at various strain rates.

Table 2. Parameters associated with the rate- and temperature-dependent plasticity model.

C	$\dot{\varepsilon}_0$ (s^{-1})	T_r (K)	T_m (K)	m	η_k	C_p (J/kg K)	$\dot{\bar{\varepsilon}}_a$ (s^{-1})
0.01366	0.00116	293	1673.7	0.921	0.9	420	1.379

5. Results and Discussions

The credibility of numerical predictions was examined by comparing the experimental force-displacement curves with the numerical simulation results and the observed fracture pattern with the numerically predicted failure mode. Note that the sudden drop in the force-displacement curves corresponds to the fracture initiation in the vicinity of the notch. Prediction of the post-failure response and crack growth path were considered of secondary importance as the adopted fracture model is suitable only for predicting the onset of a fracture and models the crack growth in an indirect way by successive deletion of failed elements.

5.1. Quasi-Static Impact Test

The experimental and numerical force-displacement curves were compared, as shown in Figure 7 for the quasi-static test and simulations. A good qualitative match was observed because the numerical results were within the experimentally observed scatter. The predictions made using the material model parameters of DH36 (upped bound flow stress curve) showed close agreement with the test results featuring the highest peak force. The stiffness, peak force, and instance of fracture were estimated accurately, but the force levels in the decay region were slightly larger than the test results. The predictions with the AH36 parameters (lower bound flow stress curve) resulted in a much lower peak force and a larger displacement at fracture compared to the experiments. (Therefore, it was deemed that the upper bound curve represents the test structural material more closely, and further analyses were conducted with the upper bound curve parameters.)

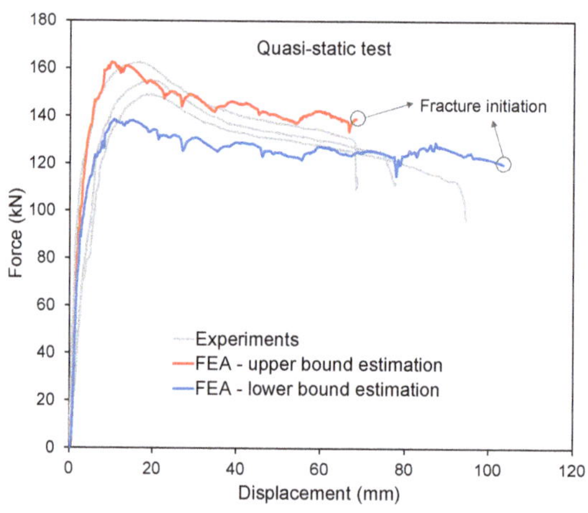

Figure 7. Force-displacement curves for quasi-static test.

In the tests, the fracture initiated close to the notch and propagated diagonally towards the contact point between the indentor and the target model. In addition, there were no surface cracks on the top flange of the test model around the folded regions. Figure 8 shows the predicted failure mode and compares the results with the post-mortem specimen. Here, the localized necking indicator, N, and ductile fracture indicator, D, were used as the plotting contour. Note that localized necking occurs around the notch, but the condition for surface cracking, which is $D = 1$ at any through-thickness integration point, was not satisfied on the top flange of the SHS. (Therefore, similar to the test observations, surface cracking was not predicted in the numerical simulations.)

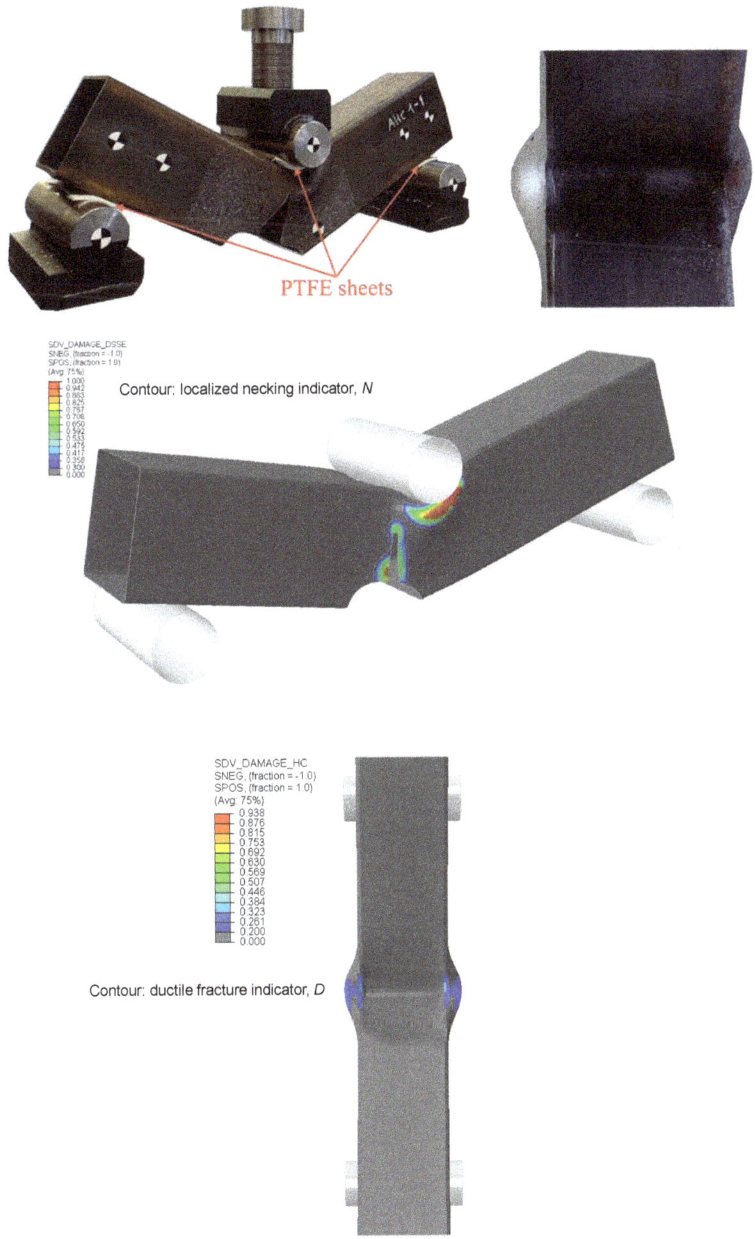

Figure 8. Predicted failure mode and comparison with the post-mortem test specimen.

5.2. Dynamic Impact Test

Next, by using the upper bound curve parameters and rate-dependent material parameters listed in Table 2, dynamic impact tests were simulated. Figure 9 shows the force-displacement curves for the low-velocity impact tests. Although the numerical simulation overestimated the force levels in the decay region, the peak force and the instant of fracture initiation were captured with reasonable accuracy. The agreement between the test scatter and the numerical predictions support the choice of assumed values for the material model parameters. Note that the numerical results were not filtered and display some oscillations, which were also present in the actual, unfiltered test results [37]. The oscillations can be attributed to the stress waves travelling back and forth, and the temporal variations of contact forces. In the dynamic tests, the failure mode was again localized necking around the notch, and no fractures were observed in the top flange of the SHS models. Figure 10 presents the finite element model of the test model at the final instant of the simulation with localized necking and ductile fracture indicator plots. The numerical simulations mimicked the observed physical test results. In agreement with the test results, surface cracking was not predicted in the simulations. It should be emphasized that the ductile failure model employed in the present work is applicable for predicting of the onset of ductile failure, not the ductile crack growth. The predicted crack growth appears to be in a straight line rather than curving as seen in the experimental results. The technique of successive deletion of failed elements yields rather an unrealistic crack shape, which also shows a significant dependence on mesh configuration.

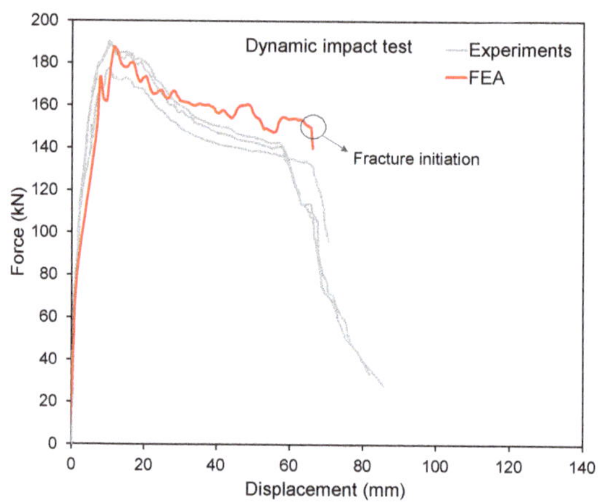

Figure 9. Comparison of the experimental and numerical force-displacement curves for the dynamic impact test.

To gain further insights into how the adopted fracture model handles different fracture mechanisms and how state variables change over the course of loading, the loading paths of the element that failed first in the simulations are plotted in Figure 11. Figure 12 shows the location of the element along with the strain rate, temperature increase, and stress state (in terms of stress triaxiality) plots at the instance of first element deletion. The loading paths were extracted from the top and bottom through-thickness integration points. It was noted that the loading paths were slightly non-proportional, that is, the stress state is not constant all the way to the onset of the fracture. Close to the notch, the elements predominantly experienced membrane stretching (as indicated by the value of stress triaxiality which remains in the biaxial tension zone $-1/3 < \eta < 2/3$), and failure was triggered by localized necking as the accumulated necking damage in all through-thickness integration points reach unity. In addition, a considerable temperature increase was estimated on the folded corners at the contact surface between the indentor and test model, and in the vicinity of the notch following Equation (12), where localized large straining occurred.

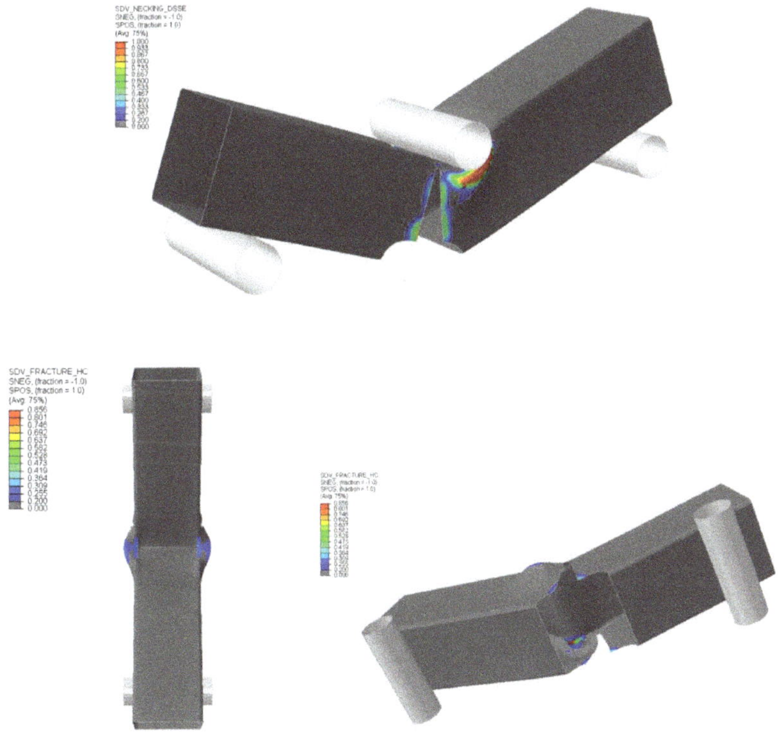

Figure 10. Predicted deformations and fracture in the simulation of the dynamic impact test.

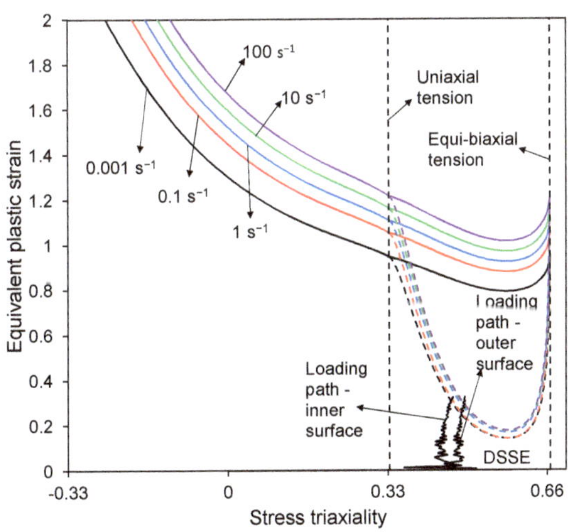

Figure 11. Loading paths of the top and bottom through-thickness integration points in the first failed element.

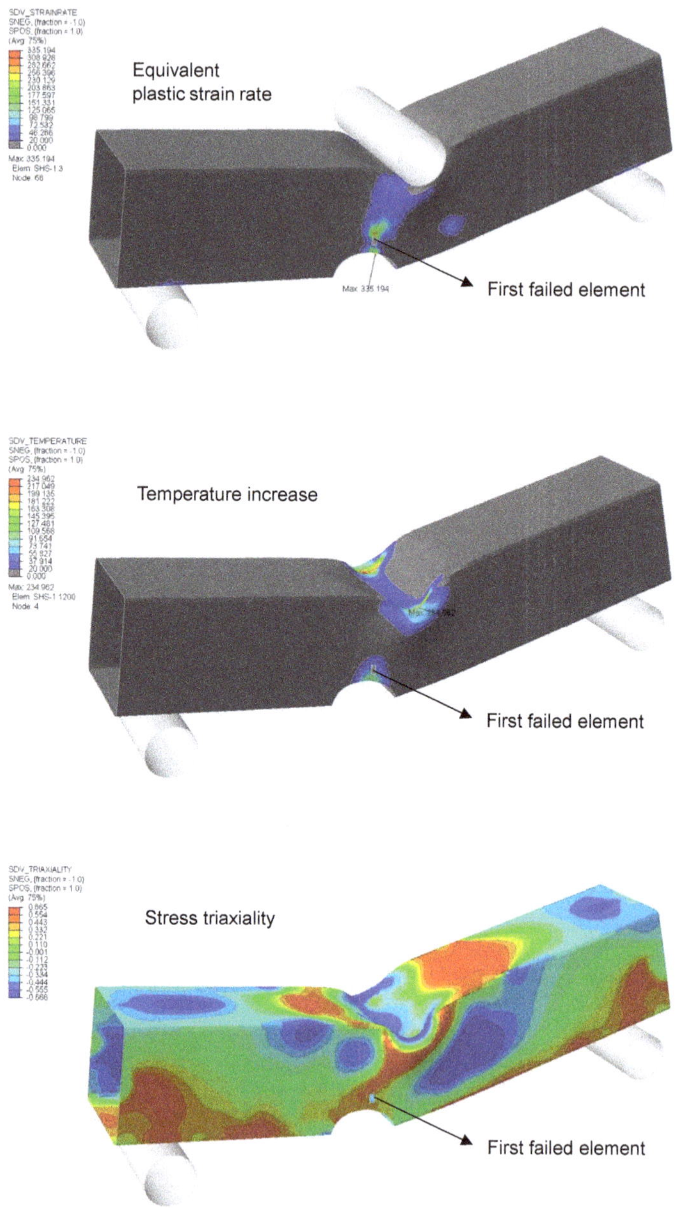

Figure 12. State variables at the initiation of fracture.

5.3. Sensitivity to Inclusion of Rate Effects

At this junction, it may be important to check whether the included rate effects play a significant role in the predicted impact response of the target structure under low-velocity mass impact. The sensitivity of the fracture predictions to the dynamic effects was checked by running the simulations, excluding the strain-rate and thermal-softening effects. Figure 13 shows the resulting force-displacement curves. It was observed that the exclusion of both the strain-rate hardening and thermal softening yielded a lower peak force associated with the plastic collapse of the SHS tube. Fracture initiation was delayed slightly, but the predicted failure mode and location were the same. On the other hand, omitting only the thermal softening led to a stiffer response. Eventually, the fracture did not occur before the kinetic energy of the indentor was fully transformed to the strain energy of the struck model.

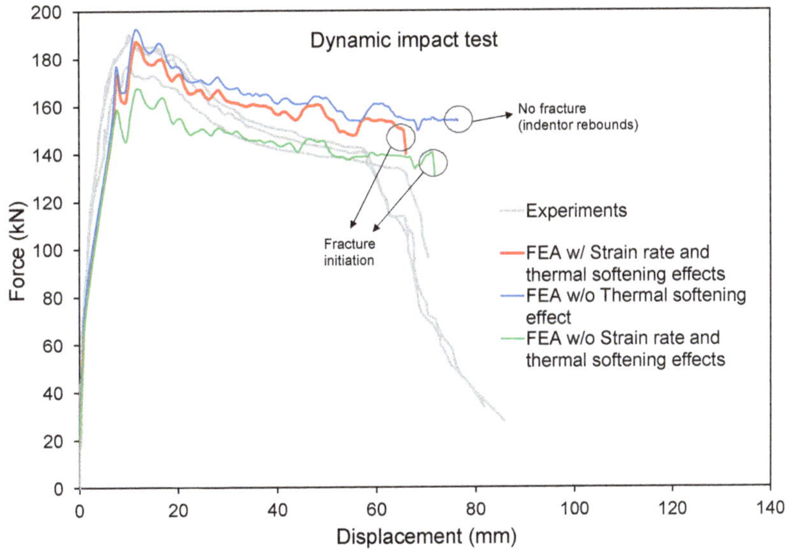

Figure 13. Force-displacement curves obtained considering or omitting dynamic effects.

5.4. Comparison with Industry Practice

Lastly, the simulations were ran using a constant fracture strain value of 0.2 to clarify the necessity of using two different "failure" indicators, which is one of the distinguishing properties of the proposed model evaluated in the previous section. The value of constant fracture strain was chosen close to the equivalent plastic strain associated with localized necking under plane strain tension and reflects the practice used in ship collision analysis [18]. The strain-rate effect (strain-rate hardening) was considered using the Cowper–Symonds equation with the commonly used coefficients for high-tensile steel ($D_{CS} = 3200\ \text{s}^{-1}$ and $q = 5$) [20]. Figures 14 and 15 show the resulting force-displacement curve and the predicted failure modes, respectively. The deleted elements and failed through-thickness integration points on the top flange of the model caused a rapid decrease in structural resistance and premature failure compared to the experiments. According to the simple failure model employed, the stress tensor components of the IP were set to zero when the equivalent plastic strain reached the critical failure strain value (in this case, 0.2). An element is deleted if all the through-thickness integration points have zero stress, i.e., assumed to have failed following the conditions above. This caused an undesirable simulation outcome in terms of both the predicted deformation and fracture patterns and force-displacement response. A larger critical failure strain would delay failure of the top flange elements and fracture initiation around the notch. It is apparent that the material model adopted in this study overcomes this issue in a computationally convenient manner by considering the basic failure modes separately.

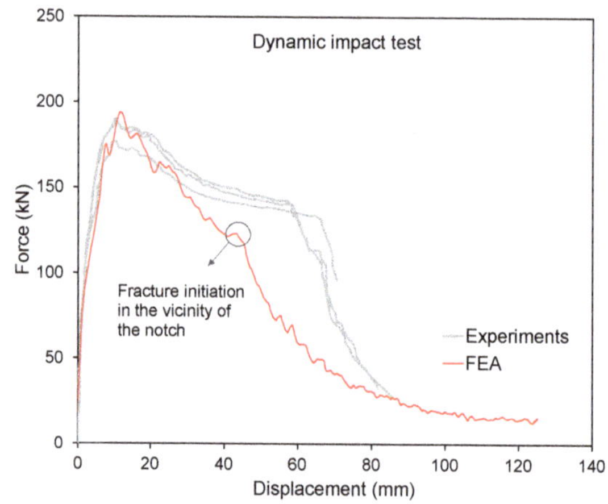

Figure 14. Force-displacement curve obtained using a constant failure strain.

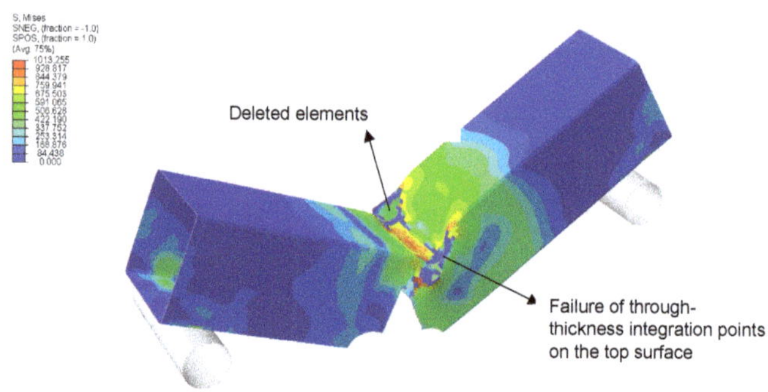

Figure 15. Equivalent stress plot of the post-mortem test model as predicted by FEA using a constant failure strain.

6. Conclusions

The present study simulated a fracture in steel-plated structures under low-velocity impact using shell elements by adopting an advanced material model that considers the strain-rate effect. The major conclusions drawn from the present study are as follows:

- It was observed that the results from both the quasi-static three-point bending and the dynamic impact simulations agreed well with the test results. The credibility of rate-dependent DSSE–HC model was confirmed as it successfully captured the overall deformation and failure behaviour of the square hollow section with an asymmetric notch.
- The necessity of distinguishing the fracture with and without localized necking when simulating a fracture with shell elements was confirmed. It was apparent that a single-failure model, i.e., a failure model for both localized necking and ductile fracture involving surface cracking or shear-induced failure, was prone to errors associated with regularizing the fracture locus (scaling the ductility limits based on mesh size).
- The adequacy of localized necking-based failure models in conjunction with the rate-dependent flow stress curves to describe the necking locus in an evolving manner was verified.
- Temperature effects were found to be significant on the absorbed energy and fracture initiation. It was noted that the inclusion of temperature effects may increase the absorbed impact energy because of increased ductility.

Author Contributions: Conceptualization, B.C.C. and J.C.; methodology, B.C.C. and J.C.; software, B.C.C.; validation, B.C.C. and J.C.; formal analysis, B.C.C. and J.C.; investigation, B.C.C. and J.C.; resources, J.C.; data curation, B.C.C.; writing—original draft preparation, B.C.C. and J.C.; writing—review and editing, B.C.C. and J.C.; visualization, B.C.C.; supervision, J.C.; project administration, J.C.; funding acquisition, J.C. All authors have read and agreed to the published version of the manuscript.

Funding: This work was supported by the Korea Environment Industry & Technology Institute funded by Korea Ministry of Environment (No. 146836) and by the Korea Energy Technology Evaluation and Planning funded by the Ministry of Trade, Industry and Energy of Korea (No. 20203030020230).

Institutional Review Board Statement: Not applicable.

Informed Consent Statement: Not applicable.

Data Availability Statement: Not applicable.

Conflicts of Interest: The authors declare no conflicts of interest. The funders had no role in the design of the study; in the collection, analyses, or interpretation of data; in the writing of the manuscript; or in the decision to publish the results.

Nomenclature

The following symbols are used in this manuscript:

α	Weighting factor in combined Swift–Voce hardening law
β	Voce hardening law parameter
γ	Fracture strain rate sensitivity parameter
$\dot{\bar{\varepsilon}}_a$	Reference strain rate for adiabatic condition
$\bar{\varepsilon}^{pr}_{DSSE}$	Localized necking strain under proportional loading
$\bar{\varepsilon}^{pr}_f$	Fracture strain under proportional loading
$\dot{\bar{\varepsilon}}_{it}$	Reference strain rate for isothermal condition
ε_0	Pre-strain in Swift hardening law
$\dot{\bar{\varepsilon}}_0$	Reference strain rate
$\bar{\varepsilon}_p$	Equivalent plastic strain
$\dot{\bar{\varepsilon}}_p$	Equivalent plastic strain rate
η	Stress triaxiality
η_k	Taylor–Quinney coefficient
$\bar{\theta}$	Lode angle parameter
ρ	Material density
$\boldsymbol{\sigma}$	Cauchy stress tensor
$\bar{\sigma}$	von Mises equivalent stress
ω	Regulating term for transition from isothermal to adiabatic condition
A	Swift law parameter
C	Strain-rate hardening sensitivity parameter
C_p	Material specific heat
D	Ductile fracture indicator
D_{CS}	Cowper–Symonds model constant
E	Young's modulus
J_2	Second invariant of deviatoric stress tensor
N	Localized necking indicator
Q	Voce law parameter
T	Temperature
T_m	Material melting temperature
T_r	Reference temperature
a, b, c	Hosford–Coulomb model parameters
f_1, f_2, f_3	Lode angle-dependent functions in Hosford–Coulomb model
g_1, g_2	Stress triaxiality-dependent functions in DSSE model
k	Deformation resistance function

k_{ε}	Strain hardening function
$k_{\dot{\varepsilon}}$	Strain-rate hardening function
k_0	Voce hardening law parameter
k_S	Swift hardening law
k_T	Thermal softening function
k_V	Voce hardening law
m	Thermal softening exponent
n	Swift law exponent
n_f	Hosford–Coulomb model transformation coefficient
p_f	DSSE model exponent
q	Cowper–Symonds model exponent

Abbreviations

The following abbreviations are used in this manuscript:

DOF	Degree of freedom
DSSE	Domain of shell-to-solid equivalence
HC	Hosford-Coulomb
IP	Integration point (through-thickness)
NLFEA	Non-Linear Finite Element Analysis
PTFE	Polytetrafluoroethylene
SHS	Square hollow section

References

1. Liu, B.; Guedes Soares, C. Recent developments in ship collision analysis and challenges to an accidental limit state design method. *Ocean Eng.* **2023**, *270*, 113636. [CrossRef]
2. Samuelides, M. Recent advances and future trends in structural crashworthiness of ship structures subjected to impact loads. *Ships Offshore Struct.* **2015**, *10*, 1–10. [CrossRef]
3. Storheim, M.; Alsos, H.S.; Amdahl, J. Evaluation of nonlinear material behavior for offshore structures subjected to accidental actions. *J. Offshore Mech. Arct. Eng.* **2018**, *140*. [CrossRef]
4. Calle, M.; Alves, M. A review-analysis on material failure modeling in ship collision. *Ocean Eng.* **2015**, *106*, 20–38. [CrossRef]
5. Storheim, M.; Amdahl, J.; Martens, I. On the accuracy of fracture estimation in collision analysis of ship and offshore structures. *Mar. Struct.* **2015**, *44*, 254–287. [CrossRef]
6. Liu, B.; Villavicencio, R.; Zhang, S.; Guedes Soares, C. A simple criterion to evaluate the rupture of materials in ship collision simulations. *Mar. Struct.* **2017**, *54*, 92–111. [CrossRef]
7. Cerik, B.C.; Park, S.J.; Choung, J. Use of localized necking and fracture as a failure criterion in ship collision analysis. *Mar. Struct.* **2020**, *73*, 102787. [CrossRef]
8. Cerik, B.C.; Lee, K.; Choung, J. Evaluation of localized necking models for fracture prediction in punch-loaded steel panels. *J. Mar. Sci. Eng.* **2021**, *9*, 1–13. [CrossRef]
9. Jones, N. The credibility of predictions for structural designs subjected to large dynamic loadings causing inelastic behaviour. *Int. J. Impact Eng.* **2013**, *53*, 106–114. [CrossRef]
10. Liu, B.; Pedersen, P.T.; Zhu, L.; Zhang, S. Review of experiments and calculation procedures for ship collision and grounding damage. *Mar. Struct.* **2018**, *59*, 105–121. [CrossRef]
11. Morin, D.; Kaarstad, B.; Skajaa, B.; Hopperstad, O.; Langseth, M. Testing and modelling of stiffened aluminium panels subjected to quasi-static and low-velocity impact loading. *Int. J. Impact Eng.* **2017**, *110*, 97–111. [CrossRef]
12. Gruben, G.; Sølvernes, S.; Berstad, T.; Morin, D.; Hopperstad, O.; Langseth, M. Low-velocity impact behaviour and failure of stiffened steel plates. *Mar. Struct.* **2017**, *54*, 73–91. [CrossRef]
13. Ohtsubo, H.; Kawamoto, Y.; Kuroiwa, T. Experimental and numerical research on ship collision and grounding of oil tankers. *Nucl. Eng. Des.* **1994**, *150*, 385–396. [CrossRef]
14. Noh, M.H.; Cerik, B.C.; Han, D.; Choung, J. Lateral impact tests on FH32 grade steel stiffened plates at room and sub-zero temperatures. *Int. J. Impact Eng.* **2018**, *115*, 36–47. [CrossRef]
15. Liu, K.; Liu, B.; Villavicencio, R.; Wang, Z.; Guedes Soares, C. Assessment of material strain rate effects on square steel plates under lateral dynamic impact loads. *Ships Offshore Struct.* **2018**, *13*, 217–225. [CrossRef]
16. Liu, B.; Guedes Soares, C. Effect of strain rate on dynamic responses of laterally impacted steel plates. *Int. J. Mech. Sci.* **2019**, *160*, 307–317. [CrossRef]
17. Cowper, G.; Symonds, P. *Strain Hardening and Strain-Rate Effects in the Impact Loading of Cantilever Beam*; Brown University Division of Applied Mathematics: Providence, RI, USA, 1957.
18. Storheim, M.; Amdahl, J. On the sensitivity to work hardening and strain-rate effects in nonlinear FEM analysis of ship collisions. *Ships Offshore Struct.* **2017**, *12*, 100–115. [CrossRef]

19. Choung, J.; Nam, W.; Lee, J.Y. Dynamic hardening behaviors of various marine structural steels considering dependencies on strain rate and temperature. *Mar. Struct.* **2013**, *32*, 49–67. [CrossRef]
20. Paik, J.K. Practical techniques for finite element modeling to simulate structural crashworthiness in ship collisions and grounding (Part I: Theory). *Ships Offshore Struct.* **2007**, *2*, 69–80. [CrossRef]
21. Ko, Y.G.; Kim, S.J.; Sohn, J.M.; Paik, J.K. A practical method to determine the dynamic fracture strain for the nonlinear finite element analysis of structural crashworthiness in ship–ship collisions. *Ships Offshore Struct.* **2018**, *13*, 412–422. [CrossRef]
22. Cerik, B.C.; Lee, K.; Park, S.J.; Choung, J. Simulation of ship collision and grounding damage using Hosford-Coulomb fracture model for shell elements. *Ocean Eng.* **2019**, *173*, 415–432. [CrossRef]
23. Cerik, B.C.; Ringsberg, J.W.; Choung, J. Revisiting MARSTRUCT benchmark study on side-shell collision with a combined localized necking and stress-state dependent ductile fracture model. *Ocean Eng.* **2019**, *187*, 106173. . 2019.106173. [CrossRef]
24. Stoughton, T.B.; Yoon, J.W. A new approach for failure criterion for sheet metals. *Int. J. Plast.* **2011**, *27*, 440–459. [CrossRef]
25. Hooputra, H.; Gese, H.; Dell, H.; Werner, H. A comprehensive failure model for crashworthiness simulation of aluminium extrusions. *Int. J. Crashworthiness* **2004**, *9*, 449–464. [CrossRef]
26. Pack, K.; Mohr, D. Combined necking & fracture model to predict ductile failure with shell finite elements. *Eng. Fract. Mech.* **2017**, *182*, 32–51. [CrossRef]
27. Cerik, B.C.; Park, B.; Park, S.J.; Choung, J. Modeling, testing and calibration of ductile crack formation in grade DH36 ship plates. *Mar. Struct.* **2019**, *173*, 27–43. [CrossRef]
28. Park, S.J.; Lee, K.; Cerik, B.C.; Choung, J. Ductile fracture prediction of EH36 grade steel based on Hosford–Coulomb model. *Ships Offshore Struct.* **2019**, *14*, 219–230. [CrossRef]
29. Park, S.J.; Cerik, B.C.; Choung, J. Comparative study on ductile fracture prediction of high-tensile strength marine structural steels. *Ships Offshore Struct.* **2020**, *15*, S208–S219. [CrossRef]
30. Cerik, B.C.; Choung, J. Ductile fracture behavior of mild and high-tensile strength shipbuilding steels. *Appl. Sci.* **2020**, *10*, 7034. [CrossRef]
31. Costas, M.; Morin, D.; Hopperstad, O.S.; Børvik, T.; Langseth, M. A through-thickness damage regularisation scheme for shell elements subjected to severe bending and membrane deformations. *J. Mech. Phys. Solids* **2019**, *123*, 190–206. [CrossRef]
32. Cerik, B.C.; Choung, J. On the prediction of ductile fracture in ship structures with shell elements at low temperatures. *Thin-Walled Struct.* **2020**, *151*, 106721. [CrossRef]
33. Park, S.J.; Choung, J. Punching fracture experiments and simulations of unstiffened and stiffened panels for ships and offshore structures. *J. Ocean Eng. Technol.* **2020**, *34*, 155–166. [CrossRef]
34. Cerik, B.C.; Choung, J. Rate-dependent combined necking and fracture model for predicting ductile fracture with shell elements at high strain rates. *Int. J. Impact Eng.* **2020**, *146*, 103697. [CrossRef]
35. Cerik, B.C.; Choung, J. Fracture Estimation in Ship Collision Analysis—Strain Rate and Thermal Softening Effects. *Metals* **2021**, *11*, 1402. [CrossRef]
36. Gruben, G.; Langseth, M.; Fagerholt, E.; Hopperstad, O. Low-velocity impact on high-strength steel sheets: An experimental and numerical study. *Int. J. Impact Eng.* **2016**, *88*, 153–171. [CrossRef]
37. Johannessen, H.; Johannessen, O.H.; Costas, M.; Clausen, A.H.; Sønstabø, J.K. Experimental and numerical study of notched SHS made of different S355 steels. *J. Constr. Steel Res.* **2021**, *182*, 106673. [CrossRef]
38. Roth, C.C.; Mohr, D. Effect of strain rate on ductile fracture initiation in advanced high strength steel sheets: Experiments and modeling. *Int. J. Plast.* **2014**, *56*, 19–44. [CrossRef]
39. Johnson, G.R.; Cook, W.H. A computational constitutive model and data for metals subjected to large strain, high strain rates and high pressures. In Proceedings of the 7th International Symposium Ballistics, Hague, The Netherlands, 19–21 April 1983; pp. 541–547.
40. Swift, H. Plastic instability under plane stress. *J. Mech. Phys. Solids* **1952**, *1*, 1–18. [CrossRef]
41. Voce, E. The relationship between stress and strain from homogenous deformation. *J. Inst. Met.* **1948**, *74*, 537–562.
42. Sung, J.H.; Kim, J.H.; Wagoner, R. A plastic constitutive equation incorporating strain, strain-rate, and temperature. *Int. J. Plast.* **2010**, *26*, 1746–1771. [CrossRef]

Journal of *Marine Science and Engineering*

MDPI

Article

Low Temperature Effect on the Mechanical Properties of EH36 with Strain Rates

Jing Zhang [1], Xuelei Kang [1], Xinghua Shi [1,2,*], C. Guedes Soares [2] and Ming Song [1]

1 School of Naval Architecture and Ocean Engineering, Jiangsu University of Science and Technology, Zhenjiang 212100, China
2 Centre for Marine Technology and Ocean Engineering (CENTEC), Instituto Superior Técnico, Universidade de Lisboa, 1049-001 Lisbon, Portugal
* Correspondence: xinghuashi@just.edu.cn; Tel.: +86-15896380832; Fax: +86-51184404433

Abstract: With the expansion of the Arctic route, the safety of ship crossing the area in light of the low temperature and ice has become of focus, especially with regards to the ship's structure. The mechanical properties of the material making up the ship's structure may not be suitable for the Arctic environment. A series of quasi-static and dynamic tests were performed to investigate the behaviour of EH36 steel, which is used to build Arctic ships, at temperatures ranging from 20 °C to −60 °C. The yield and ultimate tensile stress increased more than 10% as the temperature decreased from 20 °C to −60 °C, whereas the toughness decreased as the temperature decreased. A formula was derived to illustrate the relationship between the temperature reduction and the yield strength by fitting the experimental data. Four common constitutive rigid-perfectly plastic, elastic-perfectly plastic, bilinear elastic-plastic, and multi-linear elastic plastic models were fitted to simulate the hull structure under static loading and low temperature. Additionally, the strain rate effect of EH36 steel at low temperatures was illustrated by quasi-static and high-speed impact tests. A constitutive model including the low temperature and strain rate was introduced based on a modified Cowper-Symonds model, in which the coefficients of the constitutive model are fitted by the test results. It is improved by an iterative numerical method used to obtain more accurate coefficients using a series of numerical analyses. Detailed finite element simulations of the experiment conditions revealed that the constitutive model accurately predicts the dynamic response at low temperatures.

Keywords: constitutive model; low temperature; cowper-symonds; strain rate; dynamic behaviour

Citation: Zhang, J.; Kang, X.; Shi, X.; Guedes Soares, C.; Song, M. Low Temperature Effect on the Mechanical Properties of EH36 with Strain Rates. *J. Mar. Sci. Eng.* **2023**, *11*, 678. https://doi.org/10.3390/jmse11030678

Academic Editor: Vincenzo Crupi

Received: 23 February 2023
Revised: 14 March 2023
Accepted: 19 March 2023
Published: 22 March 2023

1. Introduction

As the air temperature has been increasing over the last several years, the Arctic shipping routes are becoming suitable for shipping for more than 4 months per year. This route reduces shipping costs significantly, especially from China to USA and Europe. However, shipping in the Arctic is a challenge for common ships, due to the threats related to low temperatures and collision with floating ice. The design and manufacture of Arctic ships are becoming an increasingly interesting topic. Some rules or guidelines were introduced for Arctic ships by the Classification Societies or other organizations [1–3]. However, the interaction mechanics of ships and ice is not fully understood, including the constitutive properties of steel at low temperature, and of the ice. While ships are crossing the ice area, repeated impacts of ice on the hull will occur, and proper methods to account for them need to be available [4]. As the basic input of the ship–ice interaction response, the constitutive relationship is one of the important factors that needs to be studied. The dynamic load and the low temperature may affect the ship's structure simultaneously. Strain rate will have a significant effect on the collision or impact response of steel or similar material with the characteristics of strain rate sensitivity [5]. It is necessary to discuss the dynamic mechanics of the ship structure's steel at low temperatures based on the quasi-static mechanical properties. A series of quasi-static and dynamic material tests should be performed. However, some special materals, such as mild steel and high-tensile steel, have been studied at low teperatures, except for EH36, which is used to build the Arctic ships, and no common constitutive model has been presented.

As the temperature is decreasing, the yield tensile stress will increase for steel, but the toughness will reduce. If the ship structure is subjected to the impact of ice, it is necessary to predict the response accurately, including the strain rate's effect on the ship's steel. The increased crashworthiness was studied in arctic conditions at sub-zero temperatures by Ehlers [6]. The steel material characteristics obtained by the quasi-static tensile material standard tests were used to perform the comparative collision simulations. A test database covering low temperatures up to −60 °C was given by Paik [7] from a series of static, quasi-static, and dynamic tests for mild steel, high-tensile steel (AH32, DH32 and DH36), aluminium alloy 5083-O, and stainless steel 304 L. Except for these tests, limited research was performed on the mechanics of the ship's steel at low temperatures. Low and middle strain rates were included in the tests by Choung et al. [8] at a low temperature (−40 °C), room temperature, and a high temperature (200 °C) to investigate the hardening behaviour. The EH series high-tensile steel is another material for Arctic ships recommended by DNV [9], ABS [2], and IACS [10], which will be studied in this paper.

Numerous researchers have published their experimental results on the effect of strain rate on the behaviour of structural steel and the mechanical properties of mild steel. The mechanical testing of metals at a high strain rate (10^2–10^4 s^{-1}) is typically performed with the Kolsky bar, also widely known as the Split Hopkinson Pressure Bar (SHPB) [11]. Low ($\dot{\varepsilon} - 10^{-3}$ s^{-1}), intermediate ($\dot{\varepsilon} - 1$ s^{-1}), and high ($\dot{\varepsilon} - 10^3$ s^{-1}) strain rate experiments are performed on dog-bone and notched specimens extracted from 1.4 mm thick TRIP780 steel sheets. Experiments in the intermediate range of strain rates ($\dot{\varepsilon} - 1$~10 s^{-1}) are typically carried out in servo-hydraulic testing machines, whereas high strain rate experiments (100–1000 s^{-1}) are usually performed in the Split Hokinson Bar system [12]. Various authors have studied the effect of strain rate at room temperature; both the behaviour of aluminium and steel have been studied, as the first is not strain-sensitive and the latter is [13]. Different constitutive models have been proposed to represent the behaviour of steel [5,14].

High temperatures induce strain ageing of the metal material. The high strain increases the strength but decreases the ductility combined with the high temperatures for the low-alloy structural carbon steel S335 [15]. The mechanical behaviour of a high strain rate that induced partially damaged mild steel under various temperature test conditions has been reported by Mirmomeni et al. [16]. High temperatures are the dominant factor, as the temperature is over 450 °C. Nemat-Nasser and Guo [17] predicted the compressive mechanical behaviour of this steel within a very large strain rate (from 0.001 to 10,000/s) and temperature ranges (−196.15 °C–726.85 °C) and only investigated −196.15 °C. The low temperature of −196 °C induces the transformation of the ductile-to-brittle behaviour of the stainless steel weldments including 304 L and 316 L, but there is little decrease in the impact energy [18]. Getter et al. [19] designed a novel high-rate testing apparatus to conduct uniaxial tensile tests of two marine types of steels under strain rates ranging from 13 to 250 s^{-1}. Chen, et al. [20] performed a series of quasi-static and dynamic uniaxial tensile tests within the range of 0.001–288 s^{-1} strain rates to study the rate-dependent mechanical properties of Q420 steel using the INSTRON and Zwick/Roell HTM5020 testing machine. Su et al. [21] tested the tensile mechanical behaviour of DH 36 steel. The strain rate range was from 0.001 to 3000/s, and the initial specimen temperatures were 19.85–526.85 °C. The flow mechanism combined with the strain rate of an Ni-based superalloy was investigated by Detrois et al. [22] under high temperatures more than 1000 °C using isothermal compression testing. Flow stress models were studied for different high temperatures.

Based on the available experimental results, some constitutive models considering the influence of strain, strain rate, and temperature on the strain-hardening behaviour have been proposed by several researchers. Forni et al. [23] developed the constitutive model of S355 structural steel by using the Johnson–Cook (J-C) model and Cowper-Symonds (C-S) model [24]. Shi et al. [25] developed a revised multi-linear constitutive model of high-strength structural steel based on material property data of Q500, Q550, and Q690 steels (over 10,000 samples). The change in yield strength between high-strength structural steels inevitably led to differences in mechanical properties and dynamic behaviour. The experimental stress–strain data obtained from high strain rate compression tests using SHPB by Bobbili et al. [26] over a range of tempering temperatures (500–650 °C), strains (0.05–0.2), and strain rates (1000–5500 s^{-1}) were employed to formulate the J-C model to predict the high strain rate deformation behaviour of high-strength armour steels.

Many studies on the mechanical properties of steel consider the effects of strain rate and temperature. However, few studies focused on the behaviour of stress–strain curves at low temperatures and strain rates. The Cowper-Symonds representation of material strain rate sensitivity is commonly employed in finite element impact simulations of steel marine structures; Cowper-Symonds coefficients of $C = 40.4$ and $P = 5.00$ were employed for mild steel [27]. However, the Cowper-Symonds parameters vary widely among different studies of mild steel [28,29]. In this paper, the experiment aimed to investigate the complex behaviour of different strain rates induced by ship–ice collision at low temperatures. The material mechanical properties from low to high strain ranging from 2×10^{-4} to 2000 s^{-1} were obtained from quasi-static tests by material testing machine (MTS) and dynamic material tests using the split Hopkinson pressure bar (SHPB) for EH36. A constitutive model was presented with low temperature and strain rate by the test and numerical data. The stress–strain curves provide a reliable material behaviour, covering strain rates ranging from 2×10^{-4} to 2000 s^{-1} and temperatures from 20 °C to −60 °C.

2. Experimental Program

2.1. Test Specimen

No rules or guidelines for the test specimen at low temperatures have yet been published now. Paik [7] performed the tension test at low temperatures, according to the standard of the tension testing of metallic materials by ASTM [30]. However, the test methods in the standard only cover the testing of metallic materials at room temperature (10–38 °C). Low temperatures increase the yield strength of the steel during the quasi-static tension process. The same specimen may be used at low temperatures.

A specimen was designed for the quasi-static test at low temperatures, as shown in Figure 1a. The sheet-type standard specimen was of 6 mm thickness, 12.5 mm width in the middle, and 20 mm width in the grip section. The length of the grip section was 50 mm. The test was performed in a low-temperature box, with one end fixed by wedge grips and the other by a pin. It was a bone type with a wider end compared with the standard specimen at normal temperature. There was a hole with a diameter of 10 mm on the centre line in the wider end to connect the specimen end test set-up, as shown in Figure 2c. A bar with a bolt through the hole in the specimen was used to connect the specimen to the upper loading equipment. The wider end was three times wider than the other end to eliminate the cross-section reduction effect. It was also confirmed that the failure happens in the middle of the specimen, as shown in Figure 3.

The length of the dynamic specimen was 3 mm and the diameter was 6 mm. The surfaces of the specimen were polished to ensure their smoothness. There were no residual stress and machining defects in the specimen.

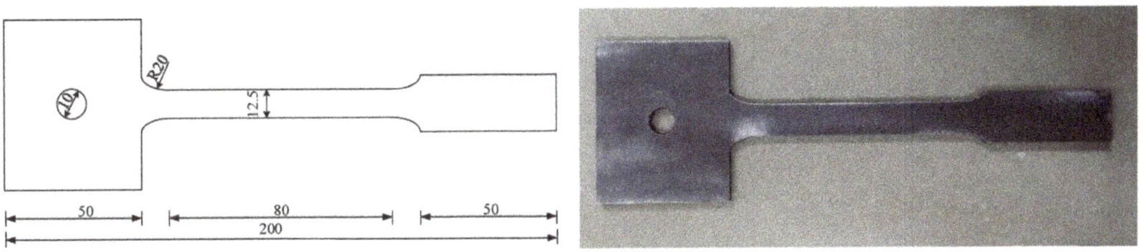

(a) Quasi-static test specimen.

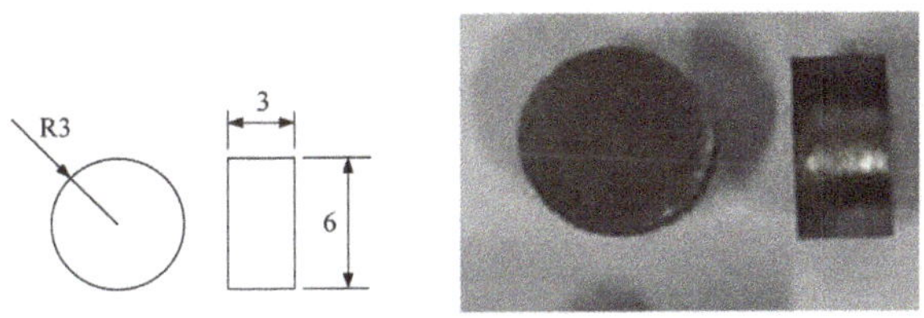

(b) Dynamic test specimen.

Figure 1. Geometry test specimen (unit: mm).

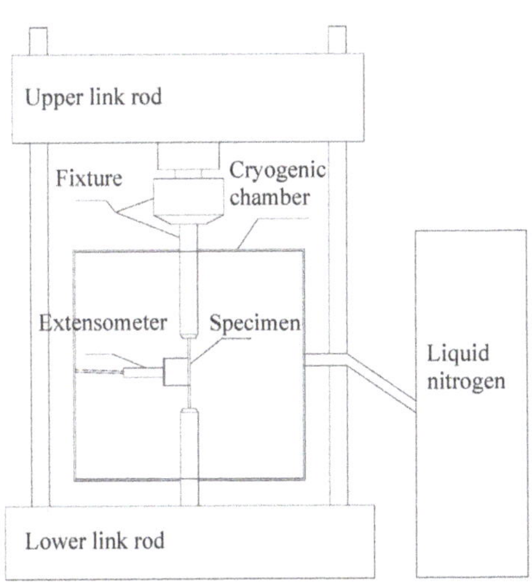

(a) Schematic of the quasi-static tensile experiment.

(b) Experimental setup for low strain rate tensile tests.

Figure 2. *Cont.*

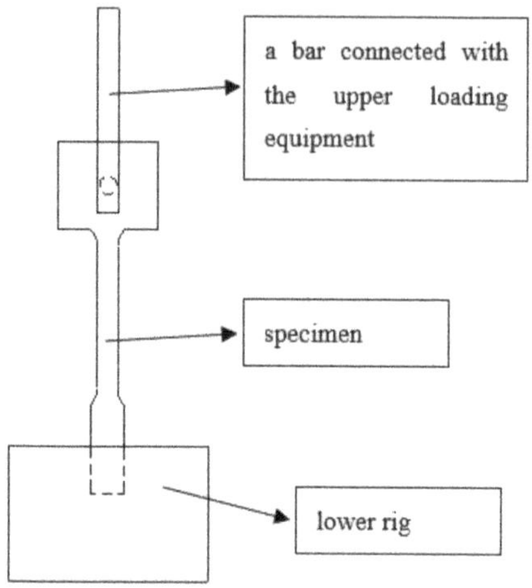

(c) Details of the specimen connection.

Figure 2. Quasi-static test setup at low temperatures.

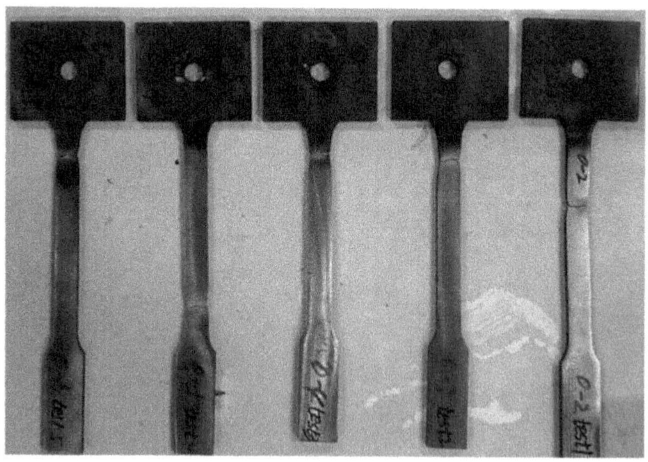

Figure 3. Failure of the quasi-static test specimen.

2.2. Test Setup

2.2.1. Quasi-Static and Low Velocity Tests

The quasi-static tension tests were performed by Material Testing Systems (MTS 810) with a maximum loading capacity of 500 KN, as the specimen was in a cryogenic chamber designed for keeping a low-temperature environment using liquid nitrogen. The arrangement of the test setup is shown in Figure 2a,b. Two holes were opened on the top and bottom surface of the chamber to install the specimen; the bottom end gripped and the top end pinned, as shown in Figure 2c. An extensometer for the temperate range of −80 to 80 °C was set on the surface of the specimen to measure the strain. The test section had to be completely immersed in the constant low-temperature chamber.

The design strain rate of quasi-static tension tests is 2×10^{-4} s^{-1}. Some low velocity tension experiments at the strain rates of 0.01, 0.1 and 1 s^{-1} were also performed by the MTS in Figure 2. The axial tension tests of two specimens were carried out at ambient temperature (room temperature, 20 °C, or low temperature). The specimen was fixed in the temperature chamber for more than 1 h until the temperature of the specimen achieved the target value. During the process, the shrinking had to be monitored by the extensometer during the cooling process. If the deformation happened as the strain, the loading bar needed to be adjusted for the MTS.

2.2.2. Dynamic Test

The experimental data of high strain rates were obtained from high-velocity impact tests using Split Hopkinson Pressure Bar (SHPB) in the Explosions and Shock Lab of Nanjing University of Science and Technology, covering a wide range of strain rates 100–2000 s^{-1}. Several high strain rate tests were performed by SHPB at room temperature and low temperatures.

The entire SHPB setup consisted of several pieces of equipment, as shown in Figure 4. Compared to the traditional equipment, a low temperature box was added only with a hole in the side of the input bar. The size of the hole was a little bigger than the input bar to ensure the impact process. The low temperature was provided by the input of ammonia.

(**a**) Experimental setup for high strain rate compressive tests

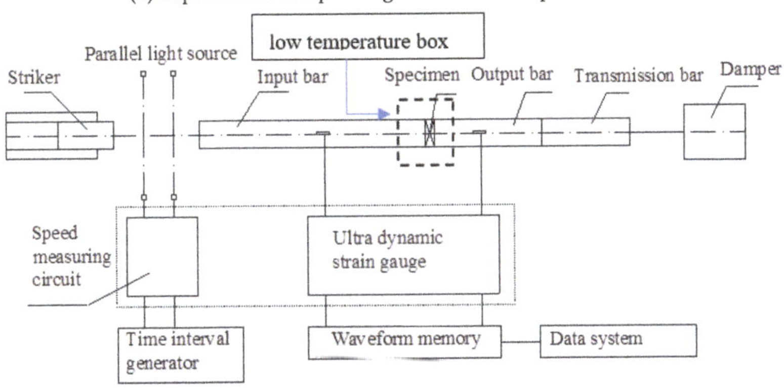

(**b**) Schematic of SHPB setup for steel testing.

Figure 4. Setup of Split Hopkinson Pressure Bar (SHPB) test system.

An SHPB system has two pressure bars, an input or incident bar, and another called output or transmitted bar, as shown in Figure 4b [31]. The yield strength of the pressure bar determines the maximum stress attainable within the deforming specimen because the cross-section of the specimen approaches that of the pressure bar during deformation. A rectangular compression wave of well-defined amplitude and length is generated in the input bar when the striker bar impacts it. As this wave reaches the specimen, part of the pressure wave is transmitted into the specimen and output bar, and part is reflected to the incident bar.

The Input wave, reflection wave, and transmission wave are measured by the ultra-dynamic strain gauge. Based on the one-dimensional and uniform assumption, the average dynamic stress, strain, and strain rate of the tested specimen are determined under a one-dimensional stress state:

$$\sigma(t) = \frac{A}{2A_s}E(\varepsilon_i + \varepsilon_r + \varepsilon_t) = E\frac{A}{A_s}\varepsilon_t \tag{1}$$

$$\varepsilon(t) = \frac{C_e}{l_0}\int_0^t (\varepsilon_i - \varepsilon_r - \varepsilon_t)dt = -\frac{2C_e}{l_0}\int_0^t \varepsilon_r dt \tag{2}$$

$$\dot{\varepsilon}(t) = \frac{C_e}{l_0}(\varepsilon_i - \varepsilon_r - \varepsilon_t) = -\frac{2C_e}{l_0}\varepsilon_r \tag{3}$$

where ε_i, ε_r, and ε_t are the strains of the input wave, reflection wave, and transmission wave, respectively. σ is the stress; ε is the strain; $\dot{\varepsilon}$ is the strain rate; E is the elastic modulus of the pressure bar; C_e is the elastic wave velocity; A is the cross-section area of the pressure bar; and l_0 and A_s are the initial length and initial cross-section of the specimen, respectively.

3. Quasi-Static Response

3.1. Test Results

A series of quasi-static tests on EH36 steel was performed by the MTS system combined with the low-temperature chamber at a strain rate of 2×10^{-4} s^{-1}. The test temperatures ranged from room temperature (20 °C) to low temperatures (0 °C, −20 °C, −40 °C, and −60 °C). The engineering and true stress–strain curves of the EH36 steel at low temperatures are shown in Figure 5. Low temperature will increase the yield strength of EH36. The average tested yield strength and ultimate tensile strength at different temperatures are shown in Figure 6, where σ_s is the yield stress and σ_u is the ultimate tensile stress. As the temperature decreased from 20 °C to −60 °C, the yield strength was increased by 11.6%, and the ultimate tensile strength was 11.3%. The nonlinearly increase trend lines were fitted by quadratic polynomial.

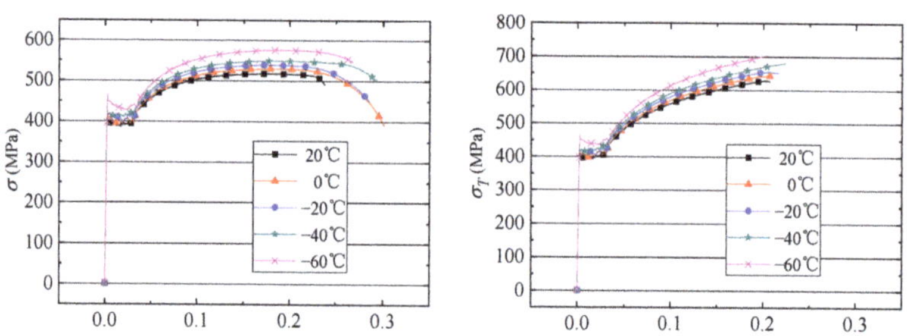

(a) Engineering stress–strain curves **(b)** True stress–strain curves

Figure 5. Stress–strain curves of EH36 at different temperatures.

The yield stress of steel can be approximated from the experimental data shown in Figure 6a as follows:

$$\sigma'_s = f_{(T'-T)}\sigma_s \tag{4}$$

where T is room temperature (20 °C), T' is the target temperature, σ'_S is the quasi-static yield stress of EH36 steel at temperature T', and $f_{(T'-T)}$ is a formula that indicates the relationship obtained from EH36 steel quasi-static tensile test at different low temperatures. Based on the fitted curve in Figure 6a, the relationship formula can be a quadratic function related to the temperature difference $(T' - T)$, $f_{(T'-T)} = a(T' - T)^2 + b(T' - T) + c$. a, b, c are the empirical coefficients and σ_s is the yield stress of quasi-static at room temperature. $a = 1.303 \times 10^{-5}$, $b = -3.071 \times 10^{-4}$, $c = 1.004$.

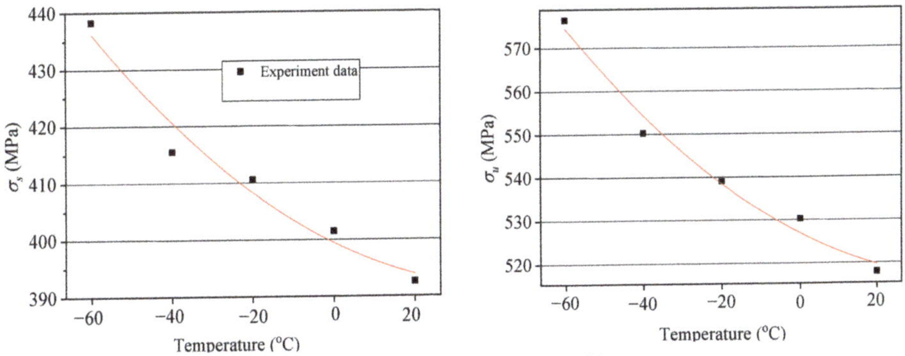

(a) Yield stress (b) Ultimate tension stress

Figure 6. Yield stress and ultimate tension stress at different temperatures (average).

Except for the yielding and ultimate tensile stress, low temperatures affects the engineering fracture strain significantly, as shown in Figure 7. The low temperatures induce a decrease in the engineering fracture strain, as well as the strain rate. Under the same temperature, the engineering fracture strain decreases with the increase in the strain rate.

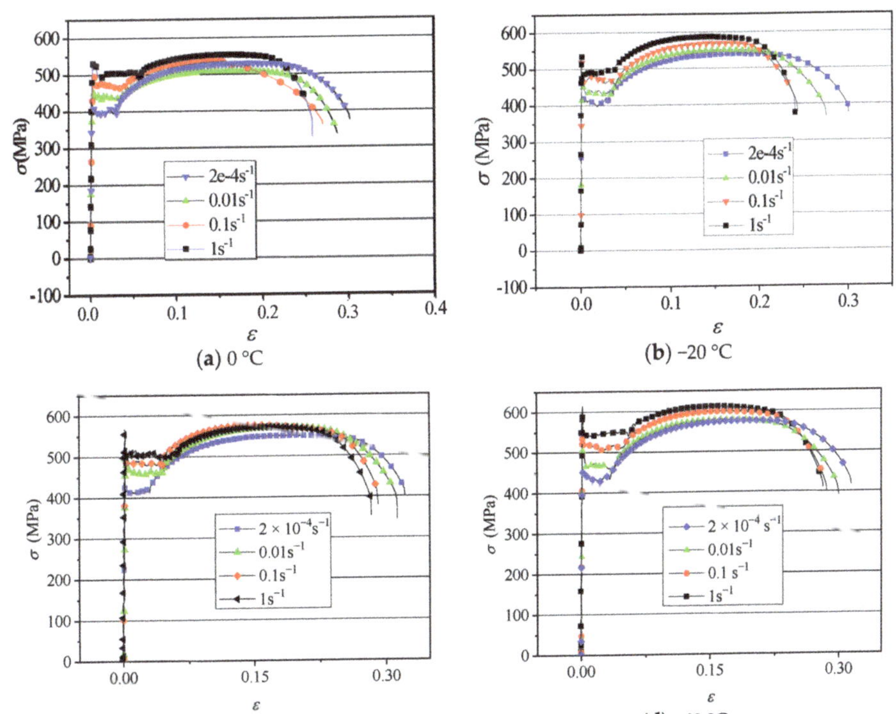

(a) 0 °C (b) −20 °C (c) −40 °C (d) −60 °C

Figure 7. Stress–strain curves of EH36 at low strain rates from −60 °C to 0 °C.

3.2. Simplified Quasi-Static Constitutive Relationship

The interaction mechanics of ice and ships or offshore structures will show strong nonlinearity, as the plastic and larger deformation will happen in the loading process for the ship's steel material. It is suitable to strengthen the structure by adding more stiffeners or increasing the plate thickness to avoid plastic deformation, according to the traditional design method. However, a flexible way involves limited plastic deformation, which can lighten the ship's structure within the design of Arctic ships, in accordance with the Rules [2]. The elastic constitutive model using linear Hooke's relationship is not suitable for the study of structural mechanics in the Arctic environment. It is necessary to study the plastic constitutive models as the input for the structural analysis.

The quasi-static constitutive relationship is discussed firstly based on the engineering stress–strain curves, according to the quasi-static test results of the EH36 steel. As shown in Figure 8, the real stress–strain curve consists of three stages: linear elastic part determined by elastic modulus E, yielding, and the hardening part. It is nonlinear and too complicated to carry out the plastic analysis. Considering the assumptions and simplifications given by the analysis objective, four constitutive models were derived as follows:

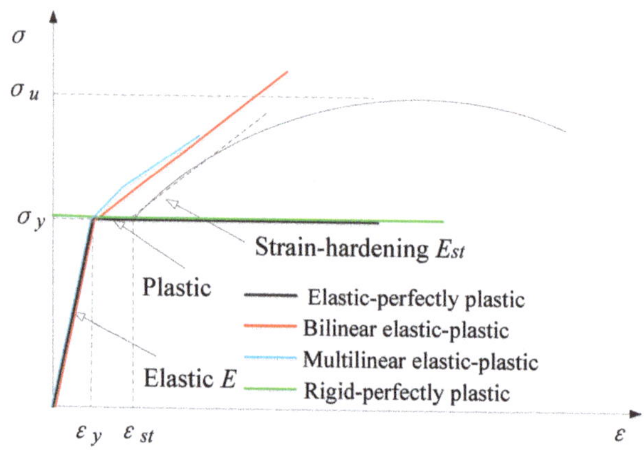

Figure 8. Quasi-static stress–strain relationship models.

(1) Elastic-perfectly plastic: as the stress reaches yield one, the material will be at the plastic stage. The yielding flow is considered, neglecting the hardening effect, when the strain is smaller than the yield strain, Hooke's Law is used to describe the linear relationship with the elastic modulus. As the strain reaches the yield strain, the stress becomes constant and equal to the yield stress, as illustrated by the black line shown in Figure 8.

(2) Bilinear elastic-plastic: the hardening effect is included but neglects the flow stress. The first relationship is linear elastic. As the stress reaches yield stress, a second linear relationship is used to describe the stress–strain curve determined by the tangent modulus E_{st}. It is shown in Figure 8 by the red line.

(3) Multi-linear elastic-plastic: an extended model of bilinear elastic hardening constitutive model. The hardening curve is defined by a multi-linear function, as shown in Figure 8 by the blue line.

(4) Rigid perfectly plastic: a simplified model of the elastic-perfectly plastic constitutive model. Elastic strain is neglected, without the plastic hardening. As the stress reaches the yielding value, the plastic flow happens.

Due to the simplification, models (1) and (4) were used to study the ultimate strength or plastic deformation analysis of structures. Models (2) and (3) were used to study the elastic and plastic response of the structure.

As the tensile stress was close to the ultimate strength, a fracture occurred in the middle part of the test section. The necking phenomenon was observed. The stress–strain curve of EH36 steel at room temperature is shown in Figure 9 as per the test. The result is a yield stress of 392.6 MPa and an ultimate tensile stress of 516.8 MPa.

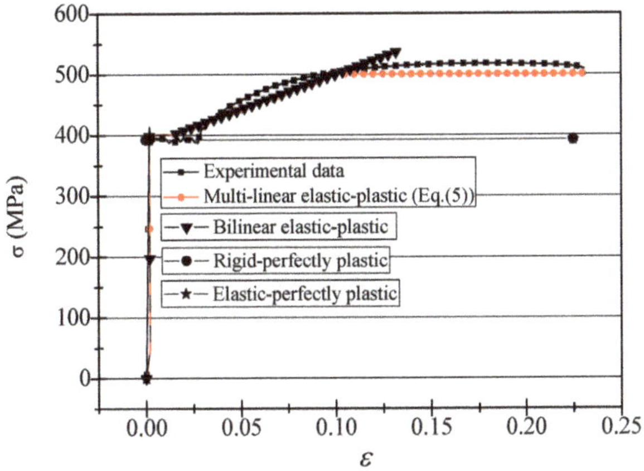

Figure 9. Fitted constitutive models of EH36 steel at room temperature.

The four constitutive models shown in Figure 8 are fitted according to the experiment data, and the multi-linear elastic-plastic relationship is fitted as follows:

$$\begin{array}{ll} \sigma = 2.07 \times 10^5 \cdot \varepsilon & (\varepsilon \leqslant 0.002) \\ \sigma = 1154 \cdot \varepsilon + 392.6 & (0.002 < \varepsilon \leq 0.1) \\ \sigma = 500 & (\varepsilon > 0.1) \end{array} \quad (5)$$

Figure 9 shows the fitting of the constitutive model curves and stress–strain experimental data at room temperature. The constitutive relationship can be derived based on the yield stress determined by Equation (4).

4. Dynamic Constitutive Model by Numerical Results

4.1. Test Results

A series of dynamic tests were performed at low temperatures including low strain rates 0.01, 0.1, and 1 s^{-1} and high strain rates ranging from 700 to 1800 s^{-1}. Dynamic stress–strain curves of EH36 steel at low strain rates were obtained using MTS at room temperature, as illustrated in Figure 10a. High strain rate tests were performed by SHPB. The stress, strain, and strain rate were calculated using Equations (1)–(3) described in Section 2.2. Figure 10b shows the stress–strain curves with varied high strain rates of 769, 932, 1030, 1312, and 1685 s^{-1} at room temperature.

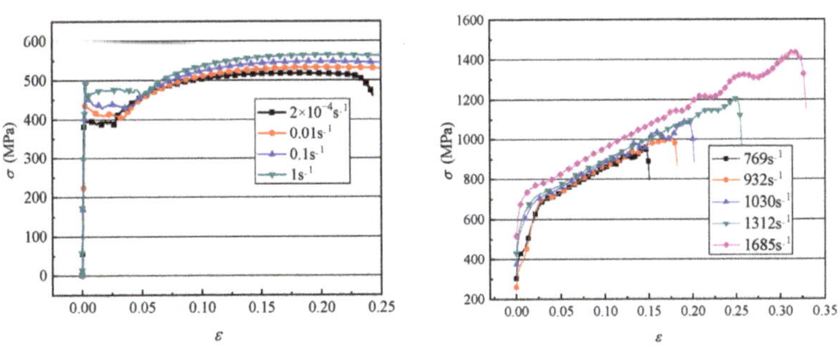

(**a**) MTS test (**b**) SHPB test

Figure 10. Stress–strain curves of EH36 at different strain rates, room temperature.

Table 1 lists the yield stress of EH36 steel from low to high strain rates. Figure 11 shows the variation of yield stress with strain rate; the experimental data were fitted by quadratic polynomial. It can be observed that the yield stress of steel increases as the strain rate increases, from 417.6 MPa at 0.01 s^{-1} to 736.26 MPa at 1685 s^{-1}. Compared with the quasi-static results, the yield stress increased by 6.4% with a strain rate of 0.01 s^{-1} and 87.5% with 1685 s^{-1}. The strain rate dependency characteristics of the test material at room temperature are indicated.

Table 1. Dynamic mechanical properties of EH36 steel at room temperature.

Strain rate/s^{-1}	0.01	0.1	1	769	932	1030	1312	1685
Yield stress/MPa	417.6	434.52	443.24	641.65	659.46	676.53	698.23	736.26

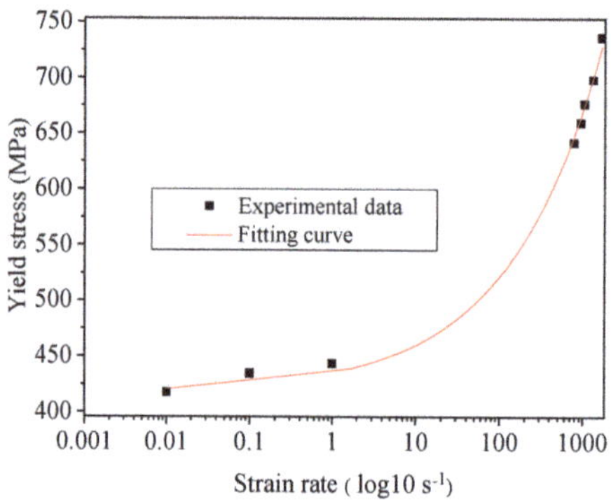

Figure 11. Variation of yield stress with strain rate.

The stress–strain curves of steel over a range of strain rates on the order of 0.01 to 1 s^{-1} from −60 °C to 0 °C are shown in Figure 7. High stain rate stress–strain curves over a range of strain rates on the order of 700 to 1900 s^{-1} from −40 °C to 0 °C are shown in Figure 12.

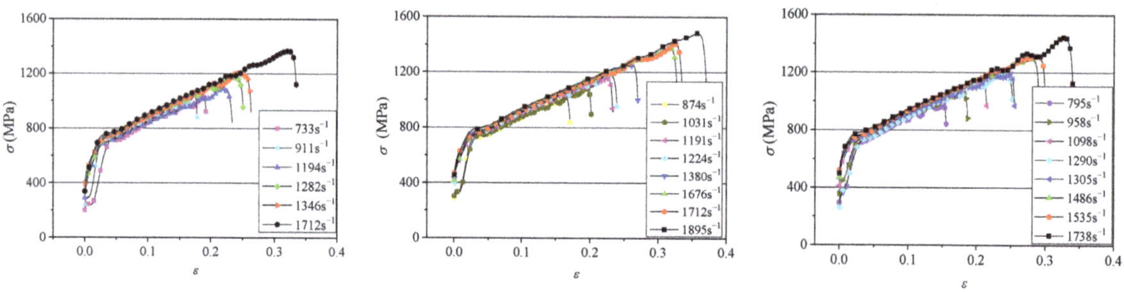

(**a**) 0 °C (**b**) -20 °C (**c**) -40 °C

Figure 12. Stress–strain curves of EH36 at high strain rate from −40 °C to 0 °C.

The temperature effect on the stress–strain curves is illustrated in Figure 13. The yield stresses increase as the temperature decreases. However, the increasing trend of the yield stress becomes slow at low temperatures compared with yield stress at room temperature, especially at the high strain rate of more than 1200 s^{-1}, as shown by the dash lines in Figure 13b.

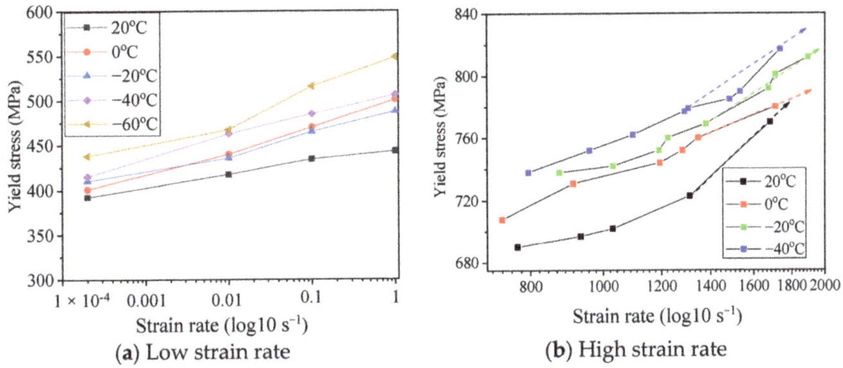

Figure 13. Yield stress at different temperatures.

4.2. Dynamic Constitutive Relationship with Strain Rate Effect and Low Temperature

The low temperature and the strain rate increases the yield stress. However, now there is no comprehensive constitutive model to describe the stress–strain relationship with the strain rate and low-temperature effect. To predict the response to the ship–ice interaction accurately, a dynamic constitutive model depending on the strain rate and low temperature may be presented as:

$$\sigma = f(\varepsilon, \dot{\varepsilon}, T) \quad (6)$$

It is difficult to establish a constitutive model combining the strain rate and low temperature directly. A function that decouples the combined effect may be written as:

$$\sigma = f_1(\varepsilon, \dot{\varepsilon}) \times f_2(\varepsilon, T) \quad (7)$$

There are several constitutive models to describe the strain rate sensitivity material. Almost all the models presented are based on experimental data. The Cowper-Symonds (CS) model is generally adopted to simulate the relationship with strain rate effect. It is written in the following form:

$$\frac{\sigma_{yd}}{\sigma_s} = 1 + \left(\frac{\dot{\varepsilon}}{C}\right)^{1/P} \quad (8)$$

where σ_{yd} is the dynamic yield stress, and C and P are material constants.

Combining the quasi-static constitutive model with the low-temperature effect, the dynamic stress depicting the effect of the strain rate and the low temperature has been expressed based on the Cowper-Symonds model. It is written as follows:

$$\sigma_{yd} = \left[1 + \left(\frac{\dot{\varepsilon}}{C}\right)^{\frac{1}{P}}\right]\left[f_{(T'-T)}\sigma_s\right] \quad (9)$$

where $f_{(T'-T)}$ is the empirical formula of the low-temperature effect, described in Equation (4), where σ_s is the yield stress of steel at room temperature.

According to the experimental results of yield stress–strain rate at room temperature in Figure 11, Equation (9) is fitted by the test data. The constitutive parameters C and P are 2370.6 and 4.00, respectively.

The constants of C and P at different low temperatures are fitted by the test data including low strain rate and high strain rate in the case of 20 °C, 0 °C, −20 °C, and −40 °C and low strain rate in the case of −60 °C, as listed in Table 2. Due to the limited sample, the values C and P fluctuate in a small range, except for −60 °C including only a low strain rate.

Table 2. C and p values at different temperatures.

Temperature	20 °C	0 °C	−20 °C	−40 °C	−60 °C
C	2370.6	2639.4	2453.6	2619.5	225.42
P	4.00	5.40	4.60	4.20	3.95

The constitutive model described in Equation (9) is the basic input of the ship–ice interaction analysis in icy waters. A series of numerical analyses was performed to validate the accuracy of the constitutive relationship under the test load conditions.

4.3. *Improved Model*

The constant values of the C and P in Equation (9) are fitted by the test results at a certain temperature, as shown in Table 2. However, only some samples are given for each temperature. The parameters may be not accurate enough as a result of the limited samples, especially when the combined effect of strain rate and low temperature is included.

An iterative approximate method was introduced to achieve the target value based on the test and numerical results, as shown in Figure 14. The steps are as follows:

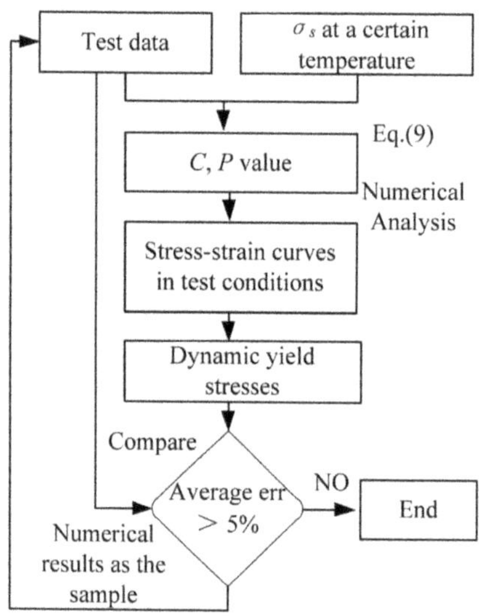

Figure 14. Flowchart of the iterative approximated method.

(1) Using the test data and the quasi-static yield stress at a certain temperature, the values of C and P in Equation (9) are fitted by the least square method.

(2) The C, P and the constitutive relationship in Equation (9) are the input to a series of numerical analyses of test conditions.

(3) The dynamic yield stress obtained by numerical analysis is compared with the one from tests. If the average error is smaller than 5%, the fitted values are accepted.

(4) If the error between the yield stress by numerical analysis and tests is large, with an average error larger than 5%, the numerical results are treated as the sample to refit the C, p values.

(5) Using the new C, p values, perform steps (2)–(4).

4.4. *Numerical Models*

The FE model of the MTS specimen is shown in Figure 15. The SHPB tests are simulated numerically by ABAQUS. The S4R element is employed in the numerical model with a mesh size of 2 mm and refined mesh in the middle area. The load is applied on one end of the specimen at a certain velocity, and the other end is fully clamped. The model defined by Equation (9) in ABAQUS was adopted as the constitutive model of EH36 steel.

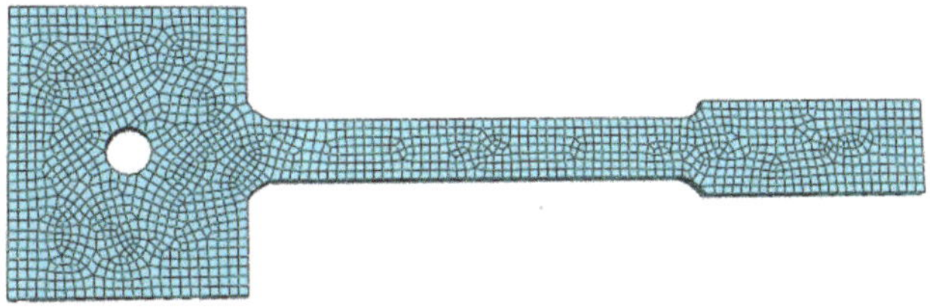

Figure 15. FE model of MTS specimen.

Figure 16 gives the FE model, simulated using solid elements C3D with a mesh size of 2 mm. The mesh size convergence was studied using a series of numerical analyses of the mild steel in [27]. The elastic modulus of the specimen was 2.07E5 MPa and Poisson ratio was 0.3. The impact bar, input bar, and output bar were also simulated by the solid element. The elastic modulus was 210 GPa and Poisson ratio was 0.3. The constitutive model defined in Equation (9) was also used.

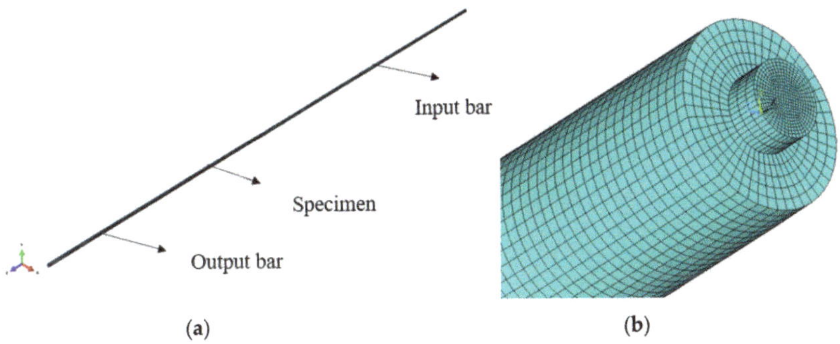

Figure 16. FE model of SHPB test.

4.5. Comparison of the Numerical and Test Results

4.5.1. Low Temperature (0 °C)

It was complicated to include the low temperature and the strain rate effect. The constitutive model at 0 °C is discussed first. As listed in Table 2, the C and p values are taken as 2639.4, 5.40, separately. A series of numerical analyses was performed in test conditions at 0 °C. Figure 17a–d gives the comparison of the test and numerical results with a low strain rate. The yield stresses obtained by the numerical method are lower than the tests in the case of 0.01, 0.1, and 1 s^{-1}, whereas the ultimate tension stresses coincide well with the tests.

The results of the high strain rate are illustrated in Figure 17e,f. The errors between the stress–strain curves of the tests and the numerical analysis are large.

Comparing the numerical results with the tests, the average error is 6.8%. The numerical results of $C = 2639.4$, and $P = 5.40$ are set as the sample data to fit the new values of C and P. The iterative approximate process is illustrated in Figure 18, with the average errors. After three iterative steps, the average error was 4.8% as the $C = 1974.7$, $P = 2.4$.

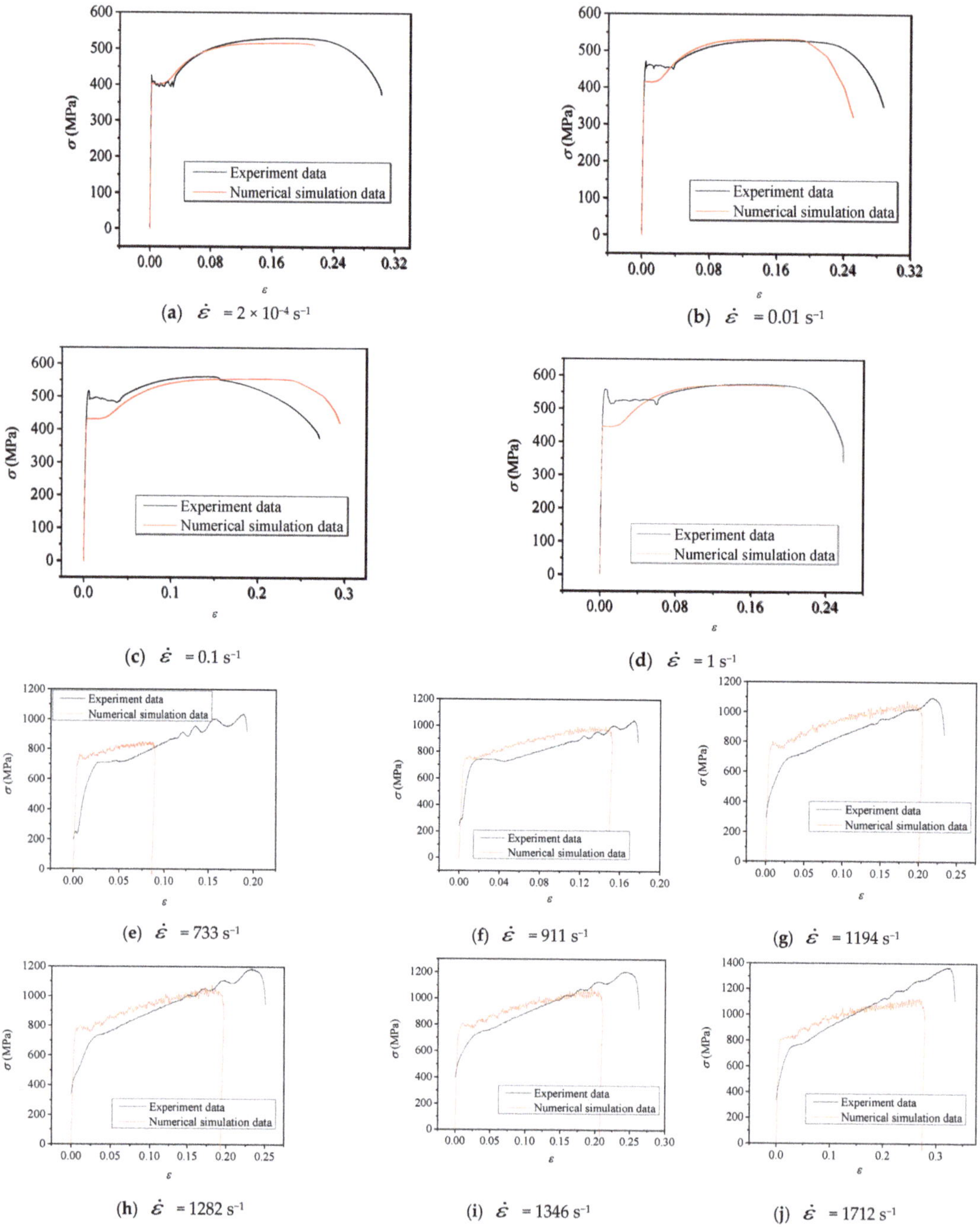

(a) $\dot{\varepsilon}$ = 2 × 10^{-4} s^{-1} (b) $\dot{\varepsilon}$ = 0.01 s^{-1}

(c) $\dot{\varepsilon}$ = 0.1 s^{-1} (d) $\dot{\varepsilon}$ = 1 s^{-1}

(e) $\dot{\varepsilon}$ = 733 s^{-1} (f) $\dot{\varepsilon}$ = 911 s^{-1} (g) $\dot{\varepsilon}$ = 1194 s^{-1}

(h) $\dot{\varepsilon}$ = 1282 s^{-1} (i) $\dot{\varepsilon}$ = 1346 s^{-1} (j) $\dot{\varepsilon}$ = 1712 s^{-1}

Figure 17. Comparison of the test and the numerical results of C = 2639.4, P = 5.40.

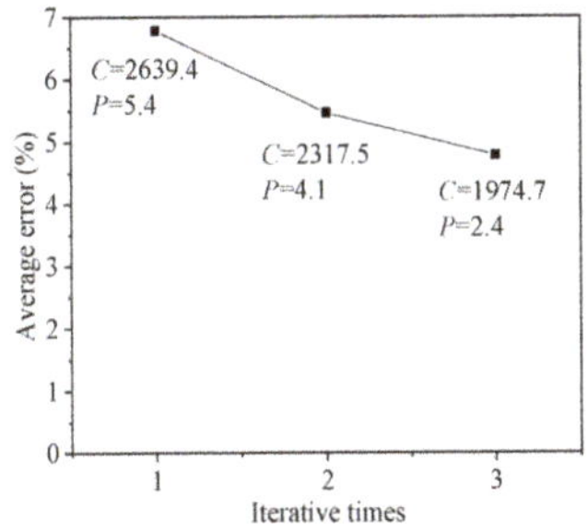

Figure 18. Average error with iterative times at 0 °C.

4.5.2. Other Temperatures

(1) Behaviour at Room Temperature (20 °C)

After four iterative calculations, the average error was 3.6%, thus less than 5%, with the constitutive coefficients $C = 3599$, $P = 2.578$. Figure 19 shows a comparison of the finite element results and experimental data at room temperature. It can be seen that the numerical results coincide with the experimental results, and the yield points are almost identical at low and high strain rates. It is shown that the constitutive model defined in Equation (9) can describe the strain rate effect of EH36 steel at room temperatures with strain rate effect accurately.

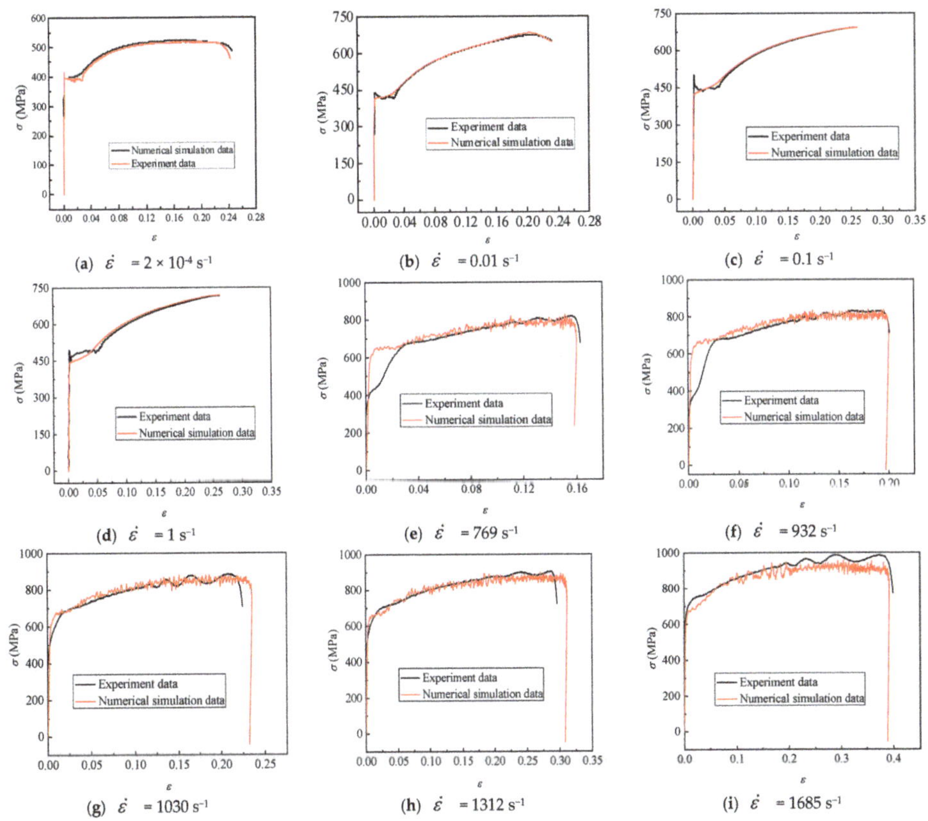

Figure 19. Comparison of numerical and test results at room temperature.

(2) Behaviour at −20 °C

Using three iterative numerical analyses, the $C = 1982.6$ and $P = 3.2$ were obtained at −20 °C, as the average error was 4.8%. The numerical results of the constitutive model with $C = 1982.6$ and $P = 3.2$ are compared with the experimental data with low and high strain rates in Figure 20.

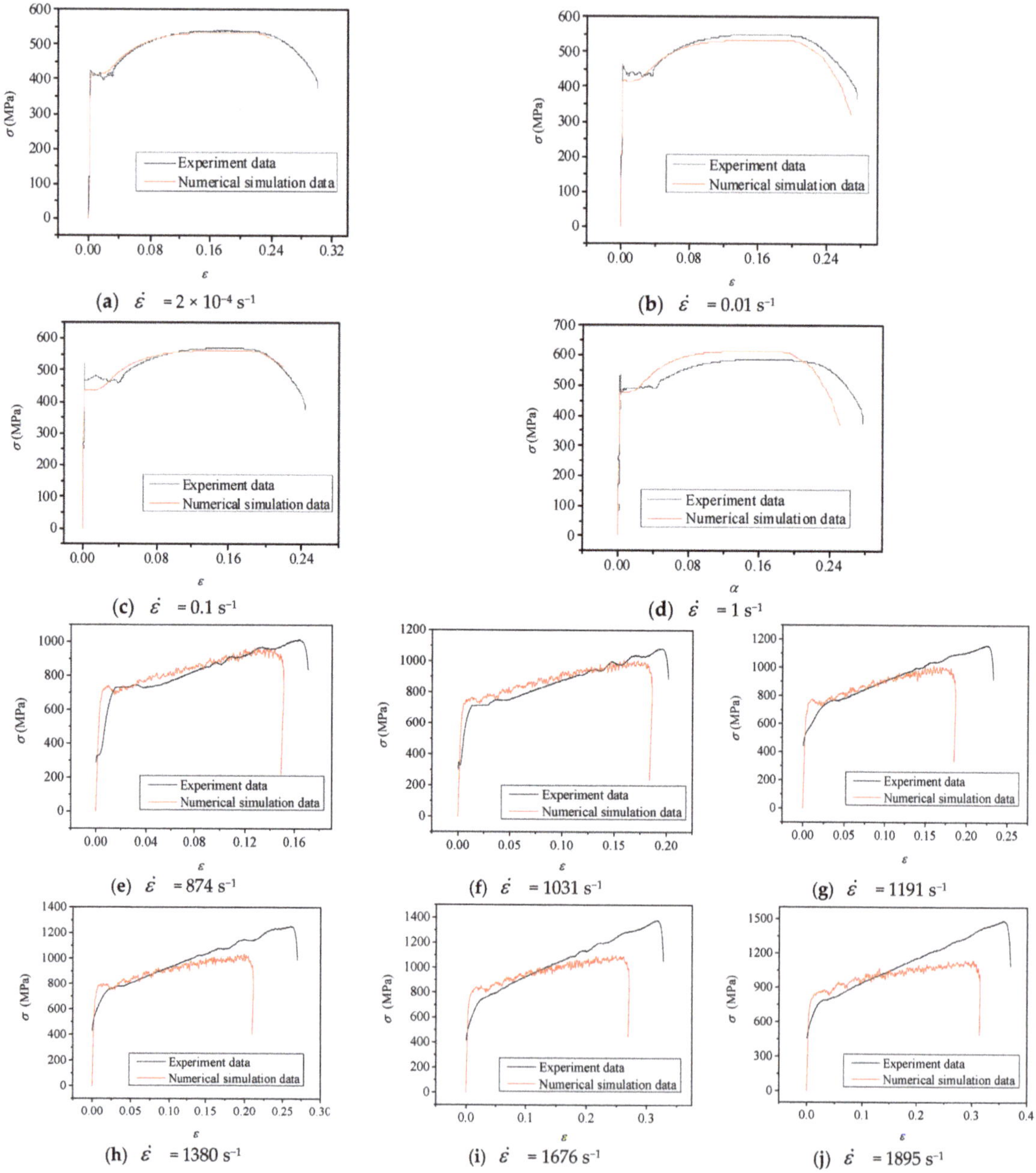

Figure 20. Comparison of numerical and test results at −20 °C with high strain rate.

(3) Behaviour at −40 °C

The constitutive coefficients $C = 2091.6$ and $P = 3.3$ were obtained by three iterations at the low temperature of −40 °C, and the average error was 4.6%. The comparison of the numerical analysis and experiment data is shown in Figure 21 with low and high strain rates.

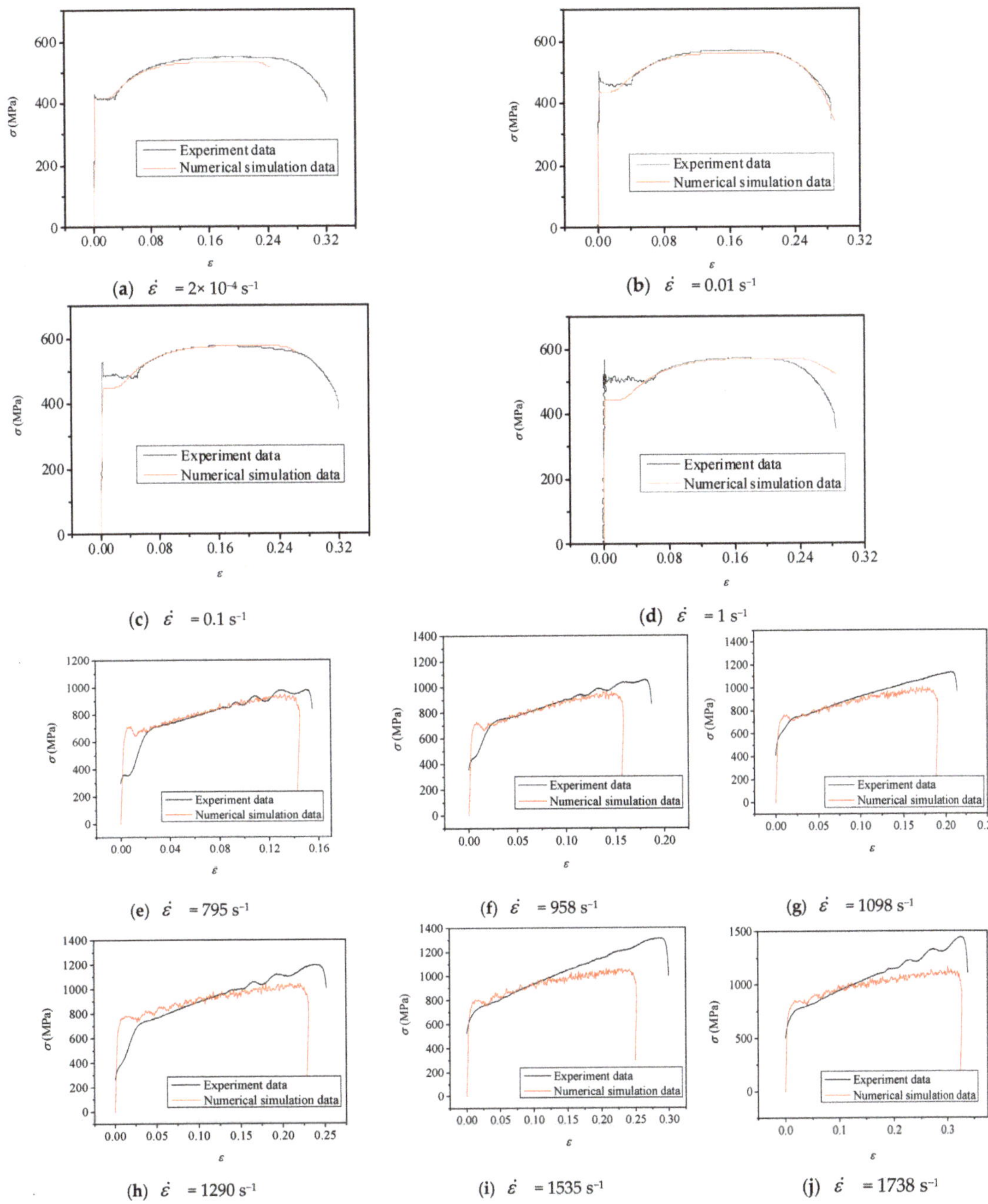

Figure 21. Comparison of numerical and test results at −40 °C.

(4) Behaviour at −60 °C

Only low strain rate test results were used to fit the constitutive coefficients, C = 225.42 and P = 3.95, with an average error of 4.9%, thus less than 5%. Figure 22 shows the comparison of numerical and experiment data including low strain rate only.

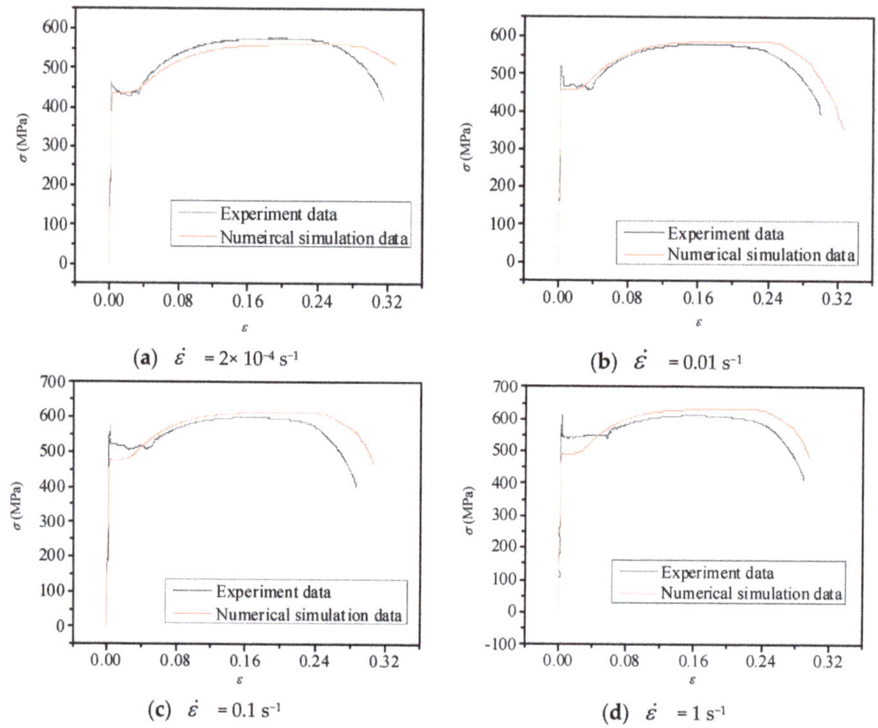

Figure 22. Comparison of numerical and test results at −60 °C.

From Figures 20–22, the stress–strain curves of the numerical results are compared with the experimental results at low temperatures −20 °C, −40 °C, and −60 °C. It can be seen that the numerical results show good agreement with the experiment considering the effect of low temperature and strain rate. Therefore, it can be concluded that the model of Equation (9) can be adopted as the stress–strain relationship of EH36 steel in the analysis of ship–ice interaction in the Arctic. However, it was found that the finite element results are conservative relative to the corresponding experimental results at a low strain rate.

5. Conclusions

In the present study, the mechanical behaviour of strain rate-sensitive steel EH36 under low-temperature test conditions were reported. The tests performed consisted of quasi-state and dynamic ones. First, the quasi-static tests ranging from room temperature to −60 °C were conducted. Then, the dynamic tests were performed on specimens at different strain rates under low temperatures. The tests of low strain rates of 0.01 s^{-1} to 1 s^{-1} were performed on an MTS testing machine and those of high strain rates of 700 s^{-1} to 2000 s^{-1} by SHPB. A constitutive model of EH36 was studied based on the Cowper-Symonds model, including the low temperature and strain rate effect. The results and conclusions can be summarized as follows:

(1) The experimental results indicate obvious strain rate effects on the nonlinear stress–strain behaviour at different strain rates of EH36 steel.

(2) EH36 steel has obvious temperature sensitivity. The yield stress of EH36 steel dramatically increased more than 10% nonlinearly with the decrease in temperature ranging from 20 °C to 60 °C, which indicated that low temperature can improve the yield stress, as well as the ultimate tensile stress. However, the toughness decreases with the temperature, shown as the decreasement of the fracture strain.

(3) An iterative approximated method was provided to fit the constants in the presented constitutive model. By performing a series of numerical analyses, a constitutive model for a certain low temperature was given by comparing the experimental and numerical results. It may describe the dynamic behaviour of EH36 at low temperatures accurately.

(4) An improved Cowper-Symonds model was established considering the effects of low temperature and strain rate by fitting the *C*, *P* constants in different temperatures ranging from 0 °C to 60 °C, which may describe the coupled effect of low temperature and strain rate.

Author Contributions: Conceptualization, J.Z., X.S. and C.G.S.; methodology, J.Z.; software, X.K.; validation, J.Z. and X.S.; formal analysis, X.K.; investigation, M.S.; resources, X.S.; data curation, J.Z.; writing—original draft preparation, J.Z.; writing—review and editing, C.G.S.; visualization, X.K.; supervision, C.G.S.; project administration, X.S. All authors have read and agreed to the published version of the manuscript.

Funding: This research was funded by National Natural Science Foundation of China (Grant No. 51809126, 51509113), the Natural Science Foundation of Jiangsu Province, China (Grant No. BK20181468, BK20200998), the Natural Science Foundation of the Higher Education Institutions of Jiangsu Province, China (Grant No. 16KJA580003).

Institutional Review Board Statement: Not applicable.

Informed Consent Statement: Not applicable.

Data Availability Statement: The data presented in this study are openly available at https://we.tl/t-hOcWbRIyb0 (accessed on 22 February 2023).

Conflicts of Interest: The authors declare no conflict of interest.

References

1. FMA. *Finnish-Swedish Ice Calss Rules*; Finnish Maritime Administration: Helsinki, Finland, 2002.
2. ABS. *Guidance Notes on Ice Class*; American Bureau of Shipping: Houston, TX, USA, 2005.
3. RMRS. *Rules for the Classification and Construction of Sea-Going Ships*; Russian Maritime Register of Shipping: Saint Petersburg, Russia, 2012.
4. He, X.; Guedes Soares, C. Pseudo-shakedown of rectangular plates under repeated impacts. *Mar. Struct.* **2022**, *85*, 103258. [CrossRef]
5. Liu, K.; Liu, B.; Villavicencio, R.; Wang, Z.; Guedes Soares, C. Assessment of material strain rate effects on square steel plates under lateral dynamic impact loads. *Ships Offshore Struct.* **2018**, *13*, 217–225. [CrossRef]
6. Ehlers, S.; Østby, E. Increased crashworthiness due to arctic conditions—The influence of sub-zero temperature. *Mar. Struct.* **2012**, *28*, 86–100. [CrossRef]
7. Paik, J.K.; Kim, K.J.; Lee, J.H.; Jung, B.G.; Kim, S.J. Test database of the mechanical properties of mild, high-tensile and stainless steel and aluminium alloy associated with cold temperatures and strain rates. *Ships Offshore Struct.* **2017**, *12* (Suppl. S1), S230–S256. [CrossRef]
8. Choung, J.; Nam, W.; Lee, J.-Y. Dynamic hardening behaviors of various marine structural steels considering dependencies on strain rate and temperature. *Mar. Struct.* **2013**, *32*, 49–67. [CrossRef]
9. DNV. *Rules for Classification for Ships Newbuildings*; Det Norske Veritas: Bærum, Norway, 2008.
10. IACS. *Requirements Concerning—POLAR CLASS*; International Association of Classification Societies: Hamburg, Germany, 2011.
11. Chen, W.W.; Song, B. *Split Hopkinson (Kolsky) Bar Design, Testing and Applications*; Springer: New York, NY, USA, 2011.
12. Dunand, M.; Mohr, D. Predicting the rate-dependent loading paths to fracture in advanced high strength steels using an extended mechanical threshold model. *Int. J. Impact Eng.* **2017**, *108*, 272–285. [CrossRef]
13. Liu, B.; Villavicencio, R.; Guedes Soares, C. On the failure criterion of aluminum and steel plates subjected to low-velocity impact by a spherical indenter. *Int. J. Mech. Sci.* **2014**, *80*, 1–15. [CrossRef]
14. Liu, B.; Guedes Soares, C. Effect of strain rate on dynamic responses of laterally impacted steel plates. *Int. J. Mech. Sci.* **2019**, *160*, 307–317. [CrossRef]
15. Forni, D.; Chiaia, B.; Cadoni, E. High strain rate response of S355 at high temperatures. *Mater. Des.* **2016**, *94*, 467–478. [CrossRef]

16. Mirmomeni, M.; Heidarpour, A.; Zhao, X.-L.; Hutchinson, C.R.; Packer, J.A.; Wu, C. Mechanical properties of partially damaged structural steel induced by high strain rate loading at elevated temperatures—An experimental investigation. *Int. J. Impact Eng.* **2015**, *76*, 178–188. [CrossRef]
17. Nemat-Nasser, S.; Guo, W.-G. Thermomechanical response of DH-36 structural steel over a wide range of strain rates and temperatures. *Mech. Mater.* **2003**, *35*, 1023–1047. [CrossRef]
18. Ibrahim, O.H.; Ibrahim, I.S.; Khalifa, T.A.F. Impact behavior of different stainless steel weldments at low temperatures. *Eng. Fail. Anal.* **2010**, *17*, 1069–1076. [CrossRef]
19. Getter, D.J.; Kantrales, G.C.; Consolazio, G.R.; Eudy, S.; Fallaha, S. Strain rate sensitive steel constitutive models for finite element analysis of vessel-structure impacts. *Mar. Struct.* **2015**, *44*, 171–202. [CrossRef]
20. Chen, J.; Li, J.; Li, Z. Experiment research on rate-dependent constitutive model of Q420 steel. *Constr. Build. Mater.* **2017**, *153*, 816–823. [CrossRef]
21. Su, J.; Guo, W.; Meng, W.; Wang, J. Plastic behavior and constitutive relations of DH-36 steel over a wide spectrum of strain rates and temperatures under tension. *Mech. Mater.* **2013**, *65*, 76–87. [CrossRef]
22. Detrois, M.; Antonov, S.; Tin, S.; Jablonski, P.D.; Hawk, J.A. Hot deformation behavior and flow stress modeling of a Ni-based superalloy. *Mater. Charact.* **2019**, *157*, 109915. [CrossRef]
23. Forni, D.; Chiaia, B.; Cadoni, E. Strain rate behaviour in tension of S355 steel: Base for progressive collapse analysis. *Eng. Struct.* **2016**, *119*, 164–173. [CrossRef]
24. Cowper, G.; Symonds, P. *Strain Hardening and Strain Rate Effects in the Loading of Cantilever Beams*; Brown University: Providence, RI, USA, 1957.
25. Shi, G.; Zhu, X.; Ban, H. Material properties and partial factors for resistance of high-strength steels in China. *J. Constr. Steel Res.* **2016**, *121*, 65–79. [CrossRef]
26. Bobbili, R.; Madhu, V.; Gogia, A.K. Neural network modeling to evaluate the dynamic flow stress of high strength armor steels under high strain rate compression. *Def. Technol.* **2014**, *10*, 334–342. [CrossRef]
27. Jones, N. *Structural Impact*; Cambridge University Press: Cambridge, MA, USA, 1997.
28. Jones, N. The credibility of predictions for structural designs subjected to large dynamic loadings causing inelastic behaviour. *Int. J. Impact Eng.* **2013**, *53*, 106–114. [CrossRef]
29. Hsu, S.S.; Jones, N. Quasi-static and dynamic axial crushing of thin-walled circular stainless steel, mild steel and aluminium alloy tubes. *Int. J. Crashworth.* **2004**, *9*, 195–217. [CrossRef]
30. *ASTM-E8*; Standard Test Methods for Tension Testing of Metallic Materials. American Society for Testing and Material: West Conshohocken, PA, USA, 2016.
31. Lin, Y.C.; Xia, Y.-C.; Chen, X.-M.; Chen, M.-S. Constitutive descriptions for hot compressed 2124-T851 aluminum alloy over a wide range of temperature and strain rate. *Comput. Mater. Sci.* **2010**, *50*, 227–233. [CrossRef]

Journal of
Marine Science and Engineering

MDPI

Article

Research on Bearing Characteristics of Gravity Anchor in Clay

Jianxing Yu [1,2], Pengfei Liu [1,2,*], Yang Yu [1,2], Xin Liu [1,2], Haoda Li [1,2], Ruoke Sun [1,2] and Xuyang Zong [3]

1 State Key Laboratory of Hydraulic Engineering Simulation and Safety, Tianjin University, Tianjin 300072, China
2 Tianjin Key Laboratory of Port and Ocean Engineering, Tianjin University, Tianjin 300072, China
3 System Engineering Research Institute, China State Shipbuilding Corporation, Beijing 100036, China
* Correspondence: m13920367969@163.com; Tel.: +86-136-2036-7969

Abstract: The applications and studies of gravity anchors in the ocean are becoming more and more extensive. Most of the research, however, has been directed toward the bearing properties of sand. Relatively less attention has been paid to the bearing properties of gravity anchors in clay. Clay is widely distributed on the seabed. The research on the bearing capacity of gravity anchors in clay is of great significance for offshore oil exploitation. Therefore, the gravity anchor was investigated by conducting reduced-scale model tests, and the bearing process of gravity anchors in clay was simulated through a 3D finite element method. Model tests and numerical simulations were used to determine the capacity curve and the *V-H* failure envelope of gravity anchors in clay. The simulation results and the test results are in good agreement. The failure form of the gravity anchor in clay was revealed by 3D finite element analysis. The effect of cohesion, internal friction angle, and mooring point height on bearing capacity have been studied. The influence of the height of the mooring point on the *V-H* failure envelope curve was explored by changing the height of the mooring point. The formula of the *V-H* failure envelope curve suitable for different mooring point heights was obtained.

Keywords: bearing characteristics; gravity anchor; clay; *V-H* failure envelope curve

Citation: Yu, J.; Liu, P.; Yu, Y.; Liu, X.; Li, H.; Sun, R.; Zong, X. Research on Bearing Characteristics of Gravity Anchor in Clay. *J. Mar. Sci. Eng.* **2023**, *11*, 505. https://doi.org/10.3390/jmse11030505

Academic Editor: Dong-Sheng Jeng

Received: 5 February 2023
Revised: 16 February 2023
Accepted: 23 February 2023
Published: 26 February 2023

1. Introduction

In recent years, oil and gas extraction has moved into increasingly deep water in search of energy [1]. The laying of pipelines is a key step in the process of offshore oil exploration [2]. In order to stabilize the pipe-laying ship, it is necessary to use an anchoring device for anchoring. Gravity anchors are the oldest mooring equipment. It provides horizontal and vertical bearing capacity through the reaction force of the soil [3] (it is the collection of soil forces reacting to the gravity anchor, including frictional forces, the adsorption of the soil, the soil resistance that arises when the soil is destroyed, etc.). Additionally, the gravity anchor is highly adaptable, meaning it can be used without too detailed soil survey data. Therefore, it adapts to a wide range of soil qualities and is widely used in engineering applications [4]. The surface of the seabed is mostly clay. To ensure safety in the process of offshore oil development, studying the bearing capacity characteristics of gravity anchors in clay is of great significance to the design work.

During the past decades, considerable efforts, including model tests and numerical studies, have been made to estimate the pull-out behavior of gravity anchors under loads. Harris et al. [5] outlined the general types of wave energy converters and discussed their anchoring requirements. Through experimental comparison to other anchoring systems, it was found that gravity anchor is the superior anchoring method. Comprehensive research was conducted by Yun G et al. experimentally and numerically [6]. Their study found that different buried depth ratios and different geometric shapes have a great influence on the vertical bearing capacity of gravity anchors. They proposed a calculation method for vertical bearing capacity. Taylor, R et al. [7] designed different mooring system options for the location of the Ocean Thermal Energy Conversion (OTEC). According to the environmental conditions, towed embedded anchors and gravity anchors with skirt wings were designed, and the selection of their size was discussed. Jim B. Peterson [8] invented a new type of articulated gravity anchor that is easy to operate and can provide greater bearing capacity. Three different anchor types—drilled shaft, gravity, and fluke anchors—were used to meet challenges for the new SR520 Evergreen Point Floating Bridge and Landings project in Seattle, Washington, by Upsall, B et al. [9]. Aiming at the movement mechanism of gravity anchors under compound loads, Li S [10] et al. studied the anti-sliding stability mechanism of gravity anchors through indoor model tests. The strip method was used to analyze the bearing capacity of gravity anchors under complex loads.

With the development of nonlinear methods, research on the numerical simulation of the bearing process of gravity anchors gradually increased. Houlsby G T [11] studied the numerical modeling of shallow circular foundations. Based on the numerical model results of shallow foundations on many calcareous soils combined with structural analysis, the working performance of the overall structure under dynamic loading conditions was predicted. Gourvenec S [12] studied the influence of different aspect ratios on the ultimate bearing capacity of gravity anchors by finite element analysis. Moreover, the influence of the aspect ratio on the critical state of gravity anchors under different loads is also studied in the following research. Gourvenec S [13] also examined the failure envelope of gravity anchors with skirts in uniform and inhomogeneous soils. The ultimate bearing mechanism of gravity anchors under general loads is proposed. Gourvenec S [14] studied the limit state and the kinematic mechanism of the failure of gravity anchors. The results show that the size and shape of failure envelopes defining the undrained capacity of shallow foundations under general loading are dependent on the embedment ratio. Three years later, Gourvenec S and Barnett S [15] improved the failure envelope of gravity anchors and developed the expression of the failure envelope. Bransby M F and Yun, G J [16] studied the influence of skirts on gravity anchors subjected to horizontal loads. Through model tests and finite element simulations, Niroumand H and Kassim K A [17] investigated the motion law of slab anchors under vertical loads in dense sand. The results show that the ultimate bearing capacity calculated by finite element is smaller than the ultimate bearing capacity of the model test. Based on the finite element limit analysis method, Mana, D S K et al. [18] studied the critical inner skirt plate spacing for the undrained failure of the skirt plate gravity anchor under plane strain conditions. The results showed that as the buried depth of the foundation increases, fewer inner skirts are required. Liu et al. [19] studied the pullout capacity and the keying process of a vertically installed OMNI-Max anchor embedded in normally consolidated clay by using three-dimensional large deformation finite element analysis. The analyses clearly showed two important processes: (1) "keying" in which the anchor rotates rapidly until reaching the best bearing capacity position; and (2) "diving" in which the anchor mainly translates with tiny rotation. The development of GIPLA's movements during its keying and its pullout capacity in clay was also investigated by Liu et al. [20]. The anchor loading angle and the anchor padeye offset angle for the diving motion were suggested, and the effects of the anchor loading angle, anchor padeye eccentricity, and soil strength profile were investigated. Li, H et al. [21] studied the effect of different types of shear bonds on the bearing capacity of gravity anchors. The results showed that the horizontal bearing capacity of the eight-key anchor is greater than that of the flat-bottomed anchor. An LDFE analysis was performed to investigate the penetration and keying of GIAs by Zhao et al. [22]. Fitted formulae were also proposed to give a quick evaluation of the performance of GIAs. A numerical model on the anchor–soil–water interaction for the dynamic installation of gravity installed anchors was proposed by Liu et al. [23]. Additionally, a theoretical model was also proposed to predict the final penetration depth of the anchor. Yang et al. [24] studied a theoretical model for three-dimensional GIAs in multi-layered clays and complicated behaviors of the OMNI-Max anchor in multi-layered clays.

In this paper, the research contents of gravity anchor bearing capacity and failure envelope are expanded on the basis of the existing research. The objectives of this study are to investigate the bearing performance of gravity anchors in clay and to analyze the factors affecting the bearing performance of gravity anchors by conducting model tests and 3D finite element analysis based on the ABAQUS software package. The damage envelope curves of gravity anchors were improved based on the damage theory for shallowly buried structures.

2. Model Test

The clay used in this model test is prepared from the soil excavated from a depth of 1500 m in the South China Sea. Figure 1 is a photo taken by the author on the spot. Since there would be some impurities in the collected soil, the clay should be dried and sieved to obtain the pure clay power that can be used to prepare the test soil. The soil is fully hydrated and mixed with the aim of obtaining a homogeneous clay. Then, the soil is fully stirred, consolidated, and settled for 3 months. The purpose of consolidation settlement for 3 months is to make the soil reach the test standard. After the soil was fully saturated and consolidated, the surface was scraped to guarantee the planeness. In addition, the soil surface was covered with plastic wrap during production. The basic physical index parameters of clay are tested before the model test. This test uses the ring knife method to measure the saturated density of the clay. The specifications of the ring knife are 61.8 × 20 mm. The density of the clay is the average of three measurements. The value is 18 kN/m^3. The shear properties of clay are measured by the direct shear test, which is a slow shear test. An appropriate amount of the configured soil sample is taken and placed in the shear box. The vertical stress is applied to the soil during the measurement, and the values are 100 kPa, 200 kPa, 300 kPa, and 400 kPa. The consolidation of the soil was completed by the application of vertical pressure. The horizontal shear stress was then applied to shear the soil slowly until the soil was completely sheared. The soil after measurement is shown in Figure 2. Therefore, the cohesive of the clay is 9.89 kPa. The average bulk density and undrained shear strength of clay soil [25] in all the tests were found to be 18 kN/m^3 and 10 kPa, respectively.

Figure 1. Field trip map of the dredger.

The grain composition of soil samples was determined by sieve analysis and hydrometer (d < 0.075 mm). The results are shown in Table 1 and Figure 3.

Table 1. Soil particle composition.

Soil Particle Content (%)			Classification of Soil
d > 0.075 mm	0.005~0.075 mm	d < 0.005 mm	fine grained soil
9.9	76.6	13.5	

(a)

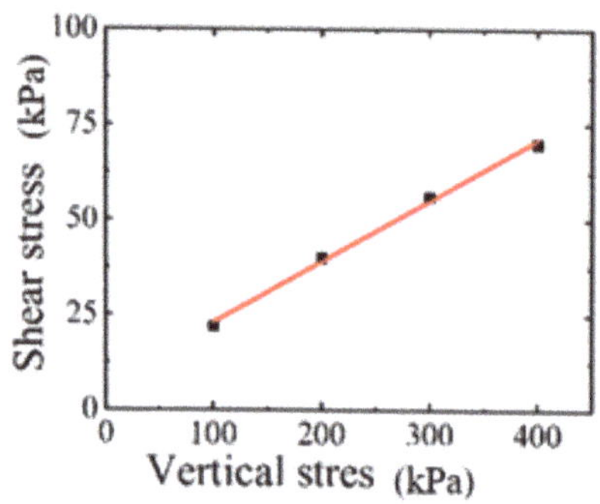

(b)

Figure 2. Soil test picture. (**a**) The shear strength curve; and (**b**) The shear strength curve.

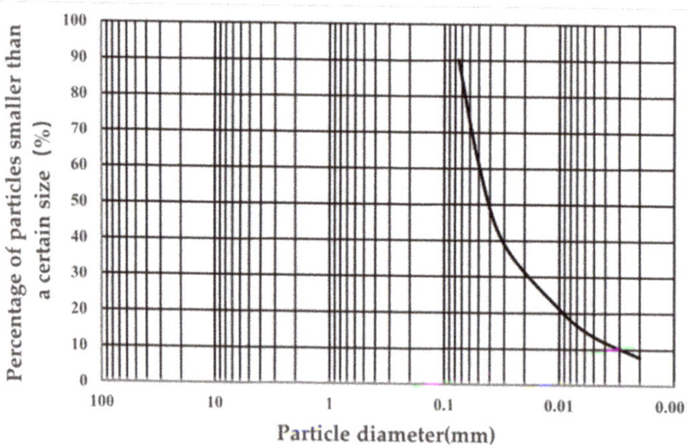

Figure 3. Particle distribution curve.

Liquid and plastic limit tests were carried out on marine soil samples. The particle size of the soil sample used in the test is less than 0.5 mm, and the combined determination method of liquid and plastic limit is adopted. When the falling cone depth of liquid and plastic limit instrument is 2 mm, the corresponding water content is the plastic limiting water content, and when the falling cone depth is 17 mm, the corresponding water content is the liquid limiting water content. The results are shown in Table 2.

Table 2. Liquid limit and plastic limit test results.

Liquid Limit (%)	Plastic Limit (%)	Plasticity Index (%)	Classification of Soil
36.2	22.6	13.6	silty clay with low liquid limit

Model tests were conducted in a test tank with a length of 1.75 m, a width of 0.6 m, and a height of 1.2 m, as shown in Figure 4. The height of the soil in the box during the test is 0.3 m. Other soil was laid 0.2 m at the bottom. The total height of the soil is 0.5 m. The height of the soil is 6 times the height of the gravity anchor. The boundary effect can be ignored. The soil is evenly and flatly filled in the test tank. In addition to the test tank, the device is mainly made up of an anchor model (the front, back, and top center are equipped with buckles that can connect to the steel strand), a height-adjustable pulley device, tension, and displacement sensors, and a travel switch (the switch is disconnected and stops loading after the anchor body moves in the model test). The height-adjustable pulley device is used to change the loading angle of the gravity anchor. The model of the tension sensor is TJL-1S, the range is 0~30 kg, and the overall accuracy is 0.02. The model of the displacement sensor is KS20-1000-01-C, the range is 0~1000 mm, and the accuracy is $\pm 0.53\%$. The shape of the gravity anchor used in the tests is shown in Figures 5 and 6. In this test, the actual gravity anchor of the project is reduced according to the ratio of 1:15. The size of the gravity anchor model in the model test is 200 mm $\times$ 200 mm $\times$ 87 mm, and the mass is 8.25 kg by using similar criterion. The anchor in the model test was made in advance. The drawing of the anchor in the actual project is shown below. The anchor is hollow and brought to a predetermined weight by adding counterweight blocks in four boxes. The oblique columns at the bottom of the anchor are the shear keys of the gravity anchor. Its function is to increase the contact area with the soil and to change the way the soil fails. These two effects increase the bearing capacity of the gravity anchor. Their exact dimensions are shown below. The height of the shear key is 260 mm, and the length is 270 mm. Its diameter is 300 mm.

Model tests include horizontal loading tests, a series of tests with different loading angles, and vertical loading tests. The angle of the oblique load test remains constant during the test. The test data revealed that the displacement of the gravity anchor was very small when damage to the soil occurred. The displacement of the gravity anchor is negligible in relation to the horizontal distance between the gravity anchor and the displacement gauge. Therefore, the angle of the oblique load test can be considered to remain constant during the test. The load angle was tested during the test. The starting position of the gravity anchor is kept constant in the model test. The force angle can be changed by varying the height of the pulley. Horizontal loading tests are first performed to obtain the load–displacement curve and the failure criteria of displacement. Then, a series of tests with different loading angles and vertical loading tests are performed based on the load-displacement curve during the tests. In the test, the AT-102 model water pump is used to add water to the water tank at a constant speed to load the gravity anchor model. The capacity of the water tank is greater than that required for the maximum test load. The switch is controlled manually. When the pump is turned on, water is poured into the tank. The weight of the injected water in the water tank is applied as a load on the gravity anchor. The loading velocity is 0.3 N/s. Therefore, the strain rate is 1.4×10^{-4}. This entire process can be considered as a quasi-static simulation.

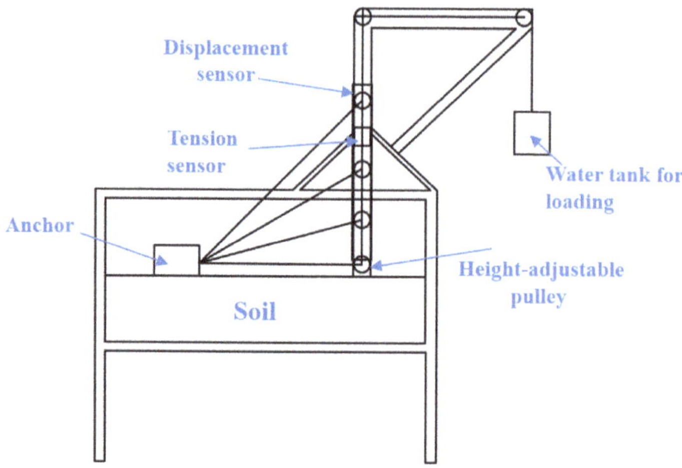

(a)

(b)

Figure 4. The model test tank. (**a**) A schematic map of the test tank; and (**b**) A physical map of the test tank.

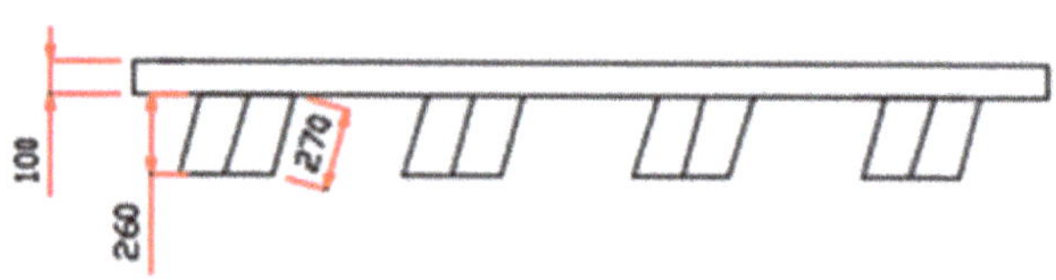

Figure 5. Gravity anchor shear key.

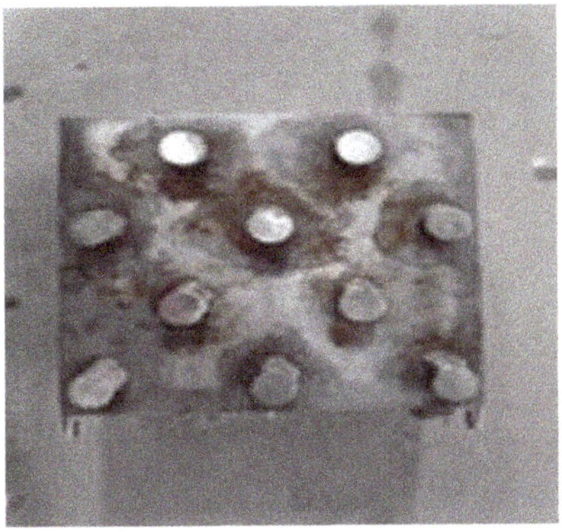

(a)

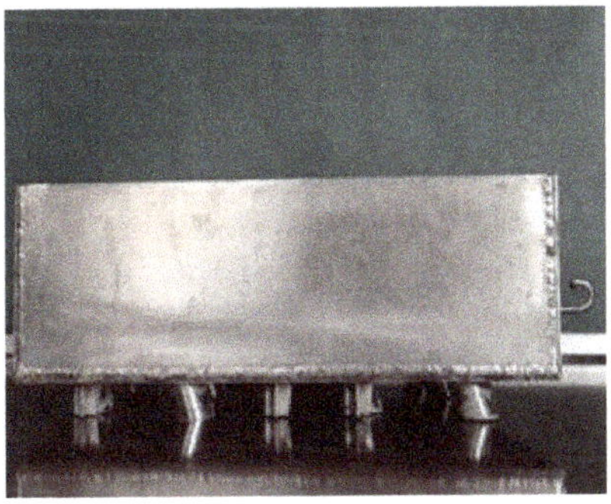

(b)

Figure 6. Model of gravity anchor. (**a**) Top view; (**b**) Front view.

The specific experimental steps are as follows. (1) Gravity anchors are slowly placed in the center of the soil. The gravity anchor is positioned 1 m from the loading end. Additionally, the gravity anchor is positioned in the middle of the test box in the width direction. It is not simply placed on the soil surface. The upper part of the gravity anchor needs to be fitted with counterweight blocks. The aim is to sink the gravity anchor to the preset position. Prior to the start of the model test, the gravity anchor was marked with a marker at the height of 10 cm. When the gravity anchor has settled exactly to the preset depth, the counterweight is removed, and the rest of the operation begins. The tension sensor is connected to the loop on the front end of the gravity anchor. Next, the displacement sensor is installed, and the readings are zeroed. The gravity anchor was left for 10 min to allow the excess pore water pressure of the soil to dissipate. (2) Start recording. Use the water pump to add water to the water tank at a constant speed to load the gravity anchor. (3) When the anchor damages the soil, stop the test. The parallel experiment was repeated 3 times. The tensile force–displacement curve was plotted by taking the average of three parallel tests.

The graph of soil deformation is shown in Figure 7. It can be observed the soil has been destroyed. The load–displacement curve can be obtained by using software to synthesize and process the data collected by the sensor. The load–displacement curve of the anchor under horizontal loads is shown in Figure 8. The load carrying capacity of a gravity anchor is the maximum force that can be provided by the gravity anchor while the anchor remains stable and does not break the soil. When the gravity anchor moves to the distance of 0.01 m, the soil has been damaged, at which point the corresponding force is the gravity anchor bearing capacity.

Figure 7. Test pictures.

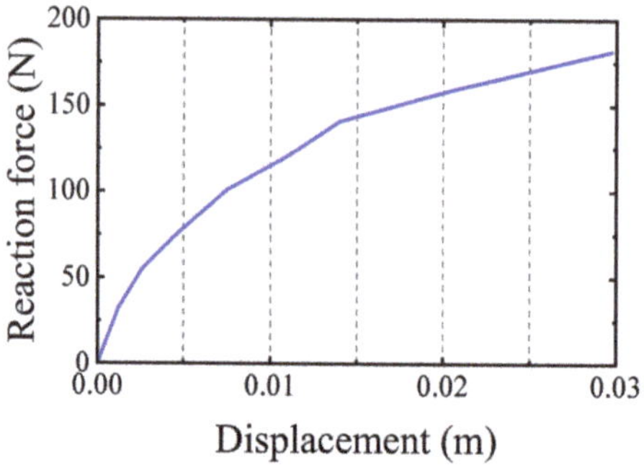

Figure 8. Load–displacement curve of the anchor under horizontal loads.

In clay, the formula for calculating bearing capacity is written as:

$$F_h = S_{uz} * A + \left[2S_{ua} * Z_s + \gamma_b * 0.5{Z_s}^2\right] * B \quad (1)$$

where F_h and S_{uz} represent horizontal bearing capacity and undrained shear strength of soil at depth Z_s, S_{ua} is the average undrained shear strength of the soil from anchor bottom to the depth Z_s, A is the area of anchor bottom, Z_s is the depth of anchor burial, B is the width of the anchor bottom, γ_b is the bulk density of the soil.

The parameters in Equation (1) are determined according to the size of the anchor in the model test, the embedded depth, and the strength of the soil. The ultimate bearing capacity were determined according to Equation (1), and the model test results are 379.5 kN and 421.9 kN. The test results are in good agreement with the calculation results based on the theoretical formula, and the difference between the two is about 10%.

3. Numerical Simulation of Gravity Anchors

To analyze horizontal bearing mechanism of the gravity anchor, the finite element method is used to analyze the gravity anchor. The model was established based on the ABAQUS software, according to the specific dimensions of the model anchor, the size of the soil domain in the model test tank, and the boundary conditions. The length and width of the gravity anchor are both 3 m, the height is 1.3 m, the weight is 80 tons, and the floating density in seawater is 5812.6 kg/m^3. The material of the gravity anchor is steel. The elastic modulus is 210 GPa, and the Poisson's ratio is 0.3. The 3D finite element model of gravity anchor is shown in Figure 9a. The gravity anchor has 10 column bases. The height of the shear key is 260 mm and the length is 270 mm. Its diameter is 300 mm. Shear keys are evenly distributed at the bottom of the gravity anchor. The calculation range of the soil body is determined according to the outer dimensions of the anchor body. To eliminate the influence of the boundary, the depth and width of the soil are both about 5 times the corresponding height and width of the anchor. Therefore, the size of the soil model is 50 × 50 × 10 m. The 3D finite element model of soil is shown in Figure 9b.

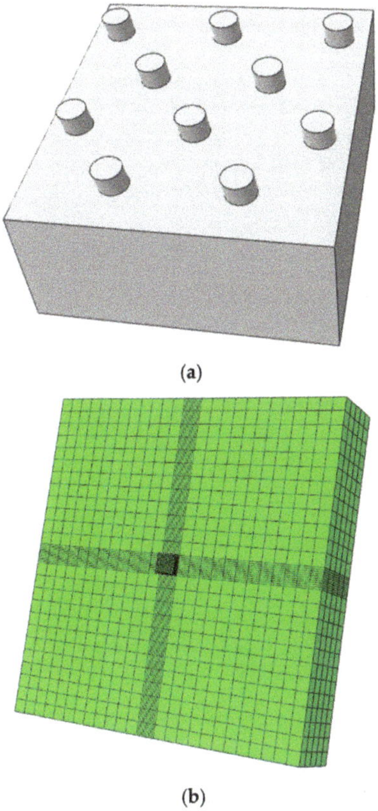

Figure 9. The 3D finite element model. (**a**) 3D finite element model of gravity anchor; (**b**) 3D finite element model of soil.

Normal horizontal constraints are applied to the vertical boundaries, and fixed constraints are applied to the bottom boundary. The top boundary is fully free. The gravity anchor model is not constrained and can move freely. The anchor is treated as a rigid body during calculations. The frictional contact is set between the model anchor and the surrounding soil. The surface of the anchor is set as a master surface, and the surface of the soil in contact with it is set as a slave surface. Separation is allowed in the normal direction of the contact surface. When the gravity anchor bears the horizontal load, the Coulomb friction contact is set between the anchor and the soil. The friction force on the contact surface of the anchor and soil is related to the normal force between the contact surface of the anchor and soil. When the gravity anchor bears an inclined load, the critical shear stress friction contact between the anchor and the soil is set. The tangential slip condition is considered to only be related to the critical shear stress. When the shear stress between the anchor–soil interface is less than the critical shear stress, the anchor and the soil are in a state of mutual bonding without relative slip; on the contrary, the anchor and the soil produce relative slip. The critical shear stress is represented by Equation (2).

$$\tau_{\max} = a s_u \tag{2}$$

where $\tau_{\max}$ is the critical shear stress; a is cohesion coefficient; s_u is undrained shear strength.

Hexahedral elements are used to divide the soil into grids, and tetrahedral elements are used to divide the gravity anchors into grids. To improve the simulation accuracy, the soil unit is divided into units by progressive encryption, as shown in Figure 9b. Mesh smoothing is used in the calculation process. Mesh smoothing is to smooth the deformed mesh without changing the connections among mesh nodes.

The contact parts of gravity anchor and soil are divided into different grids to study the grid sensitivity. The unit division is shown in the following table. When the grid number is less than 10 × 10, soil failure occurs earlier. Gravity anchors also have less carrying capacity. Therefore, the grid precision is not enough. When the grid number is greater than 10 × 10, the bearing capacity curve is basically unchanged. Considering the computational efficiency, it is better to choose a 10 × 10 grid number. The results are shown in Table 3 and Figure 10.

Table 3. Soil element number.

Serial Number	Number of Seeds
A	3 × 3
B	5 × 5
C	5 × 10
D	10 × 10
E	10 × 15
F	15 × 15

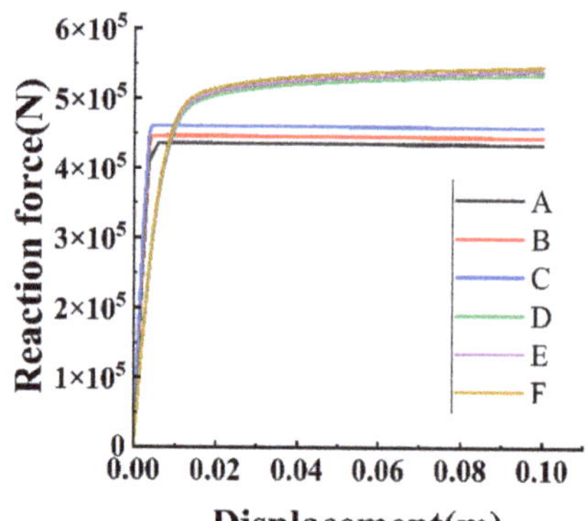

Figure 10. Bearing capacity curves of different elements.

The density of soil is 1800 kg/m^3. The floating density is 800 kg/m^3. Poisson's ratio is 0.45. Additionally, the elastic modulus is 5 MPa. The soil plastic model is the Mohr–Coulomb plastic model. The expansion angle is 0.25, and the cohesion is 10 kPa. The behaviors of the gravity anchors were analyzed by the finite element method according to three steps. (1) Geostatic: In this step, the initial ground stress of the soil is balanced so that the initial displacement of the soil is zero. (2) Contact: The purpose of this step is to establish contact between the anchor and the soil. (During the modeling process, it was assumed that the contact time between the anchor and soil was long enough at the second analysis step to ignore pore water pressure. At this point, pore water pressure is considered to have completely dissipated.) (3) Load: Load the gravity anchor to move it at this step.

Apply loads to the model and run the software. The curves of horizontal bearing capacity and horizontal displacement are extracted to draw the image shown in Figure 11. It can be found that the curve has a more obvious inflection point. The ultimate bearing capacity of the gravity anchor is generated here. The soil is destroyed at this point. The corresponding ordinate is the horizontal bearing capacity of the gravity anchor. The result of numerical simulation is 455.7 kN, with an error of 8% after comparison with the model test. The predictions are generally in agreement with the test results, which indicate the plastic strain accumulation characteristics of the soft clay surrounding the anchor can be captured by the finite element method. So, the simulation results are considered correct and can be used for comparative analysis.

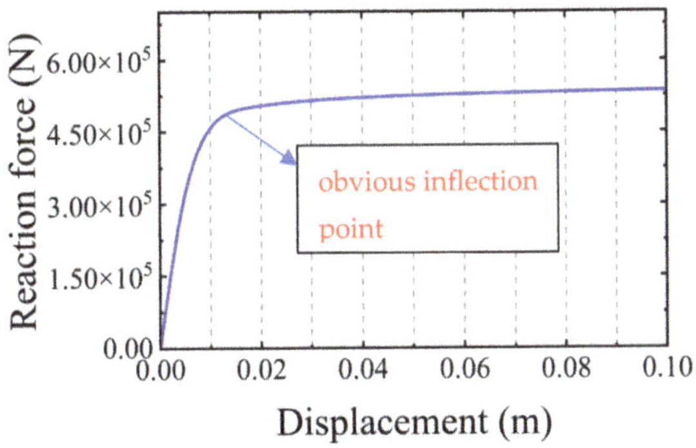

Figure 11. Numerical simulation result curve.

4. Results and Discussion

4.1. Mechanism of Sliding and Bearing Capacity of Gravity Anchor in Clay

It follows from the obtained capacity–displacement curve that the gravity anchor moves under horizontal tension in the clay. The bearing capacity tends to reach the maximum when the sliding preventer force is gradually increased to a certain level. So, the type of soil destruction is progressive destruction [26]. Sliding of the anchor occurs inside the soil, as shown in Figure 12. Figure 12 presents the deformation of the clay layer and the motion of the gravity anchor. As the displacement of the gravity anchor increases, the area and size of the soil deformation increase. The results of the model test and numerical simulations are in good agreement. The results in the figure show that the amount of soil deformation gradually increases with the motion of the gravity anchor. The settlement of the front part of the gravity anchor is larger than that of the back part. In other words, the gravity anchor will settle during the motion of the gravity anchor. The gravity anchor is a forward-tilted settlement. This is due to the fact that to balance the tensile moment, the soil body needs to provide a large vertical stress. It is accompanied by a large settlement of the front end of the soil mass. Liang J and Song X proposed that the soil in front of the anchor was formed due to the horizontal motion of soil particles. However, through the study presented in this paper, it was found that the soil in the passive zone in front of the anchor is squeezed during the tilted motion of the anchor. Due to the partially fluid nature of the clay, the clay at the bottom of the anchor will flow as it is squeezed by the gravity anchor. When a surge occurs, the forward body of the soil will move upwards after being squeezed by the clay at the bottom of the anchor. Finally, the phenomenon of soil accumulation at the front end of the anchor is formed. This is also verified in Figure 13. The displacement vector of soil indicates that accumulation of soil occurs in front of the soil. The gravity anchor is inserted into the soil. As a result, the soil in front of the gravity anchor flows and accumulates in front of the gravity anchor.

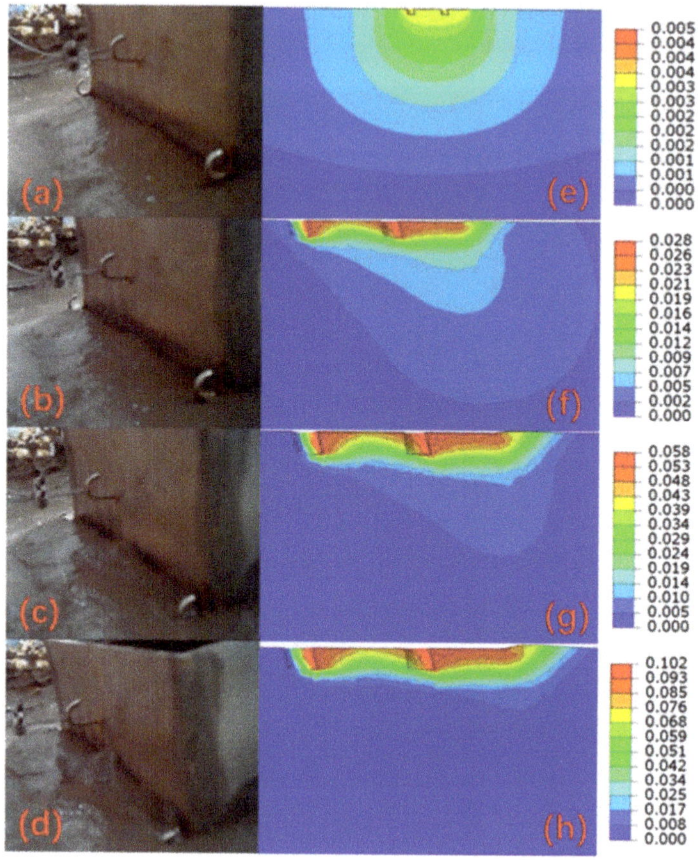

Figure 12. Soil deformation diagram of model test and numerical simulation: (**a**,**e**) are soil deformation diagrams when the gravity anchor moves 0 mm, (**b**,**f**) are soil deformation diagrams when the gravity anchor moves 25 mm, (**c**,**g**) are soil deformation diagrams when the gravity anchor moves 50 mm, and (**d**,**h**) are soil deformation diagrams when the gravity anchor moves 75 mm.

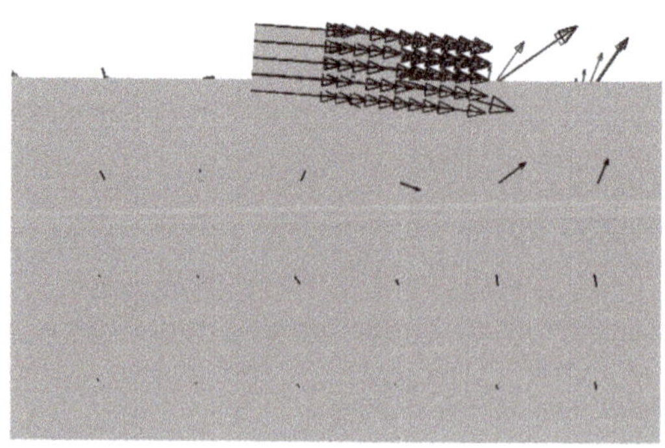

Figure 13. Soil flow mechanisms at failure.

The front soil is actively damaged by gravity anchors and is greatly deformed. The soil area behind the anchor is passive. The posterior soil is not affected by the gravity anchor. However, it can be seen from the figure that there are some deformations. This is due to the fact that the soil area at the rear end of the anchor flows as the anchor moves, and its settling and deformation is caused by its gravity. During the motion of the gravity anchor, soil deformation is concentrated around the gravity anchor. The deformation of the soil close to the gravity anchor is larger than the deformation of the soil away from the gravity anchor. The closer the soil is to the anchor, the greater the deformation. It can be seen from the Figure 14 that the plastic deformation of the soil first occurs in two small areas around the anchor. At this time, the plastic deformation is relatively small. This is caused by the gravity of the gravity anchor. As the motion process progresses, the soil around the shear key at the rear of the anchor first appears plastically deformed. So, the failure of the soil starts from the rear shear key. The soil in front of the anchor immediately undergoes plastic deformation and failure. Along with the motion of the anchor, the front shear key gradually begins to destroy the soil. Additionally, the damage extends to the soil surface. As a result, the destruction of soil is a gradual process. The plastic deformation of the soil is mainly concentrated near the shear key and around the anchor. Gravity anchors locally damage the soil during motion and then reach the ultimate bearing capacity. So, the failure mechanism of the soil is a local failure. In summary, the failure mechanism of soil mass is a local progressive failure.

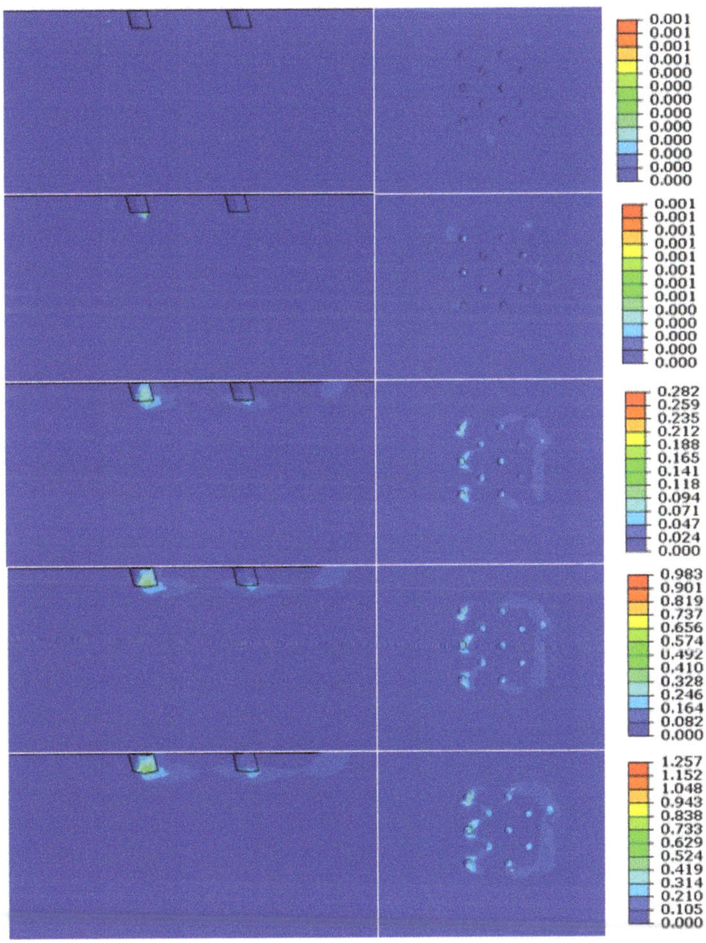

Figure 14. The plastic deformation process of the soil sections at rear of gravity achor.

4.2. Analysis of Influencing Factors of Gravity Anchor Bearing Capacity in Clay

4.2.1. Influence of Clay Properties on Bearing Capacity of Gravity Anchor

From the above analysis, it is clear that the type of soil destruction is progressive destruction and that the sliding of the anchor occurs inside the soil. Thus, the properties of the clay have a large impact on the bearing capacity of the gravity anchor. Furthermore, the influence of the friction angle and cohesion of the clay on the bearing capacity of the gravity anchor is investigated.

A 3 × 3 × 1.3 m gravity anchor model was established to study the influence of clay properties on bearing capacity. First, we changed the friction angle to study the effect of friction angle on the bearing capacity. From Figure 15, it can be found that with the increase in friction angle, the bearing capacity shows an increasing trend. The bearing capacity changes obviously when the friction angle is small. However, when the friction angle increases to 30 degrees, the change in bearing capacity tends to be stable. From the enlarged view of the end curve, it can be known the bearing capacity curves of 12.5 kPa–20 kPa coincide when the friction angle is 30 degrees. The bearing capacity reaches a fixed value, and the size of the value is 675,000 N. When the cohesion is less than 12.5 kPa, the bearing capacity increases to a greater extent with the increase in the friction angle. However, the tendency of bearing capacity to change with friction angle is small after cohesion exceeds 12.5 kPa. Therefore, as the cohesion increases, the effect of the friction angle on the bearing capacity decreases gradually.

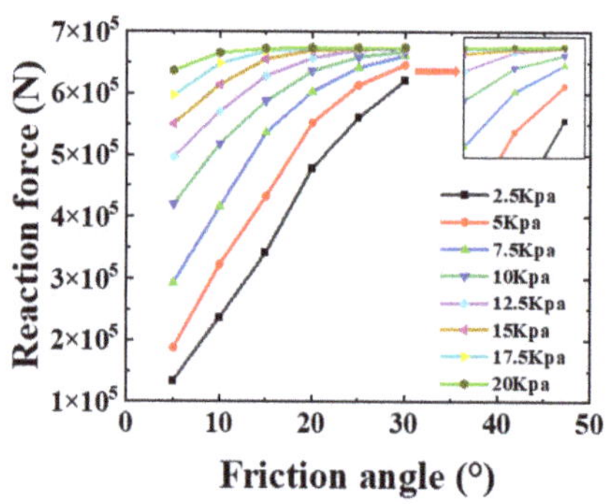

Figure 15. Friction angle–reaction force curve.

The cohesion was changed to obtain the bearing capacity curve, as shown in Figure 16. As the cohesion increases, the bearing capacity also shows an increasing trend. Additionally, with the increase in cohesion, the change in the bearing capacity curve gradually becomes gentle. It can be seen from the figure that the initial value of each bearing capacity curve increases with the increase in the friction angle. When the friction angle gradually increases to a certain value, the bearing capacity tends to remain at a certain fixed range. Therefore, as the cohesion increases, the bearing capacity curve tends to be flat. The increase in the friction angle has less and less influence on the bearing capacity.

The effects of friction angle and cohesion on the change in bearing capacity are plotted on a 3D surface plot, as shown in Figure 17. The part where the surface changes greatly is concentrated on the front side. In the range where the cohesion is less than 12.5 kPa and the friction angle is less than 15°, the bearing capacity varies greatly from 134,000 N to 629,000 N. The central slope of the surface change is 2.53. When the cohesion exceeds 12.5 kPa and the friction angle exceeds 15°, the change in the bearing capacity is small. The bearing capacity varies from 629,000 N to 675,000 N. The central slope of the surface change is 0.51. In this range, the change of cohesion and friction angle has little effect on the bearing capacity. The bearing capacity gradually becomes stable. Therefore, the bearing capacity of the gravity anchor is limited and cannot be increased indefinitely with the change of clay parameters. The bearing capacity of gravity anchors changes significantly when the clay parameters are low. When the clay parameters increase to a certain extent, since the weight of the gravity anchor does not change, the bearing capacity is limited by its weight and gradually reaches the maximum bearing capacity.

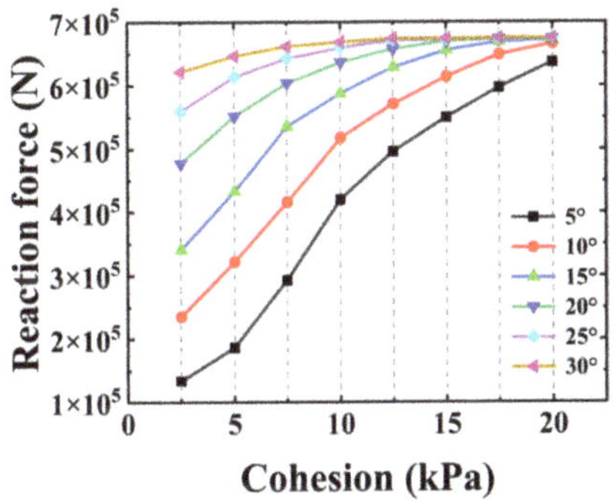

Figure 16. Cohesion–reaction force curve.

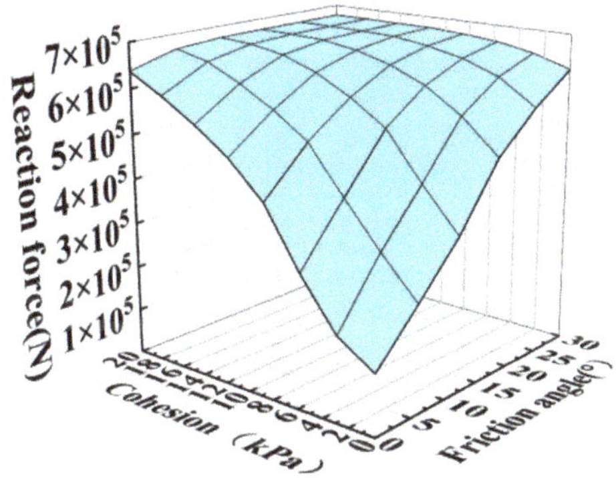

Figure 17. The 3D surface plot.

4.2.2. Influence of Gravity Anchor Bottom Area and Mooring Point Height on Bearing Capacity of Gravity Anchor

The bearing capacity of gravity anchors is mainly provided by the friction between the anchor bottom and the soil. The damage is mainly the sliding damage of the anchor bottom and the soil. The main factors affecting the friction force are the positive pressure at the bottom of the anchor and the contact area between the bottom of the anchor and the soil (i.e., the coefficient of sliding friction). When the gravity of the anchor is constant, the anchor volume is constant. When the area of the anchor bottom is changed, the vertical pressure of the anchor bottom and the contact area with the soil is changed at the same time. Therefore, the product of the two must have a maximum value. We kept other parameters unchanged and only changed the size of the anchor bottom area to study the influence of the anchor bottom area on the bearing capacity of the gravity anchor.

From Figure 18, it can be found that there is an optimal bottom area for the gravity anchor. When the area is less than the optimal area, the bearing capacity increases from 459,186 N to 710,181 N. The increase ratio is 54.7%. When the bottom area is larger than the optimal area, the bearing capacity is reduced from 710,181 N to 522,911 N. The reduction ratio is 26.4%. The soil true strain maps of gravity anchors, as shown in Figure 18, with bottom areas of 4 m^2, 9 m^2, and 16 m^2 are extracted to study the reasons for the change in bearing capacity. It is obvious that the soil mass with the optimal bottom area has the largest true strain. Additionally, the deformation area of the soil is the largest. The depth of the impact of gravity anchors on the soil is also the largest. For gravity anchors whose bottom area is less than the optimal area, the soil strain is the largest below the front edge of the anchor bottom. Therefore, the failure of the bearing capacity is mainly caused by the local failure of the soil at the front of the anchor bottom. The maximum soil strain of the gravity anchor with the bottom area larger than the optimum area appears on the surface of the subsoil. The failure of the bearing capacity is mainly caused by the damage to the surface soil at the bottom of the anchor. However, the maximum soil strain of the gravity anchor with the optimal area appears in the deep layer of the soil. The failure of the bearing capacity is mainly caused by the overall failure of the anchor bottom soil. Therefore, when the anchor bottom area is less than the optimal area, the soil failure is a local failure. When the anchor bottom area is larger than the optimal area, the soil failure is the sliding failure of the shallow layer. When the anchor bottom area is the optimal area, the soil failure is a deep failure. Therefore, the bearing capacity of the gravity anchor with the optimal area is the largest.

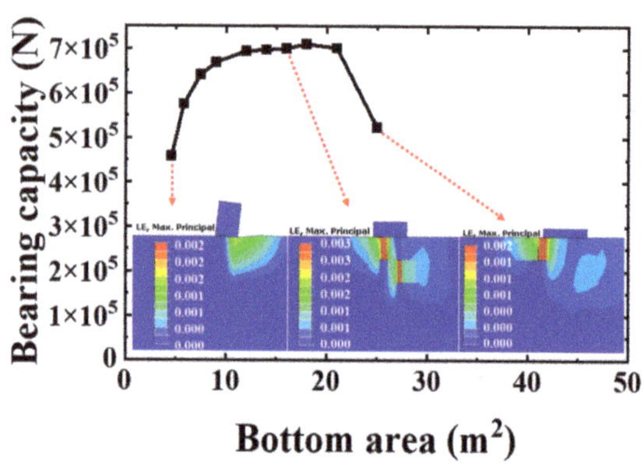

Figure 18. Anchor bottom area and bearing capacity curve.

It can be seen the optimal area is around 18 m^2. So, the optimal size is 4.3 × 4.3 × 0.633 m. According to the specification, it can be calculated that the minimum width of the anchor bottom is 2.40 m. Therefore, the finite element calculation results are within the range of the formula calculation results. It can be seen from Table 4 that when the anchor bottom area is in the range of 11–18 m^2, the horizontal bearing capacity varies little. The shapes of the bottom surfaces of the anchors studied above are all squares. In the following, rectangles with different lengths and widths, which have the same bottom area (9 m^2), are used to study the influence of the bottom shape on the bearing capacity. The calculation results are shown in Table 4. It can be found from Table 1 that the shape of the anchor bottom has little effect on the horizontal bearing capacity. The horizontal bearing capacity corresponding to the square section is slightly higher than that of the rectangular section. Therefore, a square anchor base shape is slightly better than a rectangle.

Table 4. Horizontal bearing capacity with different cross-sectional shapes.

Section Shape (m^2)	Bearing Capacity (N)
1 × 9	660,688
1.5 × 6	663,998
2 × 4.5	664,959
2.5 × 3.6	661,374
3 × 3	669,000

The location of the gravity anchor mooring point is closely related to the bearing capacity of the gravity anchor. The bearing capacity of the same gravity anchor at different mooring heights is obtained by using gravity anchor models. The calculation results are shown in Figure 19a. In each group of models, the bearing capacity of the gravity anchor decreases with the increase in mooring point height. The change trends of each group of curves in the figure are similar. The bearing capacity of the first half of the curve decreases slightly, and it can be considered that there is no change in the first half. The bearing capacity of the second half of the curve decreases greatly. In this part, the bearing capacity is greatly affected by the mooring height. It can be clearly found from Figure 20 that when $0 < D/H < 0.5$, the vertical displacement of the gravity anchor changes very little. The vertical displacement of the gravity anchor increases by a factor of 2 when D/H is between 0.5 and 1. The gravity anchor has undergone a relatively obvious flip in this interval. Stability and carrying capacity are greatly reduced. When the height of the mooring point of the gravity anchor is less than 0.5 H, the gravity anchor only translates and does not rotate under the action of tension. However, when the height of the mooring point exceeds 0.5 H, the gravity anchor will rotate. The bearing capacity of the gravity anchor will gradually decrease. The higher the height of the mooring point, the greater the rotation amplitude and the smaller the carrying capacity.

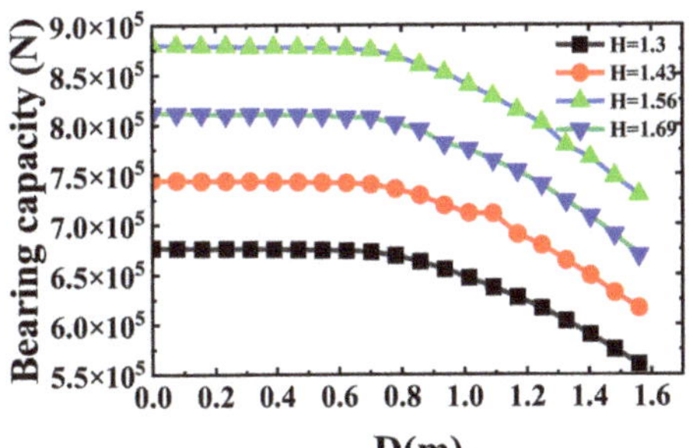

(a)

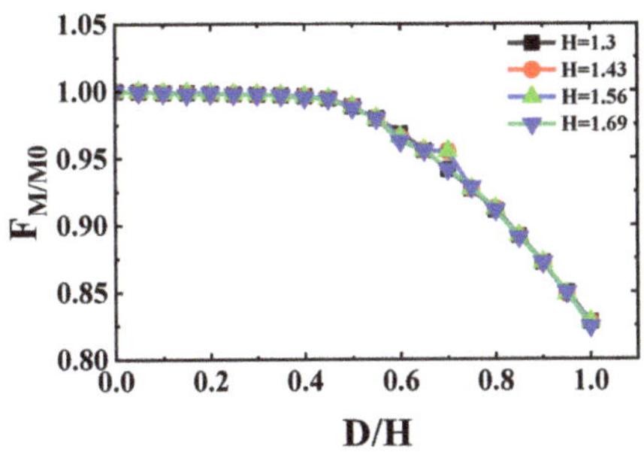

(b)

Figure 19. Relationship between mooring point height and carrying capacity. (**a**) Sensitivity analysis of mooring point height to bearing capacity; (**b**) Dimensionless relationship between mooring point height and bearing capacity.

Dimensionless processing is performed on the data in Figure 19a to obtain the dimensionless relationship between the height of the mooring point and the bearing capacity, as shown in Figure 19b. Dimensionless bearing capacity ($F_{M/M0}$) is the ratio of real-time bearing capacity (F_M) to ultimate bearing capacity (F_{M0}). Dimensionless altitude (D/H) is the ratio of real-time altitude (D) to maximum altitude (H). It can be found from the figure that the dimensionless curves of gravity anchors with different heights and weights coincide. This shows that height and weight do not affect this relationship. As the height of the mooring point increases, the rate of decrease in the bearing capacity also increases. The curve can provide a reference for engineering practice and estimate the influence of the mooring point height on the bearing capacity of gravity anchors. For the convenience of this research, the fitting formula (3) is obtained by fitting the trend curve. The fitting coefficient is R^2 = 0.99088.

$$F_{M/M0} = -0.0055\exp(\frac{D}{H}/0.28065) + 1 \quad (3)$$

The above formula can be used to estimate the influence of the height of the mooring point on the bearing capacity. When the dimensionless height increases to 0.7, the dimensionless bearing capacity drops to 0.95. While the dimensionless height increases from 0.7 to 1, the dimensionless bearing capacity drops to 0.82. The dimensionless bearing capacity decreased by 0.13 in this interval, indicating that the increase in height has a greater negative impact on bearing capacity in this interval. Therefore, it is suggested that the design height of the mooring point of the gravity anchor should not exceed 0.7 of the total height. In practical engineering, it is necessary to calculate the anti-overturning moment according to different anchor weights to ensure that gravity anchors will not overturn. Therefore, the value of *D/H* in actual engineering also needs to calculate the overturning moment to ensure the stability of the gravity anchor.

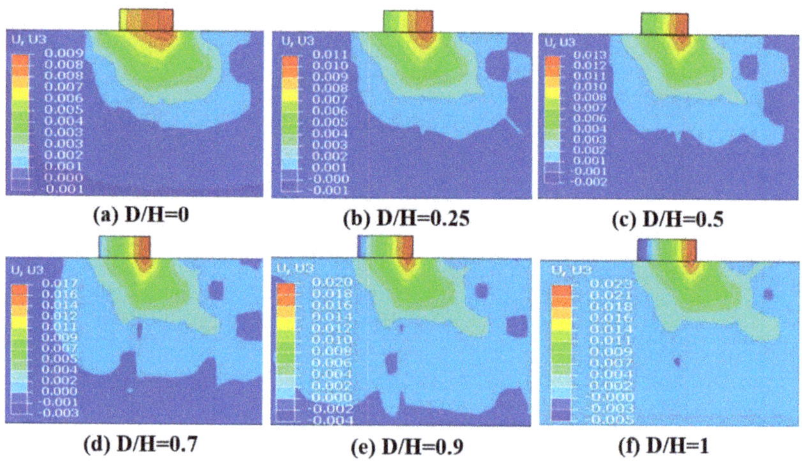

Figure 20. Soil deformation diagram at different mooring heights of gravity anchors.

4.3. *Influence of the Height of the Mooring Point on the V-H Failure Envelope Curve*

The gravity anchor model test rig mentioned in Section 2 has a height-adjustable pulley. The force angle of the anchor can be changed by adjusting the height of the pulley. We adjusted the pulley height to test and obtain the experimental data presented in Figure 21. The figure shows the bearing capacity–displacement curves of the force angles of 0°, 15°, 30°, 45°, and 90°, respectively.

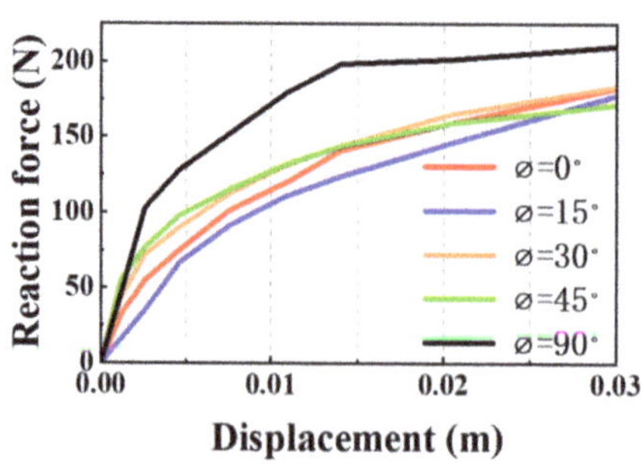

Figure 21. Bearing capacity–displacement curve at different stress angles.

The V-H failure envelope is an effective way to reasonably express the ultimate bearing capacity of shallow buried foundations. The V-H envelope of the model test is plotted in Figure 22a. $H_{simalution}$ represents the horizontal bearing capacity results in the numerical simulation. H_{Model} represents the horizontal bearing capacity results in the model test. $V_{simalution}$ represents the vertical bearing capacity results in the numerical simulation. V_{Model} represents the vertical bearing capacity results in the model test. The black curve and axis in the figure represent the failure envelope of the model test. The V-H failure envelope of the same gravity anchor is drawn by finite element modeling, as shown in Figure 22a. The red curves and axes in the figure represent the numerically simulated failure envelopes. The V-H failure envelopes of numerical simulations and model tests are normalized. The results are shown in Figure 22b. H_{ult} represents the maximum horizontal load carrying capacity. Additionally, V_{ult} represents the maximum vertical load capacity. From the figure, it can be found that the normalized V-H failure envelopes of the numerical simulation and the model test coincide. So, the results of the numerical simulation can be used for analysis.

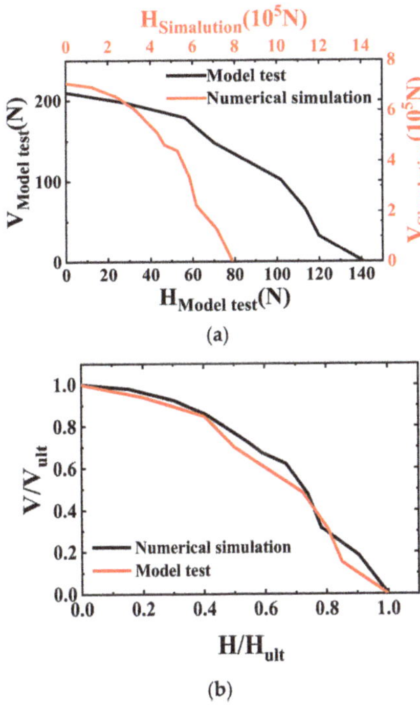

Figure 22. V-H failure envelope of gravity anchor. (**a**) The V-H envelope of the model test and numerical simulation. (**b**) Normalized V-H failure envelopes.

Next, the influence of the gravity anchor mooring point height on the V-H failure envelope is studied. The bearing capacity is obtained by changing the height of the mooring point of the gravity anchor. We plotted the normalized V-H failure envelopes for different D/H (D represents the gravity anchor mooring point height, and H represents the gravity anchor height), as shown in Figure 23. It can be known from the figure that the normalized V-H failure envelopes of gravity anchor foundations with different mooring heights are similar. However, as the mooring height increases, the failure envelope gradually shrinks inward. When the horizontal bearing capacity generated by the inclined load is less than 20% of the ultimate horizontal bearing capacity, the normalized vertical load of the anchor changes little. When the horizontal bearing capacity exceeds 20% of the ultimate horizontal bearing capacity, the normalized vertical load varies greatly.

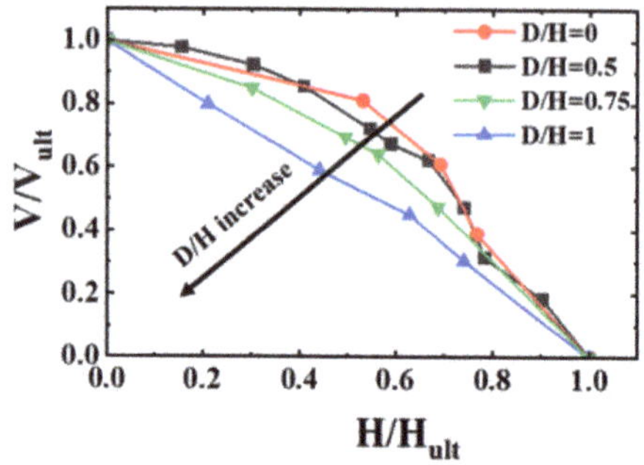

Figure 23. Normalized *V-H* failure envelope of gravity anchors at different mooring heights.

Furthermore, the mathematical expression of the normalized *V-H* failure envelope is obtained by studying the variation law of the normalized *V-H* failure envelope at different mooring heights. Bransby and Randolph [27] proposed a normalized failure envelope formula in the *V-H* stress space for shallow buried foundations.

$$\left(\frac{H}{H_{ult}}\right)^a + \left(\frac{V}{V_{ult}}\right)^b = 1 \tag{4}$$

Zdravkovic et al. [28] suggested that the values of a and b should be $a = b = 2$ when the anchor foundation with a length-to-diameter ratio $L/Di = 0.5\sim1.4$ is under an inclined load. Senders and Kay [29] suggested that the values of a and b should be $a = b = 3$ when the anchor foundation with a length-to-diameter ratio $L/Di = 2.8$ is under an inclined load. Many scholars have fitted the normalized failure envelope in the *V-H* stress space with mathematical expressions to study the values of a and b. However, the change in the height of the mooring point was not considered. Based on the content of this section, the fitting curve is obtained by fitting the above normalized *V-H* failure envelope. The fitting results are shown in Figure 24.

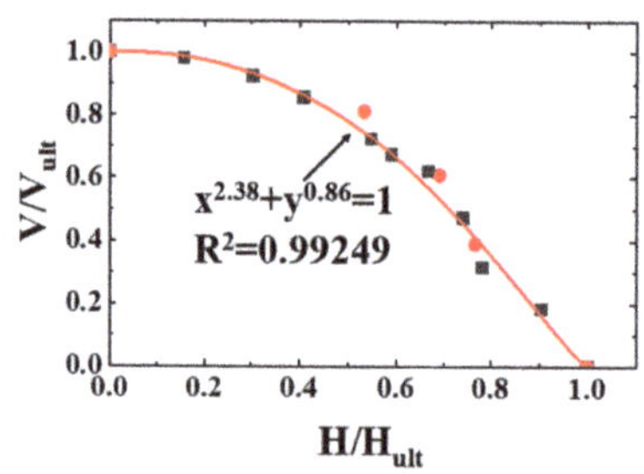

(a)

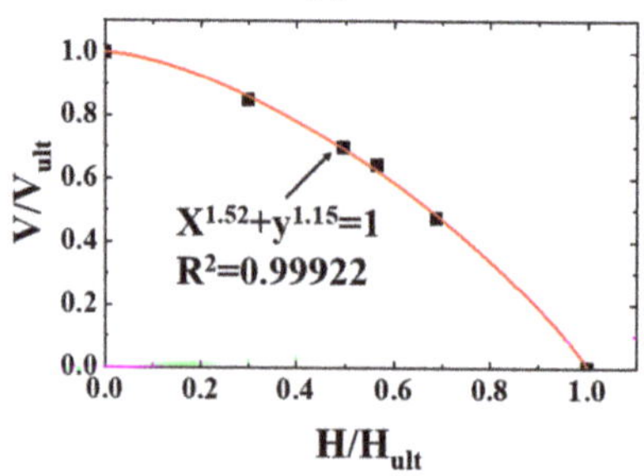

(b)

Figure 24. *Cont.*

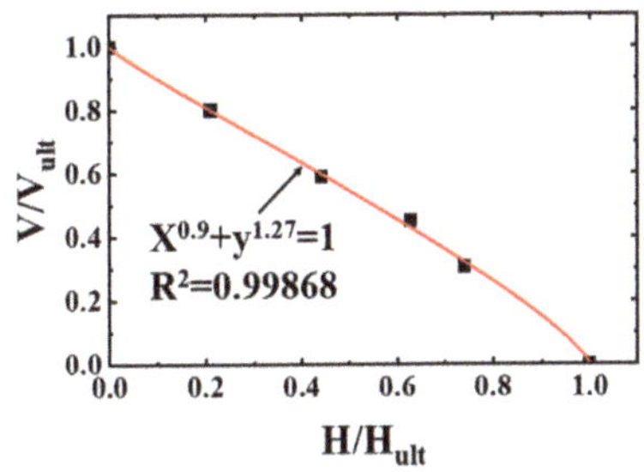

(c)

Figure 24. Fitted curve expression. (**a**) *D/H* = 0 and 0.5. (**b**) *D/H* = 0.75. (**c**) *D/H* = 1.

From the fitting result of the normalized failure envelope of the *V-H* stress space, the mathematical expression formula considering the height of the mooring point can be obtained as:

$$\left(\frac{H}{H_{ult}}\right)^a+\left(\frac{V}{V_{ult}}\right)^b=1$$
$$\begin{cases} a=\begin{cases} 2.38 & (0\le \frac{D}{H}<0.5) \\ 5.9\exp(-\frac{D}{H}/0.76)-0.7 & (0.5\le \frac{D}{H}\le 1) \end{cases} \\ b=\begin{cases} 0.86 & (0\le \frac{D}{H}<0.5) \\ 1.35-2.89\exp(-\frac{D}{H}/0.28) & (0.5\le \frac{D}{H}\le 1) \end{cases} \end{cases} \quad (5)$$

5. Conclusions

The motion behavior of gravity anchors under different angular loads is investigated by conducting model tests. Then, the bearing capacity of gravity anchors in soft clays is simulated using the finite element method based on the ABAQUS software package. The key conclusions can be summarized as follows:

(1) The type of soil destruction is progressive destruction during the movement of the gravity anchor. The damaged area of the soil is mainly concentrated near the gravity anchor. Additionally, the failure mechanism of the soil is a local failure.

(2) With the increase in friction angle and cohesion, the bearing capacity shows an increasing trend. The bearing capacity changes obviously when the friction angle and cohesion are small. However, when the friction angle and cohesion increase to a certain value, the change in bearing capacity tends to be stable.

(3) Gravity anchors have an optimal bottom area with the largest bearing capacity. When the bottom area is determined, the shape of the bottom surface has little effect on the bearing capacity.

(4) When the height of the mooring point is below 0.5 *H*, the bearing capacity of the gravity anchor is unaffected. As long as it exceeds 0.5 *H*, the bearing capacity of the gravity anchor decreases immediately. The normalized *V-H* failure envelopes of gravity anchor foundations with different mooring heights are similar. However, as the mooring height increases, the damaged envelope gradually shrinks inward.

The limitation of this study is that it only considers the bearing state of gravity anchors under a static state. The process of gravity anchor bearing under dynamic conditions was not studied further. In the future, centrifuge will be used to study the bearing process and ultimate bearing capacity of gravity anchors under dynamic conditions.

Author Contributions: Conceptualization, Y.Y. and P.L.; methodology, P.L. and J.Y.; software, P.L.; validation, P.L. and X.L.; formal analysis, P.L. and X.L.; investigation, P.L. and H.L.; resources, Y.Y. and P.L.; data curation, P.L. and X.Z.; writing—original draft preparation, P.L.; writing—review and editing, P.L.; visualization, R.S.; supervision, Y.Y.; project administration, J.Y.; funding acquisition, J.Y. All authors have read and agreed to the published version of the manuscript.

Funding: This research was funded by The National Natural Science Foundation of China (Grant No. 51879189) and National Natural Science Foundation of China (Grant No. 52071234).

Institutional Review Board Statement: Not applicable.

Informed Consent Statement: Not applicable.

Data Availability Statement: The data presented in this study are available upon request from the corresponding author.

Conflicts of Interest: The authors declare no conflict of interest.

References

1. Cheng, X.; Li, Y.; Wang, P.; Liu, Z.; Zhou, Y. Model tests and finite element analysis for vertically loaded anchors subjected to cyclic loads in soft clays. *Comput. Geotech.* **2019**, *119*, 103317. [CrossRef]
2. Liu, J.; Han, C.; Ma, Y.; Wang, Z.; Hu, Y. Experimental investigation on hydrodynamic characteristics of gravity installed anchors with a booster. *Ocean Eng.* **2018**, *158*, 38–53. [CrossRef]
3. Liang, J.; Song, X. Recent centrifuge modelling of offshore geotechnical problems at IWHR. In Proceedings of the 7th International Symposium on Deformation Characteristics of Geomaterials, Glasgow, UK, 25 June 2019; Volume 92, p. 17001. [CrossRef]
4. Zhao, X.; Cheng, L.; Zhao, M.; An, H.; He, W. Gravity anchors astride subsea pipelines subject to oscillatory and combined steady and oscillatory flows. In Proceedings of the International Conference on Offshore Mechanics and Arctic Engineering—OMAE, Rio de Janeiro, Brazil, 1–6 July 2012; pp. 597–604. [CrossRef]
5. Harris, R.E.; Johanning, L.; Wolfram, J. Mooring systems for wave energy converters: A review of design issues and choices. In Proceedings of the 3rd International Conference on Marine Renewable Energy, Blyth, UK, 1 July 2004.
6. Yun, G.; Bransby, M. The Undrained Vertical Bearing Capacity of Skirted Foundations. *Soils Found.* **2007**, *47*, 493–505. [CrossRef]
7. Taylor, R.; True, D.; Arnold, F. Anchoring and mooring considerations for an OTEC pilot plant. In Proceedings of the OCEANS 2010 MTS/IEEE Seattle, Seattle, WA, USA, 20–23 September 2010; pp. 1–8. [CrossRef]
8. Petersen, J.B.; Chellakat Satyaraj, J.-J. Cochin(IN); Ahwahnee, CA, USA, 2011. Gravity anchor: US,8,181,589B2[P/OL].2011-04-28. Available online: https://patentimages.storage.googleapis.com/a2/dc/7e/69fb28bd1979a2/US8181589.pdf (accessed on 15 January 2023).
9. Upsall, B.; Horvitz, G.; Riley, B.; Howard, T.; Olsen, K.; Struthers, J.R. Geotechnical design: Deep water pontoon mooring anchors, in: Ports 2013: Success Through Diversification. In Proceedings of the 13th Triennial International Conference, Seattle, WA, USA, 25–28 August 2013. [CrossRef]
10. Li, S.; Duan, G.; Li, H.; Huang, S.; Wang, X. Anti-sliding stability calculation of gravity anchor subjected to combined loading. *Tumu Jianzhu Yu Huanjing Gongcheng/J. Civil. Archit. Environ. Eng.* **2018**, *40*, 7. [CrossRef]
11. Houlsby, G.T. Modelling of shallow foundations for offshore structures. In Proceedings of the BGA International Conference on Foundations, Innovations, Observations, Design and Practice, Dundee, Scotland, 2–5 September 2003.
12. Gourvenec, S. Shape effects on the capacity of rectangular footings under general loading. *Geotechnique* **2007**, *57*, 637–646. [CrossRef]
13. Gourvenec, S. Failure envelopes for offshore shallow foundations under general loading. *Geotechnique* **2007**, *57*, 715–728. [CrossRef]
14. Gourvenec, S. Effect of embedment on the undrained capacity of shallow foundations under general loading. *Geotechnique* **2008**, *58*, 177–185. [CrossRef]
15. Gourvenec, S.; Barnett, S. Undrained failure envelope for skirted foundations under general loading. *Géotechnique* **2011**, *61*, 263–270. [CrossRef]
16. Bransby, M.F.; Yun, G.-J. The undrained capacity of skirted strip foundations under combined loading. *Géotechnique* **2009**, *59*, 115–125. [CrossRef]
17. Niroumand, H.; Kassim, K.A. Uplift response of square anchor plates in dense sand. *Int. J. Phys. Sci.* **2011**, *6*, 3939–3943.
18. Mana, D.S.K.; Gourvenec, S.; Martin, C.M. Critical Skirt Spacing for Shallow Foundations under General Loading. *J. Geotech. Geoenviron. Eng.* **2013**, *139*, 1554–1566. [CrossRef]
19. Liu, J.; Lu, L.; Yu, L. Large Deformation Finite Element Analysis of Gravity Installed Anchors in Clay. In Proceedings of the ASME 2014 33rd International Conference on Ocean, Offshore and Arctic Engineering, San Francisco, CA, USA, 8–13 June 2014. [CrossRef]
20. Liu, J.; Lu, L.; Hu, Y. Keying behavior of gravity installed plate anchor in clay. *Ocean Eng.* **2016**, *114*, 10–24. [CrossRef]
21. Li, H.; Huang, S.; Wang, X.; Li, S.; Yin, J. The study of horizontal bearing capacity of gravity anchor on calcareous soil with centrifugal model tests. In Proceedings of the International Offshore and Polar Engineering Conference, San Francisco, CA, USA, 25–30 June 2017.
22. Zhao, Y.; Liu, H. Toward a quick evaluation of the performance of gravity installed anchors in clay: Penetration and keying. *Appl. Ocean Res.* **2017**, *69*, 148–159. [CrossRef]
23. Liu, J.; Bu, N.; Han, C.; Yi, P. Numerical modelling on anchor-soil-water interaction for dynamic installation of gravity installed anchors. *Ocean Eng.* **2022**, *244*, 110412. [CrossRef]

24. Yang, Y.; Liu, H. Theoretical and numerical approaches to analyze gravity installed anchors in multi-layered clays. *Ocean Eng.* **2022**, *264*, 112452. [CrossRef]
25. Deb, P.; Pal, S.K. Nonlinear analysis of lateral load sharing response of piled raft subjected to combined V-L loading. *Mar. Georesources Geotechnol.* **2020**, *39*, 994–1014. [CrossRef]
26. Riyad, A.; Rokonuzzaman, M.; Sakai, T. Progressive failure and scale effect of anchor foundations in sand. *Ocean Eng.* **2019**, *195*, 106496. [CrossRef]
27. Bransby, M.F.; Randolph, M.F. Combined loading of skirted foundations. *Geotechnique* **1998**, *48*, 637–655. [CrossRef]
28. Zdravkovic, L.; Potts, D.M.; Jardine, R.J. A parametric study of the pull-out capacity of bucket foundations in soft clay. *Geotechnique* **2001**, *51*, 55–67. [CrossRef]
29. Senders, M.; Kay, S. Geotechnical suction pile anchor design in deep water soft clays. In Proceedings of the Conference Deepwater Risers Mooring and Anchorings, London, UK, 16–17 October 2002.

Journal of
Marine Science and Engineering

MDPI

Article

Coupling of Finite Element Method and Peridynamics to Simulate Ship-Ice Interaction

Renwei Liu [1], Yanzhuo Xue [2,*] and Xikui Lu [2]

1 School of Naval Architecture and Ocean Engineering, Jiangsu University of Science and Technology, Zhenjiang 212100, China
2 College of Shipbuilding Engineering, Harbin Engineering University, Harbin 150001, China
* Correspondence: xueyanzhuo@hrbeu.edu.cn

Abstract: In this work, the finite element method (PD-FEM) coupling strategy is used to simulate ship-ice interaction. Two numerical benchmark tests are selected to validate the coupling approach and its program. During the ice-breaking process simulation, the generation and propagation of radial and circular cracks in level ice are modeled and phenomena such as the shedding of wedge ice, flipping of brash ice, and cleaning of the channel are observed to be broadly consistent with experimental observation. The influence of ship speed and ice thickness on the ice load are investigated and analyzed. The ice load obtained from the numerical simulations is in general agreement with that given by Lindqvist's empirical formula. The boundary effect on the crack path can also be avoid with the current coupling method.

Keywords: level ice; ship-ice interaction; PD-FEM coupling approach; ice-breaking process; ice load

Citation: Liu, R.; Xue, Y.; Lu, X. Coupling of Finite Element Method and Peridynamics to Simulate Ship-Ice Interaction. *J. Mar. Sci. Eng.* **2023**, *11*, 481. https://doi.org/10.3390/jmse11030481

Academic Editor: Anatoly Gusev

Received: 4 February 2023
Revised: 21 February 2023
Accepted: 22 February 2023
Published: 23 February 2023

1. Introduction

In recent years, global climate change and the melting of ice in Arctic regions has raised the possibility of exploiting Arctic resources and opening an Arctic channel [1,2]. The exploitation of resources and scientific research in Arctic regions rely on icebreakers to open the necessary routes [3–5]. Therefore, it is great significant to simulate the icebreaking scenarios and calculate the ice load of ship–ice interaction, and it helps in improving the design and safe navigation of icebreakers. The ship-ice interaction scenarios are studied with full-scale tests, model tests, theoretical analyses, and numerical simulations. For full-scale testing, the results are reliable, but the associated cost is high. Model test is a promising candidate to study the ship–ice interaction [6–9]. However, compared with full-scale tests, models have many uncertainties, and can be expensive and time-consuming [10,11]. Theoretical analysis is still challenging in some cases, such as dealing with complicated structures [12]. Fortunately, numerical methods to study ship-ice interactions do not need to consider the structure complexity, and are not restricted by factors such as geography, cost, and time, and have been shown to be both efficient and accurate, both in theoretical research and engineering application [13–16]. Finite element method (FEM) was successfully applied to estimate the strength of ship structure problems [10,17,18]. The discrete element method (DEM) to calculate ice loads for offshore structures and ships [19–22]. Smoothed-particle hydrodynamics method (SPH) was adopted in the ice field to simulate the ice-structure interaction dynamics [23,24], and other methods [25,26].

In recent years, a mesh-free method of peridynamics was proposed [27]. This reformulation of the classical continuum mechanics is a non-local theory that does not assume the spatial differentiability of displacement fields. Based on integrodifferential equations, peridynamics can deal with discontinuous displacement fields. Therefore, it can simulate spontaneous crack nucleation and propagation, and can be used to simulate the ice-breaking and calculate ice loads [28–37]. However, as a non-local theory computational efficiency of peridynamics is far lower than that of FEM, especially for engineering applications like ship-ice interaction. To improve its computational efficiency, researchers have coupled peridynamics with FEM. Macek and Silling [38] proposed the PD-FEM coupling approach and implemented peridynamics in a commercial finite element analysis code, Liu et al. [39] introduced interface elements to calculate the coupling force in a PD-FEM approach, and Lee et al. [40] proposed a coupled PD-FEM approach to analyze impact fractures. To date, the advantages of combining PD with FEM have been demonstrated in applications to concrete and composite materials, but PD-FEM coupling has not been used to deal with the ship–ice interaction. In this work, the coupling strategy proposed by Liu et al. is employed for its easy to implement and robust theory foundation.

The following work is organized as, peridynamics theory and PD-FEM coupling scheme is introduced in Sections 2 and 3, respectively. The proposed coupling approach is verified with both dynamic and static cases in Section 4. The ship-ice interaction is simulated in Section 5. Conclusion is drawn in Section 6.

2. Peridynamics Framework

Peridynamics assumes that the continuum body is composed of small particles. Each particle interacts with other particles within a finite distance δ called the horizon. The pairwise interaction between two particles exists despite they are not in contact. This physical interaction is referred to as a bond, which in some way has a close analogy to a mechanical spring. In bond-based peridynamics, the kinetic equation of particle $\mathbf{x}$ in the reference configuration at time t is

$$\rho\ddot{\mathbf{u}}(\mathbf{x},t) = \int_{H_{\mathbf{x}}} \mathbf{f}(\mathbf{u}(\mathbf{x}',t) - \mathbf{u}(\mathbf{x},t), \mathbf{x}' - \mathbf{x})dV_{\mathbf{x}'} + \mathbf{b}(\mathbf{x},t)\|\mathbf{x}' - \mathbf{x}\| \leq \delta \tag{1}$$

where $H_{\mathbf{x}}$ is the domain of integration within the horizon of particle $\mathbf{x}$, $\mathbf{u}$ is the displacement vector field, and $\mathbf{b}$ is the body force density. ρ is the mass density, and $\mathbf{f}$ is a pairwise force density function defined as the force per unit volume that particle $\mathbf{x}'$ exerts on particle $\mathbf{x}$, which contains all the constitutive information of the materials.

To simplify the notation, the relative position in the initial configuration and its relative displacement are denoted as $\boldsymbol{\xi} = \mathbf{x}' - \mathbf{x}$ and $\boldsymbol{\eta} = \mathbf{u}(\mathbf{x}',t) - \mathbf{u}(\mathbf{x},t)$, respectively. Therefore, the relative position of the two interacting particles at t in the current configuration is $\boldsymbol{\xi} + \boldsymbol{\eta}$ and the pairwise force density function can be described as $\mathbf{f}(\boldsymbol{\eta}, \boldsymbol{\xi})$.

For the prototype micro-elastic brittle (PMB) material defined by Silling and Askari [41], the pairwise force density function can be expressed as

$$\mathbf{f}(\boldsymbol{\eta}, \boldsymbol{\xi}) = \frac{\boldsymbol{\xi} + \boldsymbol{\eta}}{|\boldsymbol{\xi} + \boldsymbol{\eta}|} cs\mu(t, \boldsymbol{\eta}, \boldsymbol{\xi}) \forall \boldsymbol{\eta}, \boldsymbol{\xi} \tag{2}$$

where $c = 12E/\pi\delta^4$ is the micro modulus, E is Young's modulus, and $s(\boldsymbol{\eta},\boldsymbol{\xi})$ is denoted as the stretch of the bond, which can be defined as

$$s = \frac{|\boldsymbol{\xi} + \boldsymbol{\eta}| - |\boldsymbol{\xi}|}{|\boldsymbol{\xi}|} \tag{3}$$

When the deformation stretch s exceeds a limit s_0 (described as the critical stretch for failure), the bond between the two particles breaks and no pairwise force remains. The term $\mu(t, \boldsymbol{\eta}, \boldsymbol{\xi})$ is a history-dependent scalar-valued function, which is introduced to represent the bond failure of two particles. This can be defined as

$$\mu(t, \boldsymbol{\eta}, \boldsymbol{\xi}) = \begin{cases} 1, & s < s_0 \\ 0, & s \geq s_0 \end{cases} \tag{4}$$

Accordingly, the level of damage is illustrated by the local damage at one particle, defined as

$$\varphi(\mathbf{x}, t) = 1 - \frac{\int_{H_{\mathbf{x}}} \mu(t, \boldsymbol{\eta}, \boldsymbol{\xi}) dV_{\boldsymbol{\xi}}}{\int_{H_{\mathbf{x}}} dV_{\boldsymbol{\xi}}} \tag{5}$$

When solving the elastic problem in which the damage is not considered, the critical stretch can be set to infinity. Dealing with the damage problem, the value of s_0 can be obtained from the energy release rate.

3. Coupling of PD-FEM

3.1. Coupling Scheme

The PD-FEM coupling approach proposed by Liu et al. [39] is adopted and presented. The coupling scheme is as follows: the domain to be solved is partitioned into FEM subregions, which are modeled as a non-failure area and a PD subregion containing the area expected to be damaged. An interface element is introduced to bridge from the FEM subregion to the PD subregion. The interface element contains several peridynamic nodes for calculating the coupling forces, which are the interaction forces between embedded peridynamics nodes and peridynamics nodes outside the interface element. The coupling scheme is illustrated in Figure 1.

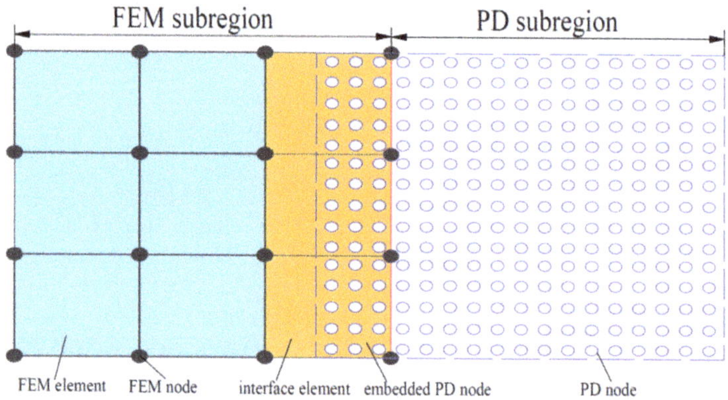

Figure 1. PD-FEM coupling scheme.

To implement the coupling scheme, interfaces between the peridynamics subregion and the FEM subregion should be defined prior to analysis. The coupling forces on embedded nodes are divided between the FEM nodes on the interface segment, as shown in Figure 2, by

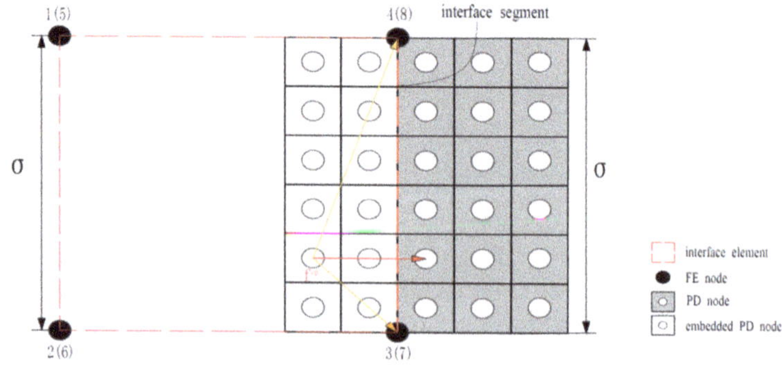

Figure 2. Coupling scheme that divides a coupling force f_p to FEM nodes on the interface segment.

$$f_{i,cp} = N_i(\xi', \eta', \psi') f_p \tag{6}$$

where $f_{i,cp}$ is the force of the FEM nodes on the interface segment, N_i is the shape function on the interface segment, f_p is the coupling force on the embedded nodes, (ξ', η', ψ') are the natural coordinates of the projection of an embedded node onto the interface segment, i is the number of FEM nodes on the interface segment, and m is the total number of embedded nodes. Note that for FEM nodes that are not on the interface segment, $f_{i,cp} = 0$.

For the FEM subregion, the equation of motion for the FEM nodes is written as

$$M_i \ddot{u}_i^n = F_{i,ext} + F_{i,int} \tag{7}$$

where $F_{i,ext}$ is the external force and the internal force is given by FEM nodes on segment FEM nodes not on segment,

$$F_{i,int} = f_{i,fem} + f_{i,cp} = \begin{cases} \left[\sum_e K^{(e)} u^{(e)}\right]_i + f_{i,cp}, & \text{FEM nodes on segment} \\ \left[\sum_e K^{(e)} u^{(e)}\right]_i, & \text{FEM nodes not on segment} \end{cases} \tag{8}$$

The displacements of the embedded peridynamics nodes are determined by

$$u_{ep} = \sum_{i=1}^{8} N_i(\xi, \eta, \psi) u_i \; i = 1, \ldots, 8 \tag{9}$$

where (ξ, η, ψ) are the natural coordinates of an embedded peridynamics node in the interface element and u_i is the nodal displacement of an interface element.

3.2. Numerical Implementation

The peridynamics equation of motion after discretization is written as

$$\rho \ddot{\mathbf{u}}_i^n = \sum_{j=1}^{m} \mathbf{f}\left(\mathbf{u}_j^n - \mathbf{u}_i^n, \mathbf{x}_j - \mathbf{x}_i\right) V_j + \mathbf{b}_i^n \tag{10}$$

For the FEM subregion, the equation of motion for the FEM nodes is written as

$$M_i \ddot{u}_i^n = F_{i,ext}^n + F_{i,\text{int}}^n \tag{11}$$

where n denotes the number of time steps. The displacement of node i can be obtained by approximating the acceleration in Equations (12) and (13) using an explicit central difference formula

$$\ddot{\mathbf{u}}_i^n = \frac{\mathbf{u}_i^{n+1} - 2\mathbf{u}_i^n + \mathbf{u}_i^{n-1}}{\Delta t^2} \tag{12}$$

$$\mathbf{u}_i^{n+1} = \begin{cases} \frac{\Delta t^2}{\rho}\left[\sum_{j=1}^{m} \mathbf{f}\left(\mathbf{u}_j^n - \mathbf{u}_i^n, \mathbf{x}_j - \mathbf{x}_i\right) V_j + \mathbf{b}_i^n\right] + 2\mathbf{u}_i^n - \mathbf{u}_i^{n-1}, & \text{for PD nodes} \\ \frac{\Delta t^2}{\rho}\left[F_{i,ext}^n - F_{i,\text{int}}^n\right] + 2\mathbf{u}_i^n - \mathbf{u}_i^{n-1}, & \text{for FEM nodes} \end{cases} \tag{13}$$

where Δt is the size of the time step. A stability condition derived by Silling and Askri [41] can be used to determine the time step size, Δt as

$$\Delta t < \sqrt{\frac{2\rho}{\sum_{j=1}^{m} \frac{c}{|(x_p - x_i)|} V_j}} \tag{14}$$

Moreover, for the PD particles, the horizon size δ has a significant influence on the accuracy of the numerical simulations. The horizon size can be selected using the scale characteristics of the simulated object. In practice, $\delta = 3\Delta x$ usually works well [42]. Therefore, the horizon size is set to $\delta = 3\Delta x$.

The numerical code for the proposed PD-FEM coupling approach is compiled using Fortran 90. A flowchart of the PD-FEM coupling approach is shown in Figure 3.

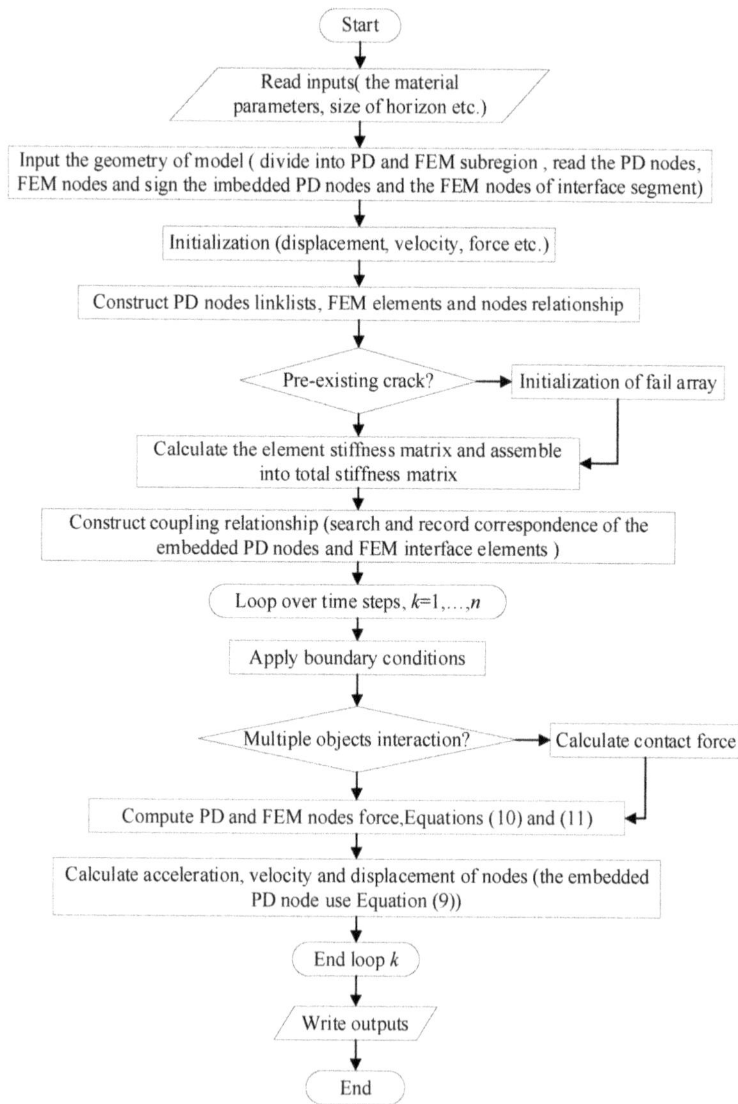

Figure 3. Flow chart of the FEM-PD coupling scheme.

4. Validation of PD-FEM Coupling Approach

4.1. Bending Deformation of Cantilever Beam

A three-dimensional cantilever beam subjected to a transverse loading of $F = 0.64$ N at the free end is examined, and the solutions given by the proposed coupling approach are compared with the FEM solutions. Because the bending deformation of a cantilever beam is a quasi-static problem, and to achieve a quantitative quasi-static calculation, the dynamic relaxation method is introduced to peridynamics [43].

The cantilever beam is 8 mm long, 2 mm wide, and 2 mm thick with Young's modulus of 1.0 GPa, Poisson's ratio of 0.25, and a density of 900 kg/m^3. Figure 4 shows the PD-FEM coupling model of this cantilever beam, which is partitioned into one FEM subregion and one PD subregion. The FEM subregion consists of 16 hexahedral elements of size 1 mm × 1 mm × 1 mm, whereas the PD subregion is discretized uniformly into 16 × 8 × 8 = 1024 particles with the grid spacing $\Delta x = 0.25$ mm and horizon size $\delta = 3\Delta x$. Three layers of peridynamics nodes are embedded in four interface elements for the coupling force calculations.

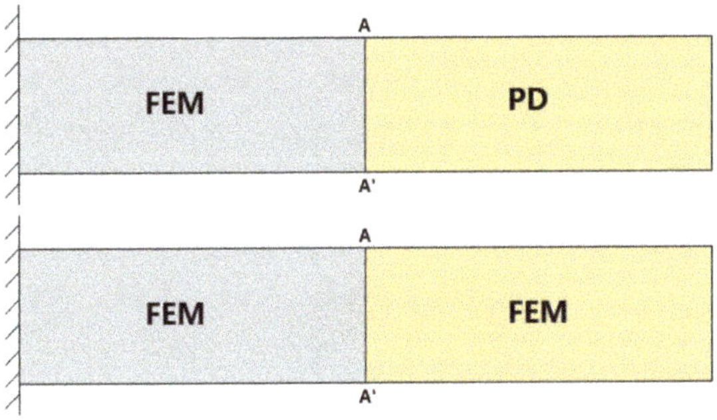

Figure 4. PD-FEM coupling model of the cantilever beamThe simulated deflection at the free end of the beam and the coupling force are presented in Table 1. The deflection and force obtained from the proposed coupling approach are very close to the FEM results (error less than 2%), which indicates that the coupling approach transfers the force accurately.

Table 1. Simulation results of bending beam.

	PD-FEM	FEM	Errors
Deflection	8.0492×10^{-5} m	8.19×10^{-5} m	1.74%
Force	−0.6327 N	−0.64 N	1.14%

Figure 5 shows the change in deflection along the central line of the beam. The displacement curve obtained by the numerical simulation is in good agreement with that of FEM, and its smoothness verifies the displacement coordination of the coupling approach. From Figure 5, it can be found that the ratio of the characteristic scale to the horizon size should be no less than 1 to obtain an acceptable result, however, more cases may be need to achieve this conclusion. The above results prove that the coupling algorithm achieves good accuracy and displacement coordination in the calculation of bending deformation, verifying the correctness of the proposed PD-FEM coupling approach in static problem.

4.2. Failure of 2D Plate with Central Crack

Mode-I crack is selected to simulate crack initiation and propagation along the plate. A 50 mm × 50 mm square plate with a 10 mm central pre-crack is stretched from both ends at a velocity of 50 mm/s, as shown in Figure 6.

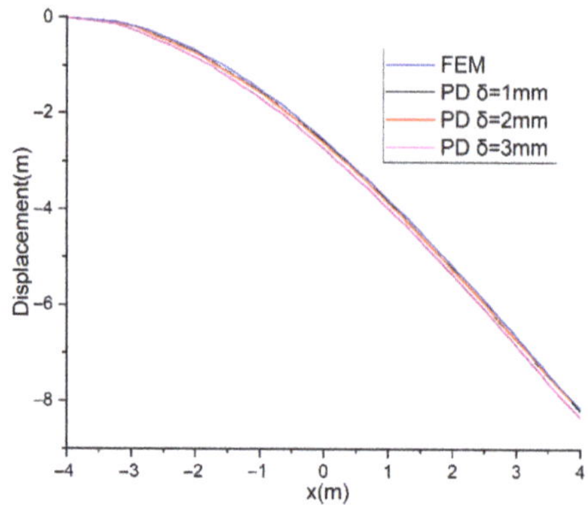

Figure 5. Vertical displacement on central axis of the beam.

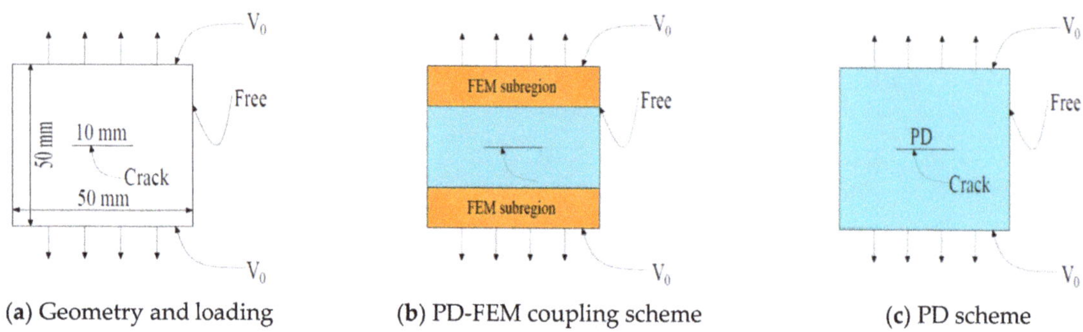

Figure 6. Geometry and loading condition of a plate with a central crack.

The PMB material properties used in this example are as follows: Young's modulus is E = 192 GPa, Poisson's ratio is v = 0.33, mass density is 8000 kg/m^3, and the critical stretch s_0 is 0.04472. These material parameters, geometry, and loading conditions are the same as the simulation example reported by Madenci and Oterkus [42]. For the coupling model, the plate is partitioned into one PD subregion and two FEM subregions (see Figure 6). The PD zone contains 100 × 200 = 20,000 particles, which are discretized regularly with a grid spacing of $\Delta x = 0.25$ mm and a horizon size of $\delta = 3\Delta x$. The FEM parts are composed of 32 four-node rectangular elements of size 6.25 mm × 6.25 mm. The interface region has three additional layers of peridynamics nodes (total of 2 × 3 × 200 = 1200 nodes). The time step is $\Delta t = 3.34 \times 10^{-8}$ s, which satisfies the stable time step condition.

For comparison, the PD solution is considered, as shown in Figure 6. The model size and load conditions are the same as for the coupling model, and the region is discretized regularly into 200 × 200 = 20,000 nodes with a grid spacing of $\Delta x = 0.25$ mm and horizon size of $\delta = 3\Delta x$. In the velocity boundary area, three virtual boundary layers are added, each with 3 × 200 = 600 nodes.

Figure 7 shows numerical simulation results of crack tracks using the PD-FEM coupling model and PD model (see Figure 7). Crack paths obtained from the proposed coupling method resemble the mode-I failure in brittle material, and are in good agreement with those obtained from the PD method. These results are similar to those reported by Madenci and Oterkus [42]. The displacements along the y-axis obtained from the PD-FEM and PD methods are plotted in Figure 8. We can find that they are in close agreement. Therefore, the PD-FEM coupling method can simulate crack propagation well, which verifies the correctness of the coupling approach in dynamic conditions.

(a) $t = 1.67\mu s$

(b) $t = 15.03\mu s$

Figure 7. Crack propagation simulations, (**left**) FEM-PD coupling model and (**right**) PD model.

(a) $t = 1.67\mu s$

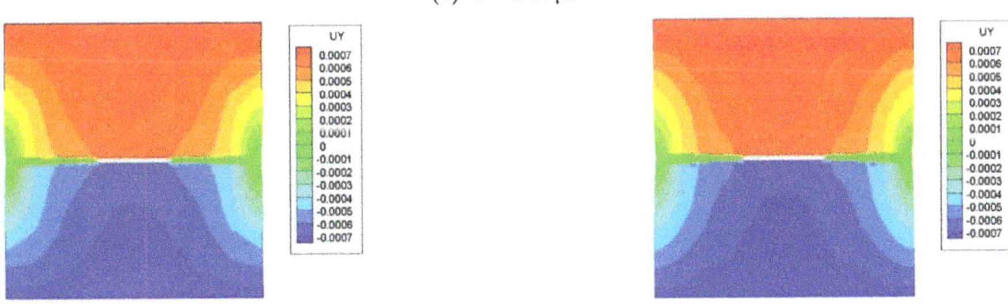

(b) $t = 15.03\mu s$

Figure 8. Vertical displacement simulations, (**left**) FEM-PD coupling model; (**right**) PD model.

5. PD-FEM Simulation of Icebreaker Navigation in Ice Level

5.1. Numerical Simulation

5.1.1. Ice Constitutive Model and Failure Criterion

Ice is a complex material that is affected by factors such as temperature, porosity, and grain size. Several laboratory tests have analyzed the brittle strength and failure patterns of ice, as well as the characteristics of ice–structure interactions [44,45]. The compressive strength and tensile strength of ice varies, with the compressive strength being around 3–4 times the tensile strength. The ductile–brittle transition of ice is another challenging topic, and ice shows different behavior at different strain rates (i.e., loading rate), as discussed by Schulson [45].

At low-speed loading rates, ice behaves as a ductile material, whereas at high-speed strain rates, the ice presents the characteristics of a linear elastic material, with a brittle mode when damage occurs. Therefore, ice is regarded as an elastic brittle material (PMB material in bond-based peridynamics) when interacting with an icebreaker, as the relatively high speed of icebreaker vessels corresponds to a high strain rate in the ice. Hence, a reasonable linear elastic constitutive model of ice is established for the bond-based peridynamics. Mechanical ice tests conducted by Schulson can be used to demonstrate the rationality of this linear elastic constitutive model.

In bond-based peridynamics, the bond force represents stress and the bond stretch represents strain. Let $s_t = s_0$ be the critical bond stretch and $|s_c| = 4s_0$ be the critical bond compression. If the bond stretch exceeds s_t in the tensile case or s_c in the compressive case, the bond will break and the bond force becomes zero.

Based on the bond-based peridynamics theory [42], the critical bond stretch in 3D cases is defined as

$$s_0 = \sqrt{\frac{5\pi G_0}{18E\delta}} \tag{15}$$

where G_0 is the energy release rate, which reflects the resistance of a material to crack propagation and can be derived from fracture mechanics. Linear elastic fracture mechanics is based on linear elastic theory and is applicable to brittle fractures. From the perspective of energy conservation, the condition for crack propagation is

$$G_0 \geq G_C \tag{16}$$

where G_C is the energy absorption rate. The primary fracture mode is tensile failures, because its compression strength is 3–4 times of tensile strength. G_0 can be expressed as

$$G_0 = \frac{K_I^2}{E} \tag{17}$$

where K_I is the fracture toughness, which reflects the resistance of a material to brittle fracture and can be measured experimentally. Therefore, the critical stretch can be calculated as

$$s_0 = \sqrt{\frac{5\pi K_I^2}{18E^2\delta}} \tag{18}$$

5.1.2. The Gravity and Buoyancy Model of Ice

The gravity and buoyancy of the ice are in balance in still water. when interacting with a ship, the ice will deviate from its equilibrium position as the gravity and buoyancy become unbalanced. To simplify the influence of gravity and buoyancy, a body force density b_z is introduced.

If a particle is completely under the waterline, we have

$$b_z = -g\rho_i + g\rho_w \tag{19}$$

If a particle is completely above the waterline, we have

$$b_z = -g\rho_i \tag{20}$$

For other particles, we have

$$b_z = -g\rho_i + g\rho_w l_w / d \tag{21}$$

where ρ_i is the density of ice, ρ_w is the density of water, l_w is the length of particles immersed in water, and d is the size of particles.

5.1.3. Ship-Ice Contact Model

The hull is modeled with FEM. The ice sheet contacting with the hull is modeled with peridynamics. Therefore, the ship–ice contact can be transformed into the interaction between peridynamics and the FEM models. In this case, the contact model developed by Liu [30] is introduced to calculate the contact force between the triangular elements of FEM and the particles of PD. To determine contact between a particle and a triangular element, some specific points must be defined (see Figure 9), namely the vertexes of the triangular element A, B, C and the particle P.

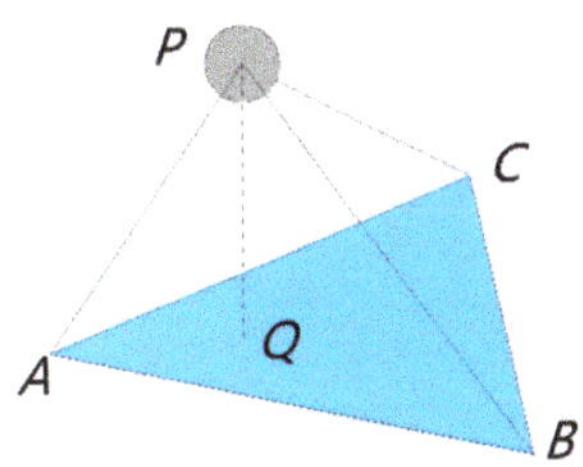

Figure 9. Diagram of ship-ice search model.

The determination of contact is divided into two stages. In the first stage, we estimate whether the distance between the particle and the plane of triangle ABC is less than the particle radius $r = \Delta x/2$. The distance is calculated with Equation (22), and the critical distance is defined by Equation (23)

$$|PQ| = PA \cdot \frac{BA \times CB}{|BA \times CB|} \tag{22}$$

$$|PQ| < r \tag{23}$$

where Q is the projection of P onto the plane of triangle ABC.

If the criterion in the first stage is not satisfied, the particle P does not contact the triangle element. When the first stage criterion is satisfied, it is necessary to further decide whether the projection point Q of P is inside the triangular region ABC, and this is the second stage. The centroid of the particle is checked to determine whether it is inside the triangle.

For any point Q in the plane ABC, the vector $\boldsymbol{d} = AQ$ can be expressed by two non-parallel vectors $\boldsymbol{a} = AC$ and $\boldsymbol{b} = AB$ in plane ABC as

$$\boldsymbol{d} = u\boldsymbol{a} + v\boldsymbol{b} \tag{24}$$

where the coefficients u and v are defined as

$$u = \frac{(\boldsymbol{b}\cdot\boldsymbol{b})(\boldsymbol{d}\cdot\boldsymbol{a}) - (\boldsymbol{b}\cdot\boldsymbol{a})(\boldsymbol{d}\cdot\boldsymbol{b})}{(\boldsymbol{b}\cdot\boldsymbol{b})(\boldsymbol{a}\cdot\boldsymbol{a}) - (\boldsymbol{b}\cdot\boldsymbol{a})(\boldsymbol{a}\cdot\boldsymbol{b})} \tag{25}$$

$$v = \frac{(\boldsymbol{a}\cdot\boldsymbol{a})(\boldsymbol{d}\cdot\boldsymbol{b}) - (\boldsymbol{a}\cdot\boldsymbol{b})(\boldsymbol{d}\cdot\boldsymbol{a})}{(\boldsymbol{a}\cdot\boldsymbol{a})(\boldsymbol{b}\cdot\boldsymbol{b}) - (\boldsymbol{a}\cdot\boldsymbol{b})(\boldsymbol{b}\cdot\boldsymbol{a})} \tag{26}$$

If the projection point Q is inside the triangular element ABC, the two coefficients must satisfy conditions, as

$$\begin{array}{l} u \geq 0 \\ v \geq 0 \\ u + v \leq 1 \end{array} \tag{27}$$

If the criteria in these two stages are satisfied, the particle P is in contact with the triangular element. The contact force is defined by the repelling short-range force. The force between two particles is given as

$$f = -\frac{PQ}{|PQ|}\min\left\{0, c_{sh}\left(\frac{|PQ|}{r} - 1\right)\right\} \tag{28}$$

where c_{sh} is the short-range force coefficient, can be choose as $c_{sh} = 5c$.

5.1.4. Numerical Model

This section describes numerical simulations of icebreaker navigation in level ice. Numerical models for the icebreaker and the level ice are illustrated in Figure 10.

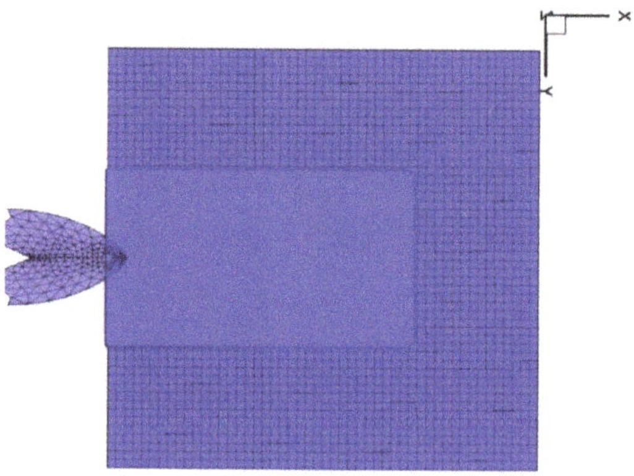

Figure 10. Numerical model of ship-ice interaction.

Xuelong icebreaker is selected and its bow is modeled. Main parameters are listed in Table 2. The FEM model of the ship's bow consists of 304 triangular elements. The ship is treated as a rigid body sailing in a straight line at a speed of 3 kn. The ice constitutive model is elastic-brittle, as described in Section 5.1.1. The ice sheet is a rectangle whose edges are fixed, except the one interacting with the icebreaker. The parameters of the ice sheet are presented in Table 3, and the critical bond stretch is calculated using Equation (18). The level ice is modeled using the proposed PD-FEM coupling approach, with PD and FEM subregions. In the PD subregion (width = 40 m, length = 70 m), the grid spacing is $\Delta x = 0.25$ m and the horizon radius is $\delta = 3\Delta x$. In the interface between the PD and FEM subregions, three layers of peridynamics nodes are embedded in each edge. Therefore, there are approximately 187,840 PD nodes. The FEM subregion is discretized into 1800 hexahedral elements of size $2 \times 2 \times 1$ m and 3936 nodes. The size of time step is $\Delta t = 1.0 \times 10^{-7}$ s, which is satisfies with the stability time step condition. The total time steps are 300,000,000 steps.

5.1.5. Numerical Result and Discussion

The numerical simulation of the ice-breaking process of an icebreaker navigating through level ice is illustrated in Figure 11. Xuelong model test is performed in the ice mechanics and ice engineering laboratory of Tianjin University to observe the failure model of level ice and the motion of broken ice. After the test is finished, Xuelong model is dragged to slowly retreat by the main trailer, and then a more complete ice breaking area can be shown on the water surface.

Table 2. Main parameters of icebreaker.

Parameter	Variable	Value
Ship length	L	166.0 m
Ship breadth	B	22.6 m
Ship depth	D	13.5 m
Bow length	l	29.6 m
Bow breadth	b	22.6 m
Draft	T	8.0 m
Stem angle	α	20°
Flooding angle	β	24°
Ship-ice friction coefficient	μ	0.15

Table 3. Parameters of level ice.

Parameter	Variable	Value
Young's modulus	E	6.83 GPa
Poisson's ratio	ν	0.25
Bending strength	σ_f	2.96 MPa
Fracture toughness	K_I	115 $kNm^{-3/2}$
Density	ρ	894 kg/m^3
Area	A	100×100 m^2
Thickness	h	1.0 m

When the icebreaker first contacts the ice, the ice is subjected to a compressive force from the bow tip, mainly along the x-axis. Once the force is greater than the ice critical strength, ice particles fall from the level ice, and a notch like the tip of the bow is formed (see Figure 11a). As the ship moves, the notch expands. The three-dimensional curved hull makes the ice bear the force in three directions. The force along the y-axis causes the ice to tear, and the notch along the profile forms a radial crack, as shown in Figure 11b.

As the contact area increases, the force along the z-axis exceeds the ice critical strength, resulting in ice sheet bending deformation and failure. Accordingly, circular cracks form, as shown in Figure 11c; these are like the circumferential cracks in model test, as shown in Figure 12.

With the ship continuous movement, the contact force increases in all three directions and radial cracks and circular cracks propagate, which is in good agreement with the ice sheet failure model observed in Xuelong model test as shown in Figure 12. The front segments of the level ice form a wedge shape, as shown in Figure 11d. The further movement of the ship causes wedge ice to form on the two sides of the bow, like the real phenomenon observed in Xuelong model test. Subsequently, ice blocks fall off, flip, push away, and pile up, as shown in Figure 11g,h.

Numerical simulation results in the generation and propagation of radial and circular cracks, as well as phenomena such as the shedding of wedge ice, flipping of brash ice, and cleaning of the channel, which are in broad agreement with experimental and real phenomena. In addition, the coupling method of finite element and perdynamics can effectively suppress the boundary effect when the level ice is failure, compared with bond-based peridynamics [28–37].

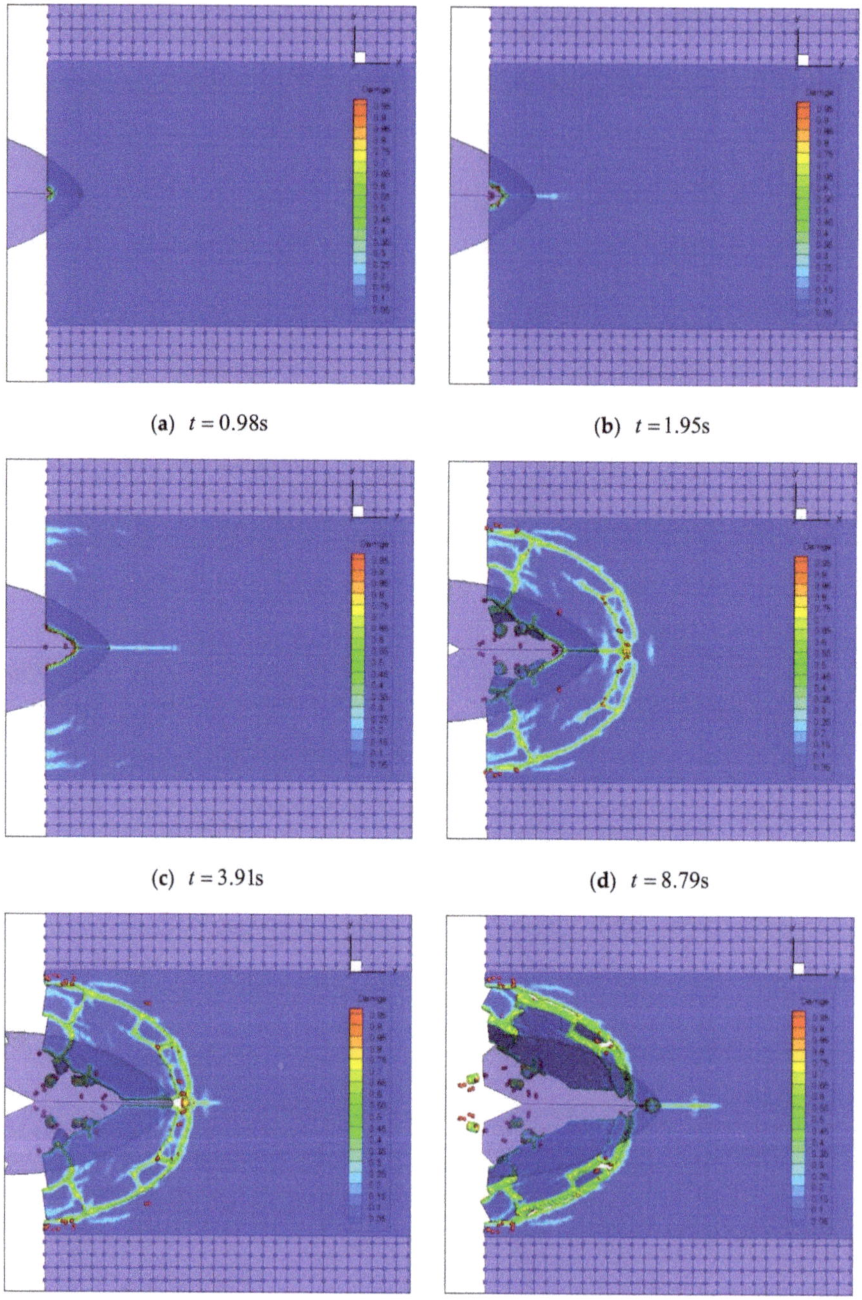

(**a**) $t = 0.98$s (**b**) $t = 1.95$s

(**c**) $t = 3.91$s (**d**) $t = 8.79$s

(**e**) $t = 10.74$s (**f**) $t = 15.63$s

Figure 11. *Cont.*

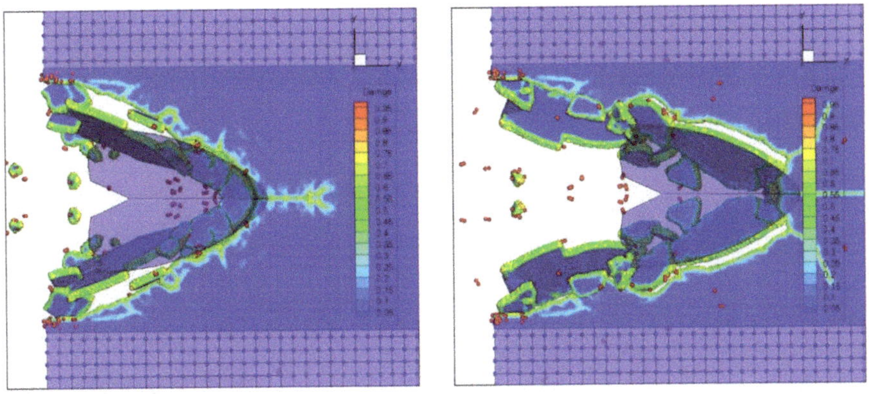

(g) $t = 20.51$s (h) $t = 29.30$s

Figure 11. Ice-breaking process of icebreaker navigating in level ice simulated by PD-FEM coupling model.

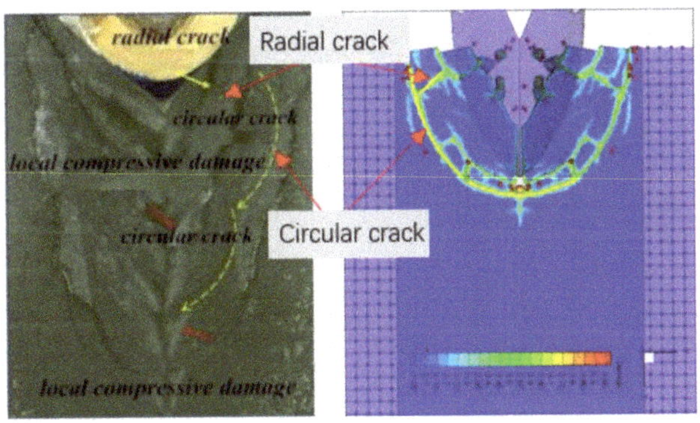

Figure 12. Picture of level ice failure in the model test of Xuelong model at Tianjin University.

To show the computational efficiency of PD-FEM coupling approach, the PD solution is considered. Except the level ice is completely modeled by PD solution, loading conditions are the same as the PD-FEM coupling approach. Table 4 shows the comparison results of PD and PD-FEM coupling method in computational time. Both two methods are compiled using Fortran 90, and use CPU_TIME function in Fortran language to calculate the program running time. In the case of the same equipment condition and 300,000 steps of the calculation time steps, the PD method needs calculate 135.47 h, while the PD-FEM coupling approach only needs 52.32 h. Compared with PD model, PD-FEM coupling model is 2.59 times more efficient and reduces the calculation time.

Table 4. Comparison of PD and PD-FEM coupling approach in computational efficiency.

Item	PD-FEM	PD
Particle number	187,840	654,400
Element number	1800	0
Total time steps	300,000	300,000
Total CPU time	52.32 h	135.47 h

5.2. Influence of Ship Speed on Ice Load

To further validate sailing speed influence on the ice load, six simulation sets are selected as speeds with 2, 3, 4, 5, 6 and 7 kn with an ice thickness of 1 m. The ice load results are shown in Figure 13. The average ice force during the period of ship-ice interaction is calculated and compared with Lindqvist's empirical formula as shown in Figure 14.

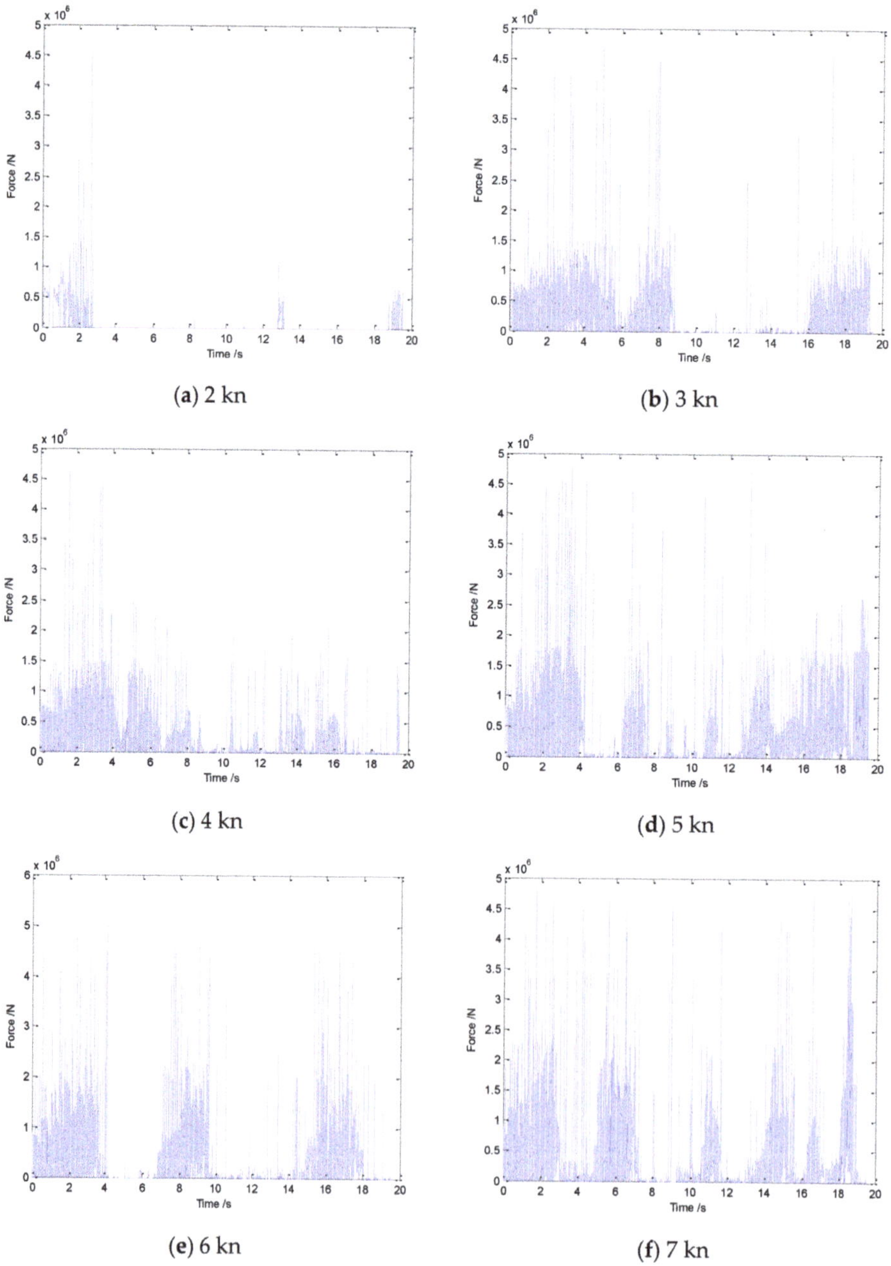

Figure 13. Time history of ice load with different ship velocity.

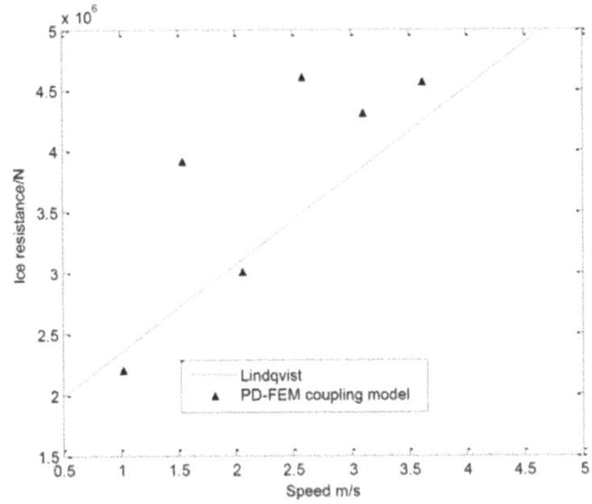

Figure 14. Ice resistance obtained by Lindqvist formula and simulated with FEM-PD coupling model for different speeds.

Lindqvist's empirical formula divides the icebreaking resistance into crushing at the stem, breaking by bending and submersion resistance. When ice is broken, the crushed ice can be cleared from both sides of the ship, and the opened channel will be larger than the ship's width. Therefore, the ice crushing at the stem and breaking by bending parts of ice resistance are mainly caused by the bow. Therefore, the ice load obtained from numerical simulation is compared with the crushing at the stem and breaking by bending resistance calculated by Lindqvist's empirical formula.

Figure 13 shows that the ice load curves (blue solid lines) are periodic. When the icebreaker interacts with the level ice, the ice load increases, but when wedge ice falls from the level ice and the icebreaker is not in contact with the ice, the ice load decreases. This is a reason for impulsive ice loads. Figure 13 also indicates that, as the ship speed increases, the period of the ice load becomes shorter and the ship velocity influences the peak of ice load. The ice load shows a rise trend as the velocity increases. From Figure 14, although there are some differences between the mean ice load results and those obtained from Lindqvist's empirical formula, they are generally in a good agreement (see Table 5 and Figure 14).

Table 5. Ice resistance calculated by PD-FEM coupling model and Lindqvist empirical formula.

Method	2 kn	3 kn	4 kn	5 kn	6 kn	7 kn
Lindqvist	2.370×10^6 N	2.743×10^6 N	3.120×10^6 N	3.490×10^6 N	3.863×10^6 N	4.237×10^6 N
PD-FEM	2.202×10^6 N	3.921×10^6 N	3.008×10^6 N	4.612×10^6 N	4.307×10^6 N	4.560×10^6 N
Errors	7.1%	42.9%	3.6%	32.1%	11.5%	7.6%

5.3. Influence of Ice Thickness on Ice Load

Numerical simulations of an icebreaker moving at a fixed speed are performed with different ice thicknesses. The ice thickness is selected as 0.5 m, 0.75 m, and 1 m, and the ship's speed is set to 3 kn.

The results of the ice load for different ice thicknesses are shown in Figure 15. It can be found that ice load presents an increasing trend in general. The mean ice load is 0.978 MN, 1.569 MN and 3.921 MN at ice thicknesses of 0.5 m, 0.75 m and 1.0 m, respectively. From Figure 16, we can find that the numerical results are in well agreement with those from Lindqvist's empirical formula.

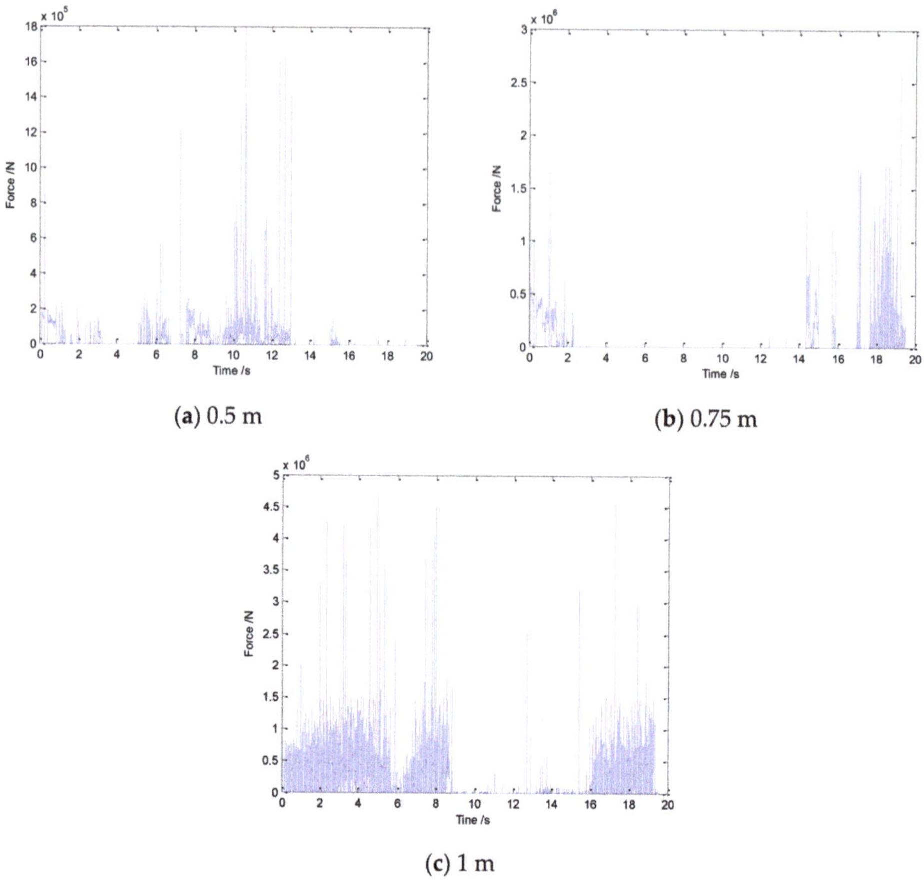

Figure 15. Time history of ice load with different ice thicknesses.

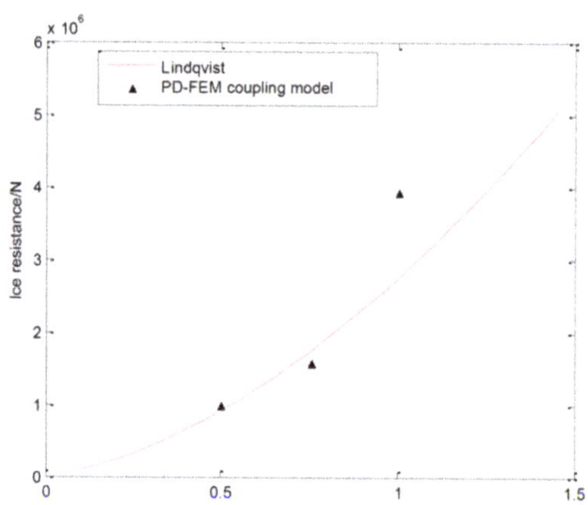

Figure 16. Ice resistance obtain by Lindqvist formula and FEM-PD coupling model for different ice thicknesses.

6. Conclusions

The coupling model of peridynamics with the finite element method is employed to simulate ship–ice interaction. The characteristics of the ice-breaking scenarios and the ice load are captured successfully. From the simulation results, the following conclusions can be drawn.

(1) The PD-FEM coupling model can successfully simulate the generation and propagation of radial and circular cracks in level ice, as well as the phenomena of wedge ice shedding, broken ice flipping, and ice cleaning of the channel during the ice-breaking process.
(2) Compared with bond-based peridynamics, the PD-FEM coupling model has better computational efficiency, and can effectively suppress the boundary effect when the level ice is failure.
(3) The ice load obtained from the PD-FEM coupling model is in good agreement with that obtained from Lindqvist's empirical formula.

Author Contributions: Conceptualization, Y.X., R.L. and X.L.; methodology, Y.X. and X.L.; program, R.L. and X.L.; validation, Y.X., R.L. and X.L.; investigation, R.L. and X.L.; writing—original draft preparation, Y.X. and R.L.; writing—review and editing, Y.X., R.L. and X.L.; supervision, Y.X. All authors have read and agreed to the published version of the manuscript.

Funding: This research was funded by the National Natural Science Foundation of China, Grant No. 51979056.

Institutional Review Board Statement: Not applicable.

Informed Consent Statement: Not applicable.

Data Availability Statement: The data presented in this study are available on request from the corresponding author. The data are not publicly available due to privacy.

Conflicts of Interest: The authors declare no conflict of interest.

References

1. Xue, Y.; Liu, R.; Li, Z.; Han, D. A review for numerical simulation methods of ship—Ice interaction. *Ocean Eng.* **2020**, *215*, 107853. [CrossRef]
2. Kujala, P.; Goerlandt, F.; Way, B.; Smith, D.; Yang, M.; Khan, F.; Veitch, B. Review of risk-based design for ice-class ships. *Mar. Struct.* **2019**, *63*, 181–195. [CrossRef]
3. Lubbad, R.; Løset, S. A numerical model for real-time simulation of ship—Ice interaction. *Cold Reg. Sci. Technol.* **2011**, *65*, 111–127. [CrossRef]
4. Raza, N.; Van Der Berg, M.; Lu, W.; Lubbad, R. Analysis of oden icebreaker performance in level ice using simulator for arctic marine structures (SAMS). In Proceedings of the International Conference on Port and Ocean Engineering under Arctic Conditions, Delft, The Netherlands, 9–13 June 2019.
5. Yu, Z.; Lu, W.; Van Den Berg, M.; Amdahl, J.; Løset, S. Glacial ice impacts: Part II: Damage assessment and ice-structure interactions in accidental limit states (ALS). *Mar. Struct.* **2021**, *75*, 102889. [CrossRef]
6. Sodhi, D.S.; Morris, C.E. Characteristic frequency of force variations in continuous crushing of sheet ice against rigid cylindrical structures. *Cold Reg. Sci. Technol.* **1986**, *12*, 1–12. [CrossRef]
7. Määttänen, M.; Marjavaara, P.; Saarinen, S.; Laakso, M. Ice crushing tests with variable structural flexibility. *Cold Reg. Sci. Technol.* **2011**, *67*, 120–128. [CrossRef]
8. Sun, J.; Huang, Y. Investigations on the ship-ice impact: Part 1. Experimental methodologies. *Mar. Struct.* **2020**, *72*, 102772. [CrossRef]
9. Sun, J.; Huang, Y. Investigations on the ship-ice impact: Part 2. spatial and temporal variations of ice load. *Ocean Eng.* **2021**, *240*, 109686. [CrossRef]
10. Li, F.; Kõrgesaar, M.; Kujala, P.; Goerlandt, F. Finite element based meta-modeling of ship-ice interaction at shoulder and midship areas for ship performance simulation. *Mar. Struct.* **2020**, *71*, 102736. [CrossRef]
11. Jou, O.; Celigueta, M.A.; Latorre, S.; Arrufat, F.; Oñate, E. A bonded discrete element method for modeling ship—Ice interactions in broken and unbroken sea ice fields. *Comput. Part. Mech.* **2019**, *6*, 739–765. [CrossRef]
12. Aksnes, V. A simplified interaction model for moored ships in level ice. *Cold Reg. Sci. Technol.* **2010**, *63*, 29–39. [CrossRef]
13. Huang, L.; Tuhkuri, J.; Igrec, B.; Li, M.; Stagonas, D.; Toffoli, A.; Cardiff, P.; Thomas, G. Ship resistance when operating in floating ice floes: A combined CFD&DEM approach. *Mar. Struct.* **2020**, *74*, 102817.

14. Luo, W.; Jiang, D.; Wu, T.; Guo, C.; Wang, C.; Deng, R.; Dai, S. Numerical simulation of an ice-strengthened bulk carrier in brash ice channel. *Ocean Eng.* **2020**, *196*, 106830. [CrossRef]
15. Kim, J.-H.; Kim, Y.; Kim, H.-S.; Jeong, S.-Y. Numerical simulation of ice impacts on ship hulls in broken ice fields. *Ocean Eng.* **2019**, *182*, 211–221. [CrossRef]
16. Ni, B.-Y.; Chen, Z.-W.; Zhong, K.; Li, X.-A.; Xue, Y.-Z. Numerical simulation of a polar ship moving in level ice based on a one-way coupling method. *J. Mar. Sci. Eng.* **2020**, *8*, 692. [CrossRef]
17. Wang, B.; Yu, H.-C.; Basu, R. Ship and ice collision modeling and strength evaluation of LNG ship structure. In Proceedings of the International Conference on Offshore Mechanics and Arctic Engineering, Estoril, Portugal, 15–20 June 2008; pp. 911–918.
18. Herrnring, H.; Ehlers, S. A finite element model for compressive ice loads based on a Mohr-Coulomb material and the node splitting technique. *J. Offshore Mech. Arct. Eng.* **2022**, *144*, 021601. [CrossRef]
19. Liu, L.; Ji, S. Comparison of sphere-based and dilated-polyhedron-based discrete element methods for the analysis of ship—Ice interactions in level ice. *Ocean Eng.* **2022**, *244*, 110364. [CrossRef]
20. Tang, X.; Zou, M.; Zou, Z.; Li, Z.; Zou, L. A parametric study on the ice resistance of a ship sailing in pack ice based on CFD-DEM method. *Ocean Eng.* **2022**, *265*, 112563. [CrossRef]
21. Zhang, J.; Zhang, Y.; Shang, Y.; Jin, Q.; Zhang, L. CFD-DEM based full-scale ship-ice interaction research under FSICR ice condition in restricted brash ice channel. *Cold Reg. Sci. Technol.* **2022**, *194*, 103454. [CrossRef]
22. Zou, M.; Tang, X.-J.; Zou, L.; Zou, Z.-J. Numerical Simulation of Ship-Ice Interaction in Pack Ice Area Based on CFD-DEM Coupling Method. In Proceedings of the International Conference on Offshore Mechanics and Arctic Engineering, Hamburg, Germany, 5–10 June 2022; p. V006T007A027.
23. Zhang, N.; Zheng, X.; Ma, Q.; Hu, Z. A numerical study on ice failure process and ice-ship interactions by Smoothed Particle Hydrodynamics. *Int. J. Nav. Archit. Ocean Eng.* **2019**, *11*, 796–808. [CrossRef]
24. Chen, Z.; He, Y.; Gu, Y.; Su, B.; Ren, Y.; Liu, Y. A novel method for numerical simulation of the interaction between level ice and marine structures. *J. Mar. Sci. Technol.* **2021**, *26*, 1170–1183. [CrossRef]
25. Herrnring, H.; Kellner, L.; Kubiczek, J.M.; Ehlers, S. Simulation of Ice-Structure Interaction with CZM-Elements. In Proceedings of the 18th German LS-Dyna Forum, Bamberg, Germany, 24 October 2018; pp. 15–17.
26. Zhang, J.; Liu, Z.; Ong, M.C.; Tang, W. Numerical simulations of the sliding impact between an ice floe and a ship hull structure in ABAQUS. *Eng. Struct.* **2022**, *273*, 115057. [CrossRef]
27. Silling, S.A. Reformulation of elasticity theory for discontinuities and long-range forces. *J. Mech. Phys. Solids* **2000**, *48*, 175–209. [CrossRef]
28. Xue, Y.; Liu, R.; Liu, Y.; Zeng, L.; Han, D. Numerical simulations of the ice load of a ship navigating in level ice using peridynamics. *Comput. Model. Eng.* **2019**, *121*, 523–550. [CrossRef]
29. Yuan, Z.; Longbin, T.; Chao, W.; Liyu, Y.; Chunyu, G. Numerical study on dynamic icebreaking process of an icebreaker by ordinary state-based peridynamics and continuous contact detection algorithm. *Ocean Eng.* **2021**, *233*, 109148. [CrossRef]
30. Liu, R.; Xue, Y.; Lu, X.; Cheng, W. Simulation of ship navigation in ice rubble based on peridynamics. *Ocean Eng.* **2018**, *148*, 286–298. [CrossRef]
31. Liu, R.; Yan, J.; Li, S. Modeling and simulation of ice—Water interactions by coupling peridynamics with updated Lagrangian particle hydrodynamics. *Comput. Part. Mech.* **2020**, *7*, 241–255. [CrossRef]
32. Ye, L.; Guo, C.; Wang, C.; Wang, C.; Chang, X. Peridynamic solution for submarine surfacing through ice. *Ships Offshore Struct.* **2020**, *15*, 535–549. [CrossRef]
33. Liu, R.; Xue, Y.; Han, D.; Ni, B. Studies on model-scale ice using micro-potential-based peridynamics. *Ocean Eng.* **2021**, *221*, 108504. [CrossRef]
34. Vazic, B.; Oterkus, E.; Oterkus, S. In-plane and out-of plane failure of an ice sheet using peridynamics. *J. Mech.* **2020**, *36*, 265–271. [CrossRef]
35. Chunyu, G.; Kang, H.; Chao, W.; Liyu, Y.; Zeping, W. Numerical modelling of the dynamic ice-milling process and structural response of a propeller blade profile with state-based peridynamics. *Ocean Eng.* **2022**, *264*, 112457. [CrossRef]
36. Vazic, B.; Oterkus, E.; Oterkus, S. Peridynamic approach for modelling ice-structure interactions. *Trends Anal. Des. Mar. Struct.* **2019**, 55–60.
37. Song, Y.; Yan, J.; Li, S.; Kang, Z. Peridynamic modeling and simulation of ice craters by impact. *Comput. Model. Eng. Sci.* **2019**, *121*, 465–492. [CrossRef]
38. Macek, R.W.; Silling, S.A. Peridynamics via finite element analysis. *Finite Elem. Anal. Des.* **2007**, *43*, 1169–1178. [CrossRef]
39. Liu, W.; Hong, J.-W. A coupling approach of discretized peridynamics with finite element method. *Comput. Method Appl. M.* **2012**, *245*, 163–175. [CrossRef]
40. Lee, J.; Liu, W.; Hong, J.-W. Impact fracture analysis enhanced by contact of peridynamic and finite element formulations. *Int. J. Impact Eng.* **2016**, *87*, 108–119. [CrossRef]
41. Silling, S.A.; Askari, E. A meshfree method based on the peridynamic model of solid mechanics. *Comput. Struct.* **2005**, *83*, 1526–1535. [CrossRef]
42. Madenci, E.; Oterkus, E. Peridynamic theory. In *Peridynamic Theory and Its Applications*; Springer: Berlin/Heidelberg, Germany, 2014; pp. 19–43.

43. Kilic, B.; Madenci, E. An adaptive dynamic relaxation method for quasi-static simulations using the peridynamic theory. *Theor. Appl. Fract. Mech.* **2010**, *53*, 194–204. [CrossRef]
44. Knott, J.F. *Fundamentals of Fracture Mechanics*; Gruppo Italiano Frattura, 1973.
45. Schulson, E.M. Brittle failure of ice. *Eng. Fract. Mech.* **2001**, *68*, 1839–1887. [CrossRef]

Journal of
*Marine Science
and Engineering*

MDPI

Article

On the Resistance of Cruciform Structures during Ship Collision and Grounding

Hewei Liu, Kun Liu *, Xiufei Wang, Zhenguo Gao and Jiaxia Wang

School of Naval Architecture and Ocean Engineering, Jiangsu University of Science and Technology, Zhenjiang 212003, China
* Correspondence: kunliu@just.edu.cn; Tel.: +86-135-1169-2085; Fax: +86-0511-8444-6543

Abstract: This paper presents an experimental, numerical, and analytical study of a novel specimen subjected to local in-plane load, to investigate its crushing deformation and resistance. The specimen was designed and fabricated to simplify the cruciform structure in double-hulled vessels subjected to external loads during collision and grounding incidents. The study results will provide reliable insights into grounding scenarios as well as side collision scenarios of double-hulled vessels. A quasi-static indentation test and related numerical research showed good agreement regarding deformation mode and force–displacement response. On the basis of the experimental and numerical results, an analytical method is proposed to derive the deformation energy, the instantaneous resistance, and the mean resistance of the deformed structure. The analytical method was verified with recorded test data and a nonlinear finite element analysis. It enables a rapid assessment of the response of the structure under accidental loads, which is a guideline for the design of crashworthy hull structures and the assessment of their crashworthiness.

Keywords: cruciform; ship's collision and grounding; model experiment; numerical simulation; analytical method

Citation: Liu, H.; Liu, K.; Wang, X.; Gao, Z.; Wang, J. On the Resistance of Cruciform Structures during Ship Collision and Grounding. *J. Mar. Sci. Eng.* **2023**, *11*, 459. https://doi.org/10.3390/jmse11020459

Academic Editor: Md Jahir Rizvi

Received: 29 December 2022
Revised: 1 February 2023
Accepted: 15 February 2023
Published: 20 February 2023

1. Introduction

Ship collision and grounding accidents may cause catastrophic consequences such as casualties, economic losses, and environmental pollution [1]. According to the statistics of the International Oil Tanker Owners Oil Pollution Federation (ITOPF), 62% of large spills (over 700 tonnes) recorded between 1970 and 2021 were caused by collisions and groundings [2]. In order to reduce losses caused by impact accidents, it is essential to develop procedures to assess double-hull tanker performance in collision and stranding accidents.

Cruciform structures in double hulls consist of joints between side stringers and transverse webs in a double side, as well as longitudinal girders and transverse frames in a double bottom. Figure 1 shows a typical collision and grounding scenario for a double-hull vessel: the bulbous bow of the collision ship and rock are simplified into rigid hemispheroid, and cruciform structures connected with outer panels suffer local axial compression load in the schematic. During the ship collision and grounding, the cruciform structure acts as one of the main structures resisting the deformation and absorbing the impact energy. Therefore, it is essential to investigate the deformation mechanism, energy absorption, and resistance of cruciform structures under in-plane loading.

In recent decades, a number of studies have been carried out focusing on the mechanical behavior of cruciforms, including model tests, numerical simulations, and simplified analytical methods. The experimental method has all along been the most reliable research means to assess the crashworthiness of hull structures. The test results provide initial insights and reliable data. Urban et al. [3] conducted crush experiments on full-size aluminum cruciform structures and the results indicated that the material fracture generated during deformation had a significant effect on its crushing behavior. Dramatic reductions were found in the energy dissipation of the specimen when fractures occurred. Abramowicz and Simonsen [4] carried out two series of crushing tests. In the first group, relatively thin steel specimens were used to investigate the relationship between structural energy absorption and the ratio of breadth to thickness (c/t). In the second group, the effect of absolute size on structural energy absorption was investigated by increasing the thickness of the plate while keeping c/t constant. The test results from Urban, as well as Abramowicz and Simonsen, became the benchmarks for later studies and provided reliable data. Zhou [5] carried out tests on three types of cruciform structures with different welding positions and lengths, and the results indicated that the energy absorption capacity of energy absorbers is tunable by adjusting the weld strength of components. Ghanbari Ghazijahani et al. [6] conducted tests to investigate the resistance and deformation mode of shell structures with different indentations.

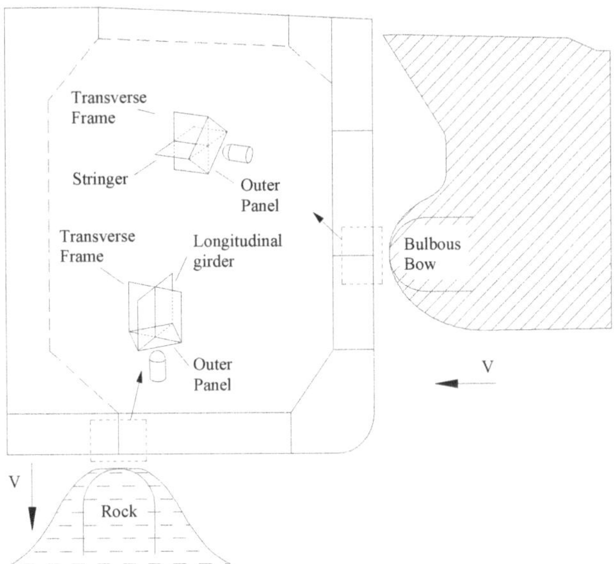

Figure 1. Typical collision and grounding scenario of double-hull amidship.

With the rapid development of high-performance computers, numerical simulation analysis is widely used to investigate the structural response of ships subjected to collision or stranding. Mohammed et al. [7] carried out numerical simulations on the basis of Simonsen's test results and the results were validated by comparison with experimental tests and DNV design code. However, material fracture and contact algorithms were not included in the simulation. Urban [8] conducted numerical simulation analyses and compared them with previous data. Various failure criteria were used in the simulation which, implemented with RTCL fracture criterion, agreed with experimental results. Haris et al. [9] also conducted numerical research based on Urban's test with LS-DYNA to validate the numerical method. Subsequently, they defined the concept of the effective cruciform width and utilized it in the crushing simulation of ship girder intersections with realistic boundary conditions. Sun et al. [10] simulated the response of the side of double-hull tanker collided at the cruciform by a rigid bow, and the numerical results were compared with a simplified method based on the study of Haris [11]. The finite element method is becoming an indispensable tool in the study of ship structures.

Generally speaking, numerical simulation calculations often require extensive modeling efforts and incur computational costs. Therefore, the analytical method remains important in structural analysis to rapidly evaluate the crashworthiness of hull structures. The simplified analytical method for crushing modeling is based on a simplified energy method called the 'kinematic method'; several scholars [12–14] have used it to investigate energy dissipation in crushed structures. Amdahl et al. [12] proposed the symmetric and asymmetric collapse modes of a cruciform structure and derived its mean crushing strength. Wierzbicki [15] studied the crushing resistance of structures consisting of different numbers of panels and proposed closed-form solutions for the mean crushing resistance of cruciforms and other basic structural elements. Hayduk and Wierzbicki [16] considered the collapse mode of cruciforms as an assembly folding mode of two basic plates, and two simplified combination fold modes were presented as upper and lower bounds of mean cruciform crushing strength. Kierkegaard et al. [17] derived a mathematical method for calculating force–indentation relations on the basis of axial crushing mechanisms for basic elements including L-, T-, and X-elements (cruciforms), and the calculated results were verified with crushing experiments of bow models provided in the previous literature [18]. Abramowicz [19] proposed formulae for calculating the axial resistance of these L-, T-, and X-shaped structures.

Table 1 lists previous research on the resistance of cruciforms. In these works, the structural responses of cruciforms subjected to flat indenters were widely studied, and the mechanical mechanisms under local indentation were less addressed. In fact, when the shape of the impacting bow or reef is more convex, the cruciform is subjected to a local compression load. This motivated us to investigate the crushing resistance and progressive deformation of cruciforms under local compress load experimentally, numerically, and analytically.

Table 1. Comparison of current study versus previous works on the resistance of cruciform structures.

Refs.	Material	Indenter	Method		
			Experimental	Analytical	Numerical
[3]	Aluminum alloy	Flat indenter	X	X	
[4]	Mild steel, aluminum alloy	Flat indenter	X		
[5]	Mild steel	Flat indenter	X		X
[7]	Mild steel	Flat indenter			X
[8]	Aluminum alloy	Flat indenter			X
[9]	Aluminum alloy	Flat and ellipsoidal indenter		X	X
[16]	Aluminum alloy	Flat indenter	X	X	
This paper	Mild steel	Hemisphere indenter	X	X	X

This paper investigates the mechanical response of cruciforms under local indentation loads. The specimen was designed and fabricated as a simplified structure of cruciform intersections in double sides or double bottoms. The analysis of the specimen was conducted via a quasi-static indentation test, nonlinear finite element simulations, and an analytical method. The details of the test and numerical simulation are provided herein, with good agreement between the recorded test data and the numerical results. In addition, a simplified analytical method was proposed to rapidly assess the instantaneous and mean crushing resistance of the cruciform structures in a double-hull structure in the event of a collision with a hemispherical object (for example, a seabed rock or a bulbous bow). The analysis was validated based on the force–displacement response of the test and the simulations, Which means that this newly proposed analytical method is an effective tool for the assessment of the crashworthiness of cruciforms in grounding or collision scenarios.

2. Experimental Details

2.1. Test Specimen

To investigate the crashworthiness of cruciform structures, a specimen was designed and manufactured. Figure 2 shows the schematic of the specimen consisting of a cruciform, a surface plate, and external frames. The specimen is simplified from a typical intersection structure on a double-hull vessel. Four 4 mm thick component panels were welded into the cruciform, and a 4 mm thick surface plate representing the outer hull panels was welded to the cruciform. The simplified hull structure was welded to the inside of a 10 mm thick support frame structure. The upper part of the external frame was reinforced with channel steels all around. The edges of the surface plate were joined to the channel steels with welded seams and the whole specimen was constructed from mild steel. The specimen was painted silver–grey and grids (20 × 20 mm) were drawn on the cruciform and surface panel so that the extent of the deformation of the specimen could be easily observed.

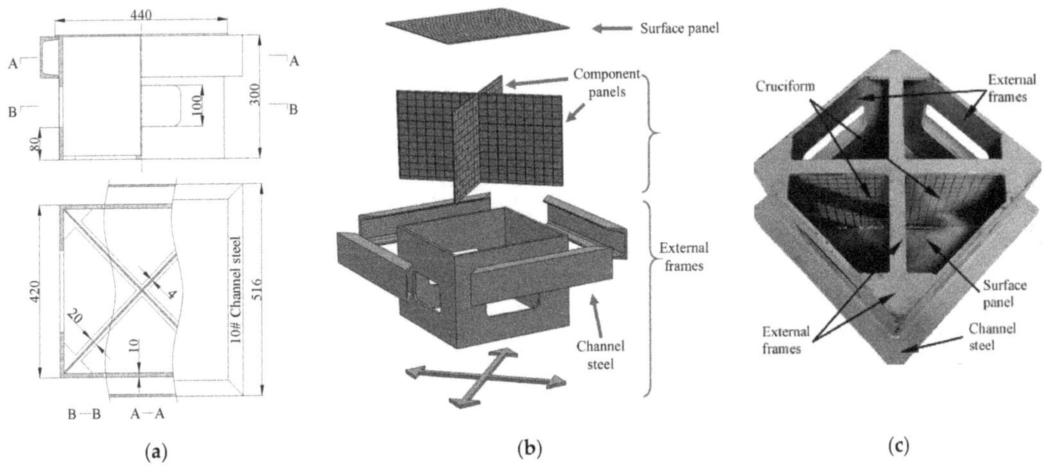

Figure 2. (**a**) The geometric dimension, (**b**) the exploded view, and (**c**) the bottom view of the specimen.

2.2. Quasi-Static Indentation Test

Since collisions and groundings of ships often occur at low speeds, as a simplified approach, quasi-static tests are applied to investigate the mechanical response of hull structures under impact [20,21]. In addition, quasi-static tests allow the continuous recording of test data during the damage process, which allows the proposal of simplified methods for the analysis of structural damage and deformation [22]. In order to investigate the deformation of the hull structure in collisions with a small rock or a bow, a hemispherical indenter with a radius of 75 mm was chosen for the test. The indenter was produced from hardened GCr15 that would not deform during loading.

The test setup is presented in Figure 3a. The indentation was executed with an Electro-hydraulic servo universal testing machine (YNS1000, CRIMS, Changchun, China) with a 1000 kN maximum load. The central axis of the specimen should be aligned with the central axis of the indenter to ensure in-plane loading of the cruciform. Before the test, the indenter was lowered until it contacted the specimen, followed by clearing the load to zero. During the test, the indenter was forced down at a constant velocity of 10 mm/min to ensure the quasi-static loading condition. The displacement and load data were recorded using the testing machine with a sampling frequency of 20 Hz.

2.3. Tensile Test

Before the indentation test, quasi-static tensile tests were performed to obtain the mechanical properties of the material. Tensile specimens were manufactured with the same material as the cruciform specimen. The tensile tests were carried out with a universal testing machine (CMT5305, MTS, Shanghai, China). The fabrication of the specimens and fulfilment of the tests were executed in accordance with standard protocols [23] and are shown in Figure 3b,c and Figure 4a. The specimen was stretched at a constant velocity of 1 mm/min. To reduce the effects of errors, the tests were carried out three times. The elongation and load were recorded with an extensometer and a force transducer, respectively, during the test. The engineering curves are plotted in Figure 4b and satisfactory repeatability is observed. The mechanical properties of the mild steel listed in Table 2 are extracted from the engineering curve. The density and the Poisson's ratio were obtained from the literature [24].

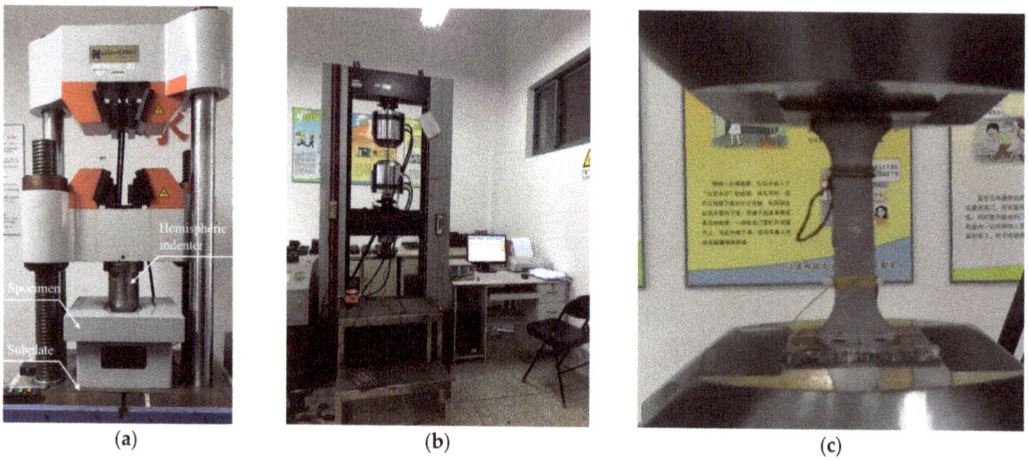

Figure 3. (**a**) The quasi-static indentation test setup, (**b**) the universal testing machine, and (**c**) the photo of tensile test.

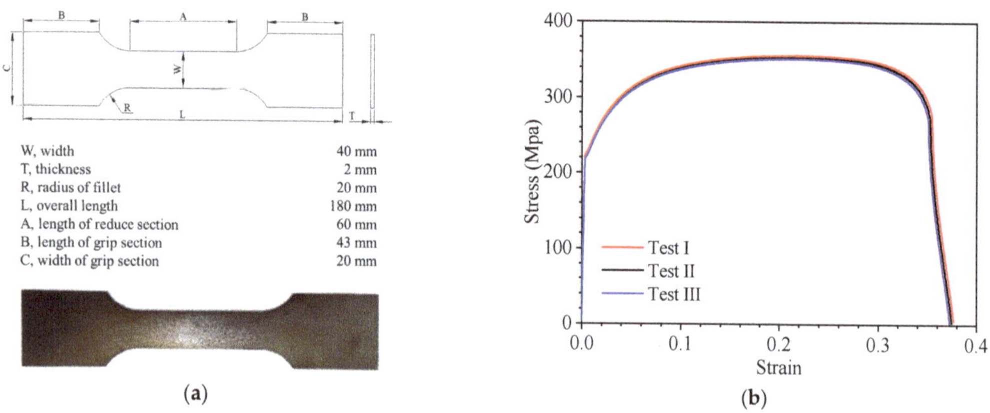

Figure 4. (**a**) Schematic of tensile specimen; (**b**) engineering stress–strain curves of three tests.

Table 2. Mechanical properties of material.

Properties	Unit	
Density	kg/m^3	7850
Young's modulus	GPa	213
Poisson's ratio	-	0.3
Yield stress	MPa	221
Ultimate tensile strength	MPa	353
Rupture strain	-	0.35

3. Numerical Simulation

Based on the structural dimensions in Figure 2, the finite element model was established using the nonlinear finite element software, Abaqus (Version 2018) [25], as shown in Figure 5a. The specimen was modeled with four-node reduced integration shell elements (S4R) from the Abaqus element library. The thickness variation of the channel profile was omitted and the equivalent thickness of the channel steel was used. The equivalent thickness of elements on both sides and in the middle of the channel steel equated to 8.5 mm and 5.3 mm, respectively. The stiffness of the indenter was large enough to ignore its deformation, so the indenter was set as a rigid body in the simulation. In addition, a fixed rigid subplate was placed under the specimen to simulate the test setup.

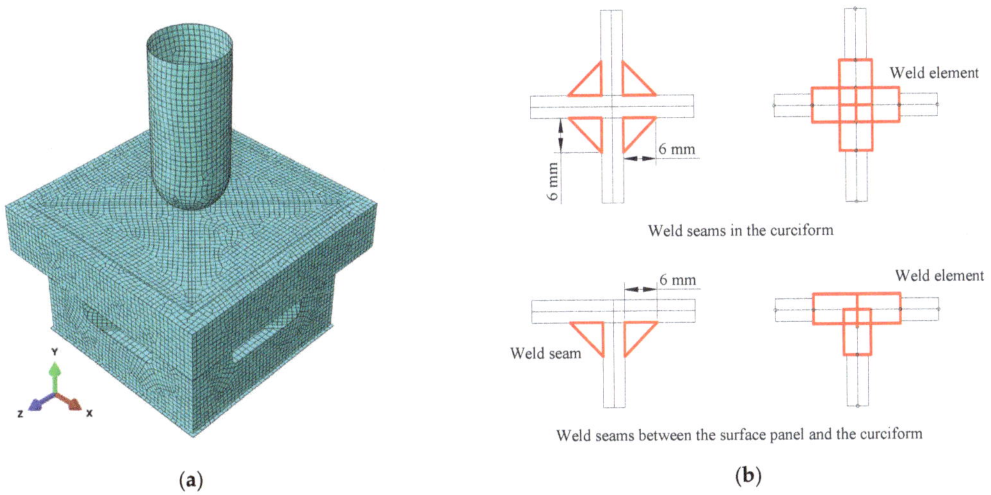

Figure 5. (**a**) Details of the finite element model; (**b**) weld elements utilized in different parts of the specimen.

Due to the specimen's small size, the reinforcement of the welds on the structure could not be ignored in the simulation. Our approach was to increase the plate thickness at the weld seams [26]. Weld elements were used on the weld seams in the cruciform, as well as those between the surface panel and the cruciform, as shown in Figure 5b. The leg width of weld seams varied from 5 mm to 7 mm. In the simulation, the width of weld elements was set to 6 mm uniformly, and the sectional area of the weld was added to the thickness of the panel on both sides. After a trial calculation, the simulation results were ideal when the thickness of both sides of the plate were increased by 1 mm. In other words, the thickness of the welding elements between the surface panel and the cruciform and inside the cross member were set to 5 mm and 6 mm, respectively. The welds at the connection between the surface plate and the channel steels were simplified with a 'tie' restraint, while other welds in other non-key areas were simplified with joint connections.

The mesh size of rigid indenter was set to 10 mm. Besides the welds and the indenter, coarse, medium, and fine elements with sizes of 15 mm, 10 mm, and 5 mm, respectively, were used in the other part of the specimen. Figure 6 shows the force–displacement curve of the three different models. Considering the calculation accuracy and efficiency, a 10 mm mesh size was selected for the current work.

The mechanical properties obtained from the tensile test was implemented in the simulation. The 'combined material' proposed by Villavicencio and Guedes Soares [27] was used to obtain the true material relationship, where the curve can be divided into three parts: part I is the true strain–stress curve before the maximum load, part II is the true strain–stress curve beyond the maximum load, and these two parts are connected with a short, straight line. Figure 7 shows the engineering and the true strain–stress curves of the material. Before the maximum load, the true stress σ_t and the true strain ε_t can be represented by the engineering stress σ_e and the engineering strain ε_e in terms of Equations (1) and (2) [28].

$$\sigma_t = \sigma_e(1 + \varepsilon_e) \tag{1}$$

$$\varepsilon_t = \ln(1 + \varepsilon_e) \tag{2}$$

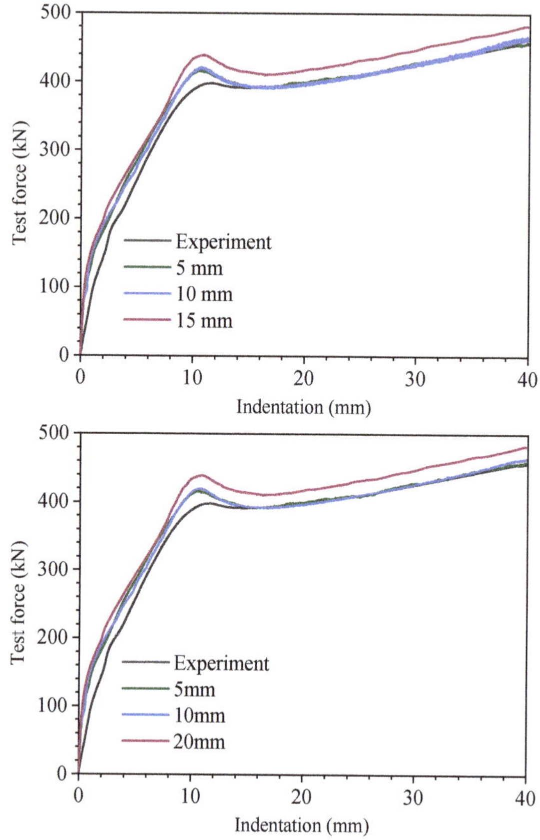

Figure 6. Force–displacement curves for coarse, medium, and fine meshed models.

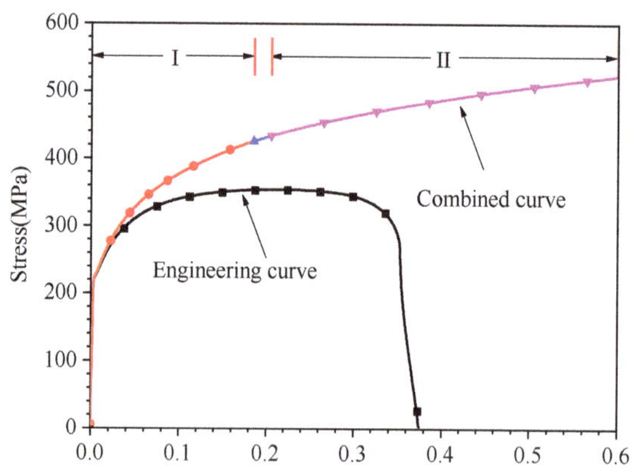

Figure 7. The combined material relationship of mild steel.

Meanwhile, the true curve beyond the maximum point can be denoted as Equation (3) [29]

$$\sigma_t = C\varepsilon_t^n \tag{3}$$

for

$$C = R_m(e/n)^n \tag{4}$$

$$n = \ln(1 + A_g) \tag{5}$$

where A_g is the maximum uniform strain related to the ultimate tensile stress R_m and e is the natural logarithmic constant. If only the ultimate stress R_m (MPa) is available, the following approximation can be used to obtain a proper A_g as

$$A_g = 1/(0.24 + 0.01395R_m) \tag{6}$$

Taking into account the computational time and efficiency, the indenter was loaded at a constant velocity of 0.01 m/s in the simulation [30]. The indenter was allowed to move in the y direction only. The interaction between the indenter and the specimen was defined by general contact, where the normal behavior was set to hard contact and the tangential behavior was set to the penalty function method with the friction coefficient set to 0.3 [30].

4. Comparative Analysis between Test and Simulation

The force–displacement responses and the deformation in the test and nonlinear finite element simulation were investigated and compared.

4.1. Force-Displacement Curve

Figure 8 shows the force–displacement curves obtained from the test and the simulation, where the whole compression process can be divided into three stages.

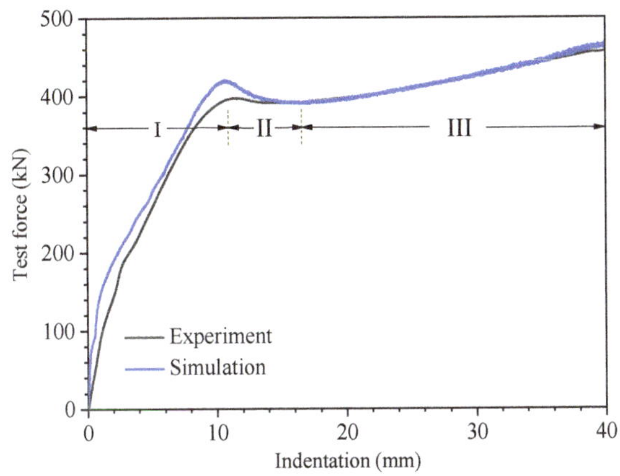

Figure 8. Comparison of crushing response between experiment and simulation.

At the beginning of stage I, the initial contact between the indenter and the specimen occurred, and indentation was generated in the local contact region. With the indenter moving downwards, the indentation region expanded, resulting in a rapid rise of the test force. The slope of the force–displacement curve in stage I appeared to vary slightly at first and then gradually decreased. As the indenter moved continually, the upper part of the cruciform yielded gradually and the nonlinear plastic property resulted in variation in the curve slope.

The test results were slightly lower than the simulation results in stage I because the whole test system was not fully fitted in the initial stage of the test. The indenter depth matched well between the simulation and the test once the peak resistance was reached. The maximum values of the test and simulated loads were 397.51 kN and 421.12 kN, respectively, with an error of 5.9%.

In stage II, the instability of the upper part of the cruciform structure led to a reduction in structural resistance. The reduction in the simulation was greater than the experimental result and these two values agreed well by the end of stage II.

In stage III, the instability area of the cruciform expanded, and besides the bending deformation, the component panels of the cruciform produced membrane tensile deformation in their length. The angle between the membrane tensile direction and the loading direction decreased as the indenter moved further downwards. No rupture was found in the specimen during the loading process and the overall structural load-carrying capacity increased steadily. The simulation was in good agreement with the recorded data.

4.2. Deformation

Figure 9 shows the deformation of the specimen in the test and the simulation. It can be seen that the cruciform produced large plastic deformation under the local compression load. A local depression was generated in the center of the surface plate by the pressure of the hemispherical indenter. The uppermost part of the cruciform constrained to the surface plate only underwent downward translation. No significant deflection was found in the mid-axis region of the cruciform, mainly because of mutual support between the four component panels of the cruciform. From the central axis outward, the deflection of the component panels rapidly increased, existing only in the upper part of the cruciform near the contact area of the indenter. The deformation was relatively small for the regions connected with the external frames and away from the loading area.

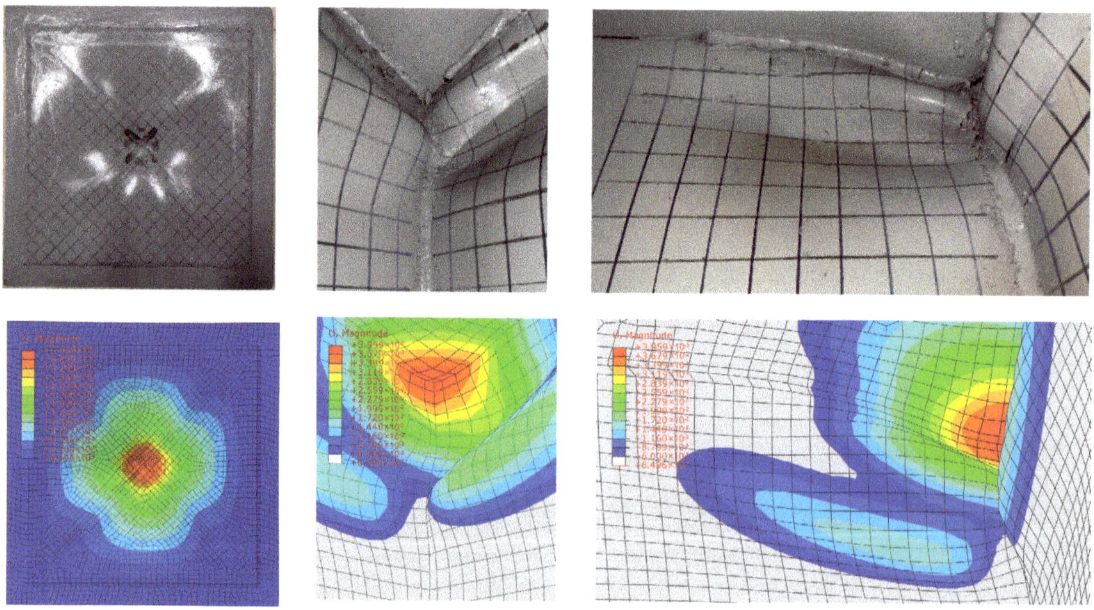

(a) (b) (c)

Figure 9. Deformation diagrams of (**a**) the surface panel, (**b**) the cruciform, and (**c**) a single component panel.

Observing Figure 9b,c it is straightforward to see that each component plate of the cruciform produced three plastic hinges that divided each plate into three small folds and a large panel at the bottom. The approximate ratio of the maximum height of each fold was 1:3:2. In addition, the deformations of the four component panels were symmetrical about the central axis, and it is straightforward to see in Figure 8b that the two adjacent panels deformed along the same rotational direction. The above deformation modes were used on the simplified deformation pattern of the cross members presented in Section 5.2.

5. Analytical Method

5.1. Upper Bound Theorem

According to the upper limit theorem, the power of work done by external forces is equal to the rate of energy dissipation within the structure, i.e.,

$$F \cdot \dot{\delta} = \dot{E}_{int} \tag{7}$$

where F is the external force, $\dot{\delta}$ is the velocity at the point applied with the external force, and $\dot{E}_{int}$ is the internal energy dissipation rate. For hull structures, the internal energy dissipation rate can be expressed as

$$\dot{E}_{int} = \int_V \sigma_{ij}\dot{\varepsilon}_{ij} dV \tag{8}$$

where σ_{ij} is the stress in the structure, $\dot{\varepsilon}_{ij}$ is the strain rate and, according to the Von Mises plastic flow theorem, the plastic energy dissipation rate is

$$\dot{E}_{int} = \int_V \sigma_0 \dot{\varepsilon}_{eq} dV \tag{9}$$

where $\dot{\varepsilon}_{eq}$ is the equivalent strain rate of the structure and σ_0 is the flow stress. The internal energy dissipation rate of a deformed structure is equal to the sum of the bending deformation energy dissipation rate and the membrane stretching deformation energy dissipation rate.

$$\dot{E}_{int} = \dot{E}_b + \dot{E}_m \tag{10}$$

The bending energy dissipation rate can be expressed as

$$\dot{E}_b = \int_A M_{\alpha\beta}\dot{k}_{\alpha\beta} dA + \sum_{i=1}^{n} M_{0,i}\dot{\theta}_i l_i,\ (\alpha, \beta = 1,2) \tag{11}$$

where $M_{\alpha\beta}$, $\dot{k}_{\alpha\beta}$, and A are the bending moment, the curvature, and the area of a plate; M_0 is the bending moment on a plastic hinge of unit length; $\dot{\theta}_i$ for the angular velocity at a plastic hinge; and l_i is the plastic hinge length. The bending deformation energy includes the continuous bending energy and plastic hinge rotational energy. In general, the energy of bending is ignored in simplified analytical methods, so Formula (12) can be derived from Formula (11).

$$\dot{E}_b = \sum_{i=1}^{n} M_{0,i}\dot{\theta}_i l_i \tag{12}$$

The membrane stretching energy dissipation rate can be expressed as

$$\dot{E}_m = \int_A N_{\alpha\beta}\dot{\varepsilon}_{\alpha\beta} dA,\ (\alpha, \beta = 1,2) \tag{13}$$

where $N_{\alpha\beta}$ is the plane membrane tensile stress tensor and $\dot{\varepsilon}_{\alpha\beta}$ is the equivalent strain rate tensor.

5.2. Deformation Mode and Analysis Method

The damage and deformation of hull structures in a collision and grounding incident are often complex. Simplified analyses of structural responses in realistic scenarios often require reasonable assumptions that can shift the deformation of the theoretical model closer to reality, which is conducive to the derivation of analytical formulae and improving the applicability of analytical methods. Based on the experimental and simulation results, the following assumptions are applied to the deformation of the cruciform structure:

- The deformation of the cruciform is divided into plastic deformation and elastic buckling zones and the dissipation of energy for elastic buckling zones is negligible. In addition, local indentation of the surface plate of the specimen is small, thus its effect on the overall structural resistance is ignored.
- It is assumed that the whole structure produces local compression deformation in the initial stages of loading. The cruciform is destabilized and folds are produced when the indentation depth reaches a threshold value.
- The deflection at the mid-axis region of the cruciform is small, and its effect on the outer part of component panels is not considered. Similarly, the deflection of the elastic buckling zones is small, therefore its effect on the deflection of the folds is neglected.
- Since the structure and the load are symmetrical about the central axis, the deformation modes of the four component panels are similar. The analytical approach in this paper assumes that the deformation of the four panels is identical and symmetrical about the central axis.
- The indenter was loaded at a low constant velocity in the test, so the effect of the material strain rate was ignored in the analytical method.

Considering the assumptions above and the deformation characteristics of the specimen, a new deformation mode for the local in-plane compression of the cruciform is proposed. The whole deformation process is divided into two stages: the initial contact stage and the fold-forming stage. The simplified deformation mode of two stages are as follows:

1. Figure 10a shows the schematic of the initial contact stage where the cruciform produces local compression deformation without folding. The area of compression is closely related to the depth of indentation and the size of the indenter.
2. Figure 10b shows the overall deformation process of a component panel of the cruciform. After the initial stage, plastic hinges and three folds are formed in the plastic deformation zone. The ratio of the maximum width of these folds from top to bottom is 1:3:2, i.e., *AB*:*BC*:*CD* = 1:3:2. The total height of the deformation zone is 6*H*. Figure 11a shows the deformation schematic of the cruciform, where the deformation of the four component panels is symmetrical about the central axis and the deformed cruciform presents a windmill shape in the top view. Figure 11b shows the fold generating process.

Based on the upper bound theorem and the simplified deformation mode, the deformation energy, the mean crushing resistance, and the instantaneous resistance of the cruciform during deformation are derived. In stage 1, it is assumed that plastic deformation in the weld location occurs at the beginning of compression. In Figure 10a, the width of the compression area in a single panel is *Lc*. It is assumed that plastic deformation in the weld location occurs at the beginning of the compression. The energy dissipation rate in the first stage is:

$$\dot{E}_{c,w,1} = 4\sigma_{0,w}t_w L_c\dot{\delta} + 2\sigma_{0,w}l^2\dot{\delta} \tag{14}$$

$$L_c = \sqrt{R^2 - (R-\delta)^2} \tag{15}$$

$$F_{w,1}(\delta) = 4\sigma_{0,w}t_w L + 2\sigma_{0,w}l^2 \tag{16}$$

where $\sigma_{0,w}$ is the flow stress; t_w denotes the thickness of the cruciform; l denotes the length of the weld, and L denotes the length of a single panel.

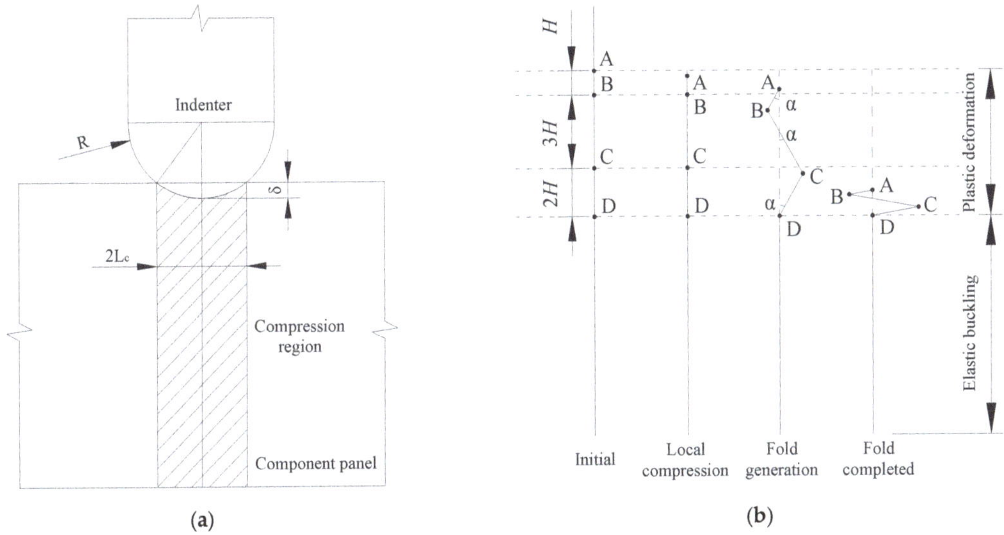

Figure 10. (**a**) Diagram of the initial contact between the indenter and the specimen; (**b**) deformation process of a component panel in the cruciform subjected to in-plane compression.

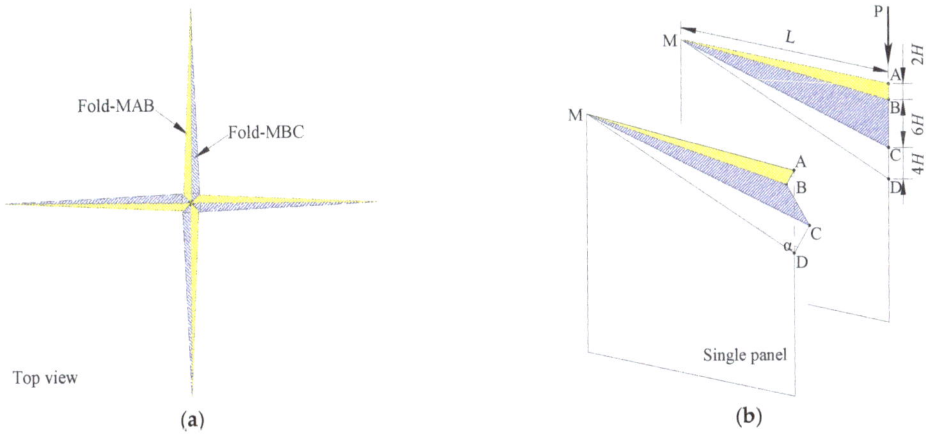

Figure 11. The deformation mode of the cruciform: (**a**) a top-view image of four symmetrical component panels; (**b**) the deformation process of a single panel.

During the formation of the folds, a single plate produces plastic deformation in three parts: *MAB*, *MBC*, and *MCD*. The fold is completed when the three parts *AB*, *BC*, and *CD* are completely flattened and the original height of the fold area is 6*H*. Assuming that the characteristic height of the fold, *H*, is significantly less than the length of the component panel, the length of the plastic hinge can be regarded as the same as the length of the panel, so that the energy dissipation rate of the plastic hinge in a single panel at this stage is:

$$\dot{E}_{b,w,2} = 6M_{0,w}\dot{\alpha}L \tag{17}$$

$$M_{0,w} = \frac{\sigma_{0,w}t_w^2}{4} \tag{18}$$

where $M_{0,w}$ is the moment on a plastic hinge per unit length and $\dot{\alpha}$ is the angular speed of rotation of the plastic hinge. According to the geometric relationship in Figure 10, the relationship between the indentation depth and the plastic hinge rotation angle is obtained as follows:

$$\delta = 6H(1 - \cos\alpha) \tag{19}$$

$$\dot{\alpha} = \frac{\dot{\delta}}{6H\sqrt{1 - \left(1 - \frac{\delta}{6H}\right)^2}} \tag{20}$$

Substituting Equation (20) into Equation (17) yields the energy dissipation rate for the plastic hinge of a single panel during fold formation.

$$\dot{E}_{b,w,2} = \frac{\dot{\delta} M_{0,w} L}{H\sqrt{1 - \left(1 - \frac{\delta}{6H}\right)^2}} \tag{21}$$

The plastic hinge angle α increases from 0 to $\pi/2$, so the total energy of the plastic hinge of the single panel is

$$E_{b,w,2} = 3\pi M_{0,w} L \tag{22}$$

When the component panels are subjected to in-plane loading, the structure is deformed in membrane tension as well as plastic hinges. Simonsen et al. [31] pointed out that the least energy dissipation occurs when the panel is deformed in tension only along its length. For this reason, only deformations along the length of the panel are considered in the analysis, and the energy dissipation rate for membrane stretching is calculated.

The energy dissipation rate from structural membrane stretching is calculated as:

$$\dot{E}_{m,w,2} = \iint_S N_{0,w} \dot{\varepsilon}_w dS \tag{23}$$

$$N_{0,w} = \sigma_{0,w} t_w \tag{24}$$

where $N_{0,w}$ is the tensile force generated by a unit thickness plate during plastic flow, S represents the plastic deformation area; $\dot{\varepsilon}_w$ is the average strain rate in the length direction of the plate. The average strain at the plastic hinge MA is

$$\varepsilon_{MA} = \left(\frac{\delta}{L}\right)^2 \tag{25}$$

The corresponding strain rate is

$$\dot{\varepsilon}_{MA} = 2\frac{\delta\dot{\delta}}{L^2} \tag{26}$$

Furthermore, MD is the intersection of the plastic deformation zone and the elastic buckling zone, where the strain rate is 0. Considering that the strain rate varies linearly from point A to D, the average strain rates at the plastic hinges MB, MC, and MD are as follows

$$\dot{\varepsilon}_{MB} = \frac{5\delta\dot{\delta}}{3L^2} \tag{27}$$

$$\dot{\varepsilon}_{MC} = \frac{2\delta\dot{\delta}}{3L^2} \tag{28}$$

$$\dot{\varepsilon}_{MD} = 0 \tag{29}$$

The average strain rate for the folds was calculated to be

$$\dot{\varepsilon}_{AD,w} = \frac{\delta\dot{\delta}}{L^2} \tag{30}$$

Substituting Equation (30) into Equation (23) yields the energy dissipation rate for the membrane stretching deformation of a single plate during fold generation as

$$\dot{E}_{m,w,2} = \frac{3\sigma_{0,w} t_w \delta \dot{\delta} H}{L} \tag{31}$$

The total energy dissipation of a single plate when the fold is flattened can be obtained by integrating the above equation.

$$E_{m,w,2} = \frac{54\sigma_{0,w} t_w H^3}{L} \tag{32}$$

The instantaneous and mean crushing resistance of the cruciform in the second stage of deformation are obtained from Equations (21) and (31) as:

$$\begin{aligned} F_{w,2}(\delta) &= 4\frac{\dot{E}_{b,w,2}+\dot{E}_{m,w,2}}{\dot{\delta}} \\ &= \frac{4M_{0,w}L}{H\sqrt{1-\left(1-\frac{\delta}{6H}\right)^2}} + \frac{12\sigma_{0,w}t_w\delta H}{L} \end{aligned} \tag{33}$$

$$F_{w,2} = \frac{4(E_{m,w,2} + E_{b,w,2})}{6H} = \frac{36\sigma_{0,w} t_w H^2}{L} + \frac{2\pi M_{0,w} L}{H} \tag{34}$$

According to the upper bound theorem, the energy dissipation of a structure is minimized at a minimum mean structural resistance.

$$\frac{\partial F_{w,2}}{\partial H} = 0 \tag{35}$$

Substituting Equation (34) into Equation (35) yields H:

$$H = \left(\frac{\pi t_w L^2}{144}\right)^{1/3} = 0.2794\left(t_w L^2\right)^{1/3} \tag{36}$$

The instantaneous and the mean structural resistance of the cruciform at the time of fold generation is obtained by substituting the characteristic height of the fold, H, into Equations (33) and (34). In the case of the indentation test herein, the characteristic height of the folds, the instantaneous structural resistance, and the average structural resistance of the specimens with the load applied at the mid-axis of the specimen are:

$$\begin{aligned} &H = 0.2794\left(t_w L^2\right)^{1/3} \\ &F_w(\delta) = \frac{4M_{0,w}L}{H\sqrt{1-\left(1-\frac{\delta}{6H}\right)^2}} + \frac{12\sigma_{0,w}t_w\delta H}{L} \\ &F_w = \frac{36\sigma_{0,w}t_wH^2}{L} + \frac{2\pi M_{0,w}L}{H} \\ &\delta \le 6H \end{aligned} \tag{37}$$

The smaller value of Equation (38) is taken as the structural resistance when the cruciform is under pressure.

$$F_w(\delta) = \begin{cases} 4\sigma_{0,w}t_wL + 2\sigma_{0,w}l^2 \\ \frac{4M_{0,w}L}{H\sqrt{1-\left(1-\frac{\delta}{6H}\right)^2}} + \frac{12\sigma_{0,w}t_w\delta H}{L} \end{cases} \tag{38}$$

5.3. Validation of the Analytical Method

The new proposed analytical method was validated with experimental and numerical results. Figure 12 shows the instantaneous resistance of the specimens obtained from the experiment, the numerical simulation, and the analytical method, in which the trend of the three curves matches well.

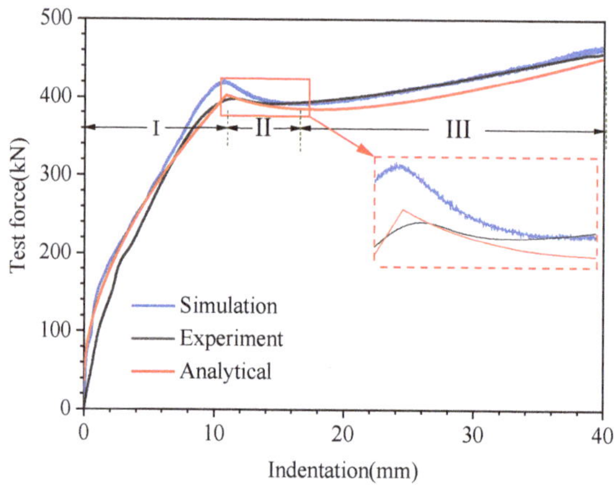

Figure 12. Comparison of crushing response between the experimental, numerical, and analytical methods.

At the initial period of stage I, the analytical result was in closer agreement with the simulation results, indicating that it is reasonable to assume a certain compression deformation of the cruciform before instability. In the later part of stage I, the analysis showed the increase in collision force slowed down; indeed, the value was smaller than in the simulation result. This is mainly because only the compressive deformation of the cruciform was considered in the analysis. In the real experiment, as the impact depth increased, the surface panel contacting the indenter generated membrane stretching deformation, making the real load higher than the analytical solution.

When the load increased to a certain level, the cruciform buckled and folded, and the load decreased. In stage II, the analytical result was similar to the experimental and simulated trends. The peak collision force in the analysis was 402.67 kN with an error of 1.3%, which was closer to the experimental value, suggesting that the analytical method can predict peak force better.

In stage III, the curves for all three methods followed the same trend. The analytical result was smaller compared to the test and the numerical results, mainly because the analytical method ignored the membrane tensile deformation of the surface panel of the specimen and its contribution to the structural load-carrying capacity.

6. Conclusions

The mechanical response of a cruciform in double-hulled hull structures during collisional grounding accidents was studied experimentally, numerically, and analytically in this paper.

The main conclusions of this study include the following:

- The results of the test and simulations, i.e., the deformation mode and the load-displacement curve of the specimen, were compared. The trend of force displacement curves matched well, and the peak forces of the test and numerical simulations were 397.51 kN and 421.12 kN, with an error of 5%.
- Observations of deformation in the test and the simulated specimen showed that the cruciform structure produced a 1:3:2 ratio of fold heights and that the deformations of four component panels were symmetrical about the central axis.
- By analyzing the deformation mode of the specimen in the test and the simulation, the deformation process of the specimen could be divided into two stages: the initial contact stage and the fold-forming stage. On this basis, a two-stage simplified deformation process of cruciform was proposed. By analyzing the energy dissipation mechanism

in each stage, we obtained analytical formulae for predicting the instantaneous and mean resistance of the cruciform under in-plane loading.

- A comparison of the analytical results with the experimental and simulation results showed that the analytical method proposed in this paper can effectively predict the structural resistance of cruciform structures under in-plane loads. The analytical peak resistance was 402.67 kN, closer to the experimental data, indicating that the analytical method provides a more accurate solution.

The test results and the numerical simulation can be of assistance to similar hull structure collisions or groundings, and the analytical method is suitable for the primary design of hull crashworthiness and the rapid evaluation of structural crashworthiness performance.

Author Contributions: Conceptualization, H.L. and X.W.; Writing—original draft, H.L.; Writing—review and editing, Z.G. and K.L.; Visualization, K.L. and J.W. All authors have read and agreed to the published version of the manuscript.

Funding: This research was funded by the National Natural Science Foundation of China (Grant No. 52171311; Grant 52271279).

Institutional Review Board Statement: Not applicable.

Informed Consent Statement: Not applicable.

Data Availability Statement: Not applicable.

Conflicts of Interest: The authors declare no conflict of interest.

Nomenclature

The following abbreviations are used in this manuscript:

A	plate area
A_g	maximum uniform strain
C	$R_m(e/n)^n$
e	natural constant
$\dot{E}_b/\dot{E}_{b,w,2}$	bending energy dissipation rate
$\dot{E}_m/\dot{E}_{m,w,2}$	membrane stretching energy dissipation rate
$\dot{E}_{int}$	internal energy dissipation rate
F	external force
F_w	average structural resistance
$F_w(\delta)$	instantaneous structural resistance
H	characteristic height of a fold
L	length of a single panel
L_c	compression width in a single panel
l	weld length
l_i	plastic hinge length
$M_{\alpha\beta}$	bending moment
$M_{0,w}$	moment on plastic hinge per unit length
$N_{\alpha\beta}$	plane membrane tensile stress tensor
$N_{0,w}$	tensile force generated by a unit thickness plate
n	$\ln(1 + A_g)$
R_m	ultimate stress
S	plastic deformation area
t_w	cruciform thickness
Greek symbols	
α	rotation angle of plastic hinge
$\dot{\alpha}/\dot{\theta}_i$	angular velocity
δ	indentation depth
$\dot{\delta}$	velocity
ε_e	engineering strain

ε_t	true strain
ε_{MA}	average strain at the plastic hinge *MA*
$\dot{\varepsilon}_{MA}$	average strain rate at the plastic hinge *MA*
$\dot{\varepsilon}_{ij}$	strain rate
$\dot{\varepsilon}_{\alpha\beta}/\dot{\varepsilon}_{eq}$	equivalent strain rate
$\dot{\varepsilon}_w$	average strain rate of a plate in lengthways
$\sigma_0/\sigma_{0,w}$	flow stress
σ_e	engineering stress
σ_t	true stress

References

1. Chang, S.E.; Stone, J.; Demes, K.; Piscitelli, M. Consequences of oil spills: A review and framework for informing planning. *Ecol. Soc.* **2014**, *19*, 26. [CrossRef]
2. Oil Tanker Spill Statistics 2021. Available online: https://www.itopf.org/knowledge-resources/data-statistics/statistics/ (accessed on 12 September 2022).
3. Urban, J.; Pedersen, P.T.; Bo, C.S. Collision Risk Analysis for HSC. In Proceedings of the 5th International Conference on Fast Sea Transportation, Seattle, WA, USA, 1 January 1999.
4. Abramowicz, W.; Simonsen, B. Effect of fracture on crushing of ship structures. *J. Ship Res.* **2003**, *47*, 194–207. [CrossRef]
5. Zhou, C.; Li, T.; Ming, S.; Song, Z.; Wang, B. Effects of welding on energy absorption of kirigami cruciform under axial crushing. *Thin Walled Struct.* **2019**, *142*, 297–310. [CrossRef]
6. Ghazijahani, T.G.; Jiao, H.; Holloway, D. Experiments on dented cylindrical shells under peripheral pressure. *Thin Wall Struct.* **2014**, *84*, 50–58. [CrossRef]
7. Mohammed, A.K.; Amdahl, J.; Skallerud, B. Collapse analysis of stiffened panels during accidental conditions. In Proceedings of the Offshore Mechanics and Arctic Engineering, Rio de Janeiro, Brazil, 3–8 June 2001.
8. Urban, J. Crushing and Fracture of Lightweight Structures. Ph.D. Thesis, Technical University of Denmark, Copenhagen, Denmark, 2003.
9. Haris, S.; Amdahl, J. Crushing resistance of a cruciform and its application to ship collision and grounding. *Ships Offshore Struc.* **2012**, *7*, 185–195. [CrossRef]
10. Sun, C.; Zhou, J.; Wu, W. Study on Collision of Ship Side Structure by Simplified Plastic Analysis Method. In Proceedings of the 2017 International Conference on Structural, Mechanical and Materials Engineering, Seoul, Republic of Korea, 13–15 July 2017.
11. Haris, S.; Amdahl, J. An analytical model to assess a ship side during a collision. *Ships Offshore Struc.* **2012**, *7*, 431–448. [CrossRef]
12. Amdahl, J. Energy Absorption in Ship-Platform Impacts. Ph.D. Thesis, The Norwegian Institute of Technology, Trondheim, Norway, 1983.
13. Wierzbicki, T.; Abramowicz, W. On the Crushing Mechanics of Thin-Walled Structures. *J. Appl. Mech-T ASMEs* **1983**, *50*, 727–734. [CrossRef]
14. Jones, N.; Wierzbicki, T. *Structural Crashworthines*; Butterworths: London, UK, 1983.
15. Wierzbicki, T. Crushing behavior of plate intersections. In *Structural Crashworthiness*; Jones, N., Wierzbicki, T., Eds.; Butterworths: London, UK, 1983; pp. 66–95.
16. Hayduk, R.J.; Wierzbicki, T. Extensional collapse modes of structural members. *Comput. Struct.* **1984**, *18*, 447–458. [CrossRef]
17. Kierkegaard, H. Ship bow response in high energy collisions. *Mar. Struct.* **1993**, *6*, 359–376. [CrossRef]
18. Hagiwara, K.; Takanabe, H.; Kawano, H. A proposed method of predicting ship collision damage. *Int. J. Impact. Eng.* **1983**, *1*, 257–279. [CrossRef]
19. Abramowicz W. *Crushing Resistance of T, Y and X Sections*; Massachusetts Institute of Technology: Boston, MA, USA, 1994.
20. Liu, K.; Zong, S.; Li, Y.; Wang, Z.; Hu, Z.; Wang, Z. Structural response of the U-type corrugated core sandwich panel used in ship structures under the lateral quasi-static compression load. *Mar. Struct.* **2022**, *84*, 103198. [CrossRef]
21. Park, S.-J.; Choung, J. Punching Fracture Experiments and Simulations of Unstiffened and Stiffened Panels for Ships and Offshore Structures. *J. Ocean. Eng. Technol.* **2020**, *34*, 155–166. [CrossRef]
22. Cerik, B.C.; Shin, H.K.; Cho, S.-R. On the resistance of steel ring-stiffened cylinders subjected to low-velocity mass impact. *Int. J. Impact. Eng.* **2015**, *84*, 108–123. [CrossRef]
23. *GB/T228.1-2010*; Metallic Materials-Tensile Testing-Part 1: Method of Test at Room Temperature. China Standard Press: Beijing, China, 2011.
24. Dieter, G.E. Mechanical behavior under tensile and compressive loads. *Asm Handb.* **2000**, *8*, 99–108.
25. Dassault Systèmes. *Abaqus*, version 2018. Windows. Dassault Systèmes: Paris, France, 2018.
26. Alsos, H.S.; Amdahl, J.; Hopperstad, O.S. On the resistance to penetration of stiffened plates, Part II: Numerical analysis. *Int. J. Impact. Eng.* **2009**, *36*, 875–887. [CrossRef]
27. Villavicencio, R.; Soares, C.G. Numerical plastic response and failure of a pre-notched transversely impacted beam. *Ships Offshore Struc.* **2012**, *7*, 417–429. [CrossRef]
28. Dieter, G.E.; Bacon, D. *Mechanical Metallurgy*; McGraw-Hill: New York, NY, USA, 1976.

29. Zhang, L.; Egge, E.-D.; Bruhns, H. *Approval Procedure Concept for Alternative Arrangements*; Germanischer Lioyd: Hamburg, Germany, 2004.
30. Liu, K.; Lu, Y.; Wang, Z.; Wang, G. An experimental, numerical and analytical study on deformation mechanisms of web girders in a collision or grounding incident. *Ships Offshore Struc.* **2019**, *14*, 839–852. [CrossRef]
31. Simonsen, B.C.; Ocakli, H. Experiments and theory on deck and girder crushing. *Thin Wall Struct.* **1999**, *34*, 195–216. [CrossRef]

Journal of
Marine Science and Engineering

MDPI

Article

Study on the Low-Frequency and Broadband Sound Absorption Performance of an Underwater Anechoic Layer with Novel Design

Jinshun Hu [1], Yongshui Lin [1,*], Zhiwei Zhou [2,*], Xiaofei Cao [1], Qingjia Chi [1] and Weiguo Wu [3]

1 School of Science, Wuhan University of Technology, Wuhan 430070, China
2 Beijing Key Laboratory of Lightweight Multi-Functional Composite Materials and Structures, Institute of Advanced Structure Technology, Beijing Institute of Technology, Beijing 100081, China
3 Green & Smart River-Sea-Going Ship, Cruise and Yacht Research Center, Wuhan University of Technology, Wuhan 430063W, China
* Correspondence: linyongshui1226@whut.edu.cn (Y.L.); zhouzw92@bit.edu.cn (Z.Z.)

Abstract: To further improve the low-frequency broadband sound absorption capability of the underwater anechoic layer (UAL) on the surface of marine equipment, a novel sound absorption structure with cavities (NSSC) is designed by adding resonators and honeycombs to the traditional sound absorption structure with cavities (SSC). Based on the principle of shear dissipation, the original intention of the design is to allow more parts of the viscoelastic material to participate the dissipation of acoustic energy. The approximate multilayer sound absorption theoretical model based on the modified transfer matrix method is used to verify the accuracy of finite element calculations. In the frequency range of 1100 Hz–10,000 Hz, the sound absorption coefficient (α) of NSSC can reach 0.8. The effects of the presence and size of cylindrical oscillators and honeycomb structures on sound absorption are discussed in detail. The results show that expanding the effective sound absorption range of the damping area of the structure is the key to improve the wideband sound absorption effect. This design concept could guide the structural design of the UAL.

Keywords: low-frequency broadband sound absorption; air cavity; cylindrical oscillator; honeycomb

Citation: Hu, J.; Lin, Y.; Zhou, Z.; Cao, X.; Chi, Q.; Wu, W. Study on the Low-Frequency and Broadband Sound Absorption Performance of an Underwater Anechoic Layer with Novel Design. *J. Mar. Sci. Eng.* **2023**, *11*, 409. https://doi.org/10.3390/jmse11020409

Academic Editor: Erkan Oterkus

Received: 17 December 2022
Revised: 8 January 2023
Accepted: 11 January 2023
Published: 13 February 2023

1. Introduction

UALs are generally used to cover the surface of an underwater vehicle to counteract the active or passive sonar detection [1–4]. Its stealth performance for submarines and other underwater vehicles is crucial and has been widely valued by various maritime military powers. The UALs with good sound absorption performance should satisfy both important principles [5]: (1) the acoustic material is consistent with the characteristic impedance of the water as much as possible, so that the sound waves can fully enter the anechoic material and the corresponding reflection is reduced; (2) the anechoic material has a high sound wave absorption capacity. Viscoelastic materials such as rubber meet the above requirements and are widely employed for underwater sound absorption [6–13].

UALs require high sound absorption capabilities at a wide frequency band, but most homogeneous materials often fall short of the requirement. To overcome the limitation, researchers proposed the polymer mixed sound absorbing materials [14], polymer foam sound absorbing materials [15], gradient sound absorbing materials [16], and filled sound absorbing materials [17]. Among them, the more mature studies of UALs mainly focus on composite structures with cavities or resonators [18].

Viscoelastic materials with cavities can absorb more energy under sound wave incidence owing to the principle of shear dissipation and the resonance of cavities [19,20]. Ivansson et al. [21] compared the sound absorption properties of an attenuated coating containing cylindrical holes with different cross-sectional shapes (circular, oval, and super-ellipse) and obtained the optimal solution using the Markov chain Monte Carlo method.

Ye et al. [22] studied the effects of different shapes of cavities and found that the horn holes have the best sound absorption performance at lower frequencies. These cavity structures are often called Alberich anechoic materials [23]. The backing material also significantly impacts the sound absorption of Alberich anechoic materials. Alberich anechoic materials are usually soaked in steel-backed water. Zhao et al. [24] studied the optimization and mechanism of Alberich anechoic materials on steel plates. It is found that laying different coatings on the two surfaces of the steel plate can effectively improve the low-frequency sound absorption and the sound absorption bandwidth. Locally resonant structures usually consist of a heavy core and soft viscoelastic material. Near the resonant frequency, the vibrations of the structure enhance the friction between the molecules, then the dissipation of sound energy rises. Gao et al. [25] introduced a broadband sound absorbing underwater metamaterial composed of viscoelastic rubber, conical cavities, cylindrical oscillators, and backing steel. Compared with pure viscoelastic rubber and traditional sound absorbing structures with cavities, the structure exhibits excellent broadband sound absorption performance below 10 kHz. Jin et al. [26] proposed a novel composite acoustic metamaterial made of periodic multi-resonators and cavities. Numerical calculations show that the coupled resonance generated by multiple resonators and cavities can effectively broaden the sound absorption frequency band. In addition, previous works have not discussed the sound absorption valley in the mid-high frequency of UALs with cavities. Compared with pure viscoelastic material, the appearance of the cavities makes UAL appear at the sound absorption peak at low frequency. Nevertheless, the α cannot maintain a high value in the subsequent mid-to-high frequency bands [7,25–30].

In this paper, NSSC is proposed to improve the low-frequency and broadband sound absorption capability of UAL. Compared with SSC, cylindrical oscillators and periodic hexagonal steel honeycomb are added. The theoretical calculation verifies the correctness of the finite element method. The validated finite element model is then used to explore the sound absorption characteristic of the new structure and to quantify the geometry dependence of the sound absorption.

For the first time, the structure of UAL (NSSC) proposed in this paper combines the viscoelastic material, cavities, cylindrical oscillator, and steel honeycomb to achieve low-frequency broadband sound absorption. Among them, the cavity plays a role in improving overall sound absorption. The cylindrical oscillators significantly rise the low-frequency sound absorption. The honeycomb enhances the dissipation of sound energy in most of the damping area of UAL, improves the sound absorption of mid- and high-frequency, and eliminates the sound absorption valley.

2. Methods

2.1. Geometric Model and Material

As displayed in Figure 1, the lateral dimension of the UAL is assumed to be infinite and consisted of a periodic arrangement of hexagonal cells. To facilitate the theoretical verification, the hexagonal prism is approximated as a cylinder based on the principle that the volume fraction of the cavity in the cell remains unchanged. This approximation has been proven to be sufficiently accurate [31]. The geometric parameters and material parameters of the UAL are listed in Tables 1 and 2. l_1 is the height from the top of the UAL to the cavity, l_2 is the height of the cavity, r_1 is the radius of the cavity, r_2 is the cell radius, r_3 is the radius of the cylinder, t is the half of thickness of the honeycomb, h_1 is the height of the honeycomb, h_2 is the height from the bottom of the cylinder to the backing steel, h_3 is the height of the cylinder.

Table 1. The geometric parameters of the UAL (mm).

l_1	l_2	r_1	r_2	r_3	t	h_1	h_2	h_3
49	1	10	15	2	0.5	45	5	10

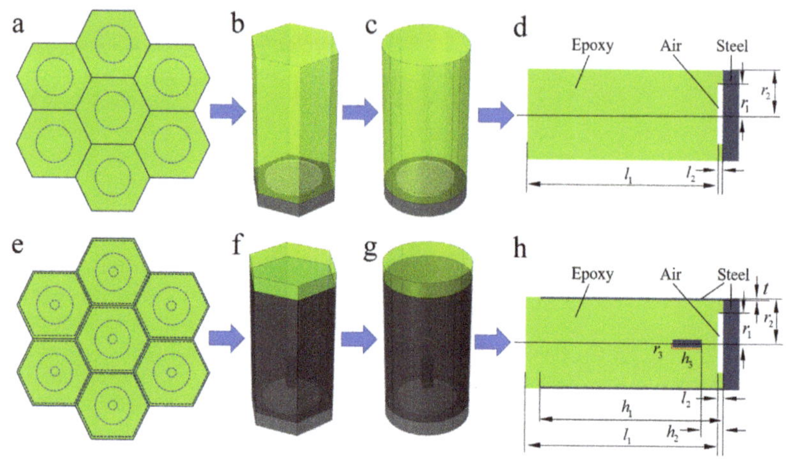

Figure 1. (**a–d**) are the top view, hexagonal cell, cylindrical cell, and cross-sectional of SSC; (**e–h**) are the top view, hexagonal cell, cylindrical cell and cross-sectional of NSSC.

Table 2. Material parameters "Adapted with permission from Ref. [26]. 2020, Elsevier Ltd.".

Solid Medium	Density (kg/m^3)	Elastic Modulus (GPa)	Poisson's Ratio	Loss Factor
Epoxy	1100	0.027	0.49	0.6
Steel	7980	210	0.28	/

2.2. Theoretical α of SSC

The α of SSC was theoretically calculated via the transfer matrix method. Figure 2 exhibits the working condition of a typical cell. The transverse dimension of the UAL is infinite. The vertically incident plane sound wave from the water passes through the UAL and reaches the back steel plate which could be regarded as the hard boundary. Then, the sound wave is reflected to the UAL and eventually back into the water.

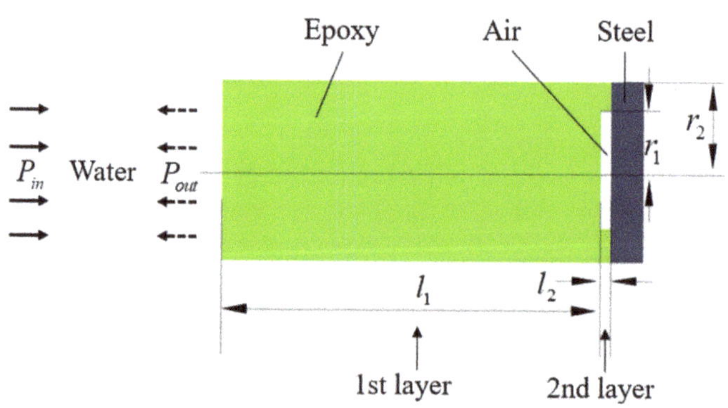

Figure 2. Structural parameters and working conditions of typical cell of SSC.

The SSC is divided into two layers. The first layer is a uniform viscoelastic material, and the sound velocity [27] is

$$c_1 = \sqrt{\frac{E_1(1-v)}{\rho_1(1+v)(1-2v)}} \tag{1}$$

$$E_1 = E_s(1 + j\eta) \tag{2}$$

where E_1 and ρ_1 are the complex elastic modulus and density of the first layer, v and η are the Poisson's ratio and loss factor of the viscoelastic material. E_s is the elastic modulus and j is the imaginary unit. The complex wavenumber of the first layer is

$$k_1 = \omega / c_1 \tag{3}$$

in which $\omega = 2\pi f$ is the circular frequency, f is the frequency.

The second layer is a viscoelastic material with a cavity, and its approximate complex wavenumber [27] is expressed as

$$k_2 = k_1 \sqrt{1 + \frac{\lambda}{\mu} \frac{\varepsilon^2}{1 - \varepsilon^2}} \tag{4}$$

with

$$\lambda = \frac{vE_1}{(1 + v)(1 - 2v)} \tag{5}$$

$$\mu = \frac{E_1}{2(1 + v)} \tag{6}$$

$$\varepsilon = r_1 / r_2 \tag{7}$$

where λ and μ are the Lame coefficients of t/he viscoelastic material, and ε is the ratio of the inner diameter to the outer diameter of the second layer. Due to the existence of the cavity, the equivalent density of the second layer is

$$\rho_2 = \rho_a \varepsilon^2 + \rho_1 \left(1 - \varepsilon^2\right) \tag{8}$$

in which ρ_a is the air density.

The sound velocity of the second layer is

$$c_2 = \omega / k_2 \tag{9}$$

For a multi-layer structure, the transfer matrix of the i-th layer is written as

$$\mathbf{T_i} = \begin{bmatrix} \cos(k_i l_i) & j\rho_i c_i \sin(k_i l_i) \\ j\sin(k_i l_i)/\rho_i c_i & \cos(k_i l_i) \end{bmatrix} \tag{10}$$

where k_i is the complex wavenumber of the i-th layer, l_i is the thickness of the i-th layer, ρ_i is the equivalent density of the i-th layer, and c_i is the sound velocity of the i-th layer. Previous studies [32] have shown that the low-frequency sound absorption performance in such cylindrical pipes is mainly affected by the first axisymmetric wavenumber associated with the first propagation mode. Therefore, only the first wavenumber k_i is considered here, and the transfer matrix formula of Equation (10) can be used. In this paper, the low-frequency range (less than 10 kHz) is considered, in which the acoustic wave wavelength far exceeds the thickness of the anechoic layer, so the transfer matrix method is appropriate.

The total transmission matrix of the anechoic layer is the successive multiplication of the transmission matrix of each layer as

$$\mathbf{T} = \begin{bmatrix} T_{11} & T_{12} \\ T_{21} & T_{22} \end{bmatrix} = \prod_{i=1}^{n} \mathbf{T_i} \tag{11}$$

According to the transfer function method, the sound pressure and vibration velocity of the front interface and the rear interface can be related by **T** as

$$\begin{bmatrix} \rho_1 \\ v_1 \end{bmatrix} = \mathbf{T}\begin{bmatrix} T_{11} & T_{12} \\ T_{21} & T_{22} \end{bmatrix}\begin{bmatrix} \rho_2 \\ v_2 \end{bmatrix} \tag{12}$$

in which ρ_1 and v_1 are the sound pressure and vibration velocity of the front interface, ρ_2 and v_2 are the sound pressure and vibration velocity of the rear interface.

The impedances of the front and rear interfaces are equal to their corresponding sound pressure, expressed as

$$Z_1 = \rho_1/v_1 \tag{13}$$

$$Z_2 = \rho_2/v_2 \tag{14}$$

When the interface impedance tends to infinity at the rigid backing, there is

$$Z_1 = T_{11}/T_{21} \tag{15}$$

The reflection coefficient of the anechoic layer is

$$R = (Z_1 - Z_w)/(Z_1 - Z_w) \tag{16}$$

where Z_w is the impedance of the aqueous medium.

Finally, the α can be achieved as

$$\alpha = 1 - |R|^2 \tag{17}$$

2.3. Verification of Finite Element Analysis

The two-dimensional axisymmetric model in the finite element software COMSOL can perfectly fit the periodicity and symmetry of the sound absorbing layer [33,34]. As shown in Figure 3, the finite element model from left to right is the perfect matching layer, the water, and the sound absorbing layer. The perfectly matched layer absorbs all sound waves and simulates an infinite acoustic domain. A background pressure field is added to the water to simulate the plane wave incidence. The radial displacement of the outer boundary of the cylinder is 0. The quadratic Lagrange element was used in the acoustic domain and the quadratic serendipity element was used in the solid domain. The mesh size is set to 1 mm to meet the calculation accuracy. The total number of elements is 1275.

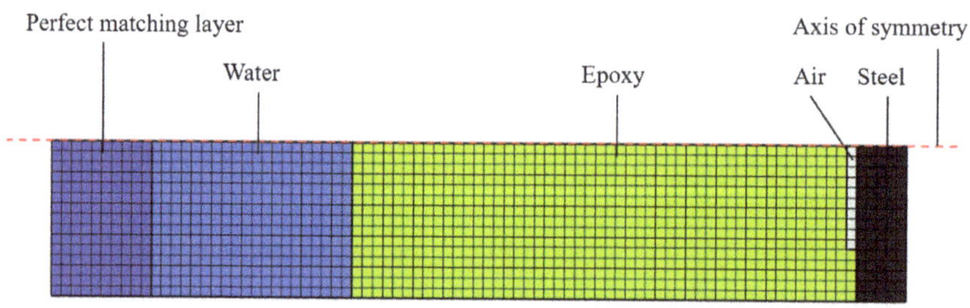

Figure 3. Finite element model of UAL of SSC.

The theoretical and the simulated α of SSC are presented in Figure 4. In the considered frequency range from 0 Hz to 10,000 Hz with the step 10 Hz, the results are in good agreement with each other. Thus, the correctness of the finite element model of the UAL is confirmed. The slight discrepancies in the α observed in the figure are a result of the approximate complex wavenumber in Equation (4).

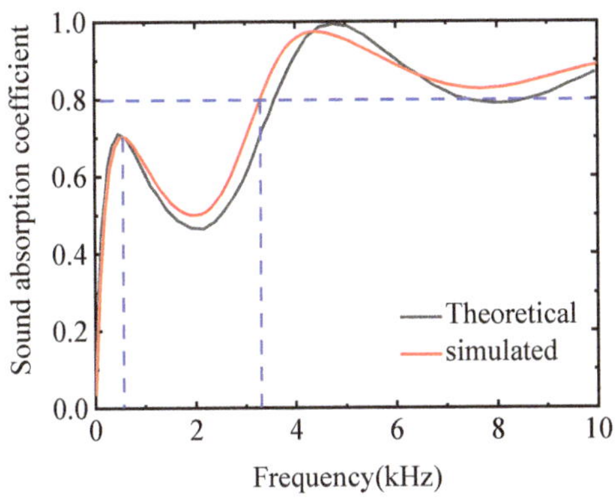

Figure 4. Comparison of α between SSC theoretical model and finite element model.

3. Acoustic Absorption Mechanism

In this paper, 0.8 is a reference value for the high α. Figure 4 shows that the SSC can achieve high sound absorption within 3300–10,000 Hz. However, the SSC also has a sound absorption valley in the mid-frequency region (around 2000 Hz), which exists widely in the work of some other researchers [7,25–30]. To improve the low-frequency and broadband sound absorption capability, a cylindrical oscillator is added and the SSC-CO (sound absorption structure with cavities and cylindrical oscillators) is formed. A periodic hexagonal steel honeycomb is added to obtain SSC-HSH (sound absorption structure with cavities and hexagonal steel honeycomb). Finally, both the cylindrical oscillator and a periodic steel honeycomb are added and the design of NSSC is gained. All structures are displayed in Figure 5.

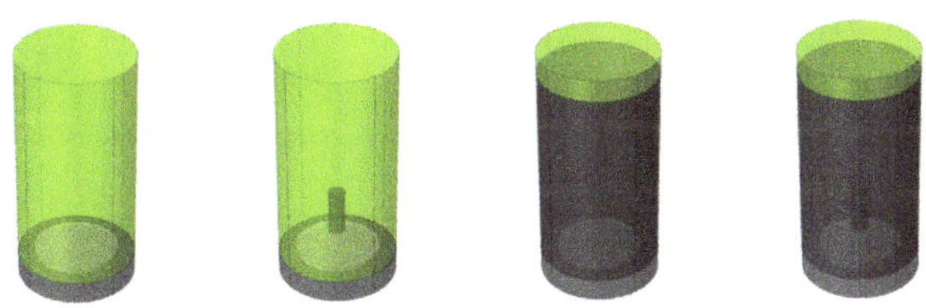

Figure 5. Sound absorption structure with cavities (SSC) and its derivatives (SSC-CO, SSC-HSH, NSSC).

3.1. The Sound Absorption Appearance of Each UAL

The comparison of the α of the above four structures is given in Figure 6. According to the trend of the α curves, it can be divided into two categories for discussion. The α curves of SSC and SSC-CO begin to decline after reaching the first absorption peak at 600 Hz and the valley of sound absorption around 2000 Hz, and then the sound absorption curve rises again. However, the α curves of SSC-HSH and NSSC have no sound absorption valley near 2000 Hz and show continuously rising trends. The NSSC has the best sound absorption ability and can achieve wide-band high sound absorption at 2000–10,000 Hz.

To further understand the sound absorption mechanism of the underwater UAL proposed in this paper, the total displacement nephogram (Figure 7) and the radial velocity nephogram (Figure 8) at 600 Hz, 2000 Hz, and 4000 Hz are selected for comparative analyses. The SSC and SSC-CO, SSC-HSH, and NSSC, respectively, have similar displacement and radial velocity distribution. This phenomenon explains that the NSSC and SSC-CO, SSC-HSH, and NSSC have similar sound absorption curves in Figure 6.

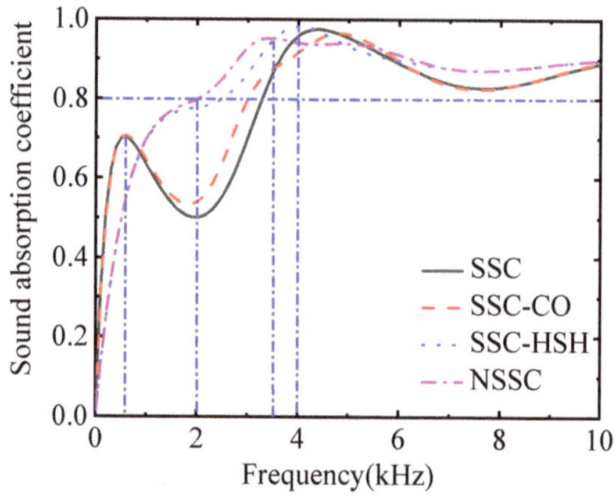

Figure 6. Comparison of α of SSC, SSC-CO, SSC-HSH, and NSSC.

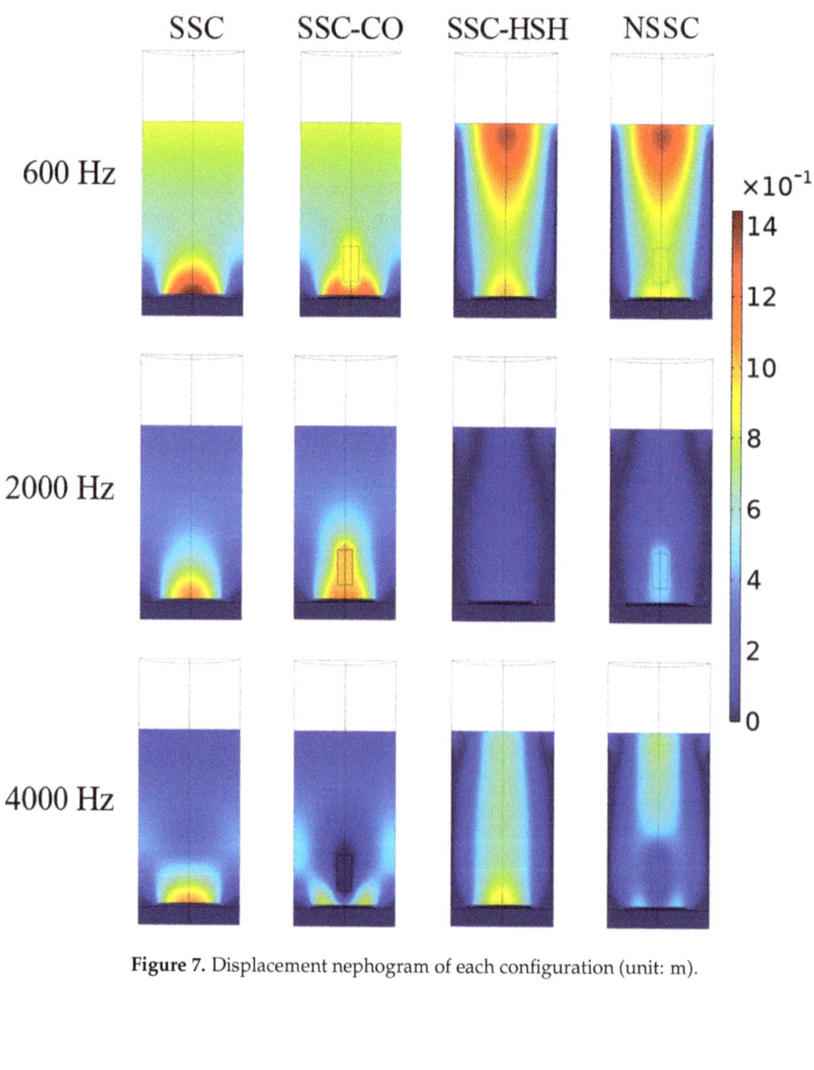

Figure 7. Displacement nephogram of each configuration (unit: m).

In Figure 7, when the UAL has no honeycomb (SSC and SSC-CO), the displacement of the damping region is relatively uniform from top to bottom, and its displacement along the radial direction is not obvious until it is close to the cavity or cylinder. When the incident wave frequency increases from 600 Hz to 2000 Hz and then to 4000 Hz, the total displacement of UAL decreases significantly. There is no displacement in the damping area near the honeycomb after the addition of honeycombs (SSC-HSH and NSSC). The displacement of UAL changes notably along the radial direction. The honeycomb weakens the displacement of the UAL. When the incident wave frequency increased from 600 Hz to 2000 Hz and then to 4000 Hz, the total displacement of UAL showed a trend of first decreasing and then increasing.

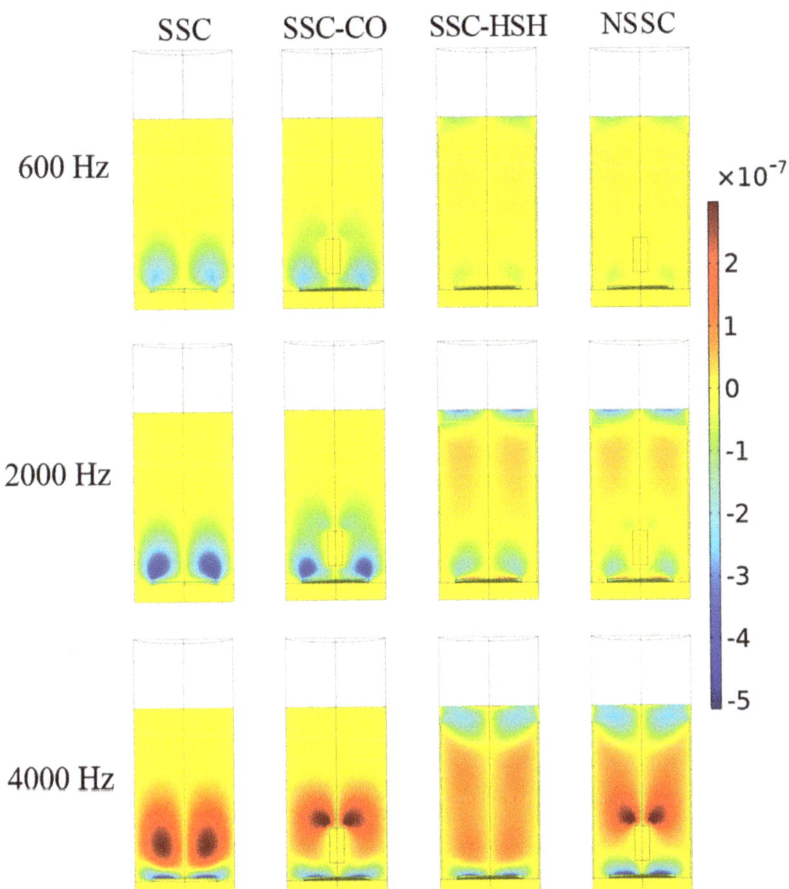

Figure 8. Radial velocity nephogram of each configuration (unit: m/s).

In Figure 8, when the UAL has no honeycomb (SSC and SSC-CO), its radial velocity is concentrated at the outer edge of the cavity, and the radial velocity is not observed in the middle and upper parts of the UAL. This is due to the nonlinear vibration of the damping area near the cavity, and the damping area in the upper middle is relatively static in the radial direction. When a honeycomb is added (SSC-HSH and NSSC), the radial velocity exists throughout the UAL due to the mutual shearing motion of the honeycomb and damping areas. Furthermore, the resonant cylinder suppresses the radial velocity of the surrounding damping region. When the incident wave frequency increases from 600 Hz to 2000 Hz and then to 4000 Hz, the radial velocity increases with the growth in frequency.

3.2. The Relationship between the Curves and the Nephograms

The radial velocity reflects the shear wave propagation. In viscoelastic media, shear waves dissipate the acoustic energy much more strongly than longitudinal waves. In Figure 8, the radial velocity of UAL is proportional to the frequency. It explains that the α of SSC-HSH and NSSC continue to rise at 600–4000 Hz. However, due to the large overall displacements of SSC and SSC-CO at 600 Hz (as shown in Figure 7), a large amount of energy is dissipated and the first sound absorption peak in the sound absorption curve is caused. At 600–4000 Hz, it shows a trend of first decreasing and then increasing. The honeycomb limits the overall displacement of SSC-HSH and NSSC at 600 Hz, which explains that the α of SSC-HSH and NSSC at 600 Hz is smaller than that of SSC and SSC-CO, and the sound absorption curve cannot reflect the first absorption peak. Compared with SSC and SSC-CO, the radial velocity distributions of SSC-HSH and NSSC are wider, and there are more parts for efficient energy dissipations of UAL. Thus, with the increase in frequency, the α of SSC-HSH and NSSC exceeds that of SSC and SSC-CO. The cylinder insertion is designed based on the contribution of the local resonant resonators to the acoustic low-frequency absorption. Consequently, before 3500 Hz, the α of UAL with the cylindrical oscillator is greater than that of UAL without the resonator cylinder (SSC-CO > SSC, NSSC > SSC-CO), but at 4000 Hz the situation is slightly opposite (SSC-CO < SSC, NSSC < SSC-CO). This can still be explained by Figures 7 and 8. At low frequencies (600 Hz), the sound absorption of the anechoic layer is determined by the overall displacement. The influence of the resonant body can be ignored, making the sound absorption curve the same. As the frequency increases, for example, when the frequency is 2000 Hz, the overall displacement of the UAL decreases, and the area near the cylindrical oscillator contributes part of the displacement and radial velocity, which makes the α of the UAL with the cylindrical oscillator larger. When the frequency increases to 4000 Hz, the region near the resonator suppresses part of the displacement and radial velocity so that the α of the UAL with the cylindrical oscillator is small.

In summary, improving the area size of structural sound energy dissipation is the key to improving the α. It is necessary to further study the influence of the size of the cylinder and honeycomb on the α.

4. The Influence of Different Parameters on the α

4.1. Influence of the Honeycomb Size on α

The analysis of the sound absorption reveals that the honeycomb confines the SSC-HSH in a vertical direction, reducing the overall displacement of the SSC-HSH. Still, the part of the SSC-HSH that can dissipate energy efficiently becomes more. Figure 9 further shows the α of the SSC-HSH under different honeycomb heights. With the increase in the height of the honeycomb, the larger the structural restraint range, the smaller the total displacement of the SSC-HSH at low frequencies (less than 600 Hz), and the correspondingly smaller α. However, at around 2000 Hz, with the increase in the height of the honeycomb, more parts of the SSC-HSH efficiently dissipate energy, and the sound absorption valley in the curve becomes smaller and smaller until it disappears. From 0 to 4000 Hz, the overall performance is that with the increase in the height of the honeycomb, the α first decreases and then increases. However, the low-frequency sacrifice is acceptable because it makes up for the defect of the valley in the α at around 2000 Hz.

Figure 10 shows the corresponding α for different honeycomb thicknesses. Changes in thickness do not substantially change the size of the energy dissipation area in SSC-HSH, which has similar α at each thickness. As the honeycomb thickness increases, the viscoelastic region of the SSC-HSH becomes more constrained. Before 1250 Hz, the larger the thickness, the smaller the α; after 1250 Hz, the larger the thickness, the larger the α. Therefore, the height of the honeycomb is the main factor affecting the α of SSC-HSH. Additionally, the sound absorption performance of SSC-HSH is proportional to the height of the honeycomb.

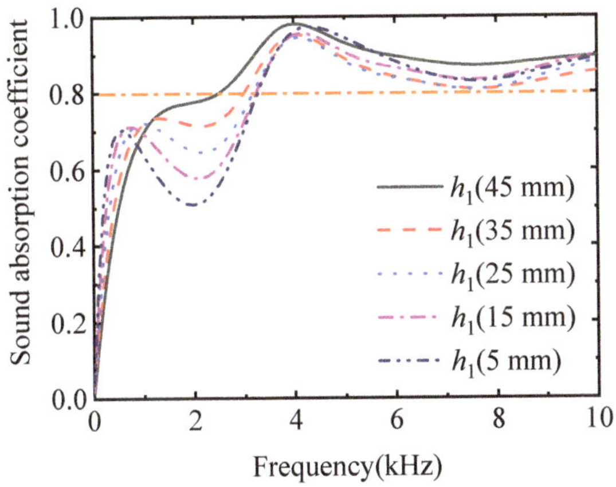

Figure 9. Comparison of α of SSC-HSH corresponding to different honeycomb heights.

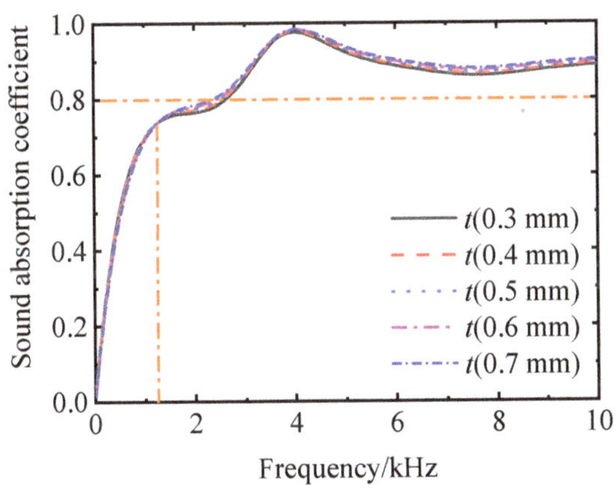

Figure 10. Comparison of α of SSC-HSH corresponding to different honeycomb thicknesses.

4.2. Influence of the Cylinder Size on α

Based on the local resonant resonator principle, the cylinder is inserted into the structure to enhance its low-frequency sound absorption. Figure 11 shows the α of SSC-CO corresponding to different cylinder radii. The radius of the cylinder has a significant effect on the sound absorption effect of SSC-CO. When the radius increases from 2 mm to 6 mm, the frequency of the sound absorption curve reaching 0.8 for the first time keeps moving to the low frequency; when the radius increases from 6 mm to 10 mm, the frequency of the sound absorption curve reaching 0.8 for the first time did not move to the low frequency significantly, and even at 10mm, the frequency reaching 0.8 for the first time moved to the high frequency instead. In addition, at around 4000 Hz, the α of SSC-CO with 8 mm and 10 mm cylinder radii dropped below 0.8. The appearance of this is due to the joint coupling between the cylinder and the bottom cavity. From the analysis of the sound absorption mechanism in Section 3.2, the displacement and radial velocity near the cavity mainly exist outside the cavity radius. When the radius of the cylinder is 10 mm, the cylinder suppresses the displacement and vibration of the viscoelastic material near the cavity and reduces the sound energy dissipation.

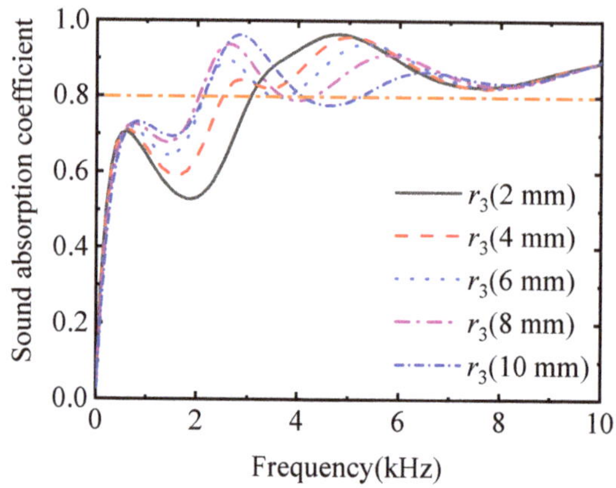

Figure 11. Comparison of α of SSC-CO corresponding to different cylinder radii.

Figure 12 shows the α of SSC-CO corresponding to different cylinder heights. Section 3.2 points out that the displacement of the viscoelastic region in the upper part of the cylinder is relatively uniform, and there is no radial velocity, so the increase in height does not significantly change the sound absorption performance of the SSC-CO. Therefore, the radius of the cylinder is the main factor affecting the α of the SSC-CO. The selection of the radius needs to consider the coupling with the cavity. In SSC-CO, the sound absorption effect is best when the radius is around 6 mm.

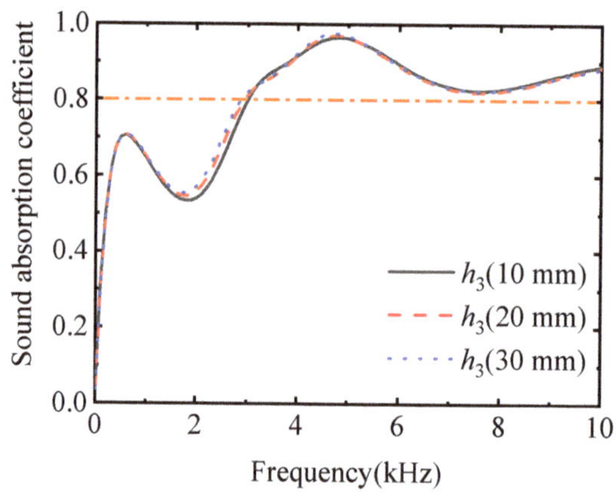

Figure 12. Comparison of α of SSC-CO corresponding to different cylinder heights.

4.3. The Effect of the Joint Action of the Honeycomb and the Cylinder on the α

In this section, the honeycomb and the cylinder are considered together to explore the impact on the sound absorption performance of NSSC. To maximize the restraint range of the honeycomb on NSSC and ensure the smoothness of the outer surface of UAL, the height of the honeycomb $h_1 = 49$ mm, and the other geometric parameters are consistent with Table 1. The influence on the α of NSSC is obtained by changing the radius of the cylinder and the height of the cylinder, respectively.

In Figure 13, as the radius of the cylinder increases, the frequency when the curve first reaches 0.8 moves to low frequencies. However, when the radius is 10 mm, the frequency at which the curve first comes 0.8 moves to the high frequency instead. This is exactly in line with the trend in Section 4.2. Due to the addition of the honeycomb, the overall sound absorption performance of NSSC has been improved to a certain extent, so there is no frequency band with a α lower than 0.8 in the follow-up. In Figure 14, the change in the cylinder height has a great influence on the sound absorption of the NSSC (refer to Section 4.2, the cylinder radius is taken as 6 mm). Unlike SSC-CO, due to the presence of the honeycomb, the damping area performs shear motion with both the honeycomb and the cylinder to dissipate more sound energy. It is shown that the higher the cylinder, the better the sound absorption capacity of NSSC.

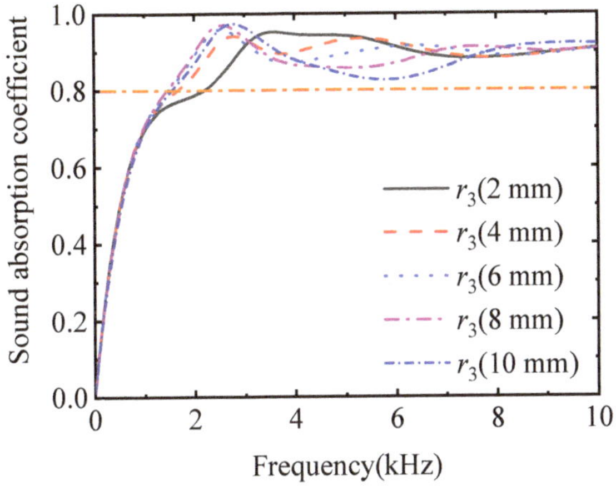

Figure 13. Comparison of α of NSSC corresponding to different cylinder radii.

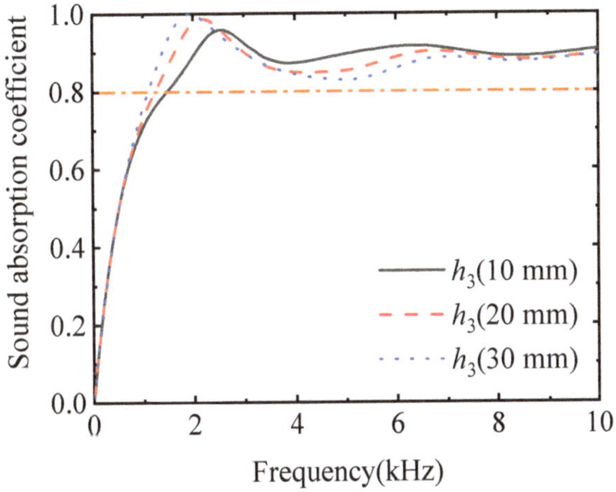

Figure 14. Comparison of α of NSSC corresponding to different cylinder heights.

4.4. Comparison with Typical UALs

For sonar detection below 10,000 Hz, it is difficult for most of the underwater anechoic layer to meet the low-frequency broadband α of 0.8 and above. To demonstrate the effectiveness of the combined structure of cavity, resonant cylinder, and honeycomb, the effective sound absorption bandwidth (α > 0.8) of NSSC is compared with the previous typical UAL (Refs. [27,28]) in Figure 15. The comparison shows that NSSC can achieve effective sound absorption in the frequency range of (1100–10,000) Hz, with the largest sound absorption bandwidth.

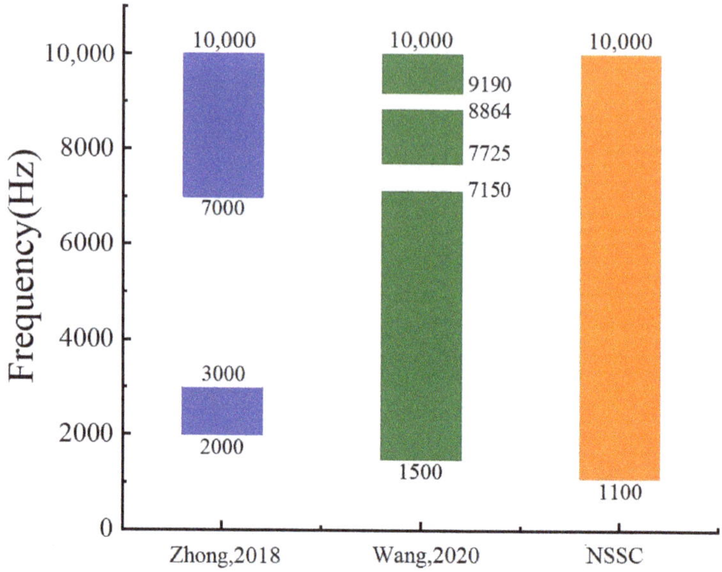

Figure 15. Comparison of absorption bandwidths with absorption coefficients above 0.8 between NSSC and other UALs in representative references [27,28].

5. Conclusions

A new type of underwater sound absorbing structure consisting of viscoelastic material, cavities, cylindrical oscillator, steel honeycomb, and backing steel was designed. The theoretical sound absorption coefficient was well matched with the simulated sound absorption coefficient, validating the reliability of the calculation. In addition, the geometry dependence of the underwater sound absorption performance of NSSC (novel sound absorption structure with cavities) was studied. The design would significantly increase the acoustic energy consumption of the structure as seen from the principle of shear dissipation. By adding honeycomb, the sound absorption valley could be effectively compensated, and the sound absorption capacity of med- and high-frequency of the underwater anechoic layer could be significantly improved. The sound absorption performance was improved as the honeycomb height increased. The low-frequency acoustic absorption capacity of the structure was improved by the addition of cylindrical oscillators. The choice of cylindrical radius and height should be determined based on the cavity radius and honeycomb height. Compared with other underwater sound absorption structures, the sound absorption coefficient of NSSC can reach 0.8 in the 1100–10,000 Hz frequency band, which has obvious low-frequency wideband sound absorption advantages. This design concept could guide the structural design of the underwater anechoic layer in the future.

Author Contributions: Conceptualization, X.C. and W.W.; Writing—original draft, J.H.; Writing—review & editing, Y.L., Z.Z. and Q.C. All authors have read and agreed to the published version of the manuscript.

Funding: This research received no external funding.

Institutional Review Board Statement: Not applicable.

Informed Consent Statement: Not applicable.

Data Availability Statement: Data can be obtained by contacting the author, Jinshun Hu (hujs@whut.edu.cn).

Conflicts of Interest: The authors declare no conflict of interest.

References

1. Ke, Y.; Zhang, L.; Zhao, X.; Tao, M. An equivalent method for predicting acoustic scattering of coated shell using identified viscoelastic parameters of anechoic coating. *Appl. Acoust.* **2021**, *179*, 108071. [CrossRef]
2. Zhang, Z.; Huang, Y.; Huang, Q. Low-frequency broadband absorption of underwater composite anechoic coating with periodic subwavelength arrays of shunted piezoelectric patches. *Compos. Struct.* **2019**, *216*, 449–463. [CrossRef]
3. Meng, H.; Wen, J.; Zhao, H.; Wen, X. Optimization of locally resonant acoustic metamaterials on underwater sound absorption characteristics. *J. Sound Vib.* **2012**, *331*, 4406–4416. [CrossRef]
4. Zhang, Z.; Li, Z.; Li, T.; Huang, Q. A novel semi-analytical approach for predicting the sound absorptions of a new underwater composite coating with transversely arranged SWCNTs. *Compos. Struct.* **2021**, *274*, 114335. [CrossRef]
5. Ross, C.T.F. A conceptual design of an underwater vehicle. *Ocean Eng.* **2006**, *33*, 2087–2104. [CrossRef]
6. Zhang, Y.; Cheng, L. Ultra-thin and broadband low-frequency underwater acoustic meta-absorber. *Int. J. Mech. Sci.* **2021**, *210*, 106732. [CrossRef]
7. Sharma, G.S.; Skvortsov, A.; MacGillivray, I.; Kessissoglou, N. Sound absorption by rubber coatings with periodic voids and hard inclusions. *Appl. Acoust.* **2019**, *143*, 200–210. [CrossRef]
8. Yu, T.; Jiang, F.; Wang, J.; Wang, Z.; Chang, Y.; Guo, C. Acoustic insulation and absorption mechanism of metallic hollow spheres composites with different polymer matrix. *Compos. Struct.* **2020**, *248*, 112566. [CrossRef]
9. Ye, C.; Liu, X.; Xin, F.; Lu, T.J. Underwater Acoustic Absorption of Composite Anechoic Layers With Inner Holes. *J. Vib. Acoust.* **2019**, *141*, 041006. [CrossRef]
10. Yang, H.; Xiao, Y.; Zhao, H.; Zhong, J.; Wen, J. On wave propagation and attenuation properties of underwater acoustic screens consisting of periodically perforated rubber layers with metal plates. *J. Sound Vib.* **2019**, *444*, 21–34. [CrossRef]
11. Fu, X.; Jin, Z.; Yin, Y.; Liu, B. Sound absorption of a rib-stiffened plate covered by anechoic coatings. *J. Acoust. Soc. Am.* **2015**, *137*, 1551–1556. [CrossRef] [PubMed]
12. Calvo, D.C.; Thangawng, A.L.; Layman, C.N., Jr.; Casalini, R.; Othman, S.F. Underwater sound transmission through arrays of disk cavities in a soft elastic medium. *J. Acoust. Soc. Am.* **2015**, *138*, 2537–2547. [CrossRef] [PubMed]
13. Wen, J.; Zhao, H.; Lv, L.; Yuan, B.; Wang, G.; Wen, X. Effects of locally resonant modes on underwater sound absorption in viscoelastic materials. *J. Acoust. Soc. Am.* **2011**, *130*, 1201–1208. [CrossRef] [PubMed]
14. Jayakumari, V.G.; Shamsudeen, R.K.; Ramesh, R.; Mukundan, T. Modeling and validation of polyurethane based passive underwater acoustic absorber. *J. Acoust. Soc. Am.* **2011**, *130*, 724–730. [CrossRef]
15. Shaid Sujon, M.A.; Islam, A.; Nadimpalli, V.K. Damping and sound absorption properties of polymer matrix composites: A review. *Polym. Test.* **2021**, *104*, 107388. [CrossRef]
16. Zhang, X.; Qu, Z.; Wang, H. Engineering Acoustic Metamaterials for Sound Absorption: From Uniform to Gradient Structures. *iScience* **2020**, *23*, 101110. [CrossRef]
17. Sharma, G.S.; Skvortsov, A.; MacGillivray, I.; Kessissoglou, N. Acoustic performance of periodic steel cylinders embedded in a viscoelastic medium. *J. Sound Vib.* **2019**, *443*, 652–665. [CrossRef]
18. Lee, T.; Iizuka, H. Heavily overdamped resonance structurally engineered in a grating metasurface for ultra-broadband acoustic absorption. *Appl. Phys. Lett.* **2018**, *113*, 101903. [CrossRef]
19. Wang, T.; Wang, G.-B.; Zhang, R.-J.; Ke, M.-Z. Low-frequency underwater sound absorption metamaterial. *Phys. Scr.* **2022**, *97*, 125706. [CrossRef]
20. Arjunan, A.; Baroutaji, A.; Robinson, J. Advances in Acoustic Metamaterials. In *Encyclopedia of Smart Materials*; Elsevier: Amsterdam, The Netherlands, 2022; pp. 1–10.
21. Ivansson, S.M. Markov-chain Monte Carlo identification of favorable design choices with application to anechoic coatings. *J. Acoust. Soc. Am.* **2014**, *135*, 3338–3351. [CrossRef]
22. Ye, C.; Liu, X.; Xin, F.; Lu, T.J. Influence of hole shape on sound absorption of underwater anechoic layers. *J. Sound Vib.* **2018**, *426*, 54–74. [CrossRef]
23. Lane, R. Absorption mechanisms for waterborne sound in Alberich anechoic layers. *Ultrasonics.* **1981**, *19*, 28–30. [CrossRef]
24. Zhao, D.; Zhao, H.; Yang, H.; Wen, J. Optimization and mechanism of acoustic absorption of Alberich coatings on a steel plate in water. *Appl. Acoust.* **2018**, *140*, 183–187. [CrossRef]
25. Gao, N.; Lu, K. An underwater metamaterial for broadband acoustic absorption at low frequency. *Appl. Acoust.* **2020**, *169*, 107500. [CrossRef]
26. Jin, G.; Shi, K.; Ye, T.; Zhou, J.; Yin, Y. Sound absorption behaviors of metamaterials with periodic multi-resonator and voids in water. *Appl. Acoust.* **2020**, *166*, 107351. [CrossRef]
27. Wang, Z.; Huang, Y.; Zhang, X.; Li, L.; Chen, M.; Fang, D. Broadband underwater sound absorbing structure with gradient cavity shaped polyurethane composite array supported by carbon fiber honeycomb. *J. Sound Vib.* **2020**, *479*, 115375. [CrossRef]
28. Zhong, J.; Zhao, H.; Yang, H.; Yin, J.; Wen, J. Effect of Poisson's loss factor of rubbery material on underwater sound absorption of anechoic coatings. *J. Sound Vib.* **2018**, *424*, 293–301. [CrossRef]
29. Shi, K.; Jin, G.; Liu, R.; Ye, T.; Xue, Y. Underwater sound absorption performance of acoustic metamaterials with multilayered locally resonant scatterers. *Results Phys.* **2019**, *12*, 132–142. [CrossRef]
30. Shi, K.; Jin, G.; Ye, T.; Zhang, Y.; Chen, M.; Xue, Y. Underwater sound absorption characteristics of metamaterials with steel plate backing. *Appl. Acoust.* **2019**, *153*, 147–156. [CrossRef]

31. Zhong, J.; Zhao, H.; Yang, H.; Yin, J.; Wen, J. On the accuracy and optimization application of an axisymmetric simplified model for underwater sound absorption of anechoic coatings. *Appl. Acoust.* **2019**, *145*, 104–111. [CrossRef]
32. Meng, T. Simplified model for predicting acoustic performance of an underwater sound absorption coating. *J. Vib. Control.* **2021**, *20*, 339–354. [CrossRef]
33. Yu, C.; Duan, M.; He, W.; Xin, F.; Lu, T.J. Underwater anechoic layer with parallel metallic plate insertions: Theoretical modelling. *J. Micromech. Microeng.* **2021**, *31*, 074002. [CrossRef]
34. Wang, T.; Liu, J.; Chen, M. Underwater sound absorption of a meta-absorption layer with double negativity. *Appl. Acoust.* **2021**, *181*, 108182. [CrossRef]

Journal of
Marine Science and Engineering

MDPI

Article

Analysis of Dynamic Response and Ultimate Strength for Box Girder under Bending Moment

Gui-Jie Shi [1,2], De-Yu Wang [1,2,*], Fu-Hua Wang [3] and Shi-Jian Cai [3]

1 State Key Laboratory of Ocean Engineering, Shanghai Jiao Tong University, Shanghai 200240, China
2 Institute of Marine Equipment, Shanghai Jiao Tong University, Shanghai 200240, China
3 Marine Design & Research Institute of China, Shanghai 200011, China
* Correspondence: dywang@sjtu.edu.cn

Abstract: The box girder can be seen as a kind of simplified ship structure that can withstand a vertical bending moment. Dynamic loads play an important role in structural safety analysis, such as ship bow slamming during harsh sea conditions. In this paper, the dynamic elastic–plastic response and ultimate strength of a box girder under a bending moment are analyzed. A box girder with the same cross-section scantlings and span length as the Nishihara experiment is selected as the analysis object. Based on the model experiment results, the non-linear FE analysis method is validated to capture the ultimate strength of a box girder under bending moment. Then, six box girders were designed to study the critical influence factors on the dynamic ultimate moment, including the model length, plate thickness, mass density and load excitation period. On the basis of structural dynamic response, an evaluation criterion of dynamic limit state for the box girder under a bending moment is proposed in this paper. Compared with the static ultimate moment, the change in the dynamic ultimate moment is discussed in detail to obtain the general principal method for dynamic strength analysis. The conclusions in this paper can provide guidance for dynamic ultimate strength evaluation.

Keywords: dynamic failure; dynamic ultimate strength; box girder; bending moment

Citation: Shi, G.-J.; Wang, D.-Y.; Wang, F.-H.; Cai, S.-J. Analysis of Dynamic Response and Ultimate Strength for Box Girder under Bending Moment. *J. Mar. Sci. Eng.* **2023**, *11*, 373. https://doi.org/10.3390/jmse11020373

Academic Editor: Erkan Oterkus

Received: 1 January 2023
Revised: 18 January 2023
Accepted: 4 February 2023
Published: 8 February 2023

1. Introduction

Recent catastrophic incidents involving large ships, such as the MSC Napoli report [1] and the MOL Comfort report [2], show that dynamic load conditions may be one of the causes of structural failure. At present, the ultimate strength of a hull girder is usually evaluated under static loading conditions, and some partial safety factors are introduced to cover the dynamic effect. The whipping response of large ships has a higher frequency than the ordinary wave response; hence, the dynamic effects may provide additional strength reserves for the ship structure [3]. A partial safety factor [4] of 0.9 is proposed to reduce the effectiveness of the whipping moment in the ultimate strength check. For the lateral dynamic load condition, the dynamic strength of the beam under lateral impact could reach up to 1.4 times that of the static strength [5]. There are some large differences related to the structural dynamic strength evaluation that require further study.

The dynamic ultimate hull girder strength of a container ship under bending moments was studied by using the non-linear FE method [6]. The ultimate rotation of the cross-section around the neutral axis was used as the evaluation criterion to estimate the dynamic ultimate strength. When the duration of the dynamic bending moment was close to the natural period of the hull girder, the maximum cross-section rotation response was 1.6 times greater than that of the same amplitude of a static bending moment. The dynamic collapse under a whipping moment for a container ship was studied [7–9] to find that the dynamic ultimate strength increased by 5.9~8.4% compared to the static conditions. Therefore, it is necessary to carry out more studies to find a reasonable justification for the evaluation criteria for dynamic strength.

Compared with static loading conditions, the inertia terms might not be negligible in the case of a dynamic response when the load duration time is shorter than the natural period of the box girder [10]. The inertia force along the beam was commonly assumed to be linear under dynamic conditions, as proposed by [11], which caused a time-varying bending moment distribution along the beam. When the load duration was very short, the box girder could not react immediately, and a time lag took place [12]. These time lags caused differences between the external applied bending moment and the internal bending moment.

For a simply supported beam [13], the beam initially deformed at the mid-span in the first impact stage, and thereafter, the impact effect propagated to the end supports. When the dynamic bending moment exceeds some degree, the beam exhibits part failures, such as plastic hinges [14]. The local failure might present during the dynamic load action period and also might present during the free vibration stage after the action of the dynamic loads [15]. It was useful to find the critical dynamic load, which induced severe damage, for the safety analysis of a box girder.

Based on the dynamic load application method, the dynamic model tests could be categorized into two categories: the water basin test and the drop hammer test. These tests can obtain the characteristics of the dynamic varying over time, which can be used to guide structural dynamic response analysis. The dynamic ultimate strength under an impact load will show the structural carrying capacity and failure mode.

In the water basin test, the girder structure should be simplified and have a lower capacity, which could be damaged by the hydrodynamic force due to basin waves. A one-degree-of-freedom beam model was used to simulate the bending stiffness and then study the collapse behavior under whipping loads by using a numerical method [16] and the water basin test [17].

The drop hammer test has often been used to study the dynamic capacity of different structures. Three-point bending tests [18] were carried out to study the dynamic and quasi-static bending behavior of thin-walled aluminum tubes. A series of drop-weight tests [19] were carried out for the double-hat structure under lateral impacts to show their crashworthiness performance. A series of repeated impact tests [20] were carried out for in situ prestressed concrete sleepers, ranging from a low drop height to the limit drop height, where the ultimate failure occurred.

Several buckling criteria were proposed to determine the dynamic critical load, such as the B-R criterion [21], the plate thickness criterion [22] and so on. The B-R criterion states that dynamic stability loss occurs when the maximum plate deflection grows rapidly at a small variation in the load amplitude. The plate thickness criterion states that the dynamic critical load corresponds to the amplitude of the impact load at which the maximum plate deflection was equal to 0.5~1 times plate thickness. The box girder had more complex structures and a stable post-buckling path. It was necessary to study the evaluation criterion applicable to the dynamic ultimate moment of the box girder.

In summary, there are some problems that require further study, including the increase in the dynamic ultimate moment compared with static conditions, the expression of the dynamic ultimate moment index, critical dynamic-moment-induced girder failure and the evaluation criterion applicable to the dynamic ultimate moment.

In this paper, the box girder with the same cross-section scantlings and span length as the experiment [23] was selected as the object of the analysis. Based on the non-linear FE method verified by experiment results, the dynamic ultimate bending moment and evaluation criterion are studied in detail. Then, the influence factors, including the model length, plate thickness, material density and load duration time, are discussed.

2. Scantlings of Box Girder with Stiffened Panel

2.1. Analysis Object

Stiffened panels are a kind of fundamental element involved in ship structures that can withstand in-plane loads and lateral loads. In the longitudinal direction, the length of a stiffened panel between two neighboring strengthened transverse frames is a span. Several stiffened panels can be assembled to form a closed cross-section. The box girder is made of some spans of closed cross-sections with stiffened panels, which can withstand many types of applied loads, including the vertical bending moment perpendicular to the box girder length direction.

Box girders are also used in experimental research due to their rich mechanical properties and easy fabrication, which can be considered to be a representation of simplified hull girders. A series of model experiments [23] for eight box girders under static bending moments were carried out. These box girders have also been selected by other researchers to study the influencing factors that affect the ultimate strength, such as initial imperfections [24], pitting corrosion and cracking [25], and the modeling method in non-linear FE analysis [26].

One of these box girders in the experiment [23] named MST-3 was selected as the analysis object in this paper in order to study its dynamic ultimate strength. The choice of this frequently used model is convenient for showing the structural response differences between static and dynamic bending moments.

In the Nishihara experiment, four points bending facilities were used to simulate a vertical bending moment condition for the mid-span of the box girder. The mid-span is the critical focused area in the experiment. At each end, the box girder was placed on a foundation that had a round steel bar on its top. Two concentrated forces were applied by the load cell on the transverse frames of the two sides of the mid-span. The two forces were applied symmetrically and simultaneously on the transverse bulkheads, which were a certain distance from the mid-span of the box girder. Therefore, the middle part of the experiment would be under pure bending conditions. More details about the experiment can be found in the reference [23].

The cross-section and scantlings of box girder MST-3 are shown in Figure 1 and Table 1. The span length between the two neighboring transverse frames of the experiment model is 540 mm. The cross-section is a square with the same scantlings of stiffened panels on four boundaries. The thickness of the plating is 3.05 mm, which is not an integer due to the actual plate measurement. The height of the stiffener web is 50 mm. A mild steel is used in the box girder with a yield strength of 278 MPa. However, the initial imperfection of the MST-3 box girder has not been mentioned in the [23] experiment. Several amplitudes of initial imperfection should be analyzed in the following FE calculation.

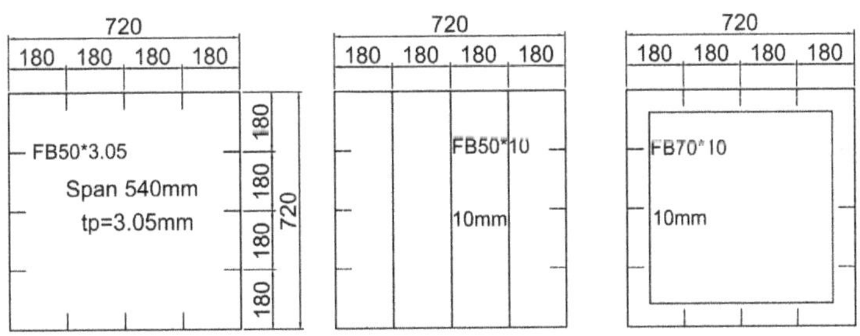

Figure 1. Cross-section of box girder MST-3 in the experiment [23]. (Span length = 540 mm; Plating thickness t_p = 3.05 mm; Stiffener FB50 × 3.05; Stiffener spacing = 180 mm).

Table 1. Scantlings of box girder MST-3 in the experiment [23].

Member	Scantlings	Yield Strength	Modulus	Span
	mm	MPa	MPa	mm
Plating	3.05	287	2.07×10^5	540
Stiffener	50×3.05	287	2.07×10^5	540

2.2. Experiment Results under Static Bending Moment

In the experiment, the upper panel under compression showed a buckling mode with several half-waves. As the bending moment increased, the side shell started to present with buckling. At the limit state, the upper panel and side shell both presented a large buckling deformation, and the bottom presented a large tension stress but no fracture. The curve of Nishihara's MST-3 model under a static bending moment is shown in Figure 2. The ultimate strength of the experiment model was about $M_{ult\_s} = 5.88 \times 10^8$ Nmm. Based on the linear beam theory, the yield moment of the girder is $M_y = 7.03 \times 10^8$ Nmm, which corresponds to the first yield point at the deck farthest from the neutral axis. The full plastic moment of the girder is $M_p = 8.24 \times 10^8$ Nmm, which corresponds to the full yielding of the whole cross-section. The yielding moment M_y and the fully plastic moment M_p for this box girder are both larger than the ultimate bending moment of M_{ult_s} obtained by the experiment. This is because the buckling strength is lower than the yielding strength of the material. Therefore, it is reasonable to use the ultimate strength in the safety assessment method.

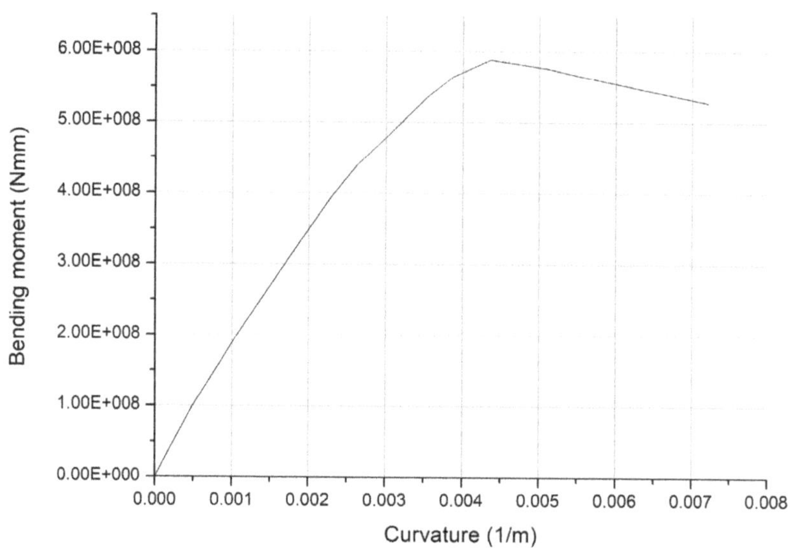

Figure 2. Load and curvature curve of box girder MST-3 under static bending moment in the experiment [23].

3. Non-Linear FE Analysis Method

The non-linear FE analysis method can be used to simulate the deformation failure process as the bending moment increases and obtain the ultimate strength of the bending moment by the extreme value of the load–deformation curve. Several parameters in the non-linear FE analysis, such as the boundary conditions, material properties, meshing density and imperfections, should be discussed and confirmed. The ABAQUS software is used in this paper.

3.1. Boundary Conditions and Loads

The coordinate system is defined as follows: X along the model length direction, Y along the model width direction, and Z along the model height direction.

The element type S4R for each element is chosen to simulate large deformation. This element has four nodes and six degrees of freedom in all nodes and a quadrilateral shape with reduced integration, hourglass control and a finite membrane strain.

The boundary conditions of the box girder are shown in Figure 3. The box girder model is assumed to be simply supported at the two ends. The aft end is constrained by the displacement degree of freedom Ux = Uy = Uz = 0 and the rotational degree of freedom URx = URz = 0 at the bottom edge. The fore-end is constrained by the displacement degree of freedom Uy = Uz = 0 and a rotational degree of freedom URx = URz = 0 at the bottom edge. Two forces, F1 = F2, along the vertical direction (Z axis), are applied as nodal forces on the upper edge of the bulkhead (BHD).

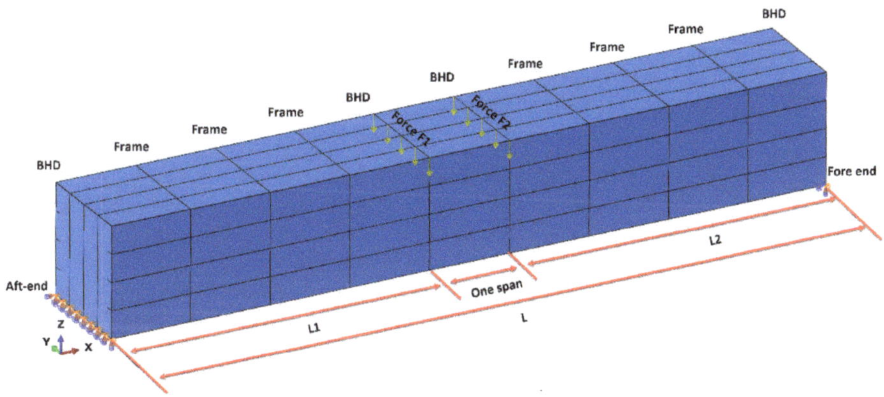

Figure 3. Boundary conditions of box girder.

The model width and height should include the whole section of the box girder. Usually, one-span or three-span are used to calculate the ultimate strength under static loading conditions. This paper will study the effect of more spans in the length direction on the dynamic ultimate strength.

3.2. Material Properties

As stated in the reference of [27], the yield strength for steel is usually higher than its nominal strength due to the positive tolerance measured by the tensile test results. The material properties based on the Nishihara test are used in this paper. The material of plates and stiffeners is modeled with Young's modulus of E = 206,000 MPa, Poisson ratio v = 0.3 and yield strength of σ_Y = 278 MPa. The material is assumed to be perfectly elastic–plastic without a strain-hardening effect.

In the case of considering the strain rate, the Cowper–Symonds model has proven to be applicable for steel structures by experiment results [28,29], as seen in the following Equation (1).

$$\frac{\sigma_d}{\sigma_s} = 1 + \left(\frac{\dot{\varepsilon}}{D}\right)^{1/q} \quad (1)$$

where σ_d denotes the dynamic stress, σ_s denotes the static stress; D and q are the material parameters. D = 40.4, q = 5 is used for the mild steel in the present study.

3.3. Meshing Density

Generally, a rectangular element with a length–width aspect ratio close to 1.0 was chosen for the non-linear FE analysis. Under the static loading condition, the meshing density of 6~10 elements in one plate width is enough to obtain a reasonable calculation result. Figure 4 shows that the calculation results of the ultimate bending moment of the box girder with 10 elements along one plate width is a little lower than that with four or six elements. The initial imperfection of two half-waves with an amplitude of 0.9 mm is applied in the analysis. When the element number exceeds 10, the calculation results of the ultimate bending moment show a stable value.

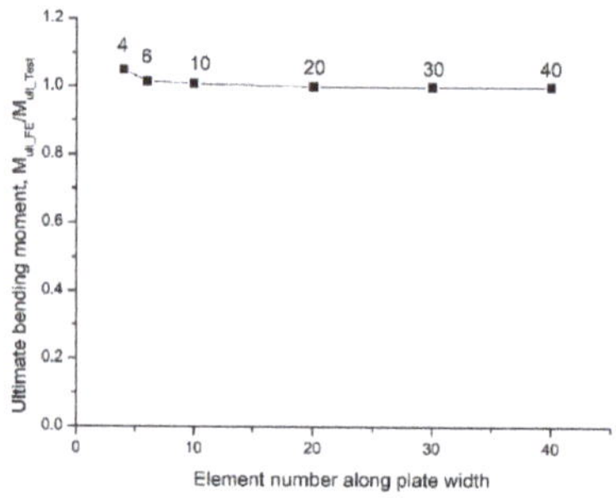

Figure 4. Ultimate bending moment of box girder with various meshing density under static load (initial imperfection assumed as 2 half-wave with the amplitude 0.9 mm).

In the dynamic explicit calculation method, the stability limit of the calculation has a relationship with the characteristic element length, i.e., the minimum element length. The shorter the element length is, the shorter the calculation time of the increase step. In order to control the calculation time of the structural response, it had better not use an overfine mesh density. In the following analysis, a mesh density with 20 elements along one plate width was used, as shown in Figure 5. The meshing size is about 9 mm for the FE analysis. There are 20 elements along the plating width, 60 elements along the plating length, six elements along the stiffener web height and eight elements along the frame web width.

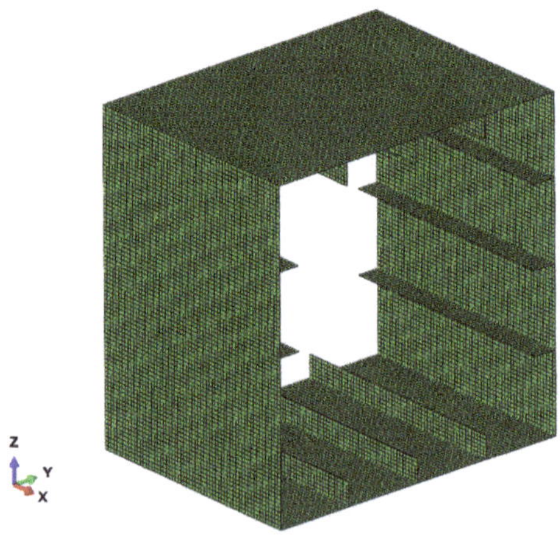

Figure 5. Mesh density of one span in the middle.

3.4. Initial Imperfections

For the simply supported plate, the buckling half-wave number m in plate length direction can be estimated by using the following Equation (2):

$$\frac{\mathrm{a}}{b} \leq \sqrt{m(m+1)} \quad (2)$$

where a is the plate length, b is the plate width and m is the buckling half-wave number.

For the box girder in this paper, a = 540 mm and b = 180 mm; therefore, the buckling half-wave number is m = 3. In order to trigger the failure mode, the initial imperfection can be assumed as the buckling mode with three half-waves in the plate length direction and one half-wave buckling mode in the plate width. This paper also considers the effect of initial imperfection with two half-waves and four half-waves in the plate length direction, as shown in Figure 6. The initial imperfection is applied to the compressed upper panel of a box girder under the bending moment.

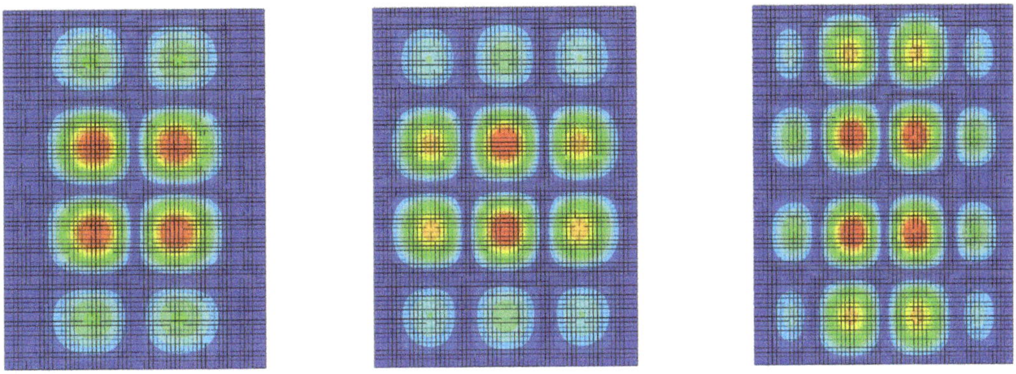

(**a**) (**b**) (**c**)

Figure 6. Initial imperfections with different half-waves along plate length and one half-wave along plate width between longitudinal stiffeners (**a**) 2 half-waves, (**b**) 3 half-waves, (**c**) 4 half-waves.

The effect of various imperfection types and various imperfection amplitudes on the ultimate bending moment of a box girder is shown in Figure 7. When the amplitude of the initial imperfection is the same, the imperfection with three half-waves gives the lowest estimation of the ultimate bending moment. For the same type of imperfection, the ultimate bending moment almost shows a linear reduction with the increase in the imperfection amplitude.

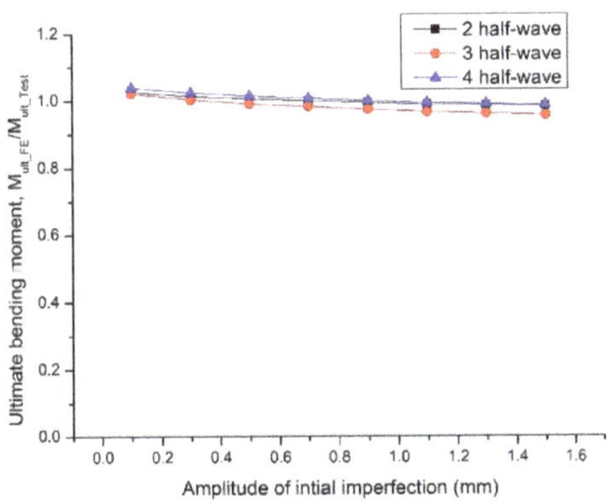

Figure 7. Ultimate bending moment of box girder with various imperfection under static load.

In the following analysis, the initial imperfection type is assumed to be in the three-half-wave buckling mode in the plate length direction, and the imperfection amplitude is assumed to be $b/200 = 180/200 = 0.9$ mm. This assumption is often adopted in ship rule research in the IACS (2014) classification societies.

The comparison of non-linear FE results and experiment results is shown in Figure 7. Through the use of FE analysis, the ultimate bending moment of the box girder is $M_{ult\_s} = 5.74 \times 10^8$ Nmm, and the displacement of the mid-span at the limit state is $W_{ult\_s} = 14.6$ mm. The value of the ultimate bending moment and displacement at the limit state by FE is almost close to the experiment result under the static condition. However, the load and displacement curves of the FE results and experiment results are not perfectly consistent. Perhaps, there are some differences between the actual imperfection in the experiment and the assumed imperfection in the FE analysis.

Under the static load conditions, the Riks method is used to carry out non-linear ultimate strength analysis. Riks' method is based on the load increment controlled by a scalar parameter of the arc length. In the following dynamic analysis, the Riks method should be replaced by the Dynamic Explicit solver, which is based on the central difference algorithm and easily obtains the convergent solution.

4. Dynamic Ultimate Strength of Box Girder under Bending Moment

The validated non-linear FE analysis method, including the boundary conditions, material properties, meshing density and initial imperfection, are extended to calculate the dynamic response of the box girder. Compared with the static load conditions, several differences should be specially considered in dynamic analysis in the time domain, including mass density, strain rate and the load change over time.

Based on the box girder of MST-3 in the Nishihara experiment, six box girder models are designed with the same cross-section scantlings and span length, as listed in Table 2. Compared with each model, the model length, plate thickness and material density have a slight difference in order to study the relevant influence factors.

Table 2. Box girder models for dynamic ultimate strength analysis.

Model	Model Length	Plate Thickness	Yielding Strength	Material Density	Static Ultimate Moment	Ultimate Displacement at Mid-Span	Natural Vibration Period
	L, mm	t_p, mm	R_{eH}, MPa	ρ, t/mm^3	M_{ult_s}, Nmm	W_{ult_s}, mm	T_0, s
M1	3780	3.05	278	7.89×10^{-9}	5.71×10^{8}	6.1	0.011
M2	5940	3.05	278	7.89×10^{-9}	5.74×10^{8}	14.6	0.025
M3	5940	4.0	278	7.89×10^{-9}	9.46×10^{8}	15.3	0.023
M4	5940	5.0	278	7.89×10^{-9}	1.29×10^{9}	17.4	0.022
M5	5940	3.05	278	1.58×10^{-8}	5.74×10^{8}	14.6	0.029
M6	5940	3.05	278	2.37×10^{-8}	5.74×10^{8}	14.6	0.034

Note: T_0 corresponds to the fundamental natural vibration mode of the box girder models.

The ultimate bending moment under the static conditions is also listed in Table 2. Models M1, M2, M5 and M6 almost have the same ultimate bending moment. That is to say, model length and material density have little effect on the calculation results under static conditions. However, the natural vibration periods of the six models are different. Therefore, the dynamic response of these models may also be different under the same dynamic loads.

4.1. Dynamic Bending Moment of Box Girder

The dynamic loads of F1 = F2 are applied to the upper edge of the bulkhead (BHD), as shown in Figure 3. Based on the water basin test of the girder under whipping loads [17] and the falling hammer test of the girder under impact loads [19], the load time history in the action period can be simplified as a half-sine shape with time. Therefore, the dynamic loads are assumed as a half-sine shape in this paper. The mid-ship bending moment can be evaluated according to the theory of simple support girder [30]. The external applied bending moment by the applied load at the mid-span of the box girder can be calculated as following Equation (3):

$$M(t) = \begin{cases} F_1 \cdot L_1 \cdot \sin\left(\frac{2\pi t}{T_e}\right) & if\, 0 \le t \le T_d \\ 0 & if\, t > T_d \end{cases} \tag{3}$$

where t is the response calculation time, T_d is the load duration time, T_e is the excitation period $T_e = 2 * T_d$, F_1 is the applied dynamic load amplitude and L_1 the distance between the applied load position and the pinned point.

As the load duration time T_d is changed, the excitation period T_e will be updated, leading to a specific ratio to the natural vibration period T_0 of the box girder.

The external applied dynamic bending moment will be withstood by the box girder deformation and inertia forces. Owing to the box girder deformation, the internal bending moment at each cross-section can be derived. Under the static load conditions, the internal bending moment is equal to the external applied bending moment at the same cross-section. However, under dynamic load conditions, the internal bending moment may be smaller than the external applied bending moment due to the inertia effect when the dynamic duration time is very short.

In the early dynamic stage, the internal moment at the mid-span is large and the upper stiffened panel is in compression, which is similar as the sagging moment. After the peak of the dynamic load, the bending moment at the mid-span reduces progressively, and then the opposite bending deformation of the box girder may be activated, leading to the opposite value of the internal bending moment, which is similar to the hogging moment. Therefore, the sagging–hogging internal moment will occur and alternately change over time, as shown in Figure 8. When the applied external bending moment reduces to 0 at a response calculation time of $t > T_d$ in Equation (3), the internal bending moment will still occur due to the girder vibration deformation.

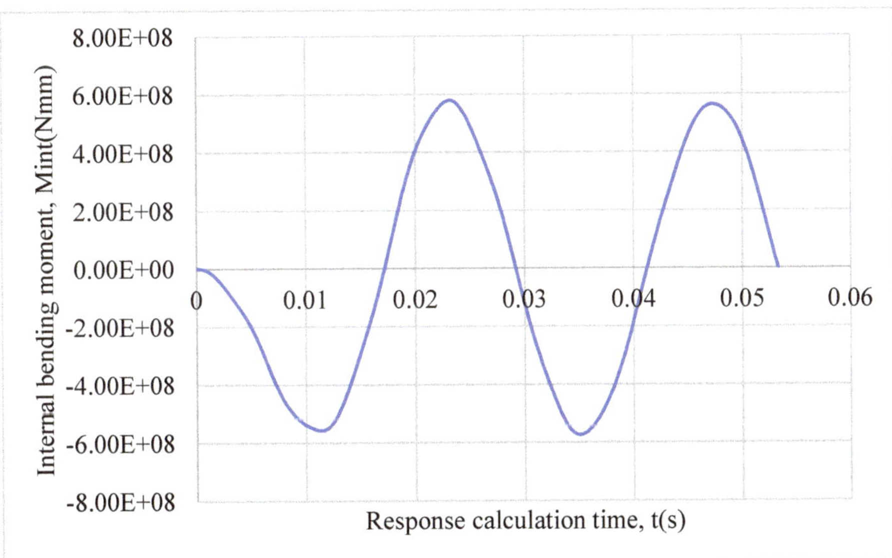

Figure 8. Internal bending moment of mid-section when the applied dynamic bending moment $M_d = 1.0M_{ult\_s}$ and the dynamic load duration time $T_d = 0.9\ T_0$.

The external applied bending moment is taken as the capacity index of the box girder in this paper, which is convenient for evaluating the external dynamic load condition.

4.2. Dynamic Response of Box Girder under Bending Moment

For the M2 box girder, a series of different amplitudes of dynamic bending moment M_d are applied as a sinusoidal shape, including $1.0M_{ult\_s}$, $2.0M_{ult\_s}$, $3.0M_{ult\_s}$ and $4.0M_{ult\_s}$. The load duration time is taken as $T_d = 0.005$ s, which is close to the drop hammer impact time in the experiment on the box girder [31]. The load duration time T_d is about 0.2 times that of the natural vibration period T_0 of the M2 box girder. Therefore, the box girder will show an apparent dynamic response under this load assumption.

The response calculation time includes the load duration time and two vibration periods, with no external applied loads to evaluate the free vibration response after the dynamic load action stage. The response calculation time is divided by half of the natural vibration period of T_0.

The vertical displacement W_d at the mid-span is derived by averaging all of the node displacements of the cross-section and then divided by the limit displacement W_{ult_s} of the same cross-section under a static bending moment. This treatment can yield a dimensionless index of vertical displacement and is convenient for comparing the dynamic and static responses.

The vertical displacement responses at the mid-span as the response calculation time for the M2 box girder are shown in Figure 9. For M_d = 1.0 M_{ult_s}, the box girder shows an elastic response, and the hogging (bending upwards)–sagging (bending downwards) vibration deformation amplitude is almost symmetrical. For M_d = 2.0 M_{ult_s}, the sagging deformation is larger than the hogging. For M_d = 3.0 M_{ult_s}, the vibration mode has been changed, and the value of dynamic deformation is always negative due to the large plastic deformation after dynamic load. For M_d = 4.0 M_{ult_s}, the vibration period T_1 (between two neighboring peaks in the free vibration stage) after dynamic load action exceeds the natural vibration period T_0 due to large failure areas induced by local buckling and plastic deformation.

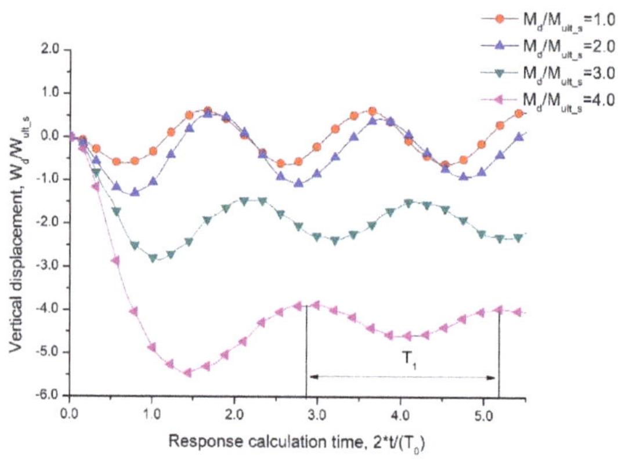

Figure 9. Vertical displacement response at mid-span as the response calculation time for box girder M2 with load duration time T_d = 0.005 s.

The Mises stress and deformation distribution at the time of the maximum vertical displacement for the mid-span of the M2 box girder are shown in Figure 8. Only in the load case of M_d = 1.0M_{ult_s} was the stress lower than the static yield strength. The maximum value of the Mises stress for the load case of M_d = 4.0M_{ult_s} is about 400 MPa, which is higher than the static yield strength. Based on Equation (1), the strain rate for the box girder is about 1/s~10/s, which shows the dynamic characteristics.

The deformed shape is also shown in Figure 10. There are three half-wave buckling modes in the upper panel of the box girder. The difference between load case M_d = 2.0M_{ult_s} and M_d = 3.0M_{ult_s} is evident. For load case M_d = 2.0M_{ult_s}, there is one cross-section with apparent deformation. However, for load case M_d = 3.0M_{ult_s}, there are three cross-sections with apparent deformation at the same time, which is different from the limit deformation under static load conditions.

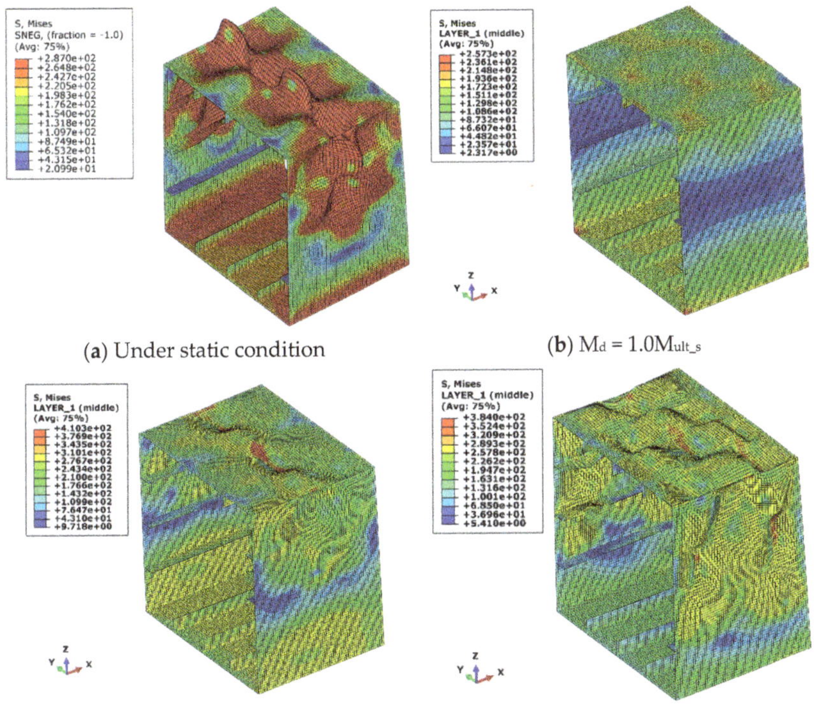

(**a**) Under static condition (**b**) $M_d = 1.0M_{ult\_s}$

(**c**) $M_d = 2.0M_{ult\_s}$ (**d**) $M_d = 3.0M_{ult\_s}$

Figure 10. Mises stress and deformation distribution at the time of maximum vertical displacement for the mid-span of box girder M2 with load duration time T_d = 0.005 s.

4.3. Evaluation Criterion of Dynamic Ultimate Strength for Box Girder

Based on the relationship between dynamic loads and structural dynamic responses, the dynamic ultimate strength evaluation is meaningful in guiding structure design and safety evaluation. However, the bending moment and vertical displacement curve under dynamic loads will not show an extreme value or sudden change because of the stable post-buckling path for the box girder. A reasonable evaluation criterion for dynamic ultimate strength is needed.

The B-R criterion [21] was proposed to evaluate the dynamic buckling of a plate. For box girders with stiffened panels, plate buckling can be regarded as one kind of local buckling. The box girder will have a stable post-buckling path, so the load–displacement will not present a severely rapid increase as the plate buckling. The B-R criterion should be revised based on the response characteristics of the box girder under a dynamic bending moment.

The internal bending moment can be derived based on the bending stresses and their locations in the cross-section of a hull girder. The curve of the internal bending moment and vertical displacement at the mid-span section is shown in Figure 11. The applied bending moment ratio (M_d/M_{ult_s}) is also plotted in Figure 11 using data labels.

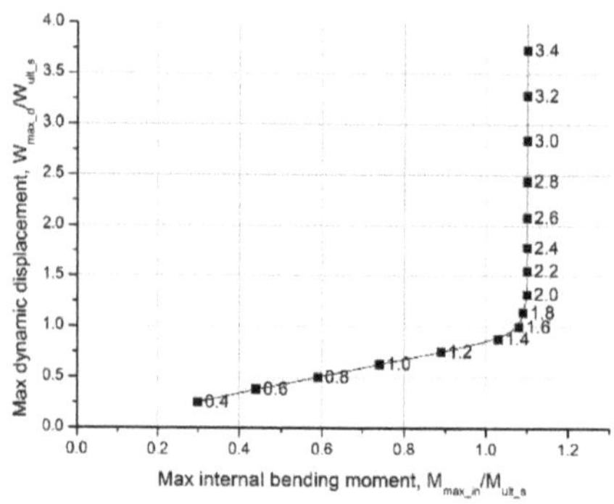

Figure 11. Internal bending moment and vertical displacement for mid-span box girder M2 at load duration time T_d = 0.005 s (Data label denotes M_d/M_{ult_s}).

When the applied bending moment ratio is small, i.e., $M_d/M_{ult\_s} < 1.4$, the vertical displacement shows an almost linear trend of increase as an increase in the bending moment. When the applied bending moment ratio is $1.4 > M_d/M_{ult\_s} < 2.0$, the curve of the internal bending moment and vertical displacement becomes non-linear, which denotes some local failure induced by buckling and yielding occurring on the box girder. When the applied bending moment ratio is $M_d/M_{ult\_s} \geq 2.0$, the internal bending moment at the mid-span of the box girder reaches its peak value. The internal bending moment will not increase as the applied bending moment, but the max displacement at the mid-span shows a rapid increase. It is an obvious signal that the box girder has reached its dynamic limit state. Therefore, the evaluation criterion of the dynamic limit state of the box girder is the curve of the internal bending moment, and vertical displacement reaches the extreme point.

For the load duration time of T_d = 0.01 s and T_d = 0.016 s, the curves of the internal bending moment and vertical displacement are shown in Figures 12 and 13. The same trend of the extreme point can be found. The proposed evaluation criterion of the dynamic limit state can be applicable to more load cases.

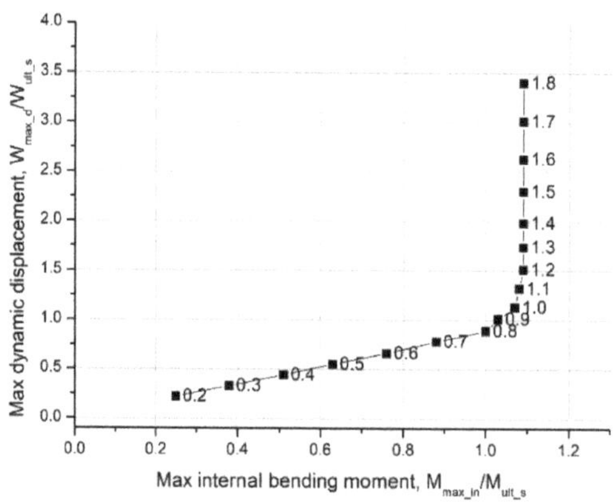

Figure 12. Internal bending moment and vertical displacement for mid-span box girder M2 at load duration time T_d = 0.01 s (Data label denotes M_d/M_{ult_s}).

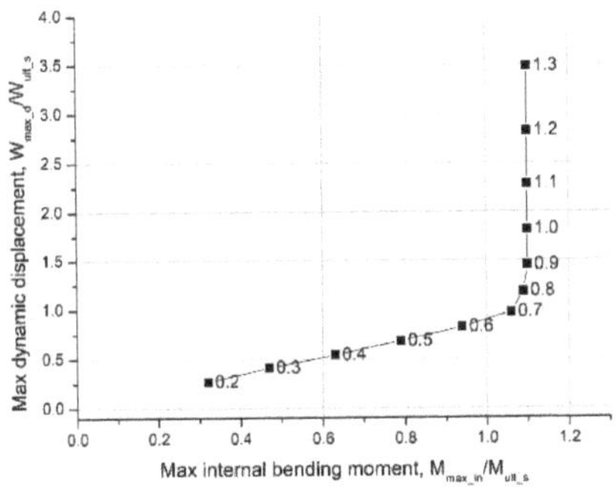

Figure 13. Internal bending moment and vertical displacement for mid-span box girder M2 at load duration time T_d = 0.016 s (Data label denotes M_d/M_{ult_s}).

The applied bending moment at the dynamic limit state for the load duration time T_d = 0.005 s, T_d = 0.01 s and T_d = 0.016 s are summarized in Table 3. As the load duration time increases, the dynamic effect for the box girder is not very obvious, and the applied bending moment at the dynamic limit state will approach the ultimate bending strength at the static limit state.

Table 3. Dynamic ultimate strength of the box girder.

Model	Load Duration Time		Applied Bending Moment at Dynamic Limit State
	T_d, s	$2T_d/T_0$	M_{ult_d}/M_{ult_s}
M2	T_d = 0.005 s	0.4	2.0
M2	T_d = 0.01 s	0.8	1.2
M2	T_d = 0.016 s	1.3	1.0

5. Influence Factors on Dynamic Ultimate Strength of Box Girder

For box girders, there are several influencing factors on dynamic ultimate strength. This paper tries to study these four factors, including model length, plate thickness, material density and load duration time.

5.1. Effect of Model Length

Each span of the box girder is strengthened by transverse frames or bulkheads at two ends. The model length is set as the product of span number and span length. The box girder M1 has seven spans and a 3780 mm model length. The box girder M2 has 11 spans and a 5940 mm model length. The ratio of model length L and span length S appears to be a non-dimensional parameter. Both models have the same cross-section scantlings and span length. These models are assumed to have the initial imperfection of the three half-wave, with an amplitude of 0.9 mm of mid-span.

As we all know, the span number in the analysis model has little effect on the static ultimate bending moment when the span number is larger [32] and the span length is kept unchanged. Therefore, two- or three-span FE models are often used in static ultimate strength analysis for box girders.

Under dynamic conditions, the curve of the internal bending moment and vertical displacement at the mid-span section for the M1 box girder with load duration times of T_d = 0.005 s and T_d = 0.016 s are shown in Figures 14 and 15. Based on the proposed evaluation criterion, the applied bending moment ratio M_{ult_d}/M_{ult_s} at the dynamic limit state is listed in Table 4. For the load duration time T_d = 0.005 s, the dynamic ultimate moment of box girder M1 (length ratio L/S = 7) is lower than that of box girder M2 (length ratio L/S = 11). However, for a load duration time of T_d = 0.016 s, there is no obvious dynamic effect, and the ultimate moment of box girder M1 and M2 is the same. It can be concluded that the model length ratio has a great influence on the dynamic ultimate moment at shorter load duration times. The longer model shows a larger dynamic carrying capacity. There are two reasons to explain this phenomenon. (1) The longer model has a greater inertia load along the model length direction, which needs a greater applied load to cause structural failure; (2) the longer model has a longer first vibration period, which shows more obvious dynamic effects for the load with duration time less than the first vibration period. It is recommended that a full-length model should be used in dynamic strength analysis instead of the simplified one-span model.

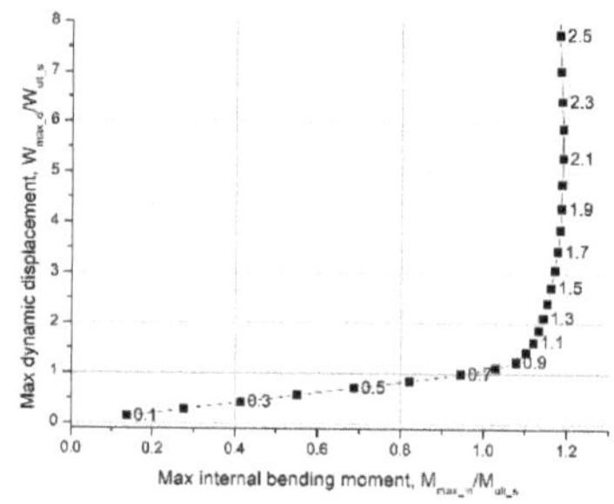

Figure 14. Internal bending moment and vertical displacement for mid-span box girder M1 at load duration time T_d = 0.005 s (Data label denotes M_d/M_{ult_s}).

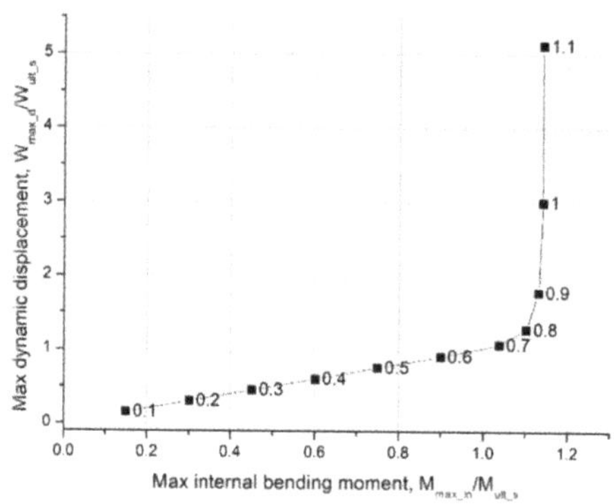

Figure 15. Internal bending moment and vertical displacement for mid-span box girder M1 at load duration time T_d = 0.016 s (Data label denotes M_d/M_{ult_s}).

Table 4. Comparison of dynamic ultimate moment for box girder with various length.

Model	Model Length	Length Ratio	Load Duration Time	Applied Bending Moment at Dynamic Limit State
	L, mm	L/S	T_d, s	M_{ult_d}/M_{ult_s}
M1	3780	7	0.005	1.7
M2	5940	11	0.005	2.0
M1	3780	7	0.016	1.0
M2	5940	11	0.016	1.0

5.2. Effect of Plate Thickness

The plate thickness is the critical parameter of structure stiffness. Three kinds of plate thickness t_p are set on the plate and stiffener web of the box girder. For box girders M2,

M3 and M4, the plate thicknesses are $t_p = 3.05$, $t_p = 4.0$ and $t_p = 5.0$, respectively. The plate slenderness ratio β can be considered a non-dimensional parameter. As the plate thickness increases, the natural vibration period T_0 has a little decrease, as shown in Table 2.

The curve of the internal bending moment and vertical displacement at the mid-span section for box girder M3 and M4 with a load duration time of $t_d = 0.005$ s are shown in Figures 16 and 17. The comparison of dynamic ultimate moment for box girder with a plate slenderness ratio of β is listed in Table 5. As the plate slenderness ratio of β changes, the applied bending moment ratio at the dynamic limit state is almost unchanged. That is to say, the dynamic effect for these girders will show a similar trend for the dynamic load with the same load duration time. It should be noted that the actual dynamic bending capacity of models M3 and M4 is higher than that of model M2 due to the increased static ultimate strength M_{ult_s}, as shown in Table 2.

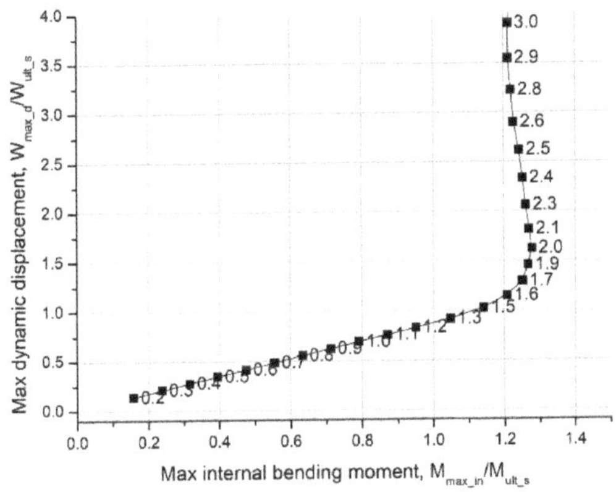

Figure 16. Internal bending moment and vertical displacement for mid-span box girder M3 at load duration time $T_d = 0.005$ s (Data label denotes M_d/M_{ult_s}).

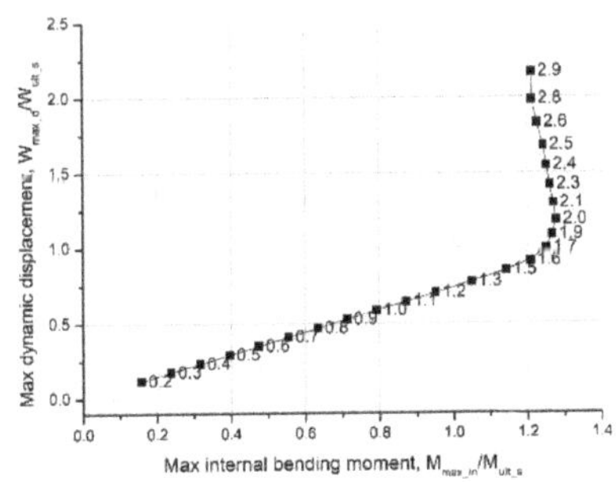

Figure 17. Internal bending moment and vertical displacement for mid-span box girder M4 at load duration time $T_d = 0.005$ s (Data label denotes M_d/M_{ult_s}).

Table 5. Comparison of dynamic ultimate moment for box girder with various plate thickness.

Model	Plate Thickness	Plate Slenderness Ratio	Load Duration Time	Applied Bending Moment at Dynamic Limit State
	t_p, mm	β	T_d, s	M_{ult_d}/M_{ult_s}
M2	3.05	2.2	0.005	2.0
M3	4	1.7	0.005	2.0
M4	5	1.3	0.005	2.0

5.3. Effect of Material Density

In actual ship structures, the box girder should withstand its structural weight and non-structure weight, including liquid/dry cargo, ballast water and fuel oil and so on. For dynamic response analysis, the changes of the mass point and density are often used to simulate various mass distribution conditions. The change in density is used in this paper to study the effect of mass distribution on dynamic ultimate moment.

In order to model the supplementary onboard distributed masses, two other kinds of equivalent material density are assumed in models M5 and M6, as shown in Table 2. The box girders of M5 and M6 are set as two times and three times of material density of M2, respectively. Although M2, M5 and M6 have the same cross-section scantlings, the box girders of M5 and M6 have longer natural vibration periods than that of M2, as shown in Table 2.

The bending moment and vertical displacement curve of the applied bending moment and vertical displacement at the mid-span section for box girders M5 and M6 with a load duration time of t_d = 0.005 s are shown in Figures 18 and 19. The relative density between the analysis model and the based model (M2) is defined. A comparison of the dynamic ultimate moment of the box girders with various densities is listed in Table 6. The larger the model's relative density, the higher the dynamic ultimate moment ratio. Material density should be assumed to be a critical factor in dynamic ultimate strength analysis.

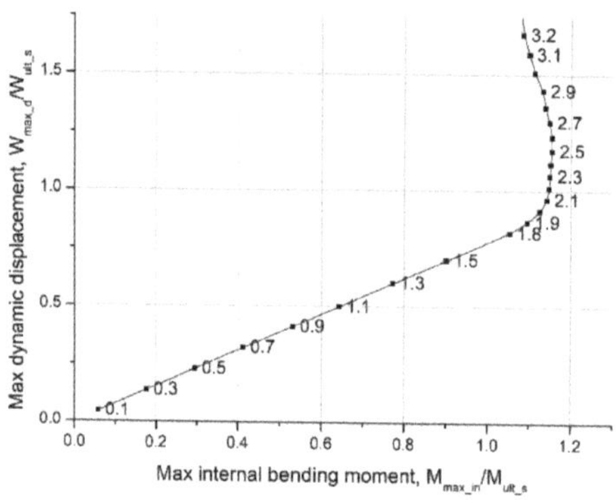

Figure 18. Internal bending moment and vertical displacement for mid-span box girder M5 at load duration time T_d = 0.005 s (Data label denotes M_d/M_{ult_s}).

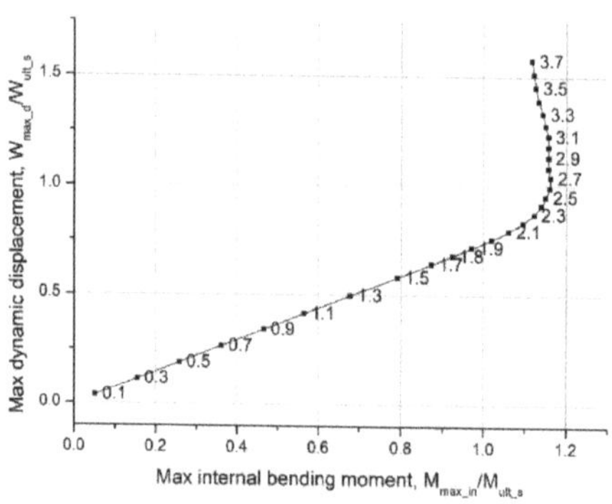

Figure 19. Internal bending moment and vertical displacement for mid-span box girder M6 at load duration time T_d = 0.005 s (Data label denotes M_d/M_{ult_s}).

Table 6. Comparison of dynamic ultimate moment of box girder with various density.

Model	Material Density	Relative Density	Load Duration Time	Applied Bending Moment at Dynamic Limit State
	ρ, t/mm³	ρ/ρ_{M2}	T_d, s	M_{ult_d}/M_{ult_s}
M2	7.89×10^{-9}	1.0	0.005	2.0
M5	1.58×10^{-8}	2.0	0.005	2.5
M6	2.37×10^{-8}	3.0	0.005	2.7

5.4. Effect of the Excitation Period

This paper assumes that the dynamic load has a half-sine shape and the start time as the beginning of the response calculation. The load duration time T_d is set from 0.004 to 0.025 s, which covers the scope of drop hammer impact on stiffened panel of steel and also the interval of box girder vibration period. The excitation period is two times that of the duration time, i.e., $T_e = 2 \times T_d$. The ratio T_e/T_0 between the excitation period and the natural vibration period can be considered a non-dimensional parameter.

The comparison of dynamic ultimate moment ratio of box girder with various load excitation period ratios T_e/T_0 is listed in Table 7. The shorter the load duration time ratio, the larger the dynamic ultimate moment. The dynamic ultimate moment ratio M_{ult_d}/M_{ult_s} is within the scope from 1.0 to 2.5 for the load duration time ratio of larger than 0.3. When the load excitation period ratio T_e/T_0 is larger than 1.3, the dynamic effect is not very obvious for box girder under bending moment.

Table 7. Comparison of applied ultimate moment ratio at dynamic limit state of box girder (Model M2) with various load duration time.

Load Excitation Period	Load Excitation Period Ratio	Applied Bending Moment at Dynamic Limit State
$T_e = 2 \times T_d$, s	T_e/T_0	M_{ult_d}/M_{ult_s}
0.008	0.3	2.3
0.010	0.4	2.0
0.016	0.7	1.4
0.020	0.8	1.2
0.026	1.1	1.1
0.032	1.3	1
0.036	1.5	1
0.050	2.0	1

Based on the FE results evaluated by proposed criterion, the relationship of dynamic ultimate moment ratio M_{ult_d}/M_{ult_s} and the load excitation period ratio T_e/T_0 is plotted in Figure 20. When the load duration time ratio increases, the dynamic ultimate moment ratio gradually decreases.

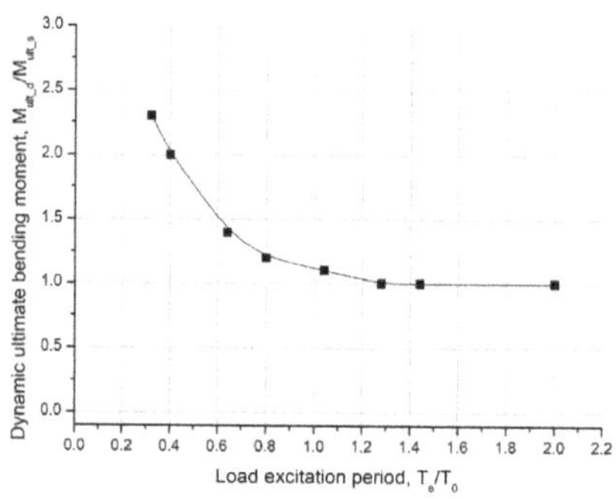

Figure 20. Effect of load excitation period on dynamic ultimate moment of box girder M2.

6. Conclusions

The main conclusions can be drawn as follows:

(1) Under larger applied dynamic moments, the box girder vibration period will be increased due to local buckling and plastic deformation. In the dynamic loading case, there are more than one cross-section with apparent deformation at the same time, which is different from only cross-section failure in the static loading case.

(2) An evaluation criterion of dynamic limit state for box girder under bending is proposed in this paper based on the curve of internal bending moment and vertical displacement, which can give reasonable prediction of dynamic ultimate strength.

(3) The model length has a great influence on the applied dynamic ultimate moment at shorter load duration times. The longer model needs a larger dynamic moment to reach the limit state. The full-length model is recommended to use in dynamic analysis.

(4) The larger the relative density of the box girder, the higher its dynamic ultimate moment ratio M_{ult_d}/M_{ult_s} due to the inertia effect. The material density should be considered a critical factor in dynamic ultimate strength analysis.

(5) When the load excitation period ratio T_e/T_0 is larger than 1.3, the dynamic effect is not very obvious for box girder bending, and the load action will be close to the quasistatic condition. For the load excitation period ratio T_e/T_0 with the scope of 0.3 and 1.3, the dynamic ultimate moment ratio M_{ult_d}/M_{ult_s} is within the scope of 2.5 to 1.0.

Author Contributions: Writing-original, G.-J.S.; Study design, D.-Y.W.; Formal analysis, F.-H.W.; Validation, S.-J.C. All authors have read and agreed to the published version of the manuscript.

Funding: This research was funded by National Natural Science Foundation of China (Grant No.51809168, Grant No.51979163, and Grant No. U2241266).

Data Availability Statement: Data will be made available on request.

Conflicts of Interest: The authors declare that they have no known competing financial interests or personal relationships that could have appeared to influence the work reported in this paper.

Nomenclature

M_{ult_s}	Bending moment at the limit state of box girder under static condition
M_{ult_d}	Bending moment at the limit state of box girder under dynamic condition
M_{max_in}	Max internal bending moment of box girder
M_d	Amplitude of applied dynamic bending moment of box girder
W_{ult_s}	Vertical displacement along Z direction of box girder mid-span section at the limit state under static condition
W_d	Vertical displacement along Z direction of box girder mid-span section during dynamic load

W_{max_d}	Maximum vertical displacement along Z direction of box girder mid-span section in the whole response calculation time (including dynamic load duration time and free vibration time after the dynamic load)
W_{p_d}	Permanent displacement after dynamic loading along Z direction of box girder mid-span section after dynamic load
t	Response calculation time
T_0	Natural vibration period of box girder
T_1	Vibration period (between two neighboring peaks in free vibration stage) of box girder after dynamic load action
T_d	Dynamic load duration time
L	Analysis model length
S	Span length between neighboring transverse frame
β	Plate slenderness ratio $\beta = b/t\sqrt{\sigma_Y/E}$
X axis	Along the box girder length direction
Y axis	Along the box girder width direction
Z axis	Along the box girder height direction

References

1. Al-Rifaie, A.; Guan, Z.W.; Jones, S.W.; Wang, Q. Lateral impact response of end-plate beam-column connections. *Eng. Struct.* **2017**, *151*, 221–234. [CrossRef]
2. Budiansky, B.; Roth, R. Axisymmetric buckling of clamped shallow spherical shells. *NASA TND* **1962**, *1510*, 597–606.
3. CLCSS. Final Report of Committee on Large Container Ship Safety. Japan Committee on Large Ship Safety. 2015. Available online: https://www.mlit.go.jp/common/001081297.pdf (accessed on 30 January 2023).
4. Derbanne, Q.; De Lauzon, J.; Bigot, F.; Malenica, S. Investigations of the dynamic ultimate strength of a ship's hull girder during whipping. In Proceedings of the PRADS, Copenhagen, Denmark, 4–8 September 2016.
5. DNV-GL. Classfication Notes No.30.12: Fatigue and Ultimate Strength Assessment of Container Ships Including Whipping and Springing. 2015. Available online: https://global.ihs.com/doc_detail.cfm?document_name=DNVGL%2DCG%2D0153&item_s_key=00663928 (accessed on 30 January 2023).
6. Duarte, I.; Vesenjak, M.; Krstulović-Opara, L. Dynamic and quasi-static bending behaviour of thin-walled aluminium tubes filled with aluminium foam. *Compos. Struct.* **2014**, *109*, 48–56. [CrossRef]
7. Hansen, A.M. Strength of midship sections. *Mar. Struct.* **1996**, *9*, 471–494. [CrossRef]
8. Hu, K.; Yang, P.; Xia, T.; Peng, Z. Residual ultimate strength of large opening box girder with crack damage under torsion and bending loads. *Ocean Eng.* **2018**, *162*, 274–289. [CrossRef]
9. Huang, Z.; Zhang, X. Three-point bending collapse of thin-walled rectangular beams. *Int. J. Mech. Sci.* **2018**, *144*, 461–479. [CrossRef]
10. Iijima, K.; Suzaki, Y.; Fujikubo, M. Scaled model tests for the post-ultimate strength collapse behaviour of a ship's hull girder under whipping loads. *Ships Offshore Struct.* **2015**, *10*, 31–38. [CrossRef]
11. Jagite, G.; Bigot, F.; Derbanne, Q.; Malenica, Š.; Le Sourne, H.; de Lauzon, J.; Cartraud, P. Numerical investigation on dynamic ultimate strength of stiffened panels considering real loading scenarios. *Ships Offshore Struct.* **2019**, *14*, 374–386. [CrossRef]
12. Jagite, G.; Bigot, F.; Malenica, S.; Derbanne, Q.; Le Sourne, H.; Cartraud, P. Dynamic ultimate strength of a ultra-large container ship subjected to realistic loading scenarios. *Mar. Struct.* **2022**, *84*, 103197. [CrossRef]
13. Jagite, G.; le Sourne, H.; Cartraud, P.; Bigot, F.; Derbanne, Q.; Malenica, Š. Dynamic ultimate strength of a container ship under sagging condition. In *International Conference on Offshore Mechanics and Arctic Engineering*; American Society of Mechanical Engineers: New York, NY, USA, 2020.
14. Jagite, G.; Le Sourne, H.; Cartraud, P.; Bigot, F.; Derbanne, Q.; Malenica, Š. Examination of the dynamic effects on the hull girder ultimate strength of ultra large container ships. *Trends Anal. Des. Mar. Struct.* **2019**, 137–148.
15. Kaewunruen, S.; Remennikov, A.M. Progressive failure of prestressed concrete sleepers under multiple high-intensity impact loads. *Eng. Struct.* **2009**, *31*, 2460–2473. [CrossRef]
16. Li, D.; Feng, L.; Huang, D.; Shi, H.; Wang, S. Residual ultimate strength of stiffened box girder with coupled damage of pitting corrosion and a crack under vertical bending moment. *Ocean Eng.* **2021**, *235*, 109341. [CrossRef]
17. Lim, H.K.; Lee, J. On the structural behavior of ship's shell structures due to impact loading. *Int. J. Nav. Archit. Ocean. Eng.* **2018**, *10*, 103–118. [CrossRef]
18. Liu, W.; Suzuki, K.; Shibanuma, K.; Pei, Z. Nonlinear Dynamic Response and Strength Evaluation of a Containership in Extreme Waves Based on Hydroelastoplasticity Method. In Proceedings of the Twenty-fourth International Ocean and Polar Engineering Conference, Busan, Korea, 15–20 June 2014.
19. Liu, X.; Jiang, D.; Liufu, K.; Fu, J.; Liu, Q.; Li, Q. Numerical investigation into impact responses of an offshore wind turbine jacket foundation subjected to ship collision. *Ocean. Eng.* **2022**, *248*, 110825. [CrossRef]
20. MAIB. *Report on the Investigation of the Structural Failure of MSC Napoli*; Marine Accident Investigation Branch: London, UK, 2008.

21. Nassirnia, M.; Heidarpour, A.; Zhao, X.-L.; Wang, R.; Li, W.; Han, L.-H. Hybrid corrugated members subjected to impact loading: Experimental and numerical investigation. *Int. J. Impact Eng.* **2018**, *122*, 395–406. [CrossRef]
22. Nishihara, S. Ultimate longitudinal strength of mid-ship cross section. *Nav. Archit. Ocean. Eng.* **1984**, *22*, 200–214.
23. Paik, J.K. *Ultimate Limit State Analysis and Design of Plated Structures*, 2nd ed; Wiley: New York, NY, USA, 2018.
24. Pham, T.M.; Hao, H. Plastic hinges and inertia forces in RC beams under impact loads. *Int. J. Impact Eng.* **2017**, *103*, 1–11. [CrossRef]
25. Sun, G.; Deng, M.; Zheng, G.; Li, Q. Design for cost performance of crashworthy structures made of high strength steel. *Thin-Walled Struct.* **2018**, *138*, 458–472. [CrossRef]
26. Volmir, S.A. Nonlinear Dynamics of Plates and Shells, Science, Moscow. 1972. Available online: https://trid.trb.org/view/17770 (accessed on 30 January 2023).
27. Wei, J.; Li, J.; Wu, C.; Liu, Z.-X.; Fang, J. Impact resistance of ultra-high performance concrete strengthened reinforced concrete beams. *Int. J. Impact Eng.* **2021**, *158*, 104023. [CrossRef]
28. Xu, M.C.; Song, Z.J.; Pan, J. Study on influence of nonlinear finite element method models on ultimate bending moment for hull girder. *Thin-Walled Struct.* **2017**, *119*, 282–295. [CrossRef]
29. Xu, M.C.; Song, Z.J.; Pan, J. Study on the similarity methods for the assessment of ultimate strength of stiffened panels under axial load based on tests and numerical simulations. *Ocean Eng.* **2020**, *219*, 108294. [CrossRef]
30. Yamada, Y. Dynamic Collapse Mechanism of Global Hull Girder of Container Ships Subjected to Hogging Moment. *J. Offshore Mech. Arct. Eng.* **2019**, *141*, 1–15. [CrossRef]
31. Yang, B.; Wang, D.-Y. Dynamic Ultimate Hull Girder Strength Analysis on a Container Ship under Impact Bending Moments. *Int. J. Offshore Polar Eng.* **2018**, *28*, 105–111. [CrossRef]
32. Zanuy, C.; Ulzurrun, G.S.; Curbach, M. Experimental determination of sectional forces in impact tests: Application to composite RC-HPFRCC beams. *Eng. Struct.* **2022**, *256*, 114004. [CrossRef]

Journal of
Marine Science and Engineering

Article

Experimental Study of the Dynamic Performance of Steel Catenary Riser within the Touchdown Zone

Yang Yu [1,2], Shengbo Xu [1,2,*], Jianxing Yu [1,2], Lixin Xu [1,2], Xin Liu [1,2] and Pengfei Liu [1,2]

1 State Key Laboratory of Hydraulic Engineering Simulation and Safety, Tianjin University, Tianjin 300072, China
2 Tianjin Key Laboratory of Port and Ocean Engineering, Tianjin University, Tianjin 300072, China
* Correspondence: edwinb@tju.edu.cn; Tel.: +86-135-0212-8143

Abstract: This study proposed a novel experimental platform to conduct dynamic loading tests of a truncated model steel catenary riser (SCR) within the touchdown zone (TDZ). The facilities of the platform, including a soil tank, a loading system, and a soil stirring system, were introduced in detail. The parameters of the test were determined through the simulation of an in situ riser. A steel pipe was adopted as the model riser, with its outer diameter equaling that of the prototype SCR. Before executing the dynamic loadings, the model riser developed its static configuration under the submerged weight and applied bending moment. Subsequently, cyclic vertical and lateral displacement loads were applied to the truncated point. The test results showed that when the vertical loading amplitude increased from 200 mm to 300 mm, the stress ranges at the front of the model riser increased by more than 100%, whereas the stress range only differed by less than 5% under different loading periods. Numerical models of the SCR were built based on the vector form intrinsic finite element (VFIFE) method. High similarities between the test and simulation results proved the reliability of the nonlinear soil model and the numerical model. During the test, a seabed trench was developed with a depth of 0.71 D and a width of 0.48 D, and its shape was similar to the in situ trench.

Keywords: steel catenary risers; touchdown zone; dynamic loading tests; vector form intrinsic finite element method; riser–seabed interaction

Citation: Yu, Y.; Xu, S.; Yu, J.; Xu, L.; Liu, X.; Liu, P. Experimental Study of the Dynamic Performance of Steel Catenary Riser within the Touchdown Zone. *J. Mar. Sci. Eng.* **2023**, *11*, 151. https://doi.org/10.3390/jmse11010151

Academic Editor: Cristiano Fragassa

Received: 5 December 2022
Revised: 28 December 2022
Accepted: 29 December 2022
Published: 8 January 2023

1. Introduction

With the depletion of onshore and shallow-water petroleum resources, the untapped hydrocarbon reserves in deep water are becoming more attractive. The risers are the conductor pipes that connect the flow lines and floating platforms, and their integrities are of vital importance. Among the various kinds of riser systems, steel catenary risers (SCR) exhibit the advantages of high bearing capacity, high cost effectiveness, and simplicity in structure, making them one of the optimum solutions used in deep water [1]. Under the cyclic movement of the floating platform and environmental loads, the SCR may suffer from fractures induced by fatigue damage, especially within the joint connected to the platform and touchdown zone (TDZ) [2]. However, accurate prediction of the fatigue life of the SCR within the TDZ is challenging, owing to the uncertainties of riser–seabed interaction and the variation of seabed profiles.

Some theoretical and numerical analyses of SCR adopt rigid or elastic seabed models, which produce unrealistic results of the fatigue performance of SCR [3,4]. Based on the pipe–soil interaction model tests, several advanced vertical riser–seabed interaction models have been proposed and integrated into the finite element models [5–8]. The degradation of soil caused by the disturbance of the riser segment can be recorded, and the suction force can be simulated. Based on the nonlinear riser–seabed interaction model proposed by Randolph and Quiggin [6], the effects of different seabed stiffness, suction ratio, and loading period on the fatigue damage of SCR were analyzed [9]. A seabed trench was developed under the oscillation of the SCR within several months [10], which also affected the mechanical behavior of the riser. The trench could be artificially inserted into the numerical model using a mathematical formulation [11], and the fatigue damage of the SCR was reduced. Using the R-Q soil model, Shiri [12] proposed a method to develop the seabed trench automatically to ensure consistency between the trench and SCR. Simulation results proved that the developed trenches have detrimental effects on the fatigue behavior of the SCR. The trenches developed using an artificial formulation, step method, and automatic method were compared and inserted into the SCR model by Randolph et al. [13], and the dynamic simulations showed that the fatigue damage of the SCR was significantly affected by the trench shape. Since the static offsets and low-frequency motions of the floating platform also contribute to the trench formation, Yu et al. [14] compared the trench shapes developed under different vessel motions and studied the trench effects on the fatigue life of the SCR using the inserting method. Shoghi and Shiri [15] and Shoghi et al. [16] simplified the trench using sloped lines and figured out how the seabed trenches influenced the fatigue damage of the SCR. These nonlinear riser–soil interaction models and seabed trenches mainly focused on determining the vertical seabed resistances; however, sophisticated lateral soil models have not been proposed and applied. Therefore, experiments should be carried out to determine the effect of horizontal riser–soil interaction on the behavior of SCR. Furthermore, the reliability of the nonlinear vertical soil models in a global SCR model still needs to be verified through experimental analysis.

Although the full-scale model tests of SCR can provide convincible results [17], the cost of a single test is considerable, and the environmental loads cannot be controlled or recorded in detail. This hinders the comparison between the test and simulation results. By contrast, laboratory tests are reliable and cost-effective alternatives. The tests can be categorized into small-scale global tests, reduced-scale cut-off model tests, and centrifuge tests. For a small-scale global test, the whole SCR was modeled using a silicone tube, whereas the seabed was substituted by dry foam, wet foam, or glass [18]. Despite the global mechanical behavior of SCR being well represented, the interaction between riser and seabed clay or sand is oversimplified, and the stress variation within the TDZ cannot be obtained accurately.

Regarding the reduced-scale cut-off model tests, the riser is truncated above the seabed, and the lower part is treated as the object to study. The truncated point of the SCR is taken as the loading point where displacement loads are applied. PVC and PE pipes were selected as the optimum choices for the model riser owing to their low rigidity. The cylinder irons stringed by steel wire ropes were input into the pipe to increase the submerged weight [19]. Both ends of the model SCR were modeled as hinged, allowing the riser to rotate freely. All these measures contributed to magnifying the deformation of SCR. The influences of heave motion-induced vibration were investigated by Wang et al. [19] and Hodder and Byrne [20], with clay and sand used as seabed soil, respectively. When cyclic sway motions were applied to the truncated point of model SCR, the seabed trench was widened and deepened [21]. After fixing the soil tank on the shaking table, the seismic loading condition can be simulated to study the dynamic stress variation of the SCR within the TDZ [22]. However, when the low-rigidity pipes were applied in the model tests, the stiffness ratio between the riser and soil was reduced. This would induce a larger stress range along the SCR and make it more difficult to determine the difference between the test result and the numerical simulation result. Once the small-diameter steel pipes were adopted in the model test, high-frequency vibrations would be significant [23].

Similar to the reduced-scale cut-off model tests, the centrifuge tests focus on part of the SCR within the TDZ. The feasibility of using a truncated model SCR in the test was verified by Bhattacharyya et al. through numerical analysis [24]. Possessing the advantages of increasing the gravity by a factor of more than 40, the experiment can simulate a full-scale riser using a small-diameter steel pipe without the need of choosing an alternative material to ensure the rule of similarity [25]. The combinations of surge and heave motions were applied to the truncated point of the model SCR, and the time history of the bending moment [26] and the fatigue life [27] were presented and analyzed. Furthermore, the clayey soil can be prepared within a shorter period using a centrifuge [28]. The limitation of the centrifuge tests is that both the mechanical facilities and the data acquisition system require a higher level of precision and safety, which is hard to realize.

Although great efforts have been made to perform a more realistic model SCR test within the TDZ, the existing model tests still have some limitations. The first limitation is that the bending moment cannot be applied to the truncated point of the model SCR. For an in situ SCR, the bending moment increases significantly before reaching the touchdown point (TDP) and contributes to the deformation and stress variation of the SCR within the TDZ. The tests can be more accurate if the bending moment is included. The second limitation is that the model SCR may not well represent the mechanical behavior of the in situ SCR. Although the reduced-scale cut-off model tests are easier to realize compared with the centrifuge tests, plastic pipes or reduced-scale steel rods may not be the optimum options to simulate the deformation and stress variation of a full-scale SCR. Furthermore, the cyclic sway motions cannot be applied in the centrifuge tests. These limitations require a novel test system for the experiment of large-scale model steel risers. The bending moment and cyclic displacement should be applied simultaneously to the model SCR.

Combining the merits of the reduced-scale cut-off model tests and centrifuge tests, a novel experimental platform was proposed and constructed to perform dynamic model tests of the SCR within the touchdown zone. The layout of the platform was presented in this study, and the structure of each component was introduced in detail. Based on the equipment, a dynamic test of the model SCR was carried out under the combination of the bending moment and cyclic displacement loads. Different loading amplitudes and periods were adopted to determine their influences on the mechanical behavior of the riser. Numerical models were built based on the vector form intrinsic finite element (VFIFE) method, and the vertical nonlinear soil model was integrated into the model. High similarities between the experiment and simulation results proved the practicability of the test platform and the reliability of the nonlinear soil model.

2. Development of Test Platform and Experimental Equipment

2.1. Layout of the Test Platform

The layout of the test platform is shown in Figure 1. The platform consists of a soil tank, a loading system, and a stirring system for clayey soil preparation. A large-scale steel pipe with a diameter of 6 inches was adopted as the truncated model SCR. Each end of the pipe was welded with a flange to connect to the facilities. The axial position of the loading facility along the soil tank was adjusted to install the pipe and fixed before the experiment. The loading end of the SCR was fastened to the bending moment loading system, and the tail end was connected to the elastic support. The riser was free to rotate at the tail end, whereas the vertical and horizontal displacements were restricted. After applying the bending moment, the model SCR developed its static configuration. The rotational angle of the loading end of the SCR was restrained by a mechanical locking device and remained constant throughout the test, and cyclic displacement loads could be applied to investigate the dynamic behavior of SCR.

2.2. Soil Tank

The length, width, and depth of the reinforced concrete soil tank were 20.05 m, 3.6 m, and 2.05 m, respectively. Under the soil tank, 6 bored cast-in-place piles were constructed and connected using the foundation beams to improve their rigidity and prevent the decline or settlement of the tank. The side wall thickness of the soil tank was 300 mm to sustain the pressure of clay. The cross section of the tank and the arrangement of reinforcements are shown in Figure 2. The plan view of the soil tank is shown in Figure 3a. Within the moving range of the loading device, the angle steel with 50 mm thickness was fixed to the surface of each side wall using bolts and shear keys, as shown in Figure 3b. For the other part of the tank, the track of the stirring system was arranged along the top of each side wall using the anchor plates. After construction, stainless-steel plates were installed on each surface of the tank for protection.

Figure 1. The layout of the test platform.

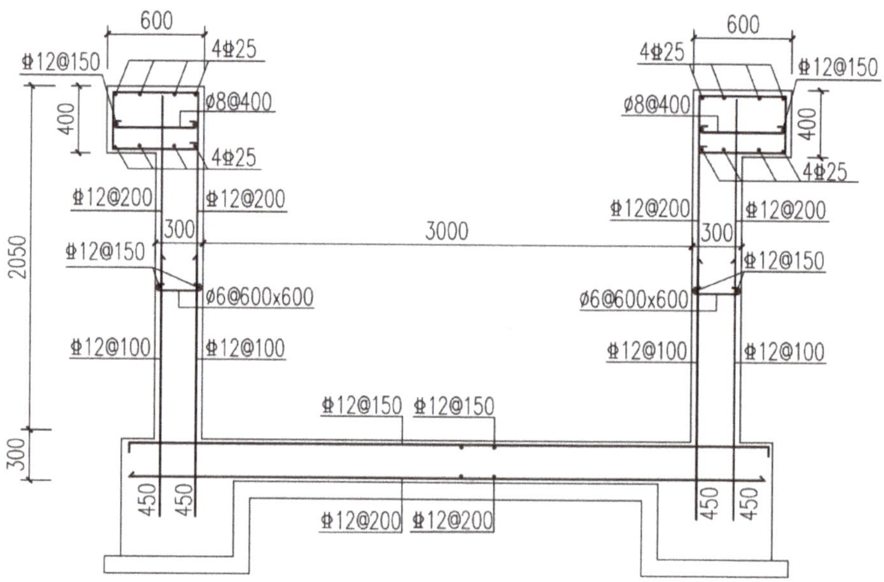

Figure 2. Reinforcement arrangement of the soil tank.

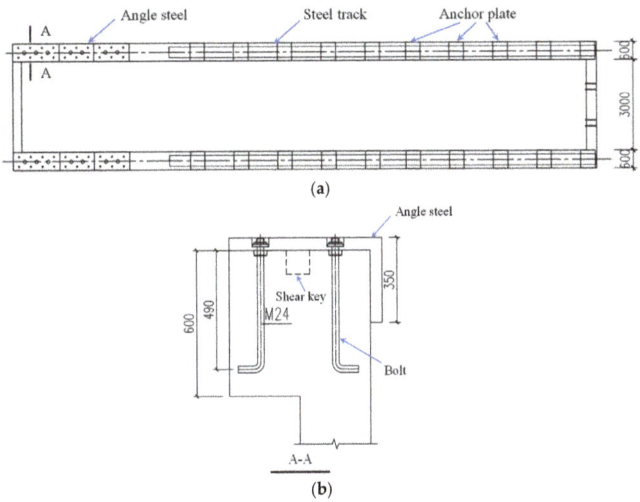

Figure 3. Plan view of the soil tank (**a**) arrangement of angle steels and steel tracks and (**b**) connection between steel angle and soil tank.

2.3. Loading System

The loading system comprises a supporting frame, a cross-sliding frame, a displacement loading system, a bending moment loading system, an elastic support, and the hydraulic power units and control system. The first four systems were assembled to apply the excitations to the SCR, as shown in Figure 4. The cross-sliding frame was fixed to the supporting frame, and its lateral movement was driven by the lateral hydraulic jack that was fixed on one side of the supporting frame. The vertical movement of the SCR was realized by the vertical hydraulic jack fastened to the top of the cross-sliding frame. The piston rod of this vertical jack was connected to the top of the bending moment loading system. A flange was installed at the front of the bending moment loading system to connect the model SCR. The bottoms of two axial hydraulic jacks were fixed on the angle steels of the soil tank, and the piston rods were connected to the supporting frame to adjust its axial position. The loading system was activated and controlled by the hydraulic power units and control system. The tail end of SCR was connected to the elastic support which is fixed to the wall of the soil tank.

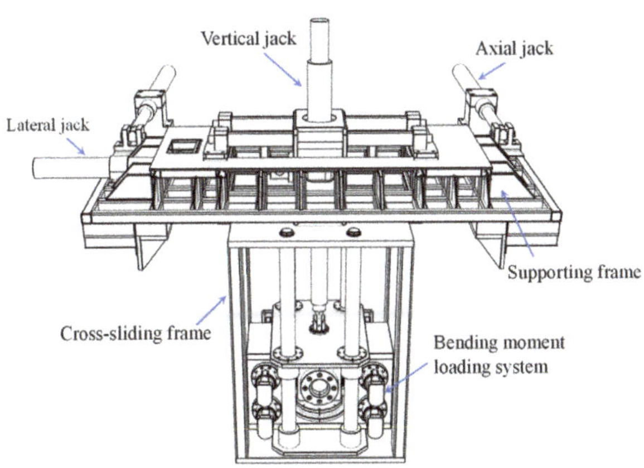

Figure 4. Assembly of the supporting frame, cross-sliding frame, and displacement and bending moment loading systems.

2.3.1. Supporting Frame

The supporting frame was constructed by welding the steel plates and steel box beams together. The three-dimensional illustration of the frame is shown in Figure 5. The purpose of this frame is to sustain the weight of the whole loading facility and the reaction force during the test. Instead of using thick steel plates, the adoption of steel box beams could reduce the self-weight and increase the flexural rigidity. Stiffeners were arranged around each side of the frame to further increase the rigidity. Since the frame should avoid any movement during the test, the electromagnets were activated to link the supporting frame and the angle steel on the soil tank.

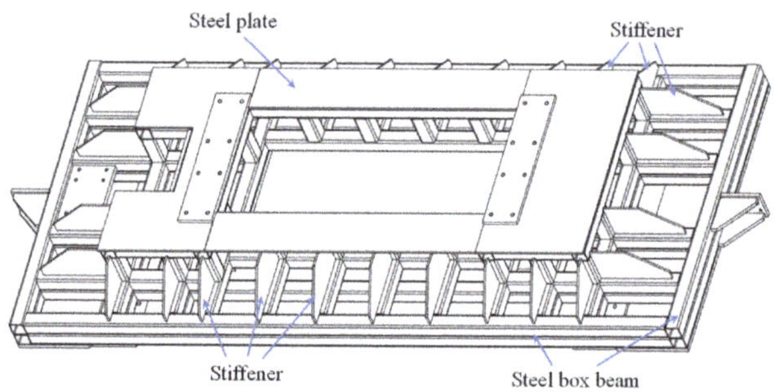

Figure 5. Schematic of the supporting frame.

2.3.2. Cross-Sliding Frame

The cross-sliding frame can accommodate the in-plane movement of the loading end of SCR. Its structure can be decomposed into the upper frame and the lower frame, as shown in Figure 6. In the upper frame, four steel rods were fastened to the supporting frame, and a sliding block was driven to slide along the rods through the lateral hydraulic jack. The vertical hydraulic jack was fixed to this sliding block. The steel plate at the top of the lower frame was also welded to the bottom of the box-shaped sliding block, and therefore, the central line of the vertical hydraulic jack would always coincide with that of the lower frame. The lower frame constituted two steel plates and four steel box columns, exhibiting the advantages of lighter self-weight and higher bending rigidity. Four steel rods were arranged within the columns as the guide of the bending moment loading system. Meanwhile, the rods also contributed to increasing the flexural rigidity of the lower frame.

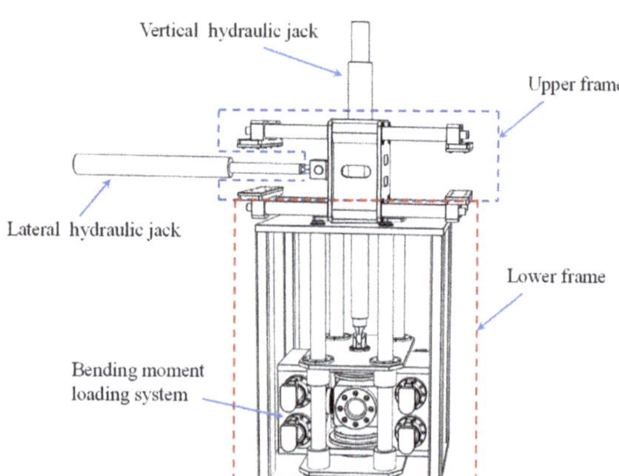

Figure 6. Schematic of the cross-sliding frame.

2.3.3. Displacement Loading System

The displacement loading system is composed of axial, lateral, and vertical hydraulic jacks. The structure of the vertical loading jack is shown in Figure 7 as an example. Since the purpose of the axial jacks was to locate the supporting frame, only low-speed displacement was allowed, whereas the other jacks could realize high-speed cyclic loading. The strokes of the axial, lateral, and vertical jacks were 960 mm, 1010 mm, and 1200 mm, respectively. For the in situ deep-water SCR, the period of short-term heave motion and sway motion is concentrated within 5~12 s. If the lowest period is reached, the moving range of the truncated point of SCR is small. The highest loading frequency of the jacks was 0.5 Hz when the loading amplitude was 100 mm, and 0.17 Hz when the loading amplitude was 300 mm. Therefore, the loading capacity of the hydraulic jacks can fulfill the aim of reproducing the moving condition of an in situ SCR.

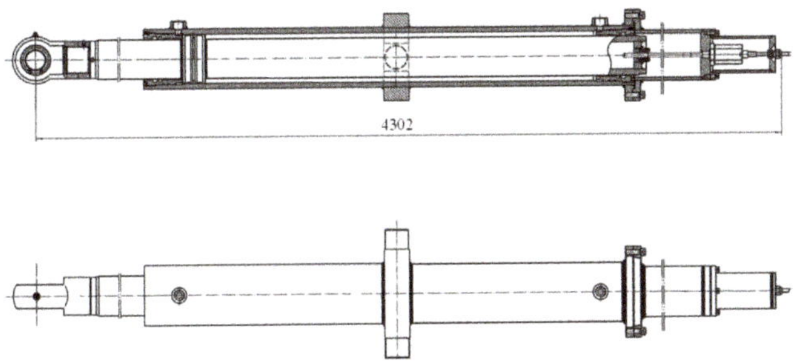

Figure 7. Structure of the vertical loading jack.

2.3.4. Bending Moment Loading System

To provide the bending moment to the loading end of SCR, two gear-rack swing hydraulic cylinders were assembled to form the bending moment loading system, as shown in Figure 8. The top and bottom of the chambers were connected through steel plates, and both output axes of the cylinders were aligned to rotate the flange simultaneously. The three-dimensional view and profile of the hydraulic cylinder are shown in Figure 9. For each cylinder, the maximum rotating angle was $\pm 15^{\circ}$ and the rated operating pressure was 21 MPa. Under this pressure, the output torque was 22.3 kN·m. Therefore, the total bending moment that could be provided to the SCR was 44.6 kN·m during the configuration-building stage. During the cyclic loading stage, the bending moment should be calculated through the stress analysis of the SCR rather than the sensor, because the rotating angle is restricted.

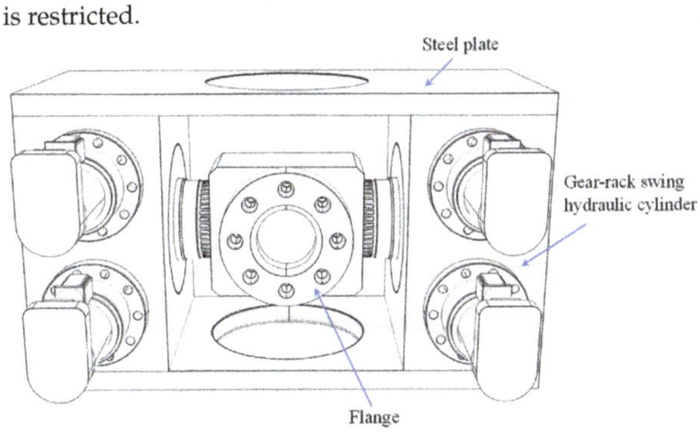

Figure 8. Schematic of the bending moment loading system.

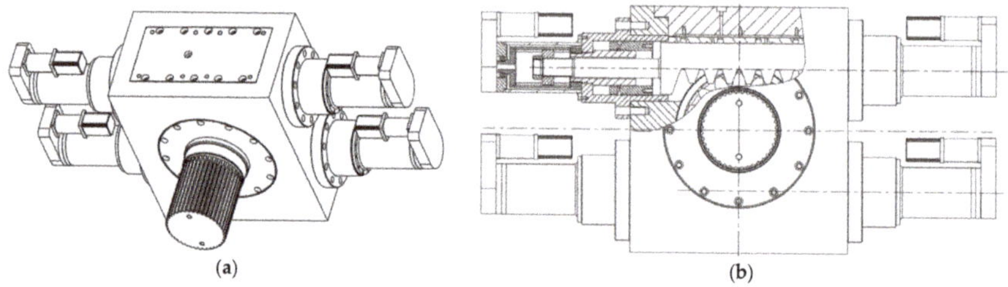

Figure 9. Structure of the gear-rack swing hydraulic cylinder: (**a**) three-dimensional view and (**b**) profile.

2.3.5. Hydraulic Power Units and Control System

The hydraulic power units contain four electric motors with a rated power of 22 kW, one electric motor of 11 kW, four hydraulic pumps, an air-cooling system, a fuel tank, and auxiliary equipment including piezometers and valves. The assembly is shown in Figure 10. The operating pressure of this system is 20 MPa, and the rate of flow is 120 L/min. To control the loading system through the electrical signal, electrohydraulic servo valves were integrated into the system. A shelf was constructed to accommodate the proportional valves, accumulators, filters, and an electrical control cabinet, as shown in Figure 11.

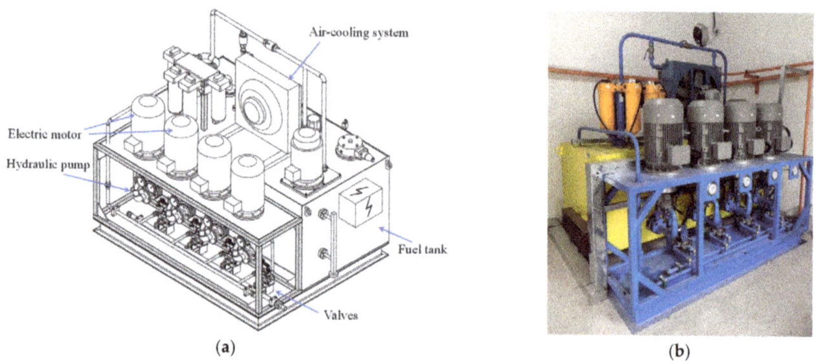

Figure 10. The hydraulic power units: (**a**) schematic illustration and (**b**) constructed assembly.

Figure 11. Assembly of the electrohydraulic servo valves and electrical control cabinet.

2.3.6. Elastic Support

The elastic support is used to simulate the constraint effect of the flowline part of the riser to the model SCR. The structure can be decomposed into two fixed supports and one movable support, as shown in Figure 12. The fixed supports were anchored to the wall of the soil tank using high-strength bolts. For each fixed support, two steel rods were arranged as the guides of the movable support. A steel plate with circular openings could slide vertically along the rods, and its movement was restricted by the vertical springs. The front steel plate of moveable support was welded perpendicularly to the sliding steel plates and could only move vertically, whereas the back steel plate shared the same vertical position with the front one, and could move axially along four steel rods. The axial stiffness was provided by two axial springs, with one end fixed to the front plate and the other fixed to the back plate. All the springs in this elastic support are changeable, allowing the stiffness to be varied according to the requirement of the test. In this study, the vertical stiffness was set to a large value to avoid the maximum penetration of SCR concentrating within this area, whereas the axial stiffness was much lower, allowing the axial movement of SCR to avoid extremely large axial force during the cyclic displacement excitations. The flange was connected to the back plate of the movable support via a spherical hinge. Therefore, the SCR could rotate freely at this end, and the bending moment could be released.

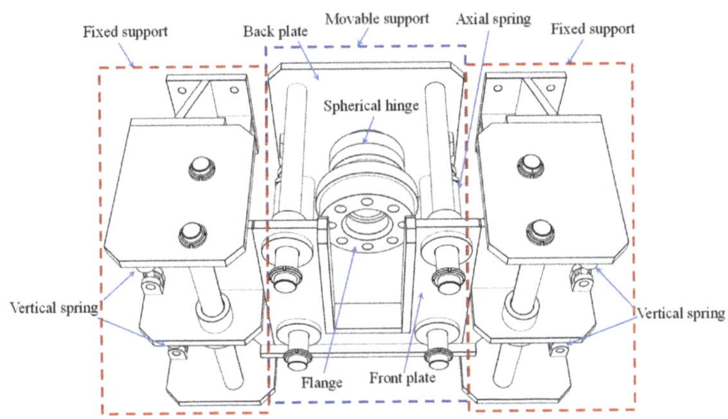

Figure 12. Schematic of the elastic support.

2.4. Soil Stirring System

The properties of the test soil can significantly affect the mechanical behavior of the model SCR. Clayey soil with an undrained shear strength similar to that in deep water should be prepared. Uniformity is another key indicator that should be guaranteed for such a large amount of soil. The clay can be prepared using kaolin clay powder, for its properties similar to the subsea soil, or by collecting near-shore or in situ soil. Since there will be some impurities in the collected soil, the clay should be dried and sieved to obtain the pure clay power. It is hard to fully saturate the clay after adding enough water and waiting for a long period, owing to its low permeability. To solve these problems, a soil stirring system was proposed. The soil powder was divided into several portions to be dropped into the soil tank. After adding each portion, the water was poured into the tank, and the stirring system was used to mix the content. When all the soil and water were thrown into the tank, the stirrer was forced to move vertically, while the frame would slide along the track on the soil tank simultaneously. This allowed all the clay to be stirred and mixed thoroughly to obtain the homogeneous test soil. Then, the soil could settle under its self-weight to be consolidated, or the vacuum preloading method could be sued to accelerate the process, as shown in Figure 13. After the soil was fully saturated and consolidated, the surface was scraped using the stirring system to guarantee the planeness.

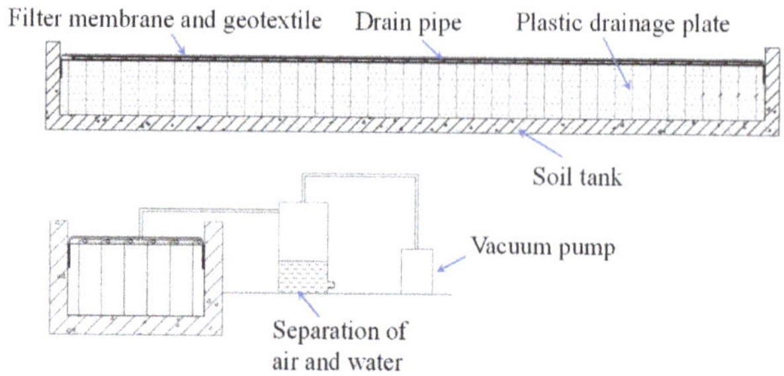

Figure 13. Vacuum preloading system.

The soil stirring system contains a portal frame, a power system, a changeable stirrer, and a scraper, as shown in Figure 14. The portal frame can travel along the track of the soil tank, whereas the vertical track can move horizontally along the lateral track that was fastened on the frame. Four electric motors were used in this system. The first one was to control the movement of the portal frame, the second one was fixed to the vertical track to drive its lateral movement, the third one was connected to the stirring shaft to control its vertical position, and the last one was installed on the top of the shaft to rotate the shaft and stirrer. The stirrer can be changed according to the property of the test soil. The scraper was adopted to guarantee the planeness of the clay, and its vertical position can also be adjusted to fit different targets of soil depth. The scraper is also detachable. Therefore, when the stirrer or the scraper was working, the other was removed to avoid interference.

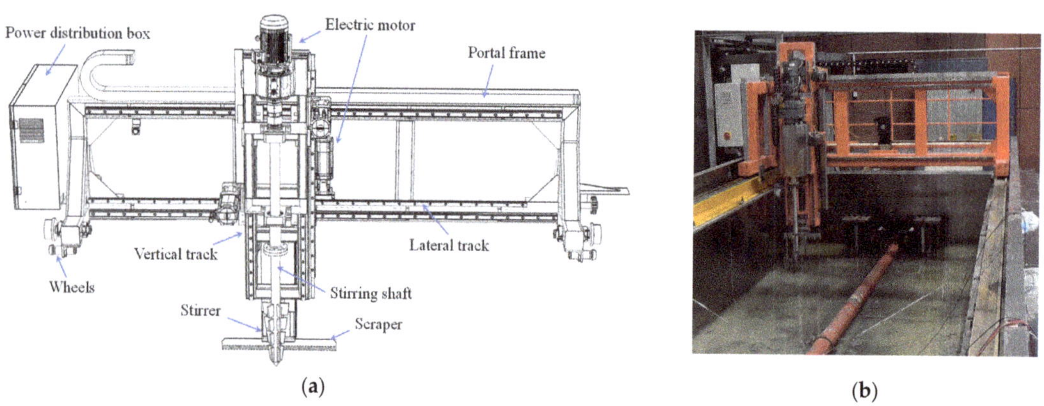

Figure 14. The soil stirring system: (**a**) schematic illustration and (**b**) constructed assembly.

3. Experiment Setup and Procedures

3.1. Determination of Experimental Parameters

To determine the parameters in the model test, a numerical model of in situ SCR was built using the vector form intrinsic finite element method (VFIFE), which is suitable to solve high-nonlinear problems. Similar to the lumped mass method adopted in the commercial software OrcaFlex, the riser was also replaced by a series of mass points. A beam element was used to connect each pair of adjacent points to transmit the internal force and provide stiffness. The position of the points represented the configuration of SCR, whereas the points' movements under the imbalanced forces were used to model the vibration of the riser. One of the major differences between the lumped mass method and the VFIFE method is that the deformations of the beam elements are calculated based on the virtual reverse motion for the VFIFE method. Detailed descriptions of the VFIFE method can be referred to in [1]. The nonlinear vertical soil model proposed by Randolph and Quiggin [6] was also integrated into this finite element model.

The prototype SCR used in this study has been put into production in the Lingshui project in China. The SCR is 2130 m long, with an outer diameter of 168 mm and a thickness of 18.3 mm. The seabed is clayey soil with an undrained shear strength of 4–8 kPa. Each end of the SCR was modeled as hinged in the numerical model. Detailed parameters used in the model are listed in Table 1. The static configuration of SCR was established under the submerged weight and soil resistance, as shown in Figure 15. The static in-plane bending moment was concentrated within the catenary part near the TDP as shown in Figure 16, with a maximum value of 17.47 kN·m. The distribution of seabed resistance and the developed seabed trench profile are shown in Figure 17. The resistance kept increasing within the trench mouth, and reached its peak value at the deepest point of the trench. Subsequently, the resistance decreased slightly along the trench tail and stabilized at the value that equaled the submerged self-weight of SCR.

Table 1. Detailed parameters used in the model.

Properties of Riser and Seabed	Value	Unit
Total length L	2130.0	m
External diameter D	168.0	mm
Wall thickness t	18.3	mm
Mudline shear strength s_{um}	6.0	kPa
Shear strength gradient k	1.5	kPa/m
Power law parameter a	6.0	
Power law parameter b	0.25	
Normalized maximum stiffness K_{max}	200	
Suction ratio f_{suc}	0.2	
Normalized suction decay distance λ_{suc}	0.5	
Normalized repenetration offset after uplift λ_{rep}	0.5	
Saturated soil density ρ_s	1500.0	kg/m^3
Soil buoyancy factor f_b	1.5	

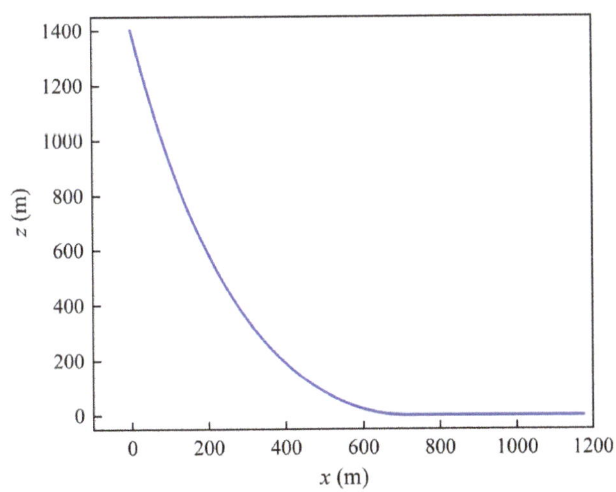

Figure 15. Static configuration of in situ SCR.

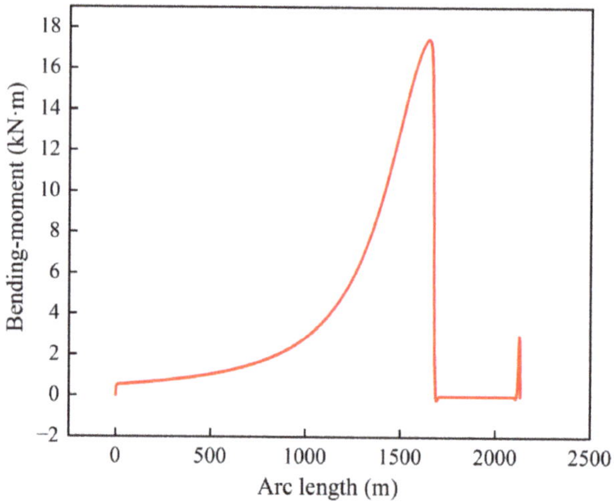

Figure 16. Distribution of bending moment.

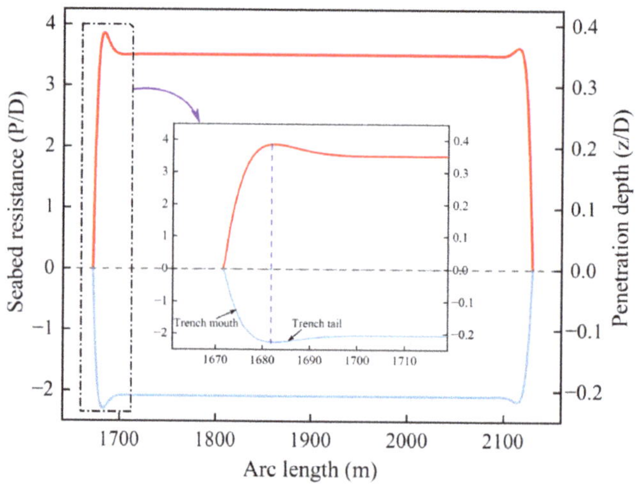

Figure 17. Distribution of seabed resistance and developed seabed trench.

A reference point P at the arc length of 1660.0 m was taken as the truncated point in the model test. The vertical clearance between point P and the seabed was 0.3 m when the static configuration of SCR was established. After applying a harmonic heave motion with an amplitude of 1.0 m and a period of 15 s, the time history of the vertical displacement of point P was obtained, as shown in Figure 18. It proves that harmonic sine-wave motions are suitable to be applied to the truncated point of the model SCR. The relationship between the displacement ranges of the top end and point P is shown in Figure 19. When the heave motion reached an amplitude of 1.5 m, the moving range of point P was 0.72 m. Therefore, the applied displacement to the truncated point lay within the capability of the vertical hydraulic loading jack. Similarly, the horizontal moving range of point P was also less than the maximum displacement of the horizontal hydraulic loading jack.

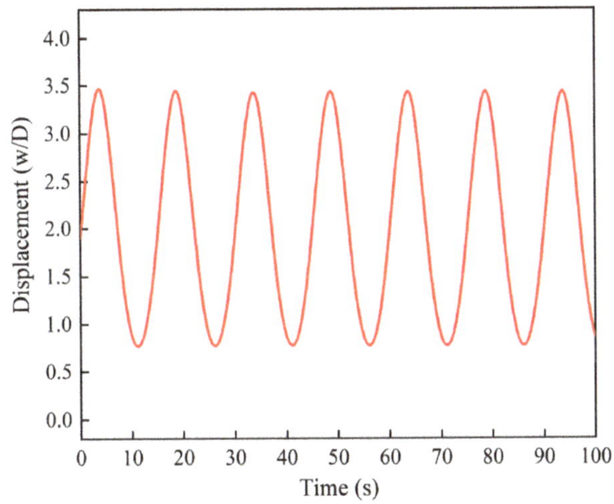

Figure 18. Displacement oscillation history of point P.

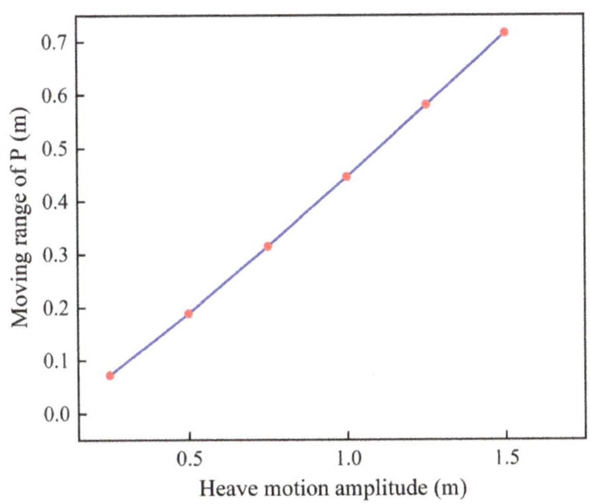

Figure 19. Relationship between heave motion amplitude and moving range of P.

In the model test of the SCR, a steel pipe with a diameter of 168 mm was adopted, similar to the in situ SCR. The wall thickness was reduced to 8 mm to magnify the deformation under the submerged weight and bending moment, since the length of the model SCR was 17.0 m. When the riser developed its configuration, the bending moment applied to the truncated point P was within the capacity of the bending moment loading system. The cyclic vertical and lateral displacement loads can be applied to the loading point of model SCR individually or simultaneously, with their amplitudes similar to the moving ranges of point P of the prototype SCR. The variation in the position of the loading end made the bending moment provided by the loading system unrealistic to precisely control during the loading stage. Therefore, the angle of the riser at the loading end was fixed, and remained constant once the static configuration of the SCR was established. It also helped to determine the boundary condition of the SCR in the numerical model. Although the stress variation may have some differences compared with that of the in situ riser, this method would not obstruct the oscillation of SCR and the interaction between the riser and seabed. Since the major target of this test is to study the influence of the riser–soil interaction on the mechanical behavior of SCR and verify the feasibility of the numerical soil model, the simplifications were deemed acceptable.

3.2. Layout of Instrumentation

To track the stress variation and determine the bending moment and axial tension distribution, 9 sets of strain gauges were arranged along the model SCR, as shown in Figure 20. For each set, four strain gauges were pasted on the same section of the pipe with an interval of 90°. Since the steel pipe remained elastic throughout the experiment, the stress can be obtained through the strain value. The tension and bending moment at each section can be determined according to Equations (1) and (2), respectively.

$$T = \sigma_{\text{ave}} A \tag{1}$$

$$M = \frac{\Delta\sigma}{D_{\text{out}}} I_z \tag{2}$$

where T is the axial force, σ_{ave} is the average value of the stresses within the section, A is the area of the cross section, $\Delta\sigma$ is the difference value of the stresses that coincides with the direction of the moment, D_{out} is the outer diameter of the pipe, and I_z is the inertia moment.

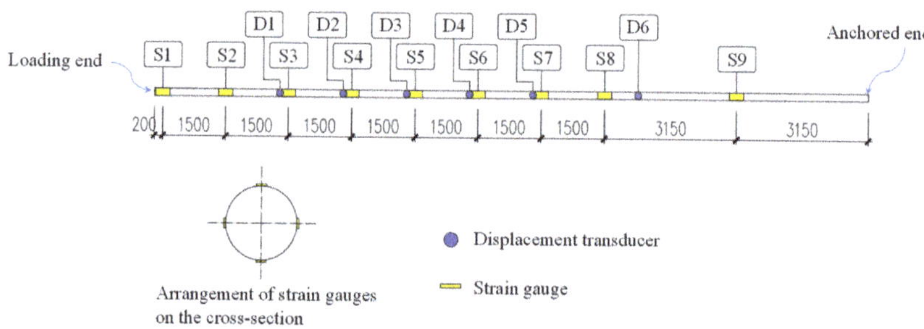

Figure 20. Arrangement of strain gauges and displacement transducers.

To obtain the displacement of the model SCR, 6 groups of draw wire displacement transducers were attached to the pipe, as shown in Figure 21. Since the riser would have in-plane and out-of-plane vibrations, the vertical and lateral movement should be recorded, respectively. Therefore, two displacement sensors were arranged diagonally in each group to realize this, as shown in Figure 22, and the horizontal and vertical displacement components could be calculated based on Equations (3) and (4), respectively.

$$\Delta H = \frac{L_1'^2 - L_2'^2}{2L_{\text{H}}} \tag{3}$$

$$\Delta V = \sqrt{L_1'^2 - \left(\frac{L_{\text{H}}^2 + L_1'^2 - L_2'^2}{2L_{\text{H}}}\right)^2} - L_{\text{v}} \tag{4}$$

where ΔH and ΔV are the vertical and horizontal displacement components, respectively, L_1' and L_2' are the displaced distance between each transducer and the model SCR, L_{H} is the horizontal distance between the transducers, and L_{v} is the vertical clearance between the transducers and the middle of the section of model SCR at the beginning of the test, and can be calculated as $L_{\text{v}} = \sqrt{L_1^2 - \frac{L_{\text{H}}^2}{2}}$, where L_1 is the initial distance between the transducer and the section of SCR. The same vertical elevations of the displacement transducers should be guaranteed when they were arranged. Since the model SCR was initially situated along the central line of the soil tank, the horizontal distances between the model SCR and each transducer in one group were the same.

Figure 21. Installation of displacement transducers.

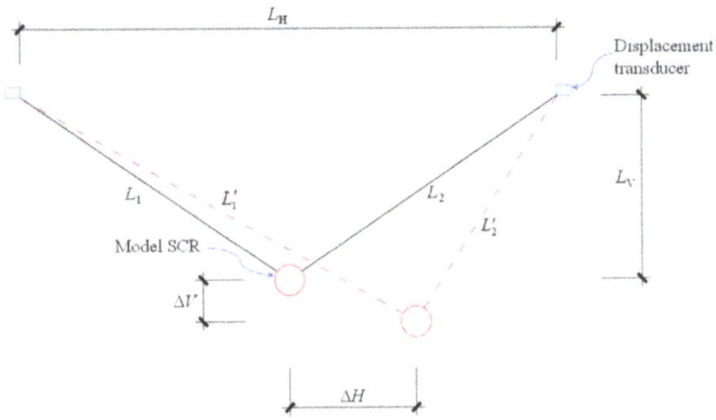

Figure 22. Calculation of the vertical and horizontal displacement.

3.3. Test Procedures

Before the experiment, the test clayey soil was prepared with the help of the soil stirring system and then consolidated via the vacuum preloading method. The undrained shear strength of the soil was obtained by the vane shear test, and its value was 6 kPa. The supporting frame was adjusted axially to make sure that the flanges of the model SCR could be connected to the bending moment loading system and the elastic support, respectively. Enough water was poured into the soil tank to ensure that the riser was submerged. The test of model SCR can be divided into three loading procedures. In the first procedure, the vertical position of the flange middle of the loading end was uplifted 300 mm above that of the hinged end, and the water was injected into the pipe to increase its self-weight. Since the wall thickness of the model SCR was decreased, the applied bending moment should be reduced according to the decrement of the section moment of inertia, allowing the stress of model SCR to be similar to that of the in situ SCR. Therefore, the bending moment of 9.2 kN·m was applied to this end. When the variation of displacement recorded by the transducers was less than 1mm, the static configuration of the model SCR was established.

Subsequently, the cyclic vertical displacements were applied to the truncated point of model SCR in the second loading stage. Different loading amplitudes and periods were selected to investigate their influences on the mechanical behavior of the SCR. The loading sequence was also an important parameter, because a severer motion would induce a deeper trench and may affect the stress distribution of SCR. The loading conditions are listed in Table 2.

Table 2. Cyclic loading conditions.

Loading Sequence	Loading Direction	Amplitude (m)	Period (s)
1	Vertical	0.1	20
2		0.2	20, 15, 10
3		0.3	20, 15, 10
4		0.2	15
5	Horizontal	0.1	15
6		0.2	15

The cyclic horizontal displacements were applied in the third stage. Although the vertical oscillation of SCR is the main factor that induces fatigue damage, the horizontal movement of SCR within the TDZ also widens the trench and affects the mechanical performance of the riser. Therefore, different amplitudes of the sway motions were considered in this test. After this stage, the water in the soil tank was also pumped out to observe the developed trench shape.

4. Experiment Results and Analysis

4.1. Static Loading Stage and Numerical Comparison

Since the riser will experience dynamic oscillations during the installation process, the seabed soil will be disturbed and the riser segments will penetrate deeper than that under its submerged weight. The overpenetration of the riser will induce larger seabed resistance when the riser oscillates laterally. To consider this effect, the model riser was placed horizontally on the seabed surface before the soil preparation process and lowered to produce a shallow trench. After the soil was fully consolidated, the distance between the seabed surface and the trench bottom was 105 mm.

After the position of the loading point was situated 0.3 m above the elastic supporting end and the bending moment was applied, the model SCR developed its configuration, as shown in Figure 23. The angle of the riser at the tail end was nonzero, because the rigidity of this steel model SCR was larger than those of the PE and PVC pipes used in the literature. The simulated configuration of the model SCR using the VFIFE method was similar to the test result, proving the correctness of the model. The distribution of the bending moment is shown in Figure 24, and high similarities exhibit the accuracy of the strain data gathered by the strain gauges. The bending moment increased gradually from the loading end and reached its peak at 7.0 m. After reaching the maximum value, the bending moment decreased to zero when approaching the tail end.

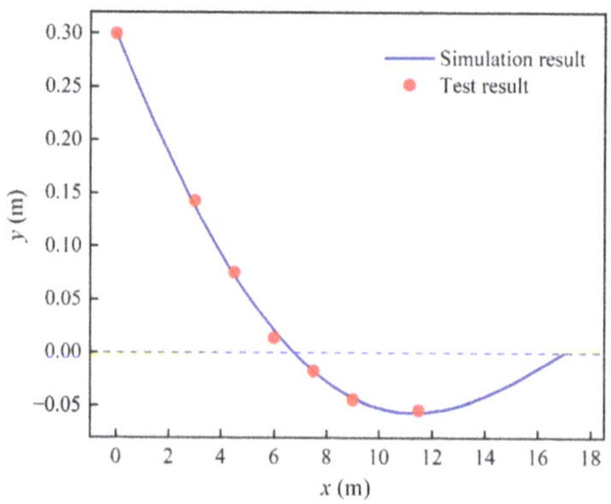

Figure 23. Static configuration of the model SCR.

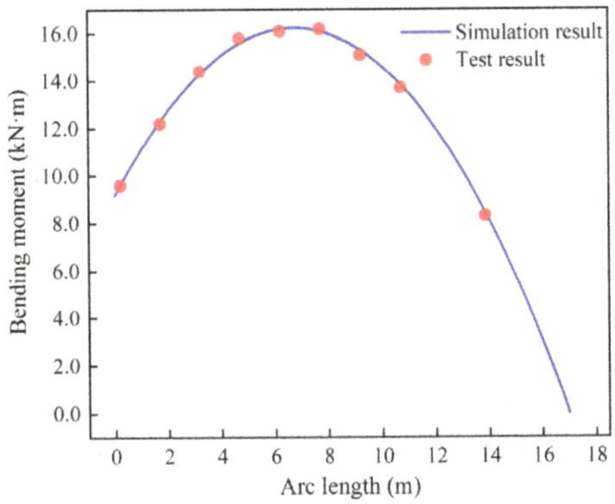

Figure 24. Distribution of bending moment of the model SCR.

4.2. Cyclic Vertical Displacement Loading

4.2.1. Effect of Increasing Loading Amplitude

The cyclic vertical displacement load with an amplitude of 100 mm and a period of 20 s was first applied to the model SCR. After several cycles, the loading amplitude increased to 200 mm while keeping the period constant. The time history of the in-plane stress values (half of the difference value of the stress obtained by the strain gauges at the bottom and top of a cross section) at points S2, S5, and S8 are shown in Figure 25. For S2, the stress variation curve is similar to a sinusoid. Once the loading end was uplifted, the stress value decreased gradually and approached the minimum.

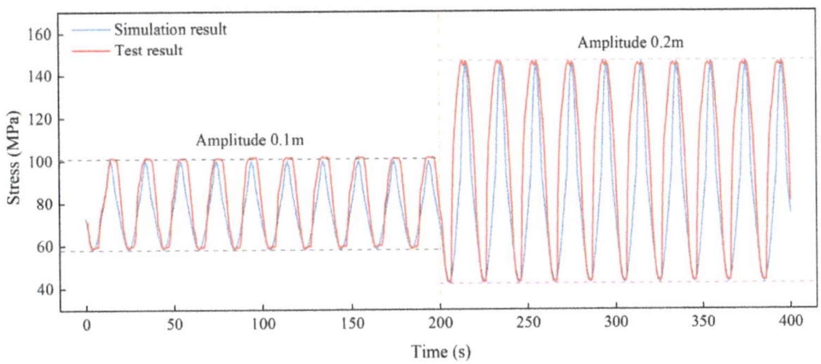

(**a**)

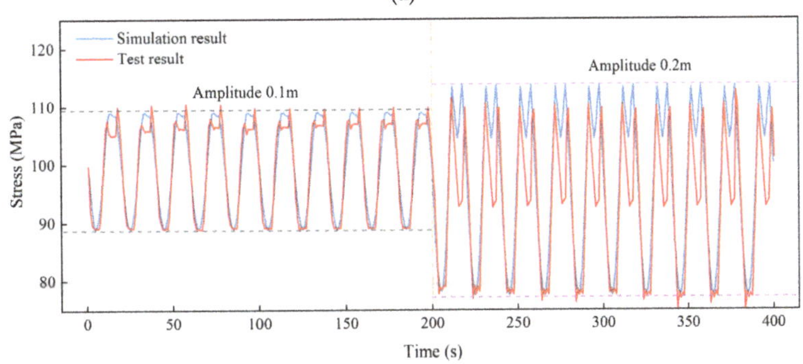

(**b**)

Figure 25. *Cont.*

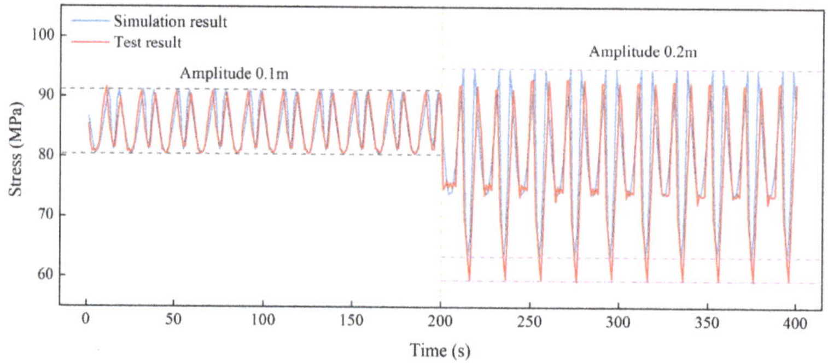

(**c**)

Figure 25. Time histories of the in-plane stress when the amplitude increased from 100 mm to 200 mm: (**a**) S2; (**b**) S5; and (**c**) S8.

Regarding S5 and S8, the time history curves of in-plane stress were no longer sinusoids, as shown in Figure 25b,c. The stress decreased when the truncated point of the model SCR approached the upper end, and vice versa. The riser segment within the middle of the SCR began to penetrate the seabed when the riser was lowered. The stress value apparently decreased during the riser penetration, since the seabed resistance became larger than the submerged weight of the riser and decreased the bending moment at the middle of the SCR. The decrease in stress was more significant for S8 than for S5. Subsequently, the riser was uplifted, and the stress increased and reached the second peak value as the seabed resistance was released. This second peak value may have been larger than the first peak because the clay seabed could provide the suction force when the riser segment began to uplift. This phenomenon was captured by both of the experiment and numerical simulation. The minimum value of stress was reached when the riser was uplifted to the top position for S5, whereas the minimum stress of S8 occurred when the SCR was lowered and a larger seabed resistance was activated. Furthermore, the variation in the minimum stress was more noticeable than the variation in the maximum stress. The stress range increased by 139.9%, 74.0%, and 203.4% for S2, S5, and S8, respectively, when the displacement amplitude increased from 100 mm to 200 mm.

The variations of time history curves of the in-plane stress exhibit similar trends when the loading amplitude increased from 200 mm to 300 mm, as shown in Figure 26. The stress range increased by 104.6%, 114.3%, and 63.6% for S2, S5, and S8, respectively. It should be mentioned that when the loading amplitude increased to 300 mm, the minimum stress of S5 was reached as the riser was lowered and penetrated deeper into the seabed.

4.2.2. Effect of Increasing the Loading Frequency

Three periods of 20 s, 15 s, and 10 s were chosen to study the influence of the loading frequency. The time history curves of stress at points S2, S5, and S8 under different loading periods are drawn in Figures 27 and 28, when the loading amplitudes were 200 mm and 300 mm, respectively. Although different loading periods were selected, the maximum and minimum stresses did not exhibit significant differences, and the trends of stress variation within different loading cycles were similar. The maximum difference in the stress ranges under different loading frequencies was less than 5%. This is attributed to the fact that the total length of the truncated model SCR is less than the in situ one, and therefore, the dynamic effect is not significantly affected by the loading period. However, for the in situ SCR, a higher loading frequency will induce a severer stress variation and larger fatigue damage, since the dynamic effect is more significant.

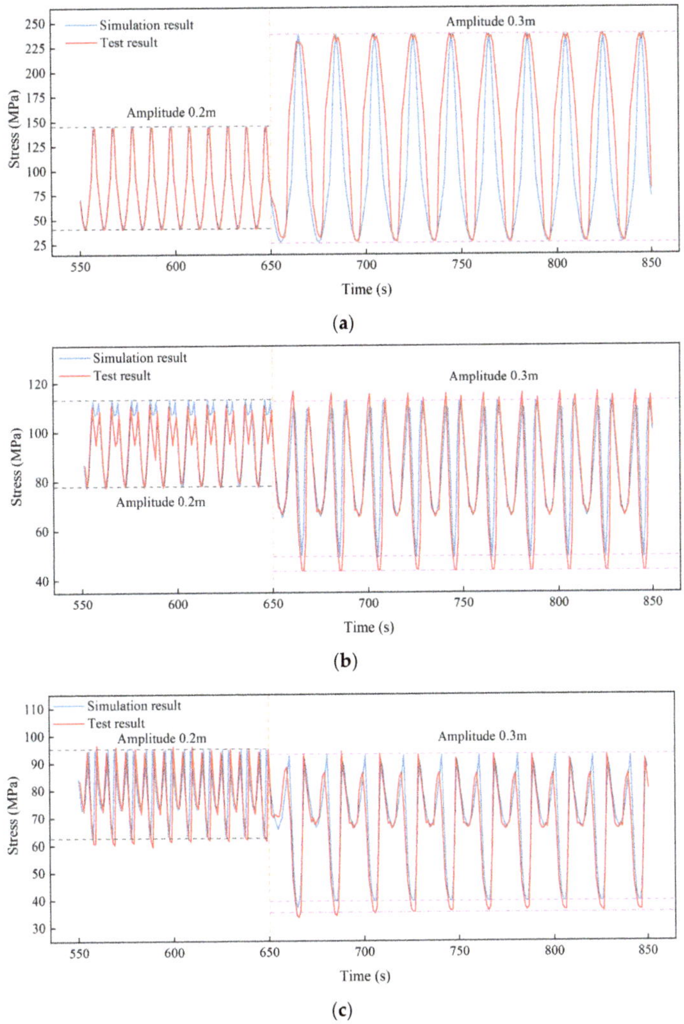

Figure 26. Time histories of the in-plane stress when the amplitude increased from 200 mm to 300 mm: (**a**) S2; (**b**) S5 and (**c**) S8.

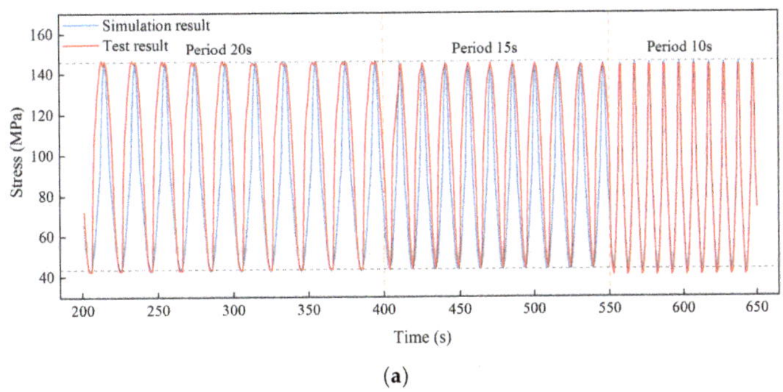

Figure 27. *Cont.*

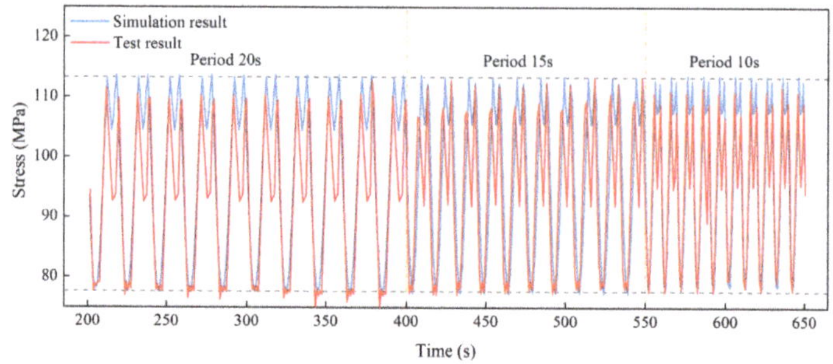

(**b**)

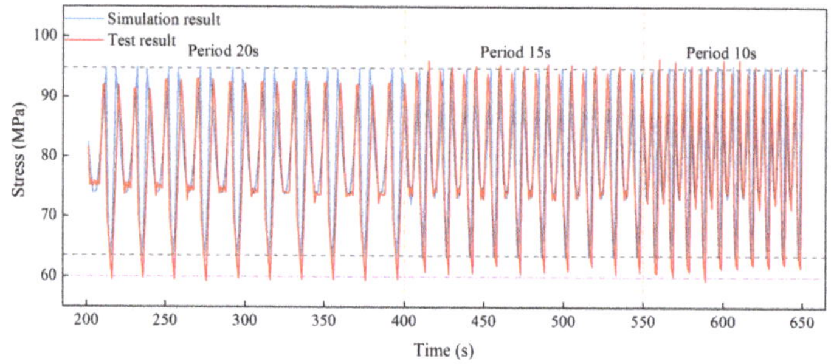

(**c**)

Figure 27. Time histories of the in-plane stress when the amplitude was 200 mm and the period decreased from 20 s to 10 s: (**a**) S2; (**b**) S5 and (**c**) S8.

4.2.3. Effect of Decreasing the Loading Amplitude

The loading amplitude decreased from 300 mm to 200 mm after executing the above loading sequences, to determine the effects of the previous severer loading conditions on the stress variation of the model SCR. The time history curves of the in-plane stresses of S2, S5, and S8 are shown in Figure 29. When comparing the stress under 0.2 m amplitude loading with those in Figure 27, the stress range decreased by 22.4%, 3.8%, and 14.1% for S2, S5, and S8, respectively. This was caused by the deformation and degradation of seabed soil. This finding also agrees with the result drawn by Yu et al. [14] that a seabed trench helps to reduce the fatigue damage of SCR. However, since the variation of trench depth during the test was far smaller than that of an in situ trench, the decrement in stress range was not so obvious.

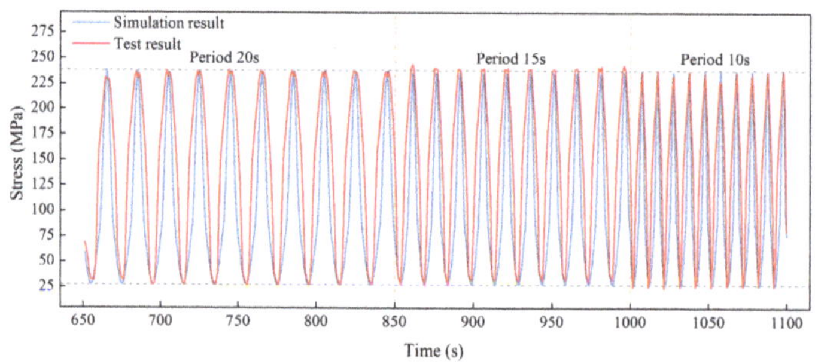

(**a**)

Figure 28. *Cont.*

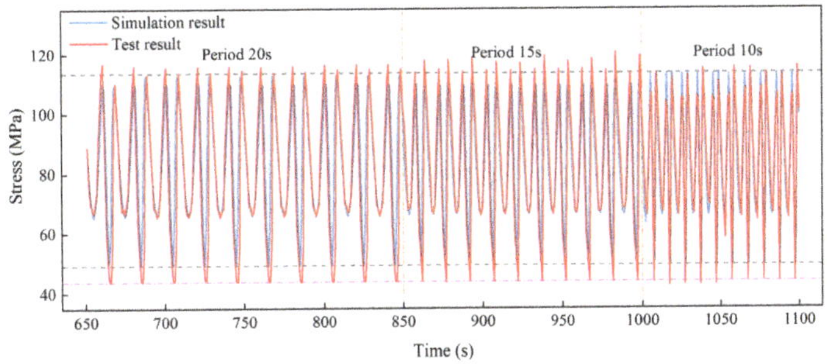

(b)

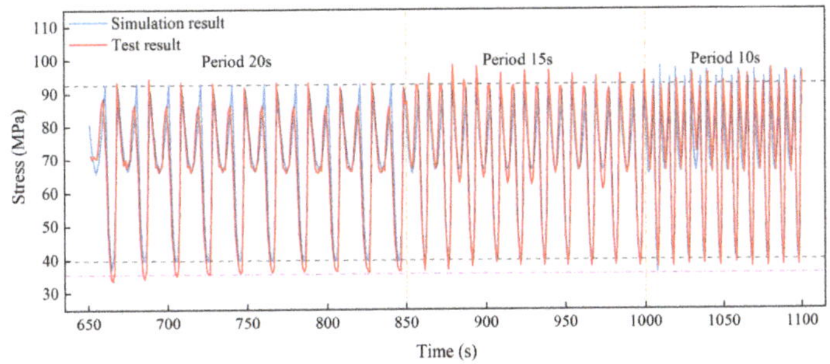

(c)

Figure 28. Time histories of the in-plane stress when the amplitude was 300 mm and the period decreased from 20 s to 10 s: (**a**) S2; (**b**) S5 and (**c**) S8.

Based on the comparisons in Figures 25–29, high similarities can be found between the simulation and experiment results. This proves the reliability of the nonlinear RQ soil model and the robustness of the VFIFE model. Furthermore, the variation in stress range was insignificant during a loading sequence, proving the degradation of soil was not severe. This supports the feasibility of using the nonlinear soil model without considering soil degradation, which can be simpler and own a higher safety factor. The effect of soil deformation and degradation can be substituted by the insertion of a proper predefined seabed trench, to give the lower bound of the fatigue damage of the SCR.

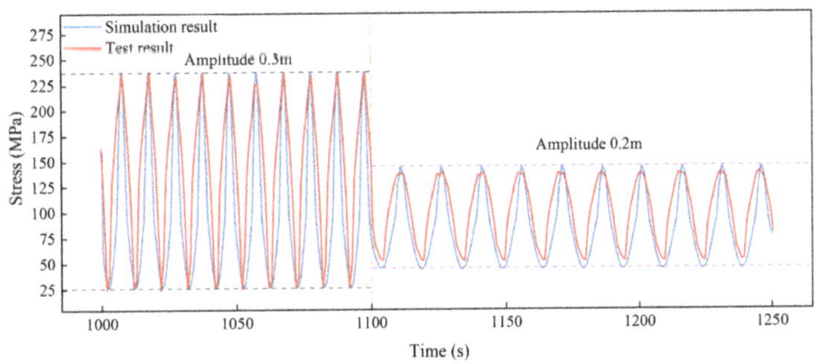

(a)

Figure 29. *Cont.*

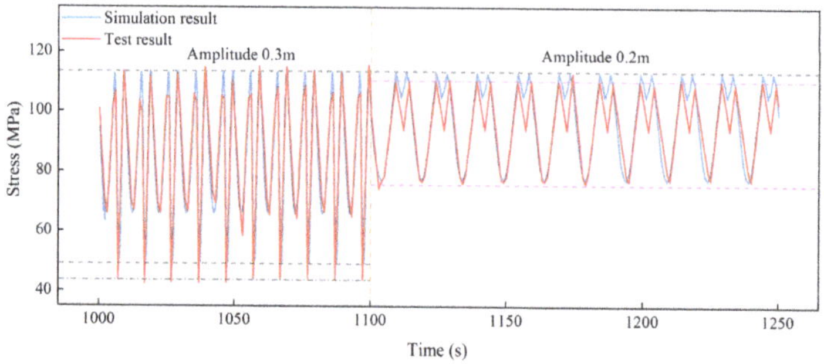

(**b**)

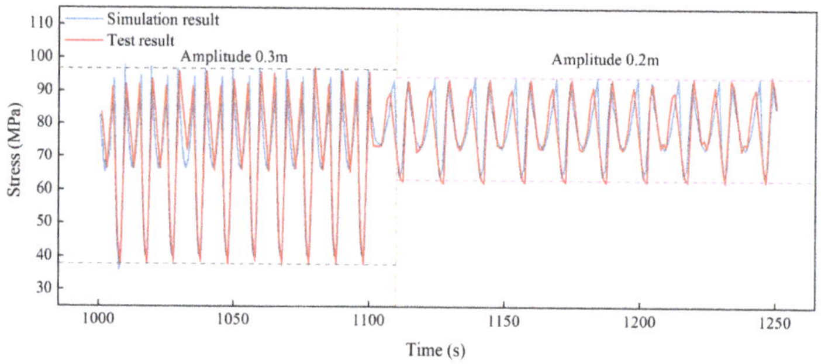

(**c**)

Figure 29. Time histories of the in-plane stress when the amplitude increased from 300 mm to 200 mm: (**a**) S2; (**b**) S5 and (**c**) S8.

4.2.4. Stress Range Analysis

Since the loading frequency would not significantly affect the stress variation, the in-plane stress range distributions along the model SCR under different loading amplitudes were analyzed. The comparisons between the test and numerical results are shown in Figure 30. The comparison between the in-plane stress ranges of the SCR under different loading amplitudes is shown in Figure 30d. When applying a severer cyclic displacement, the stress range along the whole SCR becomes larger, and the increment is more significant towards the loading end.

For the SCR with a loading amplitude of 0.1m, the stress range continued to decrease when approaching the anchored point. However, the decreasing speed has a transition point within the middle of SCR, as the slope of the curve tends to be smaller after this point. As the amplitude became larger, the stress range along the SCR began to decrease from the loading point first and then had an increment within the middle of SCR, before decreasing to zero at the anchored point. This is because the minimum stress within the middle of SCR occurred when the segment penetrated the seabed rather than uplifting to the top position. The location of the peak value of the stress range within the middle of SCR tended to move to the loading point once the loading amplitude was increased from 0.2 m to 0.3 m.

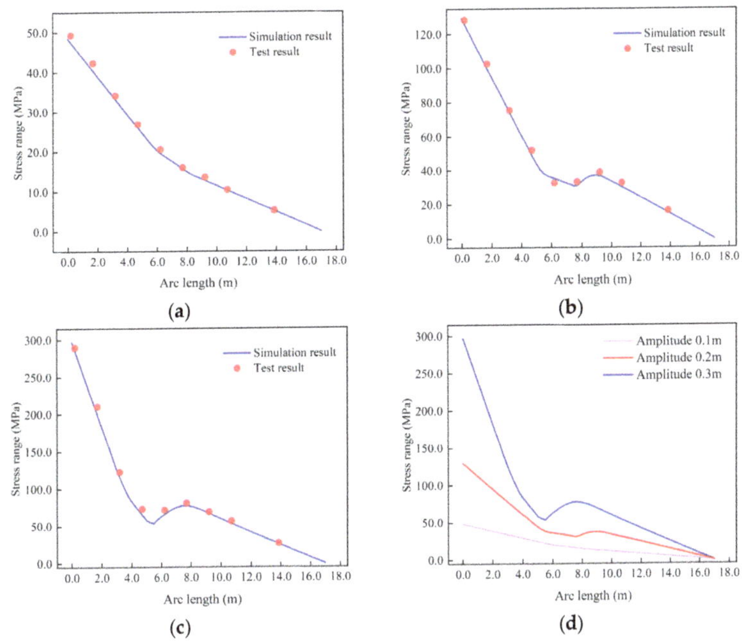

Figure 30. In-plane stress range distribution under different loading amplitudes: (**a**) 100 mm; (**b**) 200 mm; (**c**) 300 mm; (**d**) comparison of the stress ranges.

4.2.5. Seabed Trench Analysis

After executing the vertical loading sequences, the seabed was deformed under the oscillation of the model SCR. The developed trench profile can be obtained as the largest deformation of the riser during the test, and is drawn in Figure 31. A ladle-shaped trench with a maximum depth of 0.71 D was situated within the middle of SCR, and its profile was similar to that of the in situ trenches. For the centrifuge test conducted by Elliott et al. [26], although the trench shape was also similar to a ladle, its mouth was towards the anchored point. The difference in the trench directions is because the bending moment was considered in this study. This moment induced the larger deformation of the riser near the loading side, which helped to model the mechanical behavior of in situ SCR and produce a seabed trench with the correct profile.

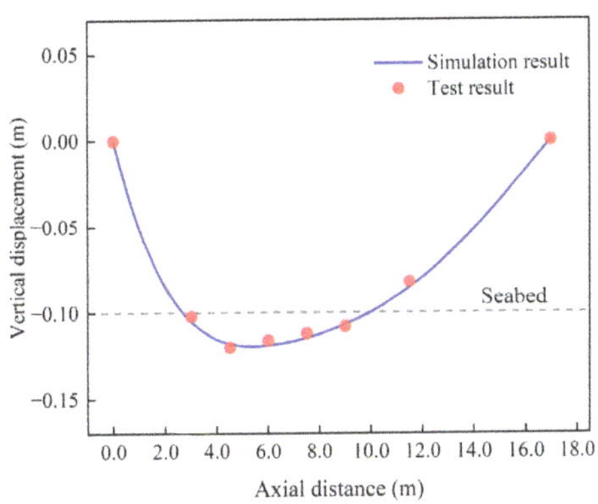

Figure 31. Developed vertical seabed trench.

4.3. Cyclic Horizontal Displacement Loading

4.3.1. Effect of Increasing Loading Amplitude

Horizontal cyclic displacement loads with amplitudes of 100 mm and 200 mm were applied to the truncated point. The vertical elevation of the loading end was adjusted to ensure that its center was 300 mm higher than the center of the tail end before the lateral loading. The loading period of 15 s was kept constant throughout this loading stage. Both the horizontal movement and the lateral seabed resistance contributed to stress variation within the model SCR. The time histories of the out-of-plane stress (half of the difference value of the stress obtained by the strain gauges at the left and right of a cross-section) at points S2, S5, and S8 are shown in Figure 32. All the curves were similar to the sinusoids, although the curve of S8 had some oscillation when the riser segment approached the initial position.

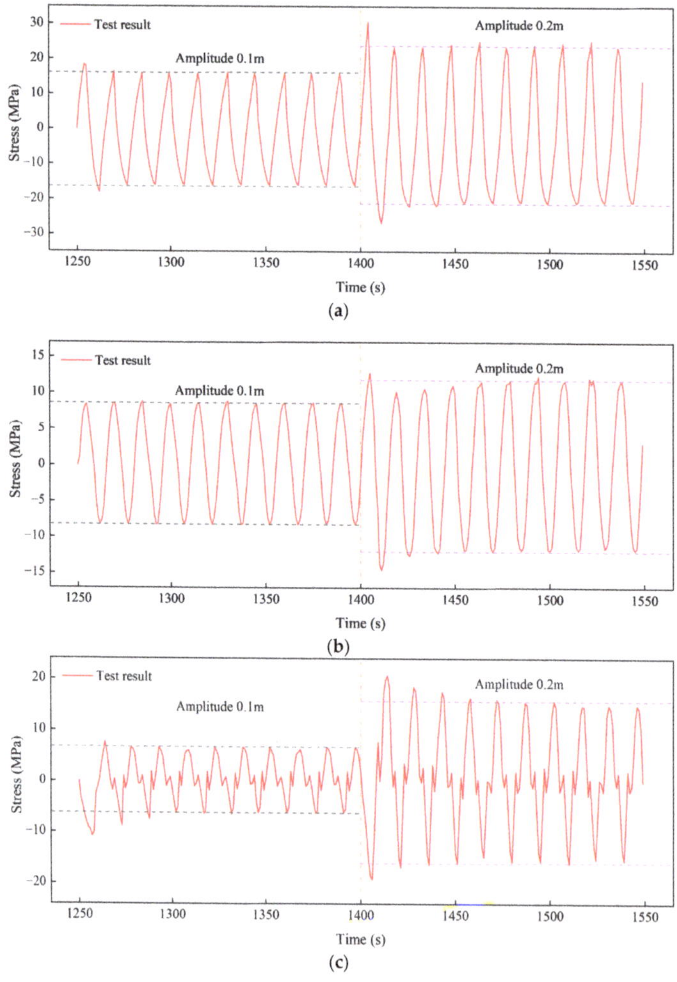

Figure 32. Time history of stress when the horizontal displacements were applied: (**a**) S2; (**b**) S5 and (**c**) S8.

For the first cycle of each loading amplitude, the stress range was larger than those of the following cycles, as the riser segment contacted the seabed and ploughed soil further. After this, the scraping thickness of the soil was reduced, and the lateral resistance provided by the seabed decreased to a stable value. When the larger amplitude was applied, the stress ranges were also enlarged. It should be mentioned that the stress of point S8 had an opposite sign compared with those of S2 and S5 at any moment, i.e., when the stress of S8 was positive, the stresses of S2 and S5 was negative and vice versa. If the lateral soil resistance is not included, the sign of stress along the model SCR should be the same. Therefore, the riser-seabed resistance can change the bending moment distribution along the model SCR and have a significant influence on the mechanical behavior of the SCR.

4.3.2. Stress Range Analysis

The out-of-plane stress range distributions along the model SCR under different lateral cyclic displacements are shown in Figure 33. The stress range decreased from the loading end and reached the valley point within the middle of the SCR. Subsequently, the sign of stress would change as discussed in the previous section, and the stress range began to increase owing to the horizontal force provided by the seabed and reached a peak value. After the peak, the stress range would decrease again when approaching the hinged end.

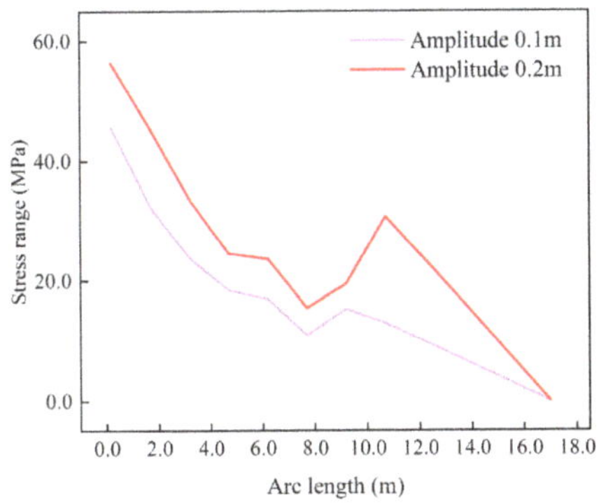

Figure 33. Out-of-plane stress range distribution along SCR.

4.3.3. Trench Plan Analysis

Considering the diameter of the riser, the shape of the trench plan can be determined according to the displacement data gathered by the transducers, and is drawn in Figure 34. The trench plan was bell-mouth-shaped with a maximum width of 0.48 D, which also coincided with the in situ trench. After the water was pumped out, the developed trench could be observed, as shown in Figure 35.

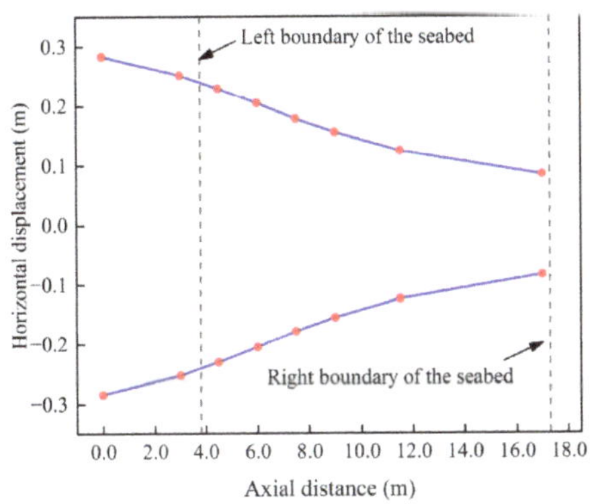

Figure 34. Lateral displacement range of the model SCR.

Figure 35. Developed trench shape.

5. Conclusions

In this study, a novel experimental platform was proposed and constructed to conduct dynamic loading tests of a truncated model SCR within the TDZ. The platform was constituted of a reinforced concrete soil tank, a loading system, and a soil stirring system. Detailed equipment and structures were introduced to make a better illustration of the feasibility of the test. The parameters adopted in this experiment were based on an in situ SCR. The layout of instrumentations and test loading conditions were designed to monitor the dynamic mechanical behavior of the model SCR and determine the influence of riser–soil interaction. Numerical simulations based on the vector form intrinsic finite element method were also conducted for comparison with the test results. According to the test and simulation results, the conclusions can be summarized as follows.

(1) The bending moment and vertical and horizontal cyclic displacement loads can be applied individually or simultaneously during the dynamic test of truncated model SCR within the TDZ. The elastic support can provide restraint to the tail end of the riser. Test clay soil can be prepared with the help of the soil stirring system and vacuum preloading system.

(2) The arrangement of the instrumentation is capable of reflecting the displacement and stress distribution along the model SCR. Based on the data gathered during the dynamic loading test, the mechanical behavior of the model SCR was analyzed, and the effect of loading amplitude and frequency were investigated. The stress range at the front of the model riser increased by more than 100% when the vertical displacement amplitude increased from 200 mm to 300 mm, whereas the loading frequency had only little influence, since the stress range only differed by less than 5%.

(3) The statical configuration, bending moment distribution, and time history of stress at different positions of SCR were compared between the numerical simulation and test results. High similarities proved the correctness and reliability of the nonlinear soil model and the VFIFE method.

(4) The maximum stress occurred when the riser was lowered. If the vertical loading amplitude was small, the minimum stress occurred when the riser was uplifted to the top. However, with the increase in vertical loading amplitude, the minimum stress within the middle of SCR happened when the riser penetrated deeper into the seabed, which may have also induced a larger stress range. The horizontal seabed resistance would vary the out-of-plane stress distribution of the SCR when the lateral displacement loads were applied.

(5) A trench with a ladle-shaped profile and a bell-mouth-shaped plan was developed during the test, and its shape was similar to the in situ trench. The maximum depth and width of the trench were 0.71 D and 0.48 D, respectively. Except for the cyclic displacement loads, the bending moment also contributed to the trench formation by increasing the deformation of SCR near the loading end.

Author Contributions: Conceptualization, Y.Y. and S.X.; methodology, S.X. and J.Y.; software, S.X.; validation, S.X., P.L. and X.L.; formal analysis, S.X. and L.X.; investigation, S.X. and P.L.; resources, Y.Y. and S.X.; data curation, S.X.; writing—original draft preparation, S.X.; writing—review and editing, S.X.; visualization, L.X.; supervision, Y.Y.; project administration, J.Y.; funding acquisition, J.Y. All authors have read and agreed to the published version of the manuscript.

Funding: This research was funded by The National Natural Science Foundation of China (Grant No. 51879189) and National Natural Science Foundation of China (Grant No. 52071234).

Institutional Review Board Statement: Not applicable.

Informed Consent Statement: Not applicable.

Data Availability Statement: The data presented in this study are available upon request from the corresponding author.

Conflicts of Interest: The authors declare no conflict of interest.

References

1. Yu, Y.; Xu, S.; Yu, J.; Xu, W.; Xu, L.; Xu, L. Dynamic Analysis of Steel Catenary Riser on the Nonlinear Seabed Using Vector Form Intrinsic Finite Element Method. *Ocean Eng.* **2021**, *241*, 109982. [CrossRef]
2. Drumond, G.P.; Pasqualino, I.P.; Pinheiro, B.C.; Estefen, S.F. Pipelines, Risers and Umbilicals Failures: A Literature Review. *Ocean Eng.* **2018**, *148*, 412–425. [CrossRef]
3. Quéau, L.M.; Kimiaei, M.; Randolph, M.F. Analytical Estimation of Static Stress Range in Oscillating Steel Catenary Risers at Touchdown Areas and Its Application with Dynamic Amplification Factors. *Ocean Eng.* **2014**, *88*, 63–80. [CrossRef]
4. Zargar, E.; Kimiaei, M.; Randolph, M. A New Hysteretic Seabed Model for Riser-Soil Interaction. *Mar. Struct.* **2019**, *64*, 360–378. [CrossRef]
5. Bai, X.; Huang, W.; Vaz, M.A.; Yang, C.; Duan, M. Riser-Soil Interaction Model Effects on the Dynamic Behavior of a Steel Catenary Riser. *Mar. Struct.* **2015**, *41*, 53–76. [CrossRef]
6. Randolph, M.; Quiggin, P. Non-Linear Hysteretic Seabed Model for Catenary Pipeline Contact. In *Volume 3: Pipeline and Riser Technology*; ASMEDC: New York, NY, USA, 1 January 2009; pp. 145–154.
7. Aubeny, C.; Biscontin, G. Interaction Model for Steel Compliant Riser on Soft Seabed. *SPE Proj. Facil. Constr.* **2008**, *3*, 1–6. [CrossRef]
8. Chen, J.; Bai, X.; Vaz, M.A. Dynamic Behavior of Steel Catenary Riser at the TDZ Considering Soil Stiffness Degeneration and Trench Development. *Ocean Eng.* **2022**, *250*, 110970. [CrossRef]
9. Elosta, H.; Huang, S.; Incecik, A. Trenching Effects on Structural Safety Assessment of Integrated Riser/Semisubmersible in Cohesive Soil. *Eng. Struct.* **2014**, *77*, 57–64. [CrossRef]
10. Bridge, C.D.; Howells, H.A. Observations and Modeling of Steel Catenary Riser Trenches. In Proceedings of the International Offshore and Polar Engineering Conference, Lisbon, Portugal, 1–6 July 2007; pp. 803–813.
11. Wang, K.; Low, Y.M. A Simple Parametric Formulation for the Seabed Trench Profile beneath a Steel Catenary Riser. *Mar. Struct.* **2016**, *45*, 22–42. [CrossRef]
12. Shiri, H. Influence of Seabed Trench Formation on Fatigue Performance of Steel Catenary Risers in Touchdown Zone. *Mar. Struct.* **2014**, *36*, 1–20. [CrossRef]
13. Mekha, B.; Randolph, M.; Bhat, S.; Jain, S. Modeling the Touchdown Zone Trench and Its Impact on SCR Fatigue Life. In Proceedings of the Offshore Technology Conference, Houston, TX, USA, 6 May 2013.
14. Yu, Y.; Xu, S.; Yu, J.; Xu, W.; Xu, L.; Xu, L. Influence of Seabed Trench on the Structural Behavior of Steel Catenary Riser Using the Vector Form Intrinsic Finite Element Method. *Ocean Eng.* **2022**, *251*, 110963. [CrossRef]
15. Shoghi, R.; Shiri, H. Modeling Touchdown Point Oscillation and Its Relationship with Fatigue Response of Steel Catenary Risers. *Appl. Ocean Res.* **2019**, *87*, 142–154. [CrossRef]
16. Shoghi, R.; Pesce, C.P.; Shiri, H. Influence of Trench Geometry on Fatigue Response of Steel Catenary Risers by Using a Boundary Layer Solution on a Sloped Seabed. *Ocean Eng.* **2021**, *221*, 108447. [CrossRef]
17. Willis, N.R.T.; West, P.T.J. Interaction between Deepwater Catenary Risers and a Soft Seabed: Large Scale Sea Trials. In Proceedings of the Offshore Technology Conference, Houston, TX, USA, 30 April 2001.
18. Bai, X.; Vaz, M.A.; Morooka, C.K.; Xie, Y. Dynamic Tests in a Steel Catenary Riser Reduced Scale Model. *Ships Offshore Struct.* **2017**, *12*, 1064–1076. [CrossRef]
19. Wang, L.; Zhang, J.; Yuan, F.; Li, K. Interaction between Catenary Riser and Soft Seabed: Large-Scale Indoor Tests. *Appl. Ocean Res.* **2014**, *45*, 10–21. [CrossRef]
20. Hodder, M.S.; Byrne, B.W. 3D Experiments Investigating the Interaction of a Model SCR with the Seabed. *Appl. Ocean Res.* **2010**, *32*, 146–157. [CrossRef]
21. Wang, L.Z.; Li, K.; Yuan, F. Lateral Cyclic Interaction between Catenary Riser and Soft Seabed. *Appl. Ocean Res.* **2017**, *63*, 11–23. [CrossRef]

22. Dai, Y.; Zhou, J. Experimental Investigations on Seismic Response of Riser in Touchdown Zone. *Int. J. Nav. Archit. Ocean Eng.* **2018**, *10*, 348–359. [CrossRef]
23. Tianfeng, Z.; Shixiao, F.; Tianhao, W. Seabed Stiffness Influences and Water Damping Effects on Vibration of the Touchdown Zone in a Steel Catenary Riser. *Int. J. Press. Vessel. Pip.* **2019**, *170*, 30–39. [CrossRef]
24. Bhattacharyya, A.; Tognarelli, M.; Li, G.; Ghosh, R.; Clukey, E.C.; Sun, Q. Simulation of SCR Behaviour at Touchdown Zone- Part I: Numerical Analysis of Global SCR Model Versus Sectional SCR Model. In Proceedings of the Offshore Technology Conference, Houston, TX, USA, 4 October 2011.
25. Elliott, B.J.; Zakeri, A.; Macneill, A.; Phillips, R.; Clukey, E.C.; Li, G. Centrifuge Modeling of Steel Catenary Risers at Touchdown Zone Part I: Development of Novel Centrifuge Experimental Apparatus. *Ocean Eng.* **2013**, *60*, 200–207. [CrossRef]
26. Elliott, B.J.; Zakeri, A.; Barrett, J.; Hawlader, B.; Li, G.; Clukey, E.C. Centrifuge Modeling of Steel Catenary Risers at Touchdown Zone Part II: Assessment of Centrifuge Test Results Using Kaolin Clay. *Ocean Eng.* **2013**, *60*, 208–218. [CrossRef]
27. Clukey, E.C.; Tognarelli, M.; Li, G.; Ghosh, R.; Phillips, R.D.; Zakeri, A.; Elliott, B.; Bhattacharyya, A.; Sun, Q. Simulation of SCR Behaviour at Touchdown Zone-Part II: Testing of a Sectional SCR Model in a Geotechnical Centrifuge. In Proceedings of the Offshore Technology Conference, Houston, TX, USA, 4 October 2011.
28. Hu, H.J.E.; Leung, C.F.; Chow, Y.K.; Palmer, A.C. Centrifuge Modelling of SCR Vertical Motion at Touchdown Zone. *Ocean Eng.* **2011**, *38*, 888–899. [CrossRef]

Journal of
Marine Science and Engineering

MDPI

Article

Effectiveness of Sacrificial Shielding for Blast Mitigation of Steel Floating Pontoons

Yasser A. Khalifa [1,2], Mohamed N. Lotfy [1,2,*] and Elsayed Fathallah [2,3,*]

1 Civil Engineering Department, McMaster University, Hamilton, ON L8S 4L8, Canada
2 Department of Civil Engineering, Military Technical College, Cairo 11865, Egypt
3 Ships and Submarines Engineering Department, Military Technical College, Cairo 11835, Egypt
* Correspondence: lotfym@mcmaster.ca (M.N.L.); saidhabib2000@hotmail.com or saidhabib2000@mtc.edu.eg (E.F.)

Abstract: Floating pontoons have played a supreme and indispensable role in crises and disasters for both civil and military purposes. Floating bridges and ferries are exposed to blast loadings in the case of wars or terrorist attacks. The protection effectiveness of sacrificial cladding subjected to a blast was numerically investigated. In this study, a steel ferry has been simulated and exposed to side explosions with different explosive charges at certain stand-off distances, according to military standards from NATO and American standard TM5. In this simulation, nonlinear three-dimensional hydro-code numerical simulation ANSYS autodyn-3d has been used. The results reported that the ferry could withstand a charge of 5 kg TNT at a stand-off distance of 1 m without failure. The main objective of this research is to achieve a design that would increase the capacity against the blast loading with minimal plastic deformation in the absence of any failure in the ferry. Therefore, an innovative mitigation system has been proposed to dissipate the blast energy of the explosion based on the scientific theory of impedance using sacrificial cladding. The new mitigation system used a specific structural system in order to install the existing pontoon structure without any distraction. The response, elastic deformations, plastic deformations and plastic failure of the ferry were illustrated in this paper. Furthermore, the results revealed that the proposed mitigation system could mitigate more than 50% of the blast waves. The new design revealed promising results, which makes it suitable for mitigating blast waves. Finally, the results were provided with a reference for the preliminary design and application of sacrificial cladding for structural protection against blast waves.

Keywords: floating; pontoon; AUTODYN; blast; mitigation; modeling; plastic deformation

Citation: Khalifa, Y.A.; Lotfy, M.N.; Fathallah, E. Effectiveness of Sacrificial Shielding for Blast Mitigation of Steel Floating Pontoons. *J. Mar. Sci. Eng.* **2023**, *11*, 96. https://doi.org/10.3390/jmse11010096

Academic Editors: Bin Liu, Kun Liu and Chenfeng Li

Received: 5 December 2022
Revised: 25 December 2022
Accepted: 27 December 2022
Published: 4 January 2023

1. Introduction

Due to accidental explosions and increasing terrorist attacks, preventive and blast-resistant techniques are needed to ensure the survivability of structures. The development of new technologies to protect structures against explosions has gained significant attention, not only for military applications but also for civilian purposes. Any design of the blast load mitigation system must incorporate influence on the mitigation efficiency and the study of attenuation mechanisms.

Techniques of protection against explosion dangers and various mitigation strategies have recently attracted the attention of many researchers [1–9]. Liu et al. [10] investigated the blast load attenuation of sandwich panels. The results demonstrated that sandwich panels could reduce the peak load by 64.69% compared with mild steel plates without a foam core. Cheng et al. [11] studied the effect of cladding sandwich panels subjected to blast loading with different configurations of tubular cores. The results illustrated that cladding panels could be used as an option for blast mitigation. Chen et al. [12] designed sandwich panels with layered-gradient aluminum foam cores subjected to blast loading.

The results illustrated that the layered-gradient sandwich panel has advantages in terms of blast resistance and energy absorption. Wang et al. [13] presented a numerical method to simulate a rocket explosion on a launch pad to study the propagation law of near-ground explosions and quantify the enhancement effect on peak overpressure. Zhou and Hao [14] demonstrated that a protective barrier could effectively reduce blast loading and protect the structures from explosion effects. Furthermore, Abdel Wahab et al. [15] presented a protection system to protect vehicles from blast effects. Alqwasmi et al. [16] studied the behavior of sacrificial sandwich steel panels with axially oriented octagonal tapered tubular cores subjected to blast loading. Additionally, it was found that the top plate and tube thickness are significantly affected by energy absorption. Additionally, Alogla et al. [17] designed protective panels to segregate building façades from the effect of an explosion. The robust design demonstrated great performance, not only in resisting the blast waves but also in providing protection against fragments. Markose and Lakshmana [18] explored a composite plate with V-shaped steel plates and coatings of hyperplastic materials, such as polyurea, to enhance blast mitigation. Moreover, Moustafa et al. [19] developed a novel blast load mitigation system employed as a protection fence, using a bent tube technique to manipulate the shockwave. The results showed that this technique could mitigate 94% of the blast waves, which means that only 6% of the blast impulse is considered an applied load on the targeted structure.

Additionally, Jin et al. [20] proposed different configurations for steel poles used as a fence for a building in order to enhance its protection by mitigating the pressure wave. As an alternative blast mitigation method, sacrificial cladding has attracted attention in recent years [21]. Sacrificial cladding generally consists of a core sandwiched with one or two face plates [22,23]. With the aid of sacrificial cladding, the load transmitted to the structure can be significantly reduced [24]. However, further investigations are still needed to provide more convincing predictions regarding the blast mitigation effectiveness of sacrificial cladding. Yasser A. Khalifa et al. [22] studied the performance of lightweight, cold-formed steel sandwich panels subjected to explosions, as well as energy absorption and scalability. The results suggested that panels with a unidirectional core configuration are preferable to mitigating blast risk. Hong yuan Zhou et al. [21] presented an experimental and numerical investigation to study the effectiveness of using a sacrificial cladding consisting of a steel face plate and an aluminum foam core for near-field blast mitigation. Additionally, Zhang et al. [25] studied the dynamic response of corrugated sandwich structures with different core arrangements to maximize the mitigation effects of blasts. Britan et al. [26] investigated the possibility of using foam materials with high deformability and energy absorption capacity in sacrificial lightweight blast walls. Taha et al. [27] studied the effect of using ultra-high-performance concrete in curved concrete barrier walls to capture the damage produced due to blast explosions. The results illustrated that using ultra-high-performance concrete in barrier walls has a good influence on mitigating the effect of blast explosions, with an average percentage of 32% compared to ordinary concrete barriers. Ehasni et al. [28] incorporated composite materials for protection against an explosive charge of 100 kg of TNT. Furthermore, McDonald et al. [29] experimentally studied the dynamic responses of high-strength and armor-grade steel subjected to explosions at different stand-off distances. Guangyong et al. [30] investigated the effect of the face–sheet configuration and materials on the blast resistance of a structure constructed from sandwich panels. Zong et al. [31] used fence walls for blast wave distraction and investigated their mitigation capabilities. The results showed that the fence walls could reduce the peak pressure and impulse. Moreover, Xia et al. [32] numerically investigated the protective performance of metallic foam cladding on a reinforced concrete slab. Wu and Sheikh [33] studied the influence of metallic foam cladding on protecting a concrete slab subjected to an explosion. Yazici et al. [34] determined the dynamic responses of a sandwich panel against blast loading. The results showed that the deflections were decreased by 50%, and the weight of the panel decreased by approximately 2.3%. Abdel Wahab et al. [35] and Markose and Rao [18] established a mitigation system (V-shape) for protecting the hull of a military vehicle subjected to an explosion. Marcel et al. [36] numerically investigated the effect of the core topologies of sandwich-structured composites in terms of dynamic responses, failure mechanisms and energy absorption. The results clarified that comparable structural characteristics could be achieved by changing the external dimensions and shapes and also by changing the core density. Lin et al. [37] designed and compared sandwich panels with different types of honeycomb cores to construct an optimized blast wall on an offshore platform. The results clarified that the sandwich plate with a concave arc honeycomb core had the best anti-blast performance.

Based on the previous studies, further investigations are needed to provide more convincing predictions about blast mitigation effectiveness. In the present study, the protective effects of sacrificial cladding were proposed. Three-dimensional numerical simulations using commercial finite element ANSYS AUTODYN have been performed to develop a blast load mitigation system that can be used as a fence to satisfy acceptable protection for a floating steel ferry. The model has first been validated using previous experimental tests. The validation process was based on a real experiment and a series of practical experiments performed by Aune et al. [38]. In addition, a new mitigation system has been proposed and simulated to attenuate the generated blast wave and significantly reduce its effect.

1.1. Blast Loading

An explosion is a very rapid release of an enormous amount of energy in a very short time accompanied by an air pressure shock wave [39,40]. This air pressure shock wave compresses the surrounding air and then decays after reaching the peak according to the type, stand-off distance and the charge mass of the explosive [41]. The pressure profiles and equivalents can be determined from Hopkinson–Cranz scaling laws, where the pressure and stand-off distance are expressed as variables of the Hopkinson scaled distance (Z), as in Equation (1) [42–45].

$$Z = R/W^{1/3} \tag{1}$$

where "R" represents the stand-off distance (meter), and "W" represents TNT equivalent charge mass (kg). The blast load depends on the explosive material and the basis of its physical state [46]. Explosions are classified as far-range explosions when the offset distance is equal to or greater than 1.2 m/kg$^{1/3}$, and near-range explosions are when the offset distance is less than 1.2 m/kg$^{1/3}$. The typical pressure–time history, which results from air explosions, is analyzed and illustrated in Figure 1 [17,47–50].

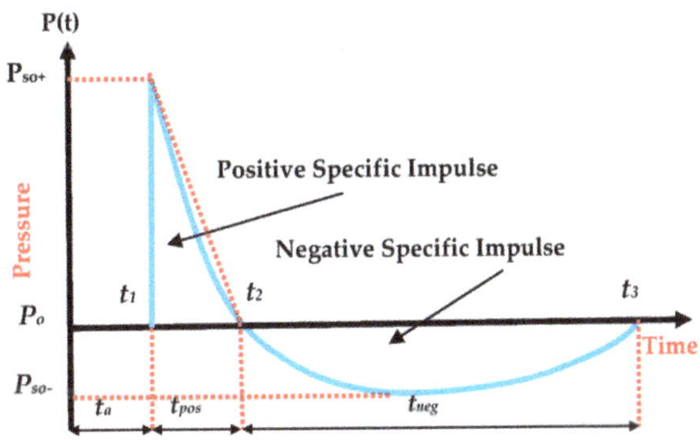

Figure 1. Typical pressure-time history from air explosions [17,47–50].

This curve represents the change in the pressure versus time at any location away from the center of the explosion. Where "P_{SO}" represents the peak incident pressure, "P_O" represents the ambient atmospheric pressure, "P_{SO}^{-}" represents the negative pressure and "t_a" is the arrival time. The blast wave reaches the structure according to the stand-off distance and the charge mass at the time equal to "t_a", and at this point, the pressure reaches its maximum and is equal to (P_{SO}). After that, the incident pressure decreases until it reaches atmospheric pressure. The elapsed time that the blast wave takes from the peak pressure to reach the atmospheric pressure is known as the positive phase duration (t_O). This is followed by a negative pressure called the negative phase duration (t_O^-). The incident impulse (i) is the integrated area under the pressure–time history curve. The pressure can be calculated at any time according to Equation (2) [17,51].

$$P(t) = P_O + P_{SO}\ (1 - \ t/t_O\ )e^{-\mu(t/t_O\ )}) \tag{2}$$

where $P(t)$ represents the pressure at any time "t", "μ" represents the pressure wave decay parameter, "P_{SO}" represents the peak incident pressure and "P_O" represents the ambient atmospheric pressure.

1.2. Strain Rate Effect

The response of the materials varies according to the type of loading. The blast and impulsive loads affect the structures with a very high velocity. The very rapid influence in a short period of time with high energy causes the nonlinearity behavior of the materials. The materials that are exposed to such high impulsive loads generate high strain rates up to (10^6 s^{-1}) [52]. This strain rate affects the strength of the materials; therefore, the simulation of these materials requires a proper understanding of the strain rate effect. Many efforts have been devoted to understanding the strain rate influence since 1960 [52–54]. Zener and Hollomon [55] concluded that most engineering structures were only affected by the strain rate in their elastic ranges, and this follows the elastic zone of the stress–strain curves, which is because of the slow loading rate of static loads. The responses of those structures have changed from the elastic zone to the plastic zone for the blast loads, affecting the high strain rate. This effect is a result of the very short duration of loading (impulse loading). Due to the importance of the strain rate in problems where plastic deformations are expected to appear, the strain rate effect must be considered in impulsive loading problems [55,56]. The Taylor impact test was performed to determine the plasticity parameters at high strain rates (up to 10^5 s^{-1}) for metals [57]. Koerber et al. [58] presented an experimental investigation of the influence of the strain rate on unidirectional carbon-epoxy characteristics, such as mechanical properties and failure strength. It was concluded that the compressive failure strain is not the strain rate sensitive to the studied strain rates.

The Johnson–Cook equation is one of the most important material models used to determine the dynamic response of materials under a high strain rate of order 10^3 s^{-1} or higher [4,59]. This model is used to represent the strength behavior of materials, typically metals that are subjected to large strains, high strain rates and high temperatures. Such behavior might arise in problems of intense, impulsive loading due to high-velocity impact. In addition, the yield stress in this model varies depending on the strain, strain rate and temperature. Equation (3) illustrates the Johnson–Cook equation [59,60].

$$\sigma = [A + B \times \varepsilon_n^p][1 + C \times \ln \varepsilon^*][1 - T_H^m] \tag{3}$$

$$\varepsilon^* = \frac{\varepsilon}{\varepsilon_o} \tag{4}$$

where "σ" represents the material strength, "ε_p^n" represents the effective plastic strain, "ε^*" is the dimensionless plastic strain for $\varepsilon_0 = 1$, "T_H^m" represents the temperature material softening and A, B, C, n, and m are five constants that depend on the characteristics of the used material [59] and are defined as follows [61–64]: Constant "A" represents the yield stress corresponding to an offset strain value of 0.2%, constant "B" and exponent "n" represents the effect of the material strain hardening, and constant "C" represents the strain rate effect. The Johnson–Cook model is represented by three terms that affect the strength of materials; the first one represents the relationship between the stress and the strain, the second represents the relationship between the stress and the strain rate while the last one represents the temperature effect on stress during plastic deformation. The proper simulation of any material in AUTODYN depends upon the definition of the used parameters according to the used equation of state, strength and failure equations.

2. Problem Explanation and Numerical Modeling

Floating bridges are very important for both civilian and military purposes [65,66]. For military usage, blast loadings have a disastrous effect on the floating ferry in the present design. Altering the design of the floating ferry requires replacing the whole number of floating pontoons already used. This problem can be solved by the implementation of an additional structure with specific characteristics to dissipate the explosive energy.

2.1. Problem Description and Methodology

The objective is to mitigate the explosion's energy to facilitate floating steel ferry/bridge operation in hazardous scenarios. For military purposes, standards and manuals (e.g., NATO Standards [67,68] and American Standards [69]) determine the levels of protection and the possible charges of several water mines. The innovative system has a simple internal configuration that can be attached to existing floating pontoons (no need to manufacture new pontoons). The simulation was performed using ANSYS AUTODYN. The reduction in the transmitted pressure waves depends on the mechanical impedance (I) of the medium [70], which is defined as the ratio between the change in the pressure (Δp) to the change in the velocity of the shock wave (Δv) traveling through the medium [71].

$$I = (\Delta p)/(\Delta v) \tag{5}$$

Using a medium with a low mechanical impedance (such as air) between two mediums with high impedance values (such as steel) attenuates the peak value of the transmitted overpressure [72,73]. In this study, an air gap is present in the mitigation system in order to dissipate the pressure wave depending on the mechanical impedance.

2.2. Uni-Float Steel Ferry

The Uni-float steel ferry is used for military purposes to accommodate the tank load of the Military Load Capacity (MLC-70). The ferry is composed of sixteen floating pontoons [74]. Figure 2 illustrates the configuration of the steel pontoons used to form the ferry.

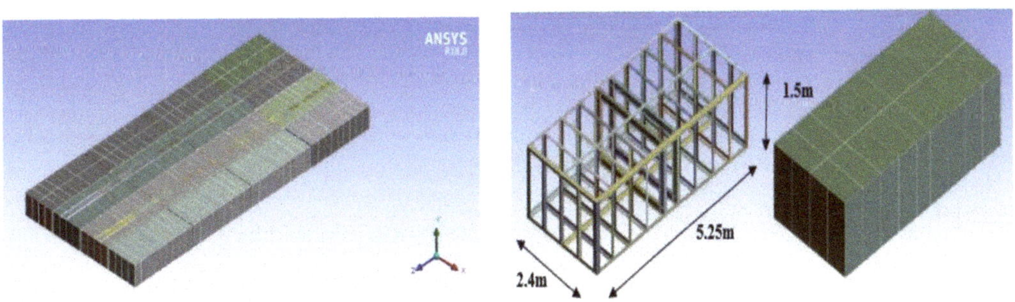

(**a**) Ferry (**b**) Steel pontoon

Figure 2. Schematic illustrating the ferry and steel pontoon configuration.

Lotfy et al. [65] illustrated the specifications of the floating steel ferry. The properties of the steel used are illustrated in Table 1 [75]. Additionally, a sensitivity analysis is carried out to obtain the sizes of the reasonable elements [65].

Table 1. The properties of the used steel (steel 37) [75].

Density	7850 (kg/m^3)	Erosion factor	1
Shear modulus	8.18×10^{10} (G Pa)	Max. eff. strain rate at failure	2.1
Elastic modulus	2.1×10^5 (G Pa)	Thermal softening exponent	1.03
Poisson's ratio	0.28	Hardening exponent	0.26
Yield stress	2.4×10^5 (kPa)	Melting temperature	1790 (K)
Hardening constant	5.1×10^5 (kPa)	Reference strain rate	1
Strain rate constant	0.014	Bulk modulus	1.59×10^5 (G Pa)

2.3. The Simulation Domain

The commercial software ANSYS AUTODYN 19.0 [76] was used for the numerical simulations and the implementation of the fluid–structure interactions (FSI). The Euler/Lagrange interaction model couples the Eulerian to the Lagrangian structural domain. The domain is defined in Euler solver as a 3D-multi-material. The volume of the domain must be determined according to the experiment's behavior.

The ferry was simulated, and the boundaries were defined to match realistic conditions, preventing the explosion from being confined; therefore, the boundaries were assigned to a flow-out type in all faces, as shown in Figure 3.

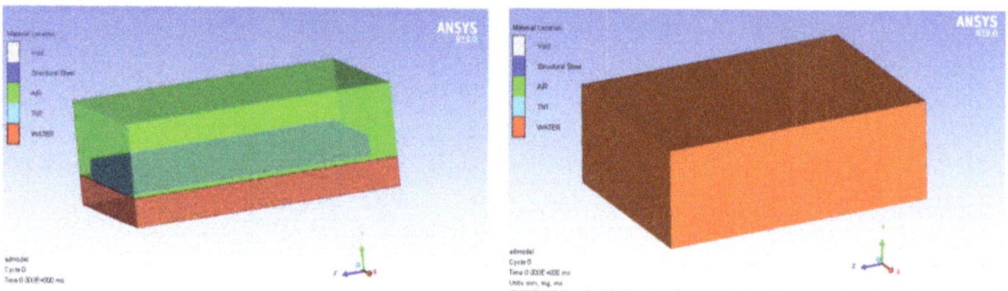

(**a**) Domain (**b**) Flow-out boundaries

Figure 3. The domain and boundaries of the ferry.

The domain was divided into the air (upper side) and water (lower side). The water surfaces were simulated without any boundary due to the effect of the water on the reflection of the pressure wave. The ferry was exposed to explosive charges of Trinitrotoluene (TNT) at a stand-off distance of one meter. The material properties of the air and TNT were retrieved from the standard AUTODYN library. The air part was modeled using 3D Euler equations, while the explosive charge (TNT) was represented according to the Jones Wilkens Lee (JWL) equation of state EOS, with a reference density of 1630 kg/m^3, according to Equation (6) [31,77,78]:

$$P = C_1\left(1 - \frac{w}{r_1 v}\right)e^{-r_1 v} + C_2\left(1 - \frac{w}{r_2 v}\right)e^{-r_2 v} + \frac{we}{v} \quad (6)$$

where "P" indicates the hydrostatic pressure, "v" denotes the specific volume and "e" indicates the specific internal energy, while C_1, r_1, C_2, r_2 and w are the material constants. The values of the constants C_1, r_1, C_2, r_2 and w have been determined for many common explosives from the dynamics. The values of the constants for a TNT explosive are available in AUTODYN, as shown in Table 2 [31,78,79].

Table 2. TNT explosive parameters.

Parameters	**C1**	**C2**	**r1**	**r2**	**w**
Value	3.7377×10^5 MPa	3.7471×10^3 MPa	4.15	0.9	0.35

In addition, the air is modeled by an ideal gas EOS, which can be expressed as in Equation (7).

$$P = C_1(\gamma - 1)\rho e \tag{7}$$

where "γ" denotes the heat-specific ratio and "ρ" denotes the density. The standard constants of the air were taken from the AUTODYN material library, with an air reference density (ρ) equal to 1.225 kg/m^3 and γ equal to 1.4. The initial internal energy of the air is assumed to be 2.068×10^5 kJ/kg [78].

The floating pontoons were exposed to side air explosions at a stand-off distance of 1 m and a height of 0.90 m from the bottom of the ferry. This height represents the mid-span of the un-submerged part of the ferry. Global erosion was performed as 1.3 geometric strains; additionally, the global cutoff of the maximum expansion was 0.2. Furthermore, the trajectory interaction between the Lagrangian part and the fully coupled Euler sub-grid and shell elements was performed. Charges of 1 kg, 10 kg, 50 kg and 100 kg were used to study their effect on the ferry in order to determine the maximum side charge that the ferry can withstand. Figure 4 illustrates the charge position for the side explosions. The gauges were installed to determine the displacement of the ferry due to the explosions. Figure 5 illustrates the gauges for the side explosion.

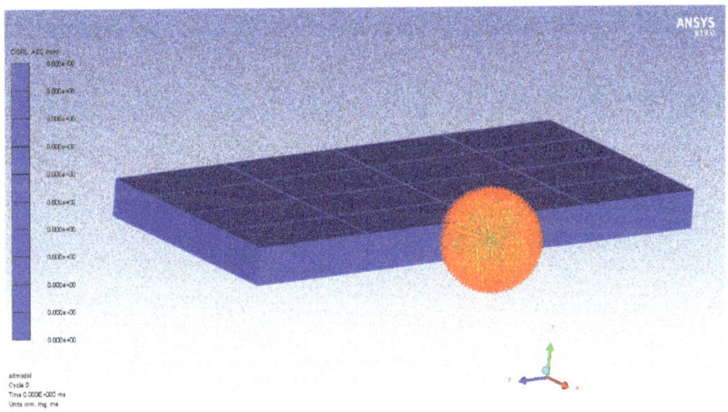

Figure 4. Side explosive charge.

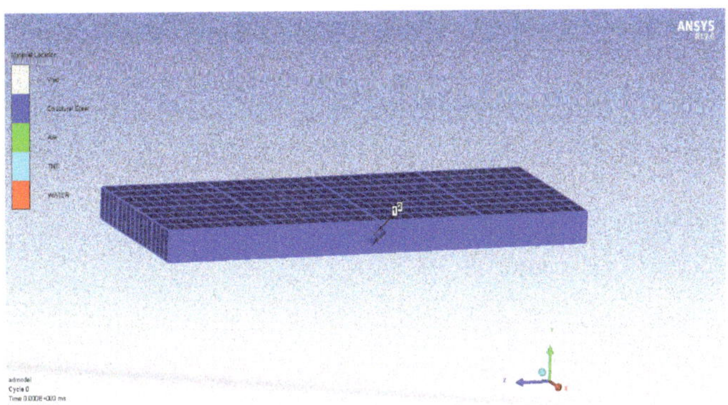

Figure 5. Gauges for the side explosion.

3. Validation

Simulating explosion in ANSYS AUTODYN requires the accurate definition of the properties of the used material, the medium that transfers the shock wave, the boundary conditions and the interaction between the different materials. The validation is divided into two parts: First, the validation of the shock wave and the pressure–time history, which affect the materials, and second, the validation of the deformation response of the used material.

3.1. Pressure-Time History

The validation of the pressure–time history depends on both the practical and numerical experiments performed by Fairlie [80]. Fairlie presented a practical example that described the air blast and determined the pressure–time history at the front of a building. Additionally, Fairlie used the scaling law to simulate the test with a scale of $1/50^{th}$. A small charge of TNT (8 g) is exploded, and the pressure is measured at the elevation of the opposite building. Figure 6 illustrates the physical and numerical experiments.

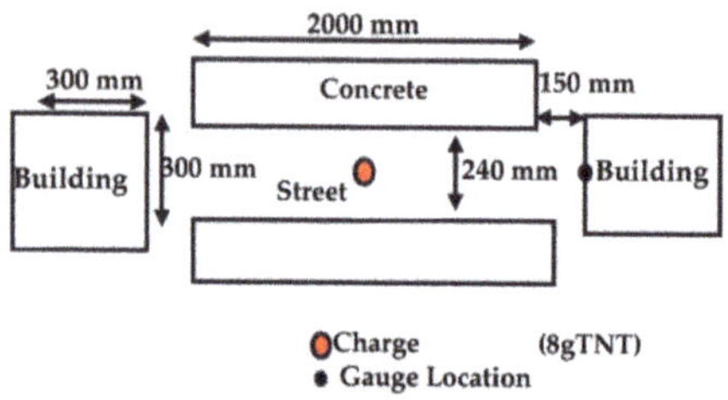

(a)

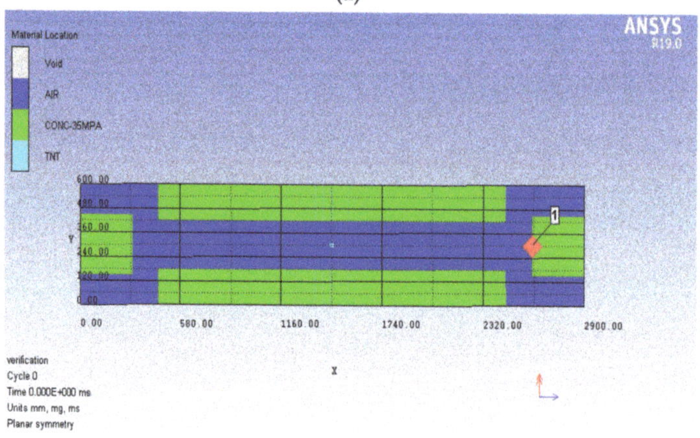

(b)

Figure 6. Details of the air blast experiment. (**a**) schematic diagram of the air blast physical [80]. (**b**) Numerical modelling of the experiment.

The maximum overpressure was measured in the field to be about 380 kPa. The maximum overpressure value from the numerical modeling equals 390 kPa. The matching percentage is 97.4%. Figure 7 illustrates the pressure–time history of both the experimental and numerical results. The pressure–time history curves are similar in both cases. The small changes in the time of arrival between the practical and numerical results are due to the change in the TNT pressure wave velocity.

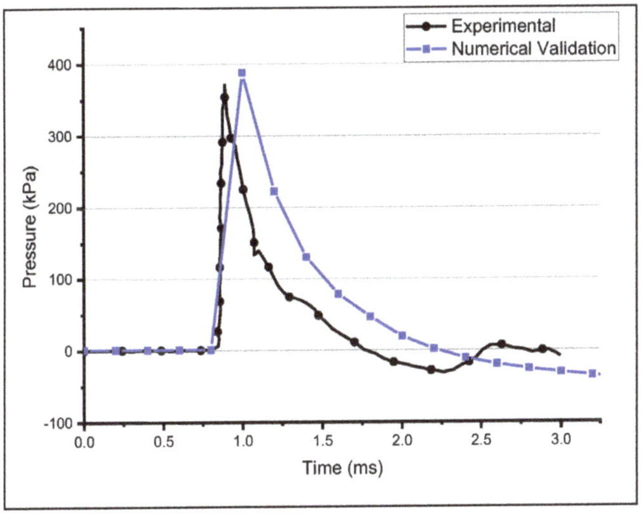

Figure 7. Pressure-time history from simulation and air blast experiment results [80].

3.2. Validation Deformation Response

The validation in this section is based on both a real experiment and a series of physical experiments performed by Aune et al. [38]. Aune et al. performed a series of physical experiments and numerical simulations to study the effect of the stand-off distance and the dynamic responses of different materials with different thicknesses. Square steel plate specimens with dimensions of 400 mm × 400 mm × 0.8 mm were tested with a TNT charge of 40.2 gm at different stand-off distances. The verified experiment is mentioned as experiment number S23. Figure 8 illustrates the setup of the experiment.

(**a**) Experiment setup (**b**) Charge of 40 gm TNT (**c**) Test plate with strain gauges

Figure 8. Field experiment [38].

The steel plate was simulated in AUTODYN with the same characteristics as that used in the experiment. The boundary conditions assigned to the plate were the same as the real experiment boundaries. The domain was simulated as an air cube around the plate with flow-out boundaries, which allowed the shockwave to propagate away and prevent its reflection. Figure 9 illustrates the simulated steel plate and the domain with the remapped TNT charge.

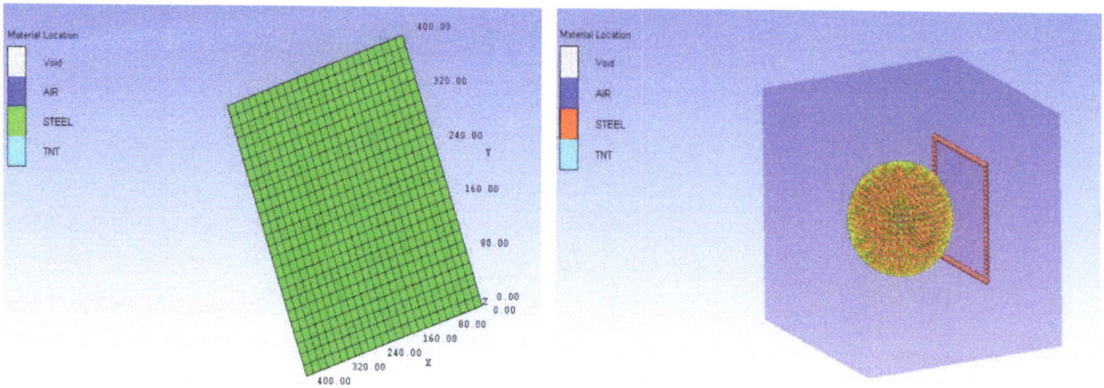

(**a**) Numerical simulation of the plate (**b**) Domain of the experiment

Figure 9. Blast test simulation.

The obtained deformation value from a spherical charge of 40 gm of TNT with a stand-off distance of 0.25 m equals 18.3 mm, as mentioned in [38]. The maximum numerical displacement value was obtained from the simulation to be 17.4 mm, with a matching percentage of about 95%. Figure 10 illustrates the displacement distribution from the numerical simulation.

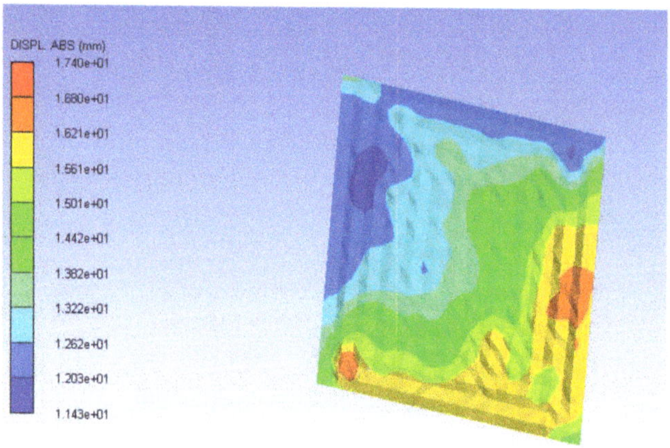

Figure 10. Numerical displacement distribution.

4. Side Blast of the Traditional Steel Ferry

The ferry was exposed to the explosion of 5 and 10 kg of TNT sequentially at a stand-off distance of 1 m. The expansion of the side explosion is illustrated in Figure 11. The explosion caused local deformations at the critical positions (in front of the charge).

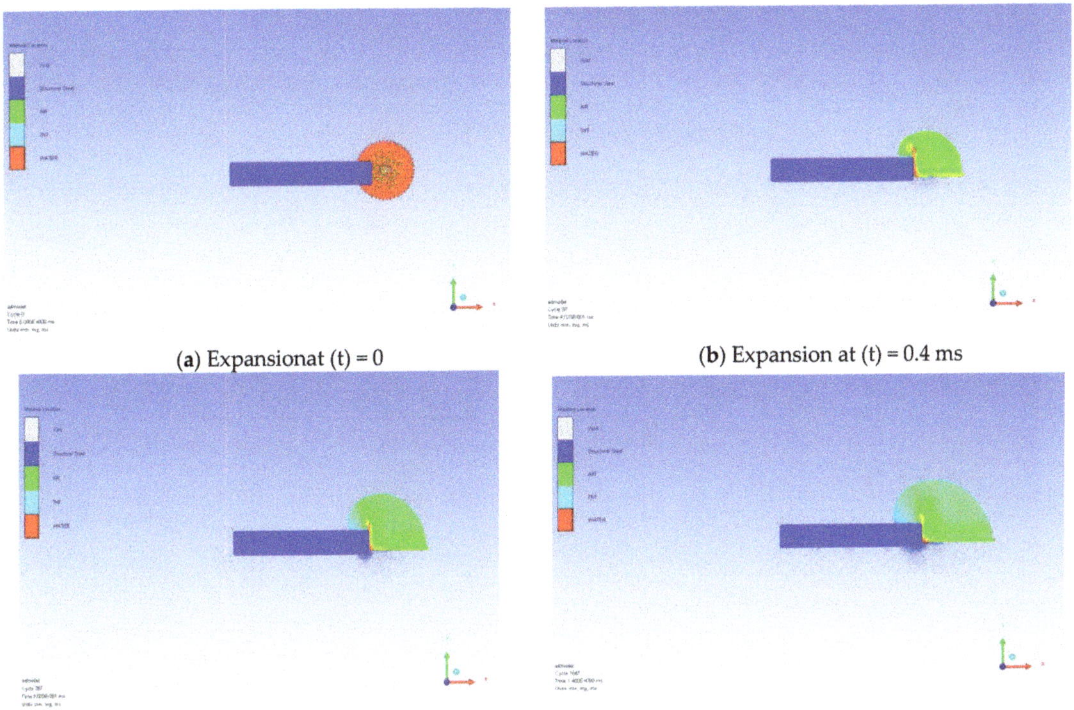

(**a**) Expansionat (t) = 0
(**b**) Expansion at (t) = 0.4 ms
(**c**) Expansion at (t) = 0.9 ms
(**d**) Expansion at (t) = 1.4 ms

Figure 11. Expansion of the side explosion.

4.1. Results for the Charge 5 kg of TNT

The explosion of a side charge of 5 kg of TNT reveals that there are local deformations at the side shells in addition to the side transition motion of the ferry. The incident pressure wave from the explosion first affects the side shell in a very short rapid time causing local deformations, and then the ferry is affected by this pressure wave. The ferry starts to gain kinetic energy from the explosion, which causes a side translation motion. This translation motion vanishes by the effect of the ferry weight and inertia. Figure 12 illustrates the displacement of the ferry during the explosion. The side translation of the ferry reaches about 8.4 cm, and the material status of the steel remains in the elastic zone.

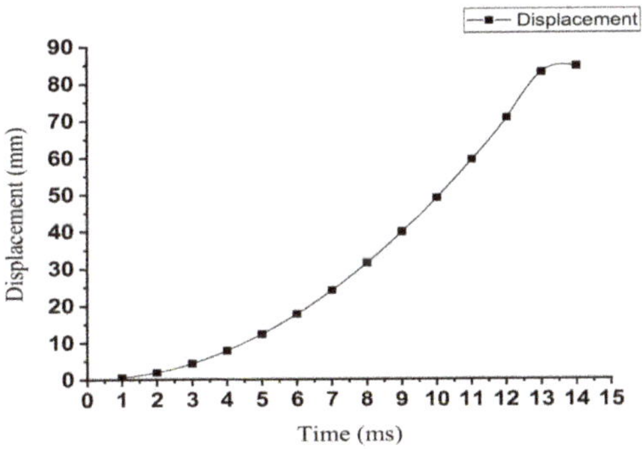

Figure 12. Displacement of the ferry from a side explosion of 5 kg TNT.

4.2. Results for the Charge 10 kg of TNT

The charge is increased to 10 kg of TNT at the same stand-off distance (1 m) to determine the maximum charge that the ferry can withstand. The explosion resulted in excessive local deformations at the critical section (in front of the charge). Figure 13 illustrates the deformed ferry due to the side explosion.

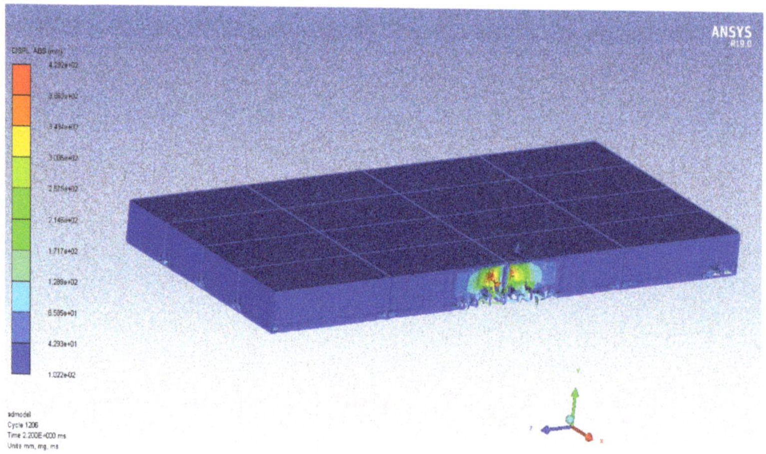

Figure 13. Deformations of the ferry.

These local deformations increase until reaching the plastic failure at the lower connection between the side and lower shells. The failure occurs at this section and not at the upper connection because of the increased rigidity in the upper connection between the upper and side shells, as the upper shell's thickness is larger than the lower shell's thickness. Plastic failure occurs at the critical section, while the plastic deformations occur at other lower connections and the side shells around the plastic failure, as illustrated in Figure 14. These deformations illustrate the plastic failure and the erosion of the ferry material.

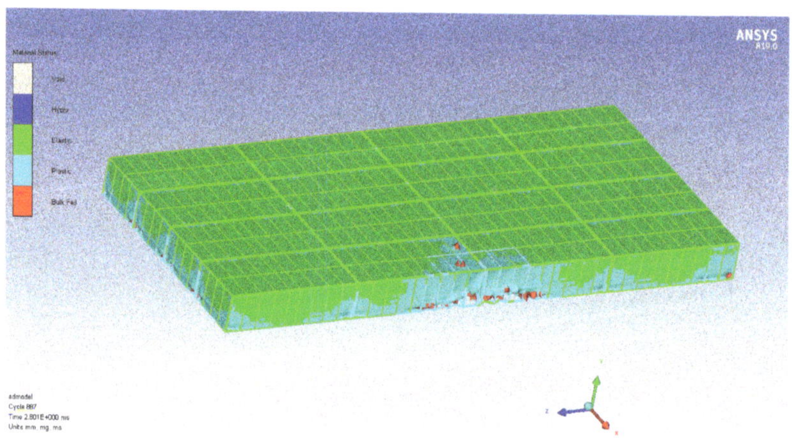

Figure 14. Material statues of the ferry.

The displacement of the ferry from a side explosion due to 10 kg of TNT is illustrated in Figure 15.

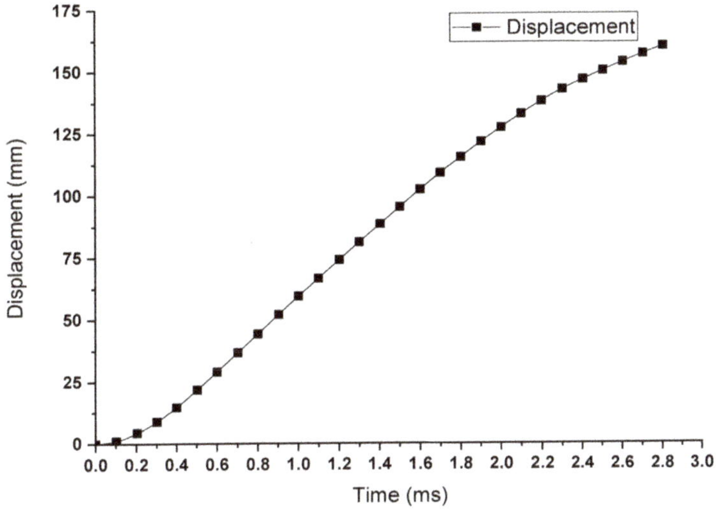

Figure 15. Deformations of the ferry from a side explosion due to 10 kg of TNT.

5. Mitigation System

The proposed mitigation system depends on the wave impedance by using an air gap, which results in the dissipation of the pressure wave; additionally, the shape of the mitigation system greatly affects the dissipation of the pressure wave. The proposed structure is a panel composed of a curved steel surface with internal stiffeners. Every panel is connected to the side of the pontoon. This panel weighs 270 kg, which is very low in comparison to the pontoon's weight (3.6 t). These panels are only added to the right and left sides of the ferry. Figure 16 illustrates the proposed mitigation system. This mitigation panel has internal stiffeners in order to increase the stiffness of the structure to withstand the explosive charge and increase the capacity against lateral buckling, as shown in Figure 17.

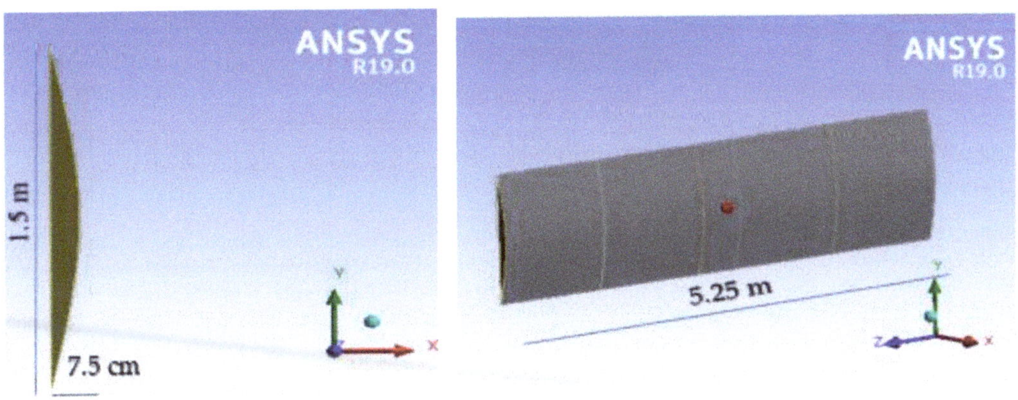

(**a**) Side view of the panel (**b**) 3D view of the panel

Figure 16. Mitigation panel.

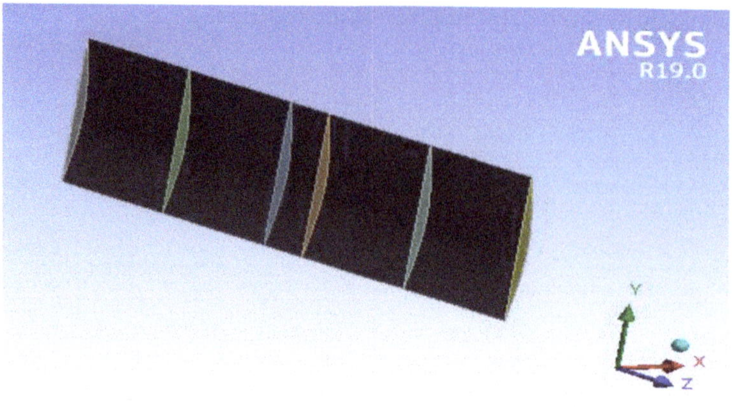

Figure 17. Internal stiffeners arrangement.

These mitigation panels are installed on two sides of the steel ferry (four panels on each side). The internal spaces between the panels and the pontoons are used as air gaps, which is the basis for using the scientific theory of impedance. Figure 18 illustrates the mitigation panels on one side of the steel ferry. The ferry is still stable on both sides after installing the mitigation panels due to the symmetry.

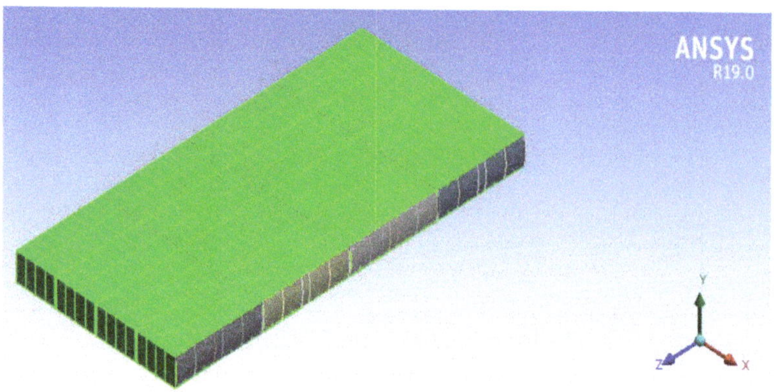

Figure 18. Mitigation panels at the side of the steel ferry.

5.1. Ferry Stability

The ferry is still stable after the installation of these new panels, whereas the centroid of the ferry is the same as before (without mitigation panels). The ferry weight is increased by about 2.15 t, but the volume of the ferry is increased, which results in almost the same values as the draft.

5.2. Side Blast Results

The fortified ferry was exposed to the same charge as the previous case (10 kg of TNT at a stand-off distance of 1 m). Figure 19 illustrates the material condition of the ferry with the new mitigation system after the explosion of the charge.

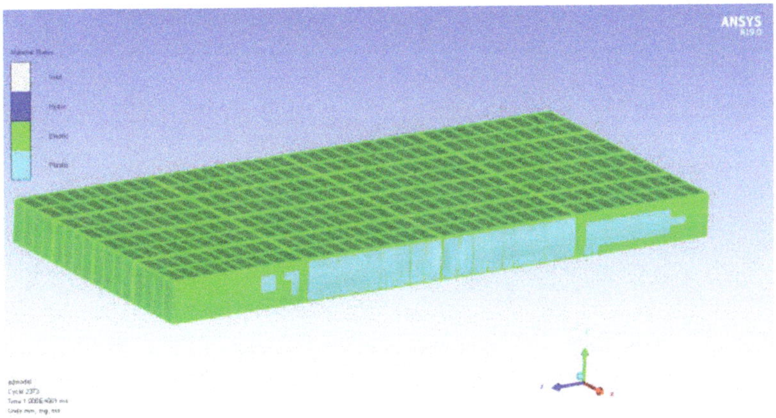

Figure 19. Deformations on the mitigation panels.

The mitigation system was found to dissipate the blast waves and reduce the effect of the explosion. The blast waves were divided into incident waves, diffracted waves and dissipated waves, which were successfully dissipated by the mitigation system.

The ferry can resist the charge without any plastic failure in the pontoons or even in the mitigation system. The installed panels are exposed to plastic deformations, while the pontoons of the ferry are not exposed to any plastic deformations or plastic failure. The effects of the explosion on the pontoons and the mitigation panels are illustrated in Figure 20.

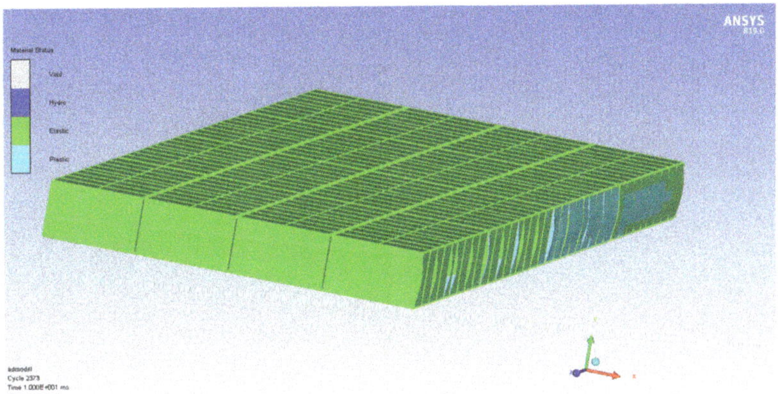

Figure 20. Material statues beyond the mitigation system.

Plastic deformations only occurred at the stiffener's positions of the mitigation panels and not in the steel pontoons. The deformations of the fortified ferry corresponding to the explosion are illustrated in Figure 21.

The maximum value of the total deformations was found to be 6.9 cm, which is lower than the used curvature of the mitigation panel (7.5 cm). Figure 22 illustrates the response of the floating ferry subjected to an explosion of 10 kg of TNT after installing the mitigation system.

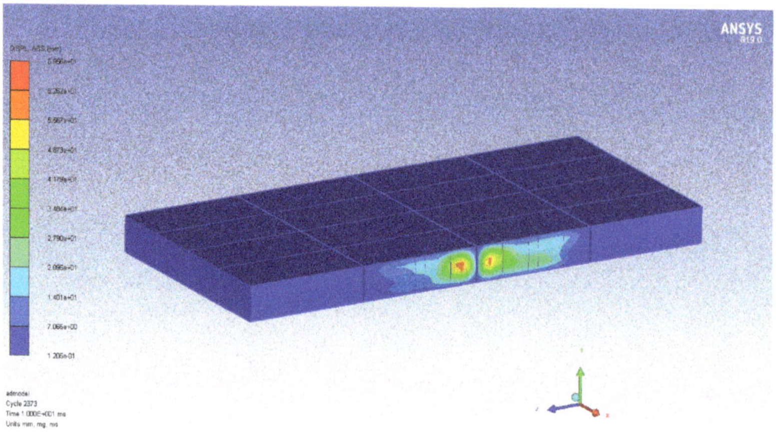

Figure 21. Total deformations of the fortified ferry.

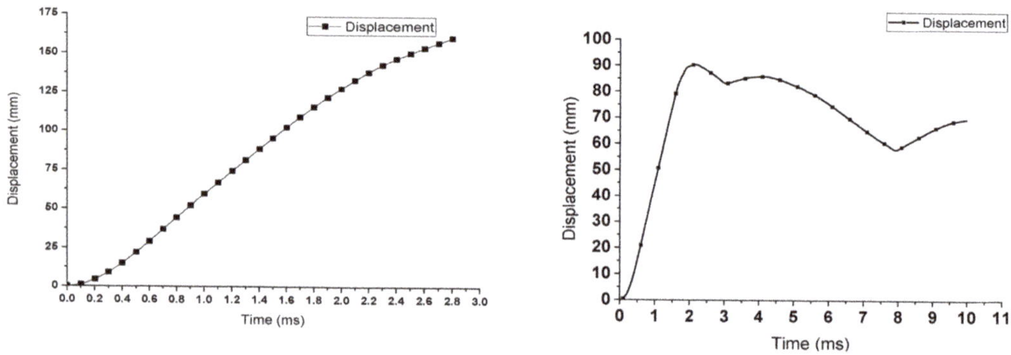

(**a**) Displacement time history without the mitigation system (**b**) Displacement time history with the mitigation system

Figure 22. The displacement response.

The maximum displacement occurs at the side shell of the ferry behind the mitigation system. The maximum deformation at the critical section behind the mitigation system equals about 8.8 cm. This deformation value occurs as a combination of the transitional movement of the ferry and the local deformations of the critical section. As a result, the ferry moved transversely in the water by about 6 cm, while the local deformations reached about 2.8 cm.

6. Conclusions

This paper presents the results of numerical analyses on steel floating pontoons exposed to side explosions with different explosive charges to achieve a design that can sustain the blast loads with minimal deformations. The fluid–structure interactions with the air and water were simulated using ANSYS AUTODYN 19.0. The ferry hull was analyzed under an explosion of 5 kg and 10 kg of TNT. For the charge of 5 kg of TNT, the maximum displacement was found to be about 8.4 cm with the elastic behavior of the ferry material, while the ferry is damaged under the explosion of 10 kg of TNT. An innovative mitigation system was proposed to resist explosion attacks from terrorists or conventional bombs. The proposed mitigation system is composed of separated mitigation steel panels. These panels increase the capacity of the floating steel ferry to withstand explosive charges until a charge of 10 kg of TNT at a stand-off distance of 1 m, while the original ferry could not withstand this charge. The mitigation panels are capable of reducing the deformation and impulse transmitted to the base structure. The maximum displacement from the explosion of 10 kg of TNT after using the mitigation panels reaches about 8.8 cm. It has been found that the mitigation system dissipates the pressure wave by using the air gap between the original pontoons and the proposed panels. The presented mitigation system showed outstanding blast wave mitigation capabilities. The results showed that the mitigation system could mitigate more than 50% of the blast waves with almost full protection for the targeted structure. Such success can only be attributed to its exceptional design. Furthermore, ANSYS AUTODYN demonstrated a great capability in studying air–water–structural interactions with a good level of accuracy.

Author Contributions: Conceptualization, E.F. and Y.A.K.; Formal analysis, E.F. and M.N.L.; Funding acquisition, E.F., Y.A.K. and M.N.L.; Methodology, E.F. and Y.A.K.; Resources, E.F. and Y.A.K.; Software, E.F. and M.N.L.; Supervision, E.F. and Y.A.K.; Writing—original draft, E.F., M.N.L. and Y.A.K.; Writing—review and editing, E.F. and Y.A.K. All authors have read and agreed to the published version of the manuscript.

Funding: This research received no external funding.

Institutional Review Board Statement: Not applicable.

Informed Consent Statement: Not applicable.

Data Availability Statement: The data presented in this study are all included in this article.

Acknowledgments: The authors are thankful to Military Technical College (Cairo, Egypt) for providing all the necessary facilities to carry out that research.

Conflicts of Interest: The authors declare no conflict of interest.

References

1. Abbas, A.; Adil, M.; Ahmad, N.; Ahmad, I. Behavior of reinforced concrete sandwiched panels (RCSPs) under blast load. *Eng. Struct.* **2019**, *181*, 476–490. [CrossRef]
2. Draganić, H.; Gazić, G.; Varevac, D. Experimental investigation of design and retrofit methods for blast load mitigation—A state-of-the-art review. *Eng. Struct.* **2019**, *190*, 189–209. [CrossRef]
3. Li, Z.; Chen, W.; Hao, H. Numerical study of sandwich panel with a new bi-directional Load-Self-Cancelling (LSC) core under blast loading. *Thin-Walled Struct.* **2018**, *127*, 90–101. [CrossRef]
4. Helal, M.M.; Elsayed, F.J.M.T. Dynamic behavior of stiffened plates under underwater shock loading. *Mater. Test.* **2015**, *57*, 506–517. [CrossRef]
5. Fathallah, E.; Qi, H.; Tong, L.; Helal, M. Numerical Simulation and Response of Stiffened Plates Subjected to Non-Contact Underwater Explosion. *Adv. Mater. Sci. Eng.* **2014**, *2014*, 1–18. [CrossRef]
6. Fathallah, E.; Qi, H.; Tong, L.; Helal, M. Numerical investigation of the dynamic response of optimized composite elliptical submersible pressure hull subjected to non-contact underwater explosion. *Compos. Struct.* **2015**, *121*, 121–133. [CrossRef]
7. Xu, J.; Liu, J.; Gu, W.; Liu, X.; Cao, T. Shock Wave Attenuation Characteristics of Aluminum Foam Sandwich Panels Subjected to Blast Loading. *Shock. Vib.* **2018**, *2018*, 1–10. [CrossRef]
8. Helal, M.; Huang, H.; Fathallah, E.; Wang, D.; ElShafey, M.M.; Ali, M.A.E.M. Numerical Analysis and Dynamic Response of Optimized Composite Cross Elliptical Pressure Hull Subject to Non-Contact Underwater Blast Loading. *Appl. Sci.* **2019**, *9*, 3489. [CrossRef]
9. Elsayed, F.; Nagy, N.M.; Lili, T. Numerical Evaluation of Composite Plates Performance under the Effect of Underwater Explosion. In Proceedings of the 15th International Conference on Aerospace Sciences & Aviation Technology, Cairo, Egypt, 28–30 May 2013; pp. 1–15.
10. Liu, H.; Cao, Z.K.; Yao, G.C.; Luo, H.J.; Zu, G.Y. Performance of aluminum foam–steel panel sandwich composites subjected to blast loading. *Mater. Des.* **2013**, *47*, 483–488. [CrossRef]
11. Wang, C.; Xu, B.; Yuen, S.C.K. Numerical analysis of cladding sandwich panels with tubular cores subjected to uniform blast load. *Int. J. Impact Eng.* **2019**, *133*, 103345. [CrossRef]
12. Chen, D.; Jing, L.; Yang, F. Optimal design of sandwich panels with layered-gradient aluminum foam cores under air-blast loading. *Compos. Part B Eng.* **2019**, *166*, 169–186. [CrossRef]
13. Wang, Y.; Wang, H.; Cui, C.; Zhao, B. Investigating Different Grounds Effects on Shock Wave Propagation Resulting from Near-Ground Explosion. *Appl. Sci.* **2019**, *9*, 3639. [CrossRef]
14. Zhou, X.Q.; Hao, H. Prediction of airblast loads on structures behind a protective barrier. *Int. J. Impact Eng.* **2008**, *35*, 363–375. [CrossRef]

15. Wahab, M.M.A.; Mazek, S.; Abada, M.M.; Atta, M.H.A. Effect of blast wave on lightweight structure performance. *J. Eng. Sci. Mil. Technol.* **2017**, *1*, 1–6. [CrossRef]
16. Alqwasmi, N.; Tarlochan, F.; Alkhatib, E.S. Study of Mild Steel Sandwich Structure Energy Absorption Performance Subjected to Localized Impulsive Loading. *Materials* **2020**, *13*, 670. [CrossRef]
17. Alogla, A.; Helal, M.; ElShafey, M.M.; Fathallah, E. Numerical Analysis for Critical Structures Protection against Blast Loading Using Metallic Panels. *Appl. Sci.* **2020**, *10*, 2121. [CrossRef]
18. Markose, A.; Rao, C.L. *Effectiveness of Polyurea Coated Steel Plates in Blast Mitigation in Vehicles*; Springer: Singapore, 2020; pp. 153–164.
19. Esa, M.; Amin, M.S.; Hassan, A. Relative performance of novel blast wave mitigation system to conventional system based on mitigation percent criteria. *Def. Technol.* **2021**, *17*, 912–922. [CrossRef]
20. Jin, M.; Hao, Y.; Hao, H. Numerical study of fence type blast walls for blast load mitigation. *Int. J. Impact Eng.* **2019**, *131*, 238–255. [CrossRef]
21. Zhou, H.; Zhang, X.; Wang, X.; Du, X.; Yu, S.; Wang, Y. Jiang Protection effectiveness of sacrificial cladding for near-field blast mitigation. *Int. J. Impact Eng.* **2022**, *170*, 104361. [CrossRef]
22. Khalifa, Y.A.; Campidelli, M.; Tait, M.J.; El-Dakhakhni, W.W. Mitigating Risk of Confined Explosion via Lightweight Sacrificial Cladding. *J. Perform. Constr. Facil.* **2020**, *34*, 04019080. [CrossRef]
23. Langdon, G.S.; Karagiozova, D.; Theobald, M.D.; Nurick, G.N.; Lu, G.; Merrett, R. Fracture of aluminium foam core sacrificial cladding subjected to air-blast loading. *Int. J. Impact Eng.* **2010**, *37*, 638–651. [CrossRef]
24. Ousji, H.; Belkassem, B.; Louar, M.A.; Reymen, B.; Martino, J.; Lecompte, D.; Pyl, L.; Vantomme, J. Air-blast response of sacrificial cladding using low density foams: Experimental and analytical approach. *Int. J. Mech. Sci.* **2017**, *128*, 459–474. [CrossRef]
25. Zhang, L.; Hebert, R.; Wright, J.T.; Shukla, A.; Kim, J.-H. Dynamic response of corrugated sandwich steel plates with graded cores. *Int. J. Impact Eng.* **2014**, *65*, 185–194. [CrossRef]
26. Britan, A.; Liverts, M.; Shapiro, H.; Ben-Dor, G. Blast Wave Mitigation by a Particulate Foam Barrier. *Transp. Porous Media* **2012**, *93*, 283–292. [CrossRef]
27. Taha, A.K.; Gao, Z.; Dahai, H.; Zahran, M.S. Response of a new structural ultra-high performance concrete barrier wall subjected to blast loading. *Aust. J. Struct. Eng.* **2020**, *21*, 154–161. [CrossRef]
28. Peña, M.E.C. Blast Loading Retrofit of Unreinforced Masonry Walls. *Struct. Perform. Mag.* **2009**, 16–20.
29. McDonald, B.; Bornstein, H.; Langdon, G.; Curry, R.; Daliri, A.; Orifici, A. Experimental response of high strength steels to localised blast loading. *Int. J. Impact Eng.* **2018**, *115*, 106–119. [CrossRef]
30. Sun, G.; Wang, E.; Zhang, J.; Li, S.; Zhang, Y.; Li, Q. Experimental study on the dynamic responses of foam sandwich panels with different facesheets and core gradients subjected to blast impulse. *Int. J. Impact Eng.* **2019**, *135*, 103327. [CrossRef]
31. Zong, R.; Hao, H.; Shi, Y. Development of a New Fence Type Blast Wall for Blast Protection: Numerical Analysis. *Int. J. Struct. Stab. Dyn.* **2017**, *17*, 1–29. [CrossRef]
32. Xia, Y.; Wu, C.; Zhang, F.; Li, Z.-X.; Bennett, T. Numerical Analysis of Foam-Protected RC Members under Blast Loads. *Int. J. Prot. Struct.* **2014**, *5*, 367–390. [CrossRef]
33. Wu, C.; Sheikh, H. A finite element modelling to investigate the mitigation of blast effects on reinforced concrete panel using foam cladding. *Int. J. Impact Eng.* **2013**, *55*, 24–33. [CrossRef]
34. Yazici, M.; Wright, J.; Bertin, D.; Shukla, A. Experimental and numerical study of foam filled corrugated core steel sandwich structures subjected to blast loading. *Compos. Struct.* **2014**, *110*, 98–109. [CrossRef]
35. Wahab, M.A.; Mazek, S.; Abada, M.; Shafy, M.A. Blast hazard impact on V-shape composite panel performance. *J. Eng. Sci. Mil. Technol.* **2018**, *2*, 90–99. [CrossRef]
36. Walkowiak, M.; Reinicke, U.; Anders, D. Numerical Investigation of Different Core Topologies in Sandwich-Structured Composites Subjected to Air-Blast Impact. *Appl. Sci.* **2022**, *12*, 9012. [CrossRef]
37. Lin, H.; Han, C.; Yang, L.; Zhang, L.; Luan, H.; Han; Xu, H.; Zhang, S. Numerical Investigation on Performance Optimization of Offshore Sandwich Blast Walls with Different Honeycomb Cores Subjected to Blast Loading. *J. Mar. Sci. Eng.* **2022**, *10*, 1743. [CrossRef]
38. Aune, V.; Fagerholt, E.; Hauge, K.O.; Langseth, M.; Børvik, T. Experimental study on the response of thin aluminium and steel plates subjected to airblast loading. *Int. J. Impact Eng.* **2016**, *90*, 106–121. [CrossRef]
39. Smith; Hetherington, J. *Blast and Ballistic Loading of Structures*, 1st ed.; Laxton's Publishing: Oxford, UK, 1994.
40. Martin, R.J.; Reza, A.; Anderson, L.W. What is an explosion? A case history of an investigation for the insurance industry. *J. Loss Prev. Process Ind.* **2000**, *13*, 491–497. [CrossRef]
41. Needham, C.E. *Blast Waves*; Springer: Berlin/Heidelberg, Germany, 2010; Volume 402.
42. Yuen, S.C.K.; Nurick, G.; Verster, W.; Jacob, N.; Vara, A.; Balden, V.; Bwalya, D.; Govender, R.; Pittermann, M. Deformation of mild steel plates subjected to large-scale explosions. *Int. J. Impact Eng.* **2008**, *35*, 684–703. [CrossRef]
43. Parviz, M.; Aminnejad, B.; Fiouz, A. Numerical simulation of dynamic response of water in buried pipeline under explosion. *KSCE J. Civ. Eng.* **2017**, *21*, 2798–2806. [CrossRef]
44. Caçoilo, A.; Mourão, R.; Belkassem, B.; Teixeira-Dias, F.; Vantomme, J.; Lecompte, D. Blast Wave Assessment in a Compound Survival Container: Small-Scale Testing. *Multidiscip. Digit. Publ. Inst. Proc.* **2018**, *2*, 540.

45. Bajić, Z.; Bogdanov, J.; Dimitrijević, R.; Jeremić, R. Investigation of Scaled Distance Influence on Shockwave Overpressure for Plastic Explosive PPE-01. In Proceedings of the 19th Seminar on New Trends in Research of Energetic Materials (NTREM), Pardubice, Czech Republic, 20–22 April 2016; pp. 361–396.
46. Goswami, A.; Adhikary, S.D. Retrofitting materials for enhanced blast performance of Structures: Recent advancement and challenges ahead. *Constr. Build. Mater.* **2019**, *204*, 224–243. [CrossRef]
47. Draganić, H.; Sigmund, V. Blast loading on structures. *Tech. Gaz.* **2012**, *19*, 643–652.
48. Iqbal, J. *Effects of an External Explosion on a Concrete Structure*; University of Engineering and Technology: Lahore, Pakistan, 2009.
49. Cooper, W. *Explosives Engineering*; John Wiley & Sons: New York, NY, USA, 2018.
50. Simmons, M.; Schleyer, G. Pulse pressure loading of aircraft structural panels. *Thin-Walled Struct.* **2006**, *44*, 496–506. [CrossRef]
51. Bulson, S. *Explosive Loading of Engineering Structures*; CRC Press: Boca Raton, FL, USA, 1997.
52. Hashmi, M. Strain rate sensitivity of a mild steel at room temperature and strain rates of up to 105 s^{-1}. *J. Strain Anal. Eng. Des.* **1980**, *15*, 201–207. [CrossRef]
53. Marsh, K.; Campbell, J. The effect of strain rate on the post-yield flow of mild steel. *J. Mech. Phys. Solids* **1963**, *11*, 49–63. [CrossRef]
54. Samanta, S.K. Resistance to dynamic compression of low-carbon steel and alloy steels at elevated temperatures and at high strain-rates. *Int. J. Mech. Sci.* **1968**, *10*, 613–636. [CrossRef]
55. Zener, C.; Hollomon, J.H. Effect of strain rate upon plastic flow of steel. *J. Appl. Phys.* **1944**, *15*, 22–32. [CrossRef]
56. NagarajaRao, N.; Lohrmann, M.; Tall, L. Effect of strain rate on the yield stress of structural steel. *ASTM J. Mater.* **1966**, *1*, 1–54.
57. Sen, S.; Banerjee, B.; Shaw, A. Taylor impact test revisited: Determination of plasticity parameters for metals at high strain rate. *Int. J. Solids Struct.* **2020**, *193*, 357–374. [CrossRef]
58. Koerber, H.; Xavier, J.; Camanho, P.P. High strain rate characterisation of unidirectional carbon-epoxy IM7-8552 in transverse compression and in-plane shear using digital image correlation. *Mech. Mater.* **2010**, *42*, 1004–1019. [CrossRef]
59. Grązka, M.; Janiszewski, J. Identification of Johnson-Cook equation constants using finite element method. *Eng. Trans.* **2012**, *60*, 215–223.
60. *ANSYS Autodyn User's Manual (Release 15.0)*; ANSYS, Inc.: Canonsburg, PA, USA, 2013.
61. Dorogoy, A.; Rittel, D. Determination of the Johnson–Cook material parameters using the SCS specimen. *Exp. Mech.* **2009**, *49*, 881. [CrossRef]
62. Milani, A.; Dabboussi, W.; Nemes, J.A.; Abeyaratne, R. An improved multi-objective identification of Johnson–Cook material parameters. *Int. J. Impact Eng.* **2009**, *36*, 294–302. [CrossRef]
63. Banerjee, A.; Dhar, S.; Acharyya, S.; Datta, D.; Nayak, N. Determination of Johnson cook material and failure model constants and numerical modelling of Charpy impact test of armour steel. *Mater. Sci. Eng. A* **2015**, *640*, 200–209. [CrossRef]
64. Wang, X.; Shi, J. Validation of Johnson-Cook plasticity and damage model using impact experiment. *Int. J. Impact Eng.* **2013**, *60*, 67–75. [CrossRef]
65. Lotfy, M.N.; Khalifa, Y.A.; Dessouki, A.K.; Fathallah, E. Dynamic Behavior of Steel and Composite Ferry Subjected to Transverse Eccentric Moving Load Using Finite Element Analysis. *Appl. Sci.* **2020**, *10*, 5367. [CrossRef]
66. Lotfy, M.; Fathallah, E.; Khalifa, Y.; Dessouki, A. Simulation and Optimization of a CFRP and a GFRP floating pontoon. *IOP Conf. Ser. Mater. Sci. Eng.* **2020**, *934*, 012037. [CrossRef]
67. Levinson, E.; Horst, J.T.; Willcocks, M. The next generation marine inertial navigator is here now. In Proceedings of the 1994 IEEE Position, Location and Navigation Symposium-PLANS'94, Las Vegas, NV, USA, 11–15 April 1994; pp. 121–127.
68. Gold, S. Military biometrics on the frontline. *Biom. Technol. Today* **2010**, *2010*, 7–9. [CrossRef]
69. US Army Engineer Waterways Experiment Station. TM5-855-1 Fundamentals of Protective Design for Conventional Weapons. In *US Army, Navy and Air Force*; US Government Printing Office: Washington, WA, USA, 1986.
70. Ma, G.; Ye, Z. Energy absorption of double-layer foam cladding for blast alleviation. *Int. J. Impact Eng.* **2007**, *34*, 329–347. [CrossRef]
71. Mahmoud, M.M.; Mahfouz, S.Y.; Farag, H.M. Effect of air gap on the structural response of aluminum foam protected reinforced concrete panels. *Int. Conf. Civ. Archit. Eng.* **2018**, *12*, 1–13. [CrossRef]
72. Cooper, G. Protection of the lung from blast overpressure by thoracic stress wave decouplers. *J. Trauma Acute Care Surg.* **1996**, *40*, 105S–110S. [CrossRef]
73. Tan, P.; Lee, B.; Tsangalis, C. FEA modelling prediction of the transmitted overpressure and particle acceleration within a frame subjected to shock tube blast loadings. In Proceedings of the 18th World IMACS/MODSIM Congress, Cairns, Australia, 13–17 July 2009.
74. Lotfy, M.N.; Khalifa, Y.A.; Fathallah, E. Simulation and Optimizationof aMetallic Ferry under MLC70Loading. *IOP Conf. Ser. Mater. Sci. Eng.* **2020**, *974*, 012008. [CrossRef]
75. Rashed, A.; Yazdani, M.; Babaluo, A.; Parvin, H. Investigation on high-velocity impact performance of multi-layered alumina ceramic armors with polymeric interlayers. *J. Compos. Mater.* **2016**, *50*, 3561–3576. [CrossRef]
76. ANSYS, Inc. *Products Release 19.0*; ANSYS, Inc.: Canonsburg, PA, USA, 2019.
77. Fedorova, N.; Valger, S.; Fedorov, A. Simulation of blast action on civil structures using ANSYS Autodyn. *AIP Conf. Proc.* **2016**, *1770*, 020016.
78. Zheng, C.; Kong, X.-S.; Wu, W.-G.; Xu, S.-X.; Guan, Z.-W. Experimental and numerical studies on the dynamic response of steel plates subjected to confined blast loading. *Int. J. Impact Eng.* **2018**, *113*, 144–160. [CrossRef]

79. Wu, C.; Lu, Y.; Hao, H. Numerical prediction of blast-induced stress wave from large-scale underground explosion. *Int. J. Numer. Anal. Methods Geomech.* **2004**, *28*, 93–109. [CrossRef]
80. Fairlie, G. The numerical simulation of high explosives using AUTODYN-2D & 3D. In *Institute of Explosive Engineers 4th Biannual Symposium*; 1998; pp. 743–751. Available online: http://www.truegrid.com/paper052f.pdf (accessed on 1 December 2022).

MDPI AG
Grosspeteranlage 5
4052 Basel
Switzerland
Tel.: +41 61 683 77 34

Journal of Marine Science and Engineering Editorial Office
E-mail: jmse@mdpi.com
www.mdpi.com/journal/jmse

www.ingramcontent.com/pod-product-compliance
Lightning Source LLC
Chambersburg PA
CBHW061551220926
51850CB00005B/309
9783725816897